The Heliosphere Near Solar Minimum
The *Ulysses* perspective

Springer
London
Berlin
Heidelberg
New York
Barcelona
Hong Kong
Milan
Paris
Santa Clara
Singapore
Tokyo

André Balogh, Richard G. Marsden and
Edward J. Smith

The Heliosphere Near Solar Minimum

The *Ulysses* perspective

Springer

Published in association with
Praxis Publishing
Chichester, UK

Professor André Balogh
Blackett Laboratory
Imperial College
London
UK

Dr Richard G. Marsden
Space Science Department of ESA ESTEC
Noordwijk
The Netherlands

Dr Edward J. Smith
Jet Propulsion Laboratory
Pasadena
California
USA

SPRINGER–PRAXIS BOOKS IN ASTROPHYSICS AND ASTRONOMY
SUBJECT *ADVISORY EDITOR*: John Mason B.Sc., Ph.D.

ISBN 1-85233-204-2 Springer-Verlag Berlin Heidelberg New York

British Library Cataloguing in Publication Data
   Balogh, A.
      The heliosphere near solar minimum : the Ulysses perspective. –
   (Springer Praxis Books in astrophysics and astronomy)
   1. Heliosphere
   I. Title II. Marsden, Richard G. III. Smith, Edward J.
   523.7
   ISBN 1-85233-204-2

Apart from any fair dealing for the purposes of research or private study, or criticism or review, as permitted under the Copyright, Designs and Patents Act 1988, this publication may only be reproduced, stored or transmitted, in any form or by any means, with the prior permission in writing of the publishers, or in the case of reprographic reproduction in accordance with the terms of licences issued by the Copyright Licensing Agency. Enquiries concerning reproduction outside those terms should be sent to the publishers.

© Praxis Publishing Ltd, Chichester, UK, 2001
Printed by MPG Books Ltd, Bodmin, Cornwall, UK

The use of general descriptive names, registered names, trademarks, etc. in this publication does not imply, even in the absence of a specific statement, that such names are exempt from the relevant protective laws and regulations and therefore free for general use.

Cover design: Jim Wilkie
Typesetting and copy-editing: Originator, Gt Yarmouth, Norfolk, UK
Printed on acid-free paper supplied by Precision Publishing Papers Ltd, UK

Contents

Preface	xi
Abbreviations	xv
List of figures, tables and plates	xvii

1 Introduction *(André Balogh, Richard G. Marsden and Edward J. Smith)* ... 1
   1.1 The heliosphere ... 1
       1.1.1 Introduction to the heliosphere ... 1
       1.1.2 An outline of the origin and properties of the heliosphere ... 3
       1.1.3 Discovery missions to the heliosphere ... 7
   1.2 *Ulysses*: scientific objectives and mission characteristics ... 8
       1.2.1 Introduction ... 8
       1.2.2 The *Ulysses* orbit ... 9
       1.2.3 Scientific payload ... 10
       1.2.4 Spacecraft and mission operations ... 13
       1.2.5 Brief history ... 15
       1.2.6 *Ulysses* data archiving ... 17
       1.2.7 *Ulysses* orbital elements ... 17
   1.3 Highlights of the scientific discoveries of the *Ulysses* mission ... 20
       1.3.1 Introduction ... 20
       1.3.2 High-latitude heliosphere ... 21
       1.3.3 Coupling between the high- and low-latitude heliosphere ... 28
       1.3.4 Interstellar constituents and the medium outside the heliosphere ... 35
   1.4 References ... 39

2 The solar-wind and heliospheric magnetic field in three dimensions *(Marcia Neugebauer)* ... 43
   2.1 Overall pattern of high- and low-speed winds – the big picture ... 43
   2.2 Solar-wind speed ... 44

2.3	Density and mass, momentum, and energy fluxes		49
2.4	Structures in the high-latitude wind		52
2.5	Heavy ions		58
	2.5.1	Ionization states of heavy ions	59
	2.5.2	Abundance of heavy ions	62
	2.5.3	Helium	65
2.6	Ion-distribution functions		68
	2.6.1	Temperatures	68
	2.6.2	High-energy tails	73
	2.6.3	Ion beams	73
	2.6.4	Anisotropies	78
2.7	Electrons		80
	2.7.1	Overview	80
	2.7.2	*Ulysses* instrumentation for electron measurements	81
	2.7.3	Radial temperature gradient and polytrope index	83
	2.7.4	Core anisotropy	86
	2.7.5	Halo electrons and the interplanetary potential	87
	2.7.6	Electron-heat flux	88
	2.7.7	Strahl	90
2.8	Heliospheric magnetic field		91
	2.8.1	Solar source and field polarity	92
	2.8.2	Field strength	93
	2.8.3	Field direction	94
2.9	Acknowledgments		98
2.10	References		99

3 Corotating and transient structures in the heliosphere *(R. J. Forsyth and J. T. Gosling)* 107

3.1	Introduction		107
3.2	Corotating structures		108
	3.2.1	CIR formation and structure	108
	3.2.2	*Ulysses* CIR observations	116
	3.2.3	Three-dimensional models of CIRs	128
3.3	Transient flows		134
	3.3.1	The origin of CMEs and their signatures in the heliosphere	134
	3.3.2	Summary of *Ulysses* ICME observations	142
	3.3.3	ICMEs observed at high latitudes	148
3.4	Summary and Conclusion		158
3.5	Acknowledgements		159
3.6	References		159

4 *Ulysses* measurements of waves, turbulence and discontinuities *(Tim S. Horbury and Bruce Tsurutani)* 167

4.1	Introduction		167
	4.1.1	Why are *Ulysses* data unique?	167

4.2	Fluctuations in the polar heliosphere		168
	4.2.1	General character of fluctuations	170
	4.2.2	Large-scale evolution in the polar heliosphere	175
	4.2.3	Small-scale turbulent processes	185
	4.2.4	Fluctuations and solar-wind structure	192
4.3	Discontinuities and Alfvén waves		195
	4.3.1	DD radial and latitudinal gradients	196
	4.3.2	Discontinuity 'thicknesses'	199
	4.3.3	The relationship between discontinuities and Alfvén waves	201
	4.3.4	Alfvénic shocks?	205
	4.3.5	North–south asymmetries?	205
	4.3.6	Evolution of non-linear Alfvén waves and rotational discontinuities	206
	4.3.7	TDs versus RDs	207
	4.3.8	Nature of tangential discontinuities at high latitudes	209
	4.3.9	MDs and magnetic holes	212
	4.3.10	Slow shocks	217
4.4	Conclusions		218
4.5	Acknowledgements		222
4.6	References		222

5 **Waves and instabilities in the three-dimensional heliosphere** *(Robert J. MacDowall and Paul J. Kellogg)* 229

5.1	Introduction		229
	5.1.1	Wave modes in the solar wind	229
	5.1.2	Instrumentation for wave observations: *Ulysses* overview	231
5.2	Radio bursts caused by solar flares		233
	5.2.1	Type III radio burst theory and observations	233
	5.2.2	Applications in remote studies of the IPM	239
5.3	Waves associated with IP shocks		240
	5.3.1	Low-frequency electromagnetic waves	241
	5.3.2	Ion acoustic waves	241
	5.3.3	Langmuir waves	244
	5.3.4	Radio waves	244
5.4	Waves in coronal mass ejections and magnetic clouds		245
5.5	Waves at IP discontinuities and magnetic holes		246
5.6	The 'quiet' solar wind		249
	5.6.1	Whistler waves and heat-flux regulation	250
	5.6.2	VLF waves associated with expanding regions of the solar wind	250

viii Contents

		5.6.3	More ion acoustic waves	250
		5.6.4	Thermal noise	251
	5.7	Summary and remaining questions		253
	5.8	References		254

6 Energetic particles in the heliosphere *(Louis J. Lanzerotti and Trevor R. Sanderson)* ... 259

	6.1	Introduction and Stage Setting	259
	6.2	Sources of energetic particles	263
		6.2.1 Solar sources	263
		6.2.2 Interplanetary sources	266
		6.2.3 Low energy anomalous cosmic rays	272
	6.3	Particle propagation and transport	275
		6.3.1 Latitudinal particle transport	276
		6.3.2 Propagation of charged particles in interplanetary structures	276
	6.4	Periodic structures in particle flux time series	278
	6.5	*Ulysses* and solar–terrestrial connections	280
	6.6	Summary	282
	6.7	Acknowledgements	282
	6.8	References	282

7 Heliospheric and interstellar phenomena revealed from observations of pickup ions *(George Gloeckler, Johannes Geiss and Lennard A. Fisk)* ... 287

	7.1	Introduction	287
	7.2	Production and measurements of pickup ions	288
	7.3	Hydrogen and helium pickup ions	291
		7.3.1 Pickup-ion velocity distributions observed in the high-speed solar wind of the polar coronal hole	292
		7.3.2 Pickup-ion velocity distributions observed in the in-ecliptic, low-speed solar wind	299
	7.4	Interstellar pickup ions with masses heavier than helium	304
	7.5	Ionization state of the local interstellar cloud	306
	7.6	Composition of the local interstellar cloud	309
	7.7	A new pickup-ion population from an extended inner source	312
	7.8	Interaction of pickup ions with the solar wind	319
	7.9	Summary and conclusions	321
	7.10	Acknowledgments	322
	7.11	References	323

8 Cosmic rays at all latitudes in the inner heliosphere *(R. Bruce McKibben)* 327

	8.1	Introduction	327
		8.1.1 Motivation for the *Ulysses* investigations	328
		8.1.2 Status of modulation and cosmic-ray composition studies prior to *Ulysses*	329
		8.1.3 Summary of available *Ulysses* measurements	331

8.2		Heliospheric modulation of cosmic rays, electrons and atypical components	334
	8.2.1	Summary of modulation theory and available modelling tools	334
	8.2.2	Observations of solar cycle modulation	336
	8.2.3	Observations of short-term, 26-day recurrent modulations	346
	8.2.4	Implications of measurements for modulation theory and models	349
8.3		Composition studies	351
	8.3.1	Instrumentation for isotopic studies	351
	8.3.2	Measurements of stable isotopes in the galactic cosmic rays	353
	8.3.3	Measurements of radioactive isotopes in the galactic cosmic rays	357
8.4		Future contributions expected from the continuing Ulysses mission	361
	8.4.1	Modulation studies	361
	8.4.2	Composition studies	363
	8.4.3	Summary outlook	364
8.5		Acknowledgments	364
8.6		References	364

9 Cosmic dust *(Eberhard Grün, Harald Krüger and Markus Landgraf)* — 373

9.1		Introduction	373
	9.1.1	Dust objectives	375
9.2		Instrumentation	376
9.3		Interplanetary dust background in the ecliptic plane and above the solar poles	378
	9.3.1	*Ulysses'* south–north traverse	380
	9.3.2	β-meteoroids	381
	9.3.3	Mass distribution of interplanetary particles	382
	9.3.4	Description of model distributions	382
	9.3.5	Comparison with zodiacal light	385
9.4		Electromagnetically interacting dust: Jupiter dust streams	385
	9.4.1	Observations	385
	9.4.2	Particle masses and speeds	389
	9.4.3	Dust source and particle acceleration in Jupiter's magnetosphere	390
9.5		Interstellar dust	393
	9.5.1	Discovery and identification of interstellar dust grains	394
	9.5.2	Mass distribution and cosmic abundances	396
9.6		Summary and conclusions	398
9.7		Acknowledgements	400
9.8		References	400

Preface

The *Ulysses* mission is one of the cornerstones of heliospheric physics, the science of the region of space dominated by the Sun. It is the first space mission which, thanks to its unique orbit over the poles of the Sun, provides us with a three dimensional view of the complex structures and dynamic phenomena that shape the heliosphere. Since the start of the space age, this region has become the laboratory in which the properties of the medium, a magnetized plasma, can be directly observed by scientific space probes. A key objective of heliospheric research is to understand the complex relationships between the Sun and its environment, and in doing so, to test our understanding of space plasma, the medium that fills space everywhere in the Universe.

This book focuses on the heliosphere as seen by *Ulysses* during the minimum of the solar activity cycle. Sunspots have been observed for several centuries. The many solar phenomena that show an approximately eleven-year periodicity, together with the sunspot numbers, have become the object of great interest in the past century, not least because of their association with terrestrial phenomena. We now know that solar activity is caused by the Sun's magnetic fields that undergo quasi-regular changes with a period of about twenty two years. The outer atmosphere of the Sun, the corona, as well as the solar wind that originates there, streaming away from the Sun and filling a volume of space that reaches beyond the orbits of the furthest planets are particularly sensitive to solar magnetic activity. Sunspots, as well as many other solar phenomena, wax and wane at twice the rate of the magnetic changes. This is called the eleven year solar activity cycle.

Ulysses was launched in October 1990, just after the maximum of solar activity. Following a journey out to Jupiter, the spacecraft's orbit was deflected to a near-polar solar orbit, the first time that such an orbit has been achieved. It reached the regions over the south pole of the Sun in the second half of 1994. Following the pass over the south pole, the spacecraft traversed across all latitudes in the first half of 1995, reaching the northern polar region in summer 1995. From mid-1992, solar activity slowly declined; solar minimum occurred in 1996. *Ulysses* was therefore well placed to observe all aspects of the three-dimensional heliosphere under condition of declining solar activity. The picture that emerged was a relatively simple one: high speed solar winds dominated the heliosphere away from the ecliptic plane, with regions of interaction between slow and fast streams restricted to near the Sun's equatorial plane. This was an ideal period to observe heliospheric phenomena associated with quiet solar conditions; the many discoveries related to the structure of the heliosphere that *Ulysses* made are the subject of this book.

We are keenly aware of the privilege of taking part in this pioneering mission. No other spacecraft had visited these regions before *Ulysses*, and no others are planned by the space agencies at present. This ensures that the *Ulysses* observations presented in this book are bound to remain unique for at least another generation as ground truth for heliospheric physics.

A remarkable aspect of the *Ulysses* mission is the breadth of disciplines encompassed by its investigations. In addition to addressing the main themes of this book – the three dimensional heliosphere – the experiments carried by *Ulysses* are helping to resolve long-standing questions concerning the nature of the heliosphere's immediate surroundings in the Galaxy, the Local Interstellar Medium. Some of these findings, relating in particular to the gas and dust components of that medium, are described in this book. Other topics of astrophysical significance include the origin of the nuclear component of cosmic rays, and studies of relevance to cosmology. An important set of measurements made by *Ulysses* not discussed in the context of this book concern cosmic gamma-ray bursts. In all of these fields, *Ulysses* has made, and continues to make, a fundamental – and often unique – contribution.

The mission was made possible by the conviction of scientists who put forward the arguments that a truly three-dimensional view of the heliosphere was an essential element in understanding our neighbourhood in space; the perseverance of the European Space Agency with the implementation of the mission; and the continued support of NASA. Many people in these agencies, as well as the project staff at Dornier Systems (now Astrium), the Prime Contractors for the spacecraft, have contributed significantly to the success of the mission. This mission has also been a great success thanks to the efforts of the *Ulysses* scientific teams that built, calibrated and have operated the nine scientific instruments making up the payload, over a period that now spans a quarter of a century. The Project Managers who led the programme to its successful launch in 1990 were Derek Eaton (ESA) and Willis Meeks (NASA); in the past years, the project has been successfully managed by Peter Wenzel (ESA) and Ed Massey (NASA). Two of the Editors of this book, Richard Marsden (ESA) and Ed Smith (NASA) are Project Scientists for the mission, responsible for coordinating the scientific exploitation of *Ulysses*. The highly successful flight- and ground-segment operations have been carried out by Peter Beech, Nigel Angold and their colleagues in the *Ulysses* Mission Operations Team.

The chapters in this book draw on the many scientific papers published by scientists associated with the mission; the book would not have been possible without the research carried out both in Europe and in the United States by the many dedicated scientists who analysed and interpreted the *Ulysses* observations and created a vast body of knowledge concerning the heliosphere. The Editors wish to thank the entire *Ulysses* team for the successful exploitation of the mission.

The first, introductory chapter by the Editors describes the heliosphere, the *Ulysses* mission and summarises its main scientific achievements. Chapter 2, by Marcia Neugebauer, provides a comprehensive description of the solar wind. The many dynamic and transient phenomena leading to structures observed in the solar wind are described in Chapter 3 by Robert Forsyth and Jack Gosling. What we have learnt about turbulence and discontinuities in the solar wind is described in Chapter 4 by Timothy Horbury and Bruce Tsurutani. Chapter 5, by Robert McDowall and Paul Kellogg deals with the many forms of waves observed in the heliospheric medium. Energetic particles associated in this phase of the solar cycle mainly with Corotating Interaction Regions are described by Louis Lanzerotti and Trevor Sanderson. The propagation of cosmic rays and their gradients in the three-dimensional heliosphere are discussed by Bruce McKibben in Chapter 8. Finally, in Chapter 9, the distribution of cosmic dust in the heliosphere is described by Eberhard Grün, Harald Krüger and Markus Landgraf.

Chapters in the book were refereed by Peter Bochsler, Nancy Crooker, Mel Goldstein, Bill Kurth, Rosine Lallement, Marty Lee, Marius Potgieter and Ian Richardson. The Editors are very grateful to these scientists who provided many constructive comments and criticism to ensure that the book is truly a sound and balanced account of the voyage of discovery by *Ulysses* in the heliosphere near solar activity minimum. The Editors wish to thank the Publisher, Clive Horwood, and both Bruce and Neil Shuttlewood of Originator for their very patient and helpful support of this book. The diplomatic and administrative support of Mrs Susan Balogh during the preparation of the book is gratefully acknowledged by the Editors.

André Balogh, Richard Marsden and Ed Smith.

List of Abbreviations

ACE	Advanced Composition Explorer (satellite)
ACR	Anomalous Cosmic Ray
BDE	Bi-Directional suprathermal Electrons
CIR	Corotating Interaction Region
CME	Coronal Mass Ejection
COSPIN	Cosmic Ray and Solar Particle Investigations
CRIS	Cosmic Ray Isotope Spectrometer
CRR	Corotating Rarefaction Region
CRRES	Chemical Release and Radiation Effects (satellite)
CD	Contact Discontinuities
DD	Directional Discontinuities
DSN	Deep Space Network
DUST	Cosmic Dust experiment
EPAC	Energetic Particle Composition experiment
ESA	European Space Agency
FES	Fast Envelope Sampler
FIP	First Ionization Potential
FIT	First Ionization Time
FLS	Fast Latitude Scan
FS	Forward Shocks
GAS	Interstellar Neutral Gas experiment
GCR	Galactic Cosmic Ray
GRB	Gamma Ray Burst experiment
HCS	Heliospheric Current Sheet
HET	High Energy Telescope
HI-SCALE	Heliosphere Intrument for Spectra, Composition and Anisotropy at Low Energies
HMF	Heliospheric Magnetic Field
IAL	Ion Acoustic-Like
IAW	Ion Acoustic Waves

ICME	Interplanetary CME
IP	Interplanetary
IPM	Interplanetary Medium
IPS	Interplanetary Scintillation
JPL	Jet Propulsion Laboratory
KET	Kiel Electron Telescope
KNLS	Kinetic Non-Linear Schrödinger
LASCO	Large Angle Spectroscopic Coronagraph
LET	Low Energy Telescope
LIC	Local Interstellar Cloud
LISM	Local Interstellar Medium
LW	Langmuir waves
MDLS	Modified Derivative Non-linear Schrödinger
MD	Magnetic Decrease
NASA	National Aeronautics and Space Administration
OESP	Out of Ecliptic/Solar Polar
OOE	Out-Of-Ecliptic
PBS	Pressure Balance Structure
PFR	Plasma Frequency Receiver
QTN	Quasithermal Noise
RAR	Radio Astronomy Receiver
RD	Rotational Discontinuities
ROID	Rate of Occurrence of Interplanetary Discontinuities
RS	Reverse Shocks
RW	Reverse Waves
SEP	Solar Energetic Particle
SLS	Slow Latitude Scan
SMM	Solar Maximum Mission
SWICS	Solar Wind Ion Composition Spectrometer
SXT	Soft X-ray Telescope
TD	Tangential Discontinuities
URAP	Unified Radio and Plasma Wave experiment
UVCS	Ultra-Violet Coronagraph Spectrograph
VHM/FGM	Vector Helium Magnetometer/Flux Gate Magnetometer
WFA	Waveform Analyzer
WMW	Whistler Mode Wave

List of figures, tables and plates

Figure 1.1	A schematic representation of the heliosphere	6
Figure 1.2	The anti-correlation between sunspot numbers and the intensity of cosmic rays	6
Figure 1.3	The *Ulysses* orbit viewed from 15° above the ecliptic plane	11
Figure 1.4	The heliographic latitude of the *Ulysses* spacecraft	11
Figure 1.5	Projection of the *Ulysses* orbit (ecliptic latitude and longitude) on the solar disc	12
Figure 1.6	The *Ulysses* spacecraft in pre-launch configuration	15
Figure 1.7	An overview of the *Ulysses* observations as a function of heliolatitude	19
Figure 2.1	Contours of solar-wind speed versus heliographic latitude and a longitude defined by a rotation rate of 29 days relative to *Ulysses*	46
Figure 2.2	Solar-rotation averages of solar-wind speed and scatter plot of solar-rotation averages	48
Figure 2.3	Median, 5%, and 95% values for selected fluid parameters over each solar rotation as a function of heliographic latitude during the *Ulysses* FLS	50
Figure 2.4	Median energy-flux densities, binned by solar rotation, versus heliographic latitude during the *Ulysses* FLS	51
Figure 2.5	Polar plot of the square root of solar-rotation averages of the momentum flux of the solar wind versus heliographic latitude	53
Figure 2.6	Power spectra of hourly averages of the radial R, tangential T, and normal N components the solar-wind proton velocity	54
Figure 2.7	Hourly averages of solar-wind speed, plasma pressure (ΣNkT), and magnetic pressure ($B^2/8\pi$) for the 1–24 April 1994, observed by *Ulysses*	55
Figure 2.8	Superposed epoch analyses of variation of 6-hour averages in several plasma parameters	57
Figure 2.9	Time series of alpha-particle speed and ionization temperature calculated from the charge states of oxygen ions	61
Figure 2.10	Coronal electron temperature versus distance from the Sun	61
Figure 2.11	Elemental abundances in the solar wind as a function of the first ionization time	63
Figure 2.12	Scatter plot of the ratio of abundances of magnesium and oxygen ions detected by *Ulysses*	64
Figure 2.13	Superposed epoch plot showing the systematic variation in solar-wind speed, ionization temperature T_O, and the magnesium to oxygen abundance ratio as a function of the phase of solar rotation	65

xviii List of figures, tables and plates

Figure 2.14	SWOOPS measurement of solar-rotation averages of the ratio of the densities of alpha particles and protons	66
Figure 2.15	Kinetic temperature ratios for fourteen pairs of heavy ions observed by *Ulysses*/SWICS at mid-latitudes	71
Figure 2.16	Mass-adjusted ratios of kinetic temperatures for oxygen and neon relative to helium as functions of solar-wind speed	72
Figure 2.17	Correlograms of speeds and kinetic temperatures of C^{6+} versus He^{2+} ions	72
Figure 2.18	Two proton distributions observed by *Ulysses*	73
Figure 2.19	Magnitude of the alpha–proton differential flow versus heliocentric distance and Averages (6-hour) of the magnitude of the alpha–proton differential flow as a function of travel time from the Sun	74
Figure 2.20	Averages (6-hour) of the ratio $V_{\alpha p}/V_A$ versus travel time from the Sun	76
Figure 2.21	Correlation between $V_{\alpha N}$ and B_N as a function of $V_{\alpha p}$	77
Figure 2.22	Distribution of the stability criterion for mirror-mode waves	79
Figure 2.23	A typical energy spectrum, summed over all angles of incidence	81
Figure 2.24	A typical quasi-thermal noise spectrum obtained by URAP	82
Figure 2.25	Median and 25 and 75 percentile values of core electron temperatures versus heliocentric distance for the *Ulysses* in-ecliptic trajectory	84
Figure 2.26	Radial variation in the core electron temperature and total electron density in the latitude ranges $\pm(40–80°)$ as determined by the URAP quasi-thermal noise technique	85
Figure 2.27	Decline in electron-heat flux with distance from the Sun	89
Figure 2.28	The electron heat flux multiplied by a factor $R^{2.9}$ to correct for dependence on distance from the Sun	89
Figure 2.29	Electron-heat flux and magnetic-field magnitude	90
Figure 2.30	Median full width at half maximum of the strahl of 77-eV electrons versus heliolatitude	91
Figure 2.31	*Ulysses* observations of daily averages of the radial component B_R of the HMF, normalized to a solar distance 1 AU, versus time and versus heliographic latitude	93
Figure 2.32	Solar-rotation averages of the azimuthal angle of the HMF expected on the basis of the Parker spiral, the *Ulysses* trajectory parameters, and the observed solar-wind speed	94
Figure 2.33	Histograms of the direction of the high-latitude HMF (ϕ_B) relative to the expected Parker spiral angle (ϕ_P)	95
Figure 2.34	Correlation between normal components of the proton velocity V_N and the magnetic field B_N along the *Ulysses* trajectory	96
Figure 2.35	A diagram demonstrating how the dragging of field lines by solar convection generates transverse fields at the solar surface	97
Figure 2.36	Projection of tracings of magnetic-field lines originating at a latitude of 70°S according to the Fisk model and the Parker spiral field	98
Figure 3.1	A plot of flow speed and pressure from a simple one-dimensional simulation showing the formation and evolution of interaction regions with distance from the Sun	109
Figure 3.2	A schematic diagram of a corotating interaction region in two dimensions in the equatorial plane	111
Figure 3.3	Schematic diagrams of the variation of the coronal magnetic field with the solar cycle	112
Figure 3.4	An example of the variation with time of selected plasma and magnetic field parameters through a CIR observed by *Ulysses*	114

List of figures, tables and plates xix

Figure 3.5	An illustration of the two-dimensional geometry of CIRs showing the shocks, stream interface and unshocked regions.	115
Figure 3.6	Solar wind speed throughout the *Ulysses* mission	117
Figure 3.7	Solar wind speed, magnetic field strength and proton number density plotted through the southern hemisphere CIRs and the northern hemisphere CIRs	118
Figure 3.8	Magnetic-field azimuth angle, magnetic-field strength, and solar wind speed for a subset of the southern hemisphere CIRs	119
Figure 3.9	Variation of oxygen and carbon freezing-in temperatures through a CIR observed by *Ulysses*	122
Figure 3.10	A summary of the corotating forward and reverse shocks observed by *Ulysses*	123
Figure 3.11	Examples of flow deflections from two CIRs	124
Figure 3.12	Schematic diagram illustrating how the tilted streamer belt leads to equatorward propagating forward shocks and poleward propagating reverse shocks	126
Figure 3.13	The magnetic field geometry associated with the counterstreaming electron beams produced by CIR shocks	128
Figure 3.14	The tilted dipole flow geometry used at the inner boundary of the three-dimensional simulations of CIRs	130
Figure 3.15	A latitude-longitude grey scale map of solar wind speed from the three dimensional CIR simulation of Pizzo and Gosling (1994)	131
Figure 3.16	A meridional plane map of the solar wind speed and pressure from the CIR simulation of Pizzo and Gosling (1994).	132
Figure 3.17	Stack plots of the latitude and longitude variation of speed (left) and pressure (right) from the CIR simulation of Pizzo and Gosling (1994)	133
Figure 3.18	An example of a coronal mass ejection observed by the LASCO coronagraph on SOHO	135
Figure 3.19	Plasma and magnetic field parameters for an ICME observed by *Ulysses*.	138
Figure 3.20	Sketch showing two possible configurations of magnetic field lines closed back to the Sun within an ICME	139
Figure 3.21	Sketch illustrating how reconnection between the legs of neighbouring magnetic field loops which are sheared relative to one another leads to the formation of a flux rope within a CME	141
Figure 3.22	X-ray images from the *Yohkoh* spacecraft	144
Figure 3.23	A diagram illustrating the successive stages of 3D reconnection	146
Figure 3.24	*Ulysses* magnetic field data from days 286 to 302 (12 to 28 October) of 1996.	147
Figure 3.25	A schematic diagram illustrating how a magnetic flux rope originating from 3-D reconnection in a helmet streamer propagates outwards as an occlusion in the heliospheric current sheet	148
Figure 3.26	Plasma and field parameters from the June 1993 and April 1994 ICMEs observed by *Ulysses*.	150
Figure 3.27	Solar wind speed and pressure plotted against heliocentric distance for a simulated ICME that has propagated out to 5 AU	151
Figure 3.28	Schematic diagram illustrating an overexpanding ICME propagating in the solar wind	153
Figure 3.29	Plasma and field parameters obtained from both *Ulysses* and *IMP-8* for an ICME observed in February 1994	155
Figure 3.30	Plots in the meridional plane of the radial velocity component, meridional velocity component, and pressure within a two-dimensional simulation of an ICME	156

xx List of figures, tables and plates

Figure 3.31	A meridional plane cross section of ICMEs propagating in a corotating fast and slow solar wind flow structure.	157
Figure 4.1	A typical day (1995 day 120, 1.4 AU and 42°N) of magnetic-field and plasma data in the polar heliosphere	169
Figure 4.2	Power spectra of total component and field-magnitude power for the same day as Figure 4.1 and spectral index of field-component fluctuations as a function of frequency	170
Figure 4.3	Time series of velocity and magnetic field fluctuations in high-speed polar wind	173
Figure 4.4	Elsässer variable power spectra for three intervals of high-speed solar wind	173
Figure 4.5	Minimum-variance directions superimposed on the magnetic-field direction	175
Figure 4.6	Turbulent energy and changes in the shape of the fluctuation power spectrum in high-speed wind	176
Figure 4.7	Distance and latitude dependence of power and spectral index for magnetic-field components as a function of wavenumber in the polar solar wind	180
Figure 4.8	Comparison of power levels measured in high-speed streams between 0.3 and 1 AU by *Helios 1* and *Ulysses* polar measurements	183
Figure 4.9	Variation of spectral breakpoint with solar distance in and out of the ecliptic	184
Figure 4.10	Structure-function values for a range of scales and moments, calculated from 5 days of polar magnetic-field data and gradients of structure functions on spacecraft scales of 20–60 s, corresponding to solar-wind scales of about 1.5×10^4 to 4.5×10^4 km.	186
Figure 4.11	Observed $g(m)$ values and a least-squares fit of the generalised p model. Estimates of the intermittency measure p and energy scaling q from several intervals of data and several time scales in polar data	188
Figure 4.12	Variation in intermittency outside the inertial range.	191
Figure 4.13	Two examples of events when near-radial magnetic fields are accompanied by dramatically reduced magnetic-field power levels.	194
Figure 4.14	Schematics of idealized rotational and tangential discontinuities.	197
Figure 4.15	The in-ecliptic rate of occurrences of interplanetary discontinuities (ROID) values from 1 to 5 AU.	197
Figure 4.16	The solar-wind plasma and magnetic field and ROID values for the TS and LB criteria.	198
Figure 4.17	The temporal and spatial 'thicknesses' of interplanetary discontinuities at 67°S latitude and 3.0 AU	200
Figure 4.18	The magnetic field at 5.2 AU at 6.0°S latitude	202
Figure 4.19	The relationship between a rotational discontinuity and slowly rotating Alfvén waves.	203
Figure 4.20	The different possible polarizations for plane waves and spherical waves	204
Figure 4.21	The normalized jumps in field magnitude, density, and temperature across rotational discontinuities	206
Figure 4.22	A comparison of discontinuity-occurrence rates over the north pole and over the south pole.	207
Figure 4.23	Discontinuity phase spaces (B_3/B_L, $\Delta|B|/B_L$) distributions for days 218–221 1993 at the north heliographic pole.	208
Figure 4.24	Several different types of MDs. These examples were taken at 80°S latitude, 2.3 AU from the Sun.	210
Figure 4.25	Minimum variance results for 129 discontinuities bounding MDs.	211
Figure 4.26	A distribution of field-magnitude changes within MDs.	211
Figure 4.27	The MD thickness distribution for different values of $\Delta|B|/B_L$	212

List of figures, tables and plates xxi

Figure 4.28	A histogram of MD thicknesses	213
Figure 4.29	An example of a linear magnetic hole.	214
Figure 4.30	An example of a TD at the edge of MD-related field decreases	215
Figure 4.31	A current-sheet-related TD on day 239 1994. *Ulysses* was at 79.3°S heliographic latitude.	216
Figure 4.32	Observed and simulated of two components of the field in minimum-variance co-ordinates.	217
Figure 4.33	A slow shock pair	219
Figure 4.34	The discontinuity normals for days 154–155 1994	220
Figure 5.1	Example of an intense IP type III radio burst and associated *in-situ* waves	234
Figure 5.2	Type III burst occurrence as a function of the solar cycle	235
Figure 5.3	After the onset of a solar flare at time t_0, electrons escaping along open field lines will form a cutoff distribution function $f(v)$ at large distances from the flare	236
Figure 5.4	High time resolution envelope of Langmuir waves during the event in Figure 5.1, detected by the *Ulysses* URAP FES.	237
Figure 5.5	Spectral cuts through the type III burst	238
Figure 5.6	Low frequency limit of type III radio bursts versus the local plasma frequency at *WIND* and *Ulysses* and histograms comparing the low frequency cutoff of strong and weak type III bursts to the local plasma frequency at *WIND* and *Ulysses*	238
Figure 5.7	Type III radio burst trajectories observed remotely from *Ulysses* at high heliographic latitudes.	240
Figure 5.8	A forward-reverse shock pair from the perspective of the *Ulysses* wave data	242
Figure 5.9	Occurrence probability of ion acoustic wave activity as a function of distance from forward and reverse, quasiparallel and quasiperpendicular shocks	243
Figure 5.10	A 24-hour interval of *Ulysses* radio data with IP type II radio burst activity	245
Figure 5.11	Solar wind, wave, and magnetic-field data for the magnetic cloud interval of 13–22 October 1996	247
Figure 5.12	Example of Langmuir, ion acoustic, and whistler waves associated with magnetic holes	248
Figure 5.13	Electric field spectrum of URAP RAR data on 22 February 1991, histogram of f_{peak}/f_{pe} for Langmuir waves associated with several type III events observed by URAP, electric field spectra of URAP RAR data on 9 December 1994 and histogram of f_{peak}/f_{pe} for 20 intervals of wave activity in magnetic holes	249
Figure 5.14	Example of ion acoustic wave activity in high-latitude, fast solar wind	251
Figure 5.15	Thermal noise spectrum from the URAP with fit to derive the solar wind velocity V, proton temperature T_p, electron density n_e, electron core temperature T_c, electron temperature ration T_h/T_c, and density ratio n_h/n_c	252
Figure 5.16	Electric and magnetic field wave activity at several frequencies along the *Ulysses* fast latitude scan in 1994–5	254
Figure 6.1	Intensity of 1.8–3.8 MeV protons.	260
Figure 6.2	Intensities of an electron and a proton channel measured by the HI-SCALE and COSPIN LET instruments	261
Figure 6.3	Association of interplanetary energetic protons with occurrence of a coronal mass ejection event	264
Figure 6.4	Differential energy spectra of energetic ions propagating at two different pitch angles along magnetic fields associated with a coronal mass ejection event	265
Figure 6.5	Proton and alpha particle fluxes as a function of time during the passage of 13 CIR events in 1993–1994	267

xxii List of figures, tables and plates

Figure 6.6	Relative timing of electron (50 keV) and proton (0.5 MeV) fluxes during 18 CIR-associated events during the southern heliosphere pass of *Ulysses*.	268
Figure 6.7	Twelve solar rotations (each of 26 days) beginning on 12 January 1993 during the southern heliosphere pass.	269
Figure 6.8	Sixteen solar rotations beginning 1 January 1996 during the northern heliosphere pass.	271
Figure 6.9	Velocity distribution functions measured by two instruments on *Ulysses* (SWICS and HI-SCALE).	272
Figure 6.10	Velocity distribution functions for four ion species measured during two different CIR events.	273
Figure 6.11	Relationship between the intensities of anomalous cosmic ray oxygen fluxes and variations in galactic cosmic ray intensities during an interval of nine CIR events	274
Figure 6.12	Anomalous cosmic ray oxygen, nitrogen, and neon fluxes for heliographic latitudes $>60°$S and $>60°$N.	275
Figure 6.13	Solar electron (42–65 keV) event observed at $\sim 74°$S heliolatitude during days 298–305, 1994.	277
Figure 6.14	Propagation of electrons in an interplanetary propagation channel.	278
Figure 6.15	Portion of a power spectrum of 38–53 keV electron fluxes calculated with eight prolate spheroidal data windows and a time-band width product of 5.0	280
Figure 7.1	Schematic representation of sources and processes leading to the creation of pickup ions in the heliosphere and their subsequent transport and acceleration to form the ACRs.	289
Figure 7.2	Phase-space density of H^+ versus W, the ion speed in the spacecraft frame of reference divided by the solar-wind speed.	293
Figure 7.3	Same as Figure 7.2 for He^{++}.	295
Figure 7.4	Same as Figure 7.3 for He^+.	296
Figure 7.5	Plot of the average triple-coincidence counts of ions and Phase-space density of $^4He^+$ and $^3He^+$.	298
Figure 7.6	Ratio of H^+ to $^4He^+$ pickup-ion fluxes versus heliocentric distance.	299
Figure 7.7	Phase-space density versus W of H^+, He^+, and He^{++} ions in the slow (375 km s^{-1} average speed), in-ecliptic solar wind.	300
Figure 7.8	Distribution functions of H^+, He^+, and He^{++} downstream of the forward shock, downstream of the reverse shock and distribution functions of H^+, He^+ and He^{++} in the region downstream of the RS.	301
Figure 7.9	Plot of the double coincidence counts of ions with normalized speed W above 1.4 per mass/charge (m/q) intervals versus the mean m/q of each interval.	305
Figure 7.10	Phase-space density of O^+ versus W averaged over 135 days from 17 February to 1 July 1992 when *Ulysses* was still in the ecliptic plane.	306
Figure 7.11	Variation of the calculated magnetic-field strength B just beyond the interstellar bow shock with distance to the termination shock R_{TS} and verage values of the electron density and ionization fractions of H and He.	310
Figure 7.12	Plot of the double coincidence counts per mass/charge (m/q) interval of ions with normalized speed W (ion speed/solar-wind speed) between 0.8 and 1.0 versus the mean m/q of each interval.	313
Figure 7.13	Phase-space density of C^+ and O^+ versus W during all of 1994.	313
Figure 7.14	Velocity distributions of pickup H^+, C^+, and O^+.	314
Figure 7.15	Plot of the double coincidence counts per mass/charge (m/q) interval of ions with W between 0.85 and 1.4 versus the mean m/q of each interval.	316

Figure 7.16	One-year averaged (30 January 1997–18 January 1998) velocity distributions of O^+, N^+, and Ne^+ when *Ulysses* was again in the ecliptic plane but at ~ 5.2 AU	317
Figure 7.17	Mass per charge distributions of heavy pickup ions in the inner source	318
Figure 7.18	The dependence of the ratio of the average distribution function observed between W of 1.8–2.0 to that observed between W of 1.6–2.0 on the magnetic field angle ψ is compared with theoretical estimates.	321
Figure 8.1	Energy ranges for electrons, protons, and helium over which COSPIN sensors provide measurements	332
Figure 8.2	Energy and nuclear charge ranges over which COSPIN sensors provide measurements for study of the elemental and isotropic composition of cosmic rays	333
Figure 8.3	*Ulysses* heliographic longitude and latitude versus time, 6-hour average solar-wind speeds from the *Ulysses* SWOOPS instrument, daily average magnetic-field strength from the *Ulysses* VHM/FGM instrument and daily average counting rates of protons $> \sim 90\text{-}100$ MeV from the *Ulysses* COSPIN HET and the University of Chicago cosmic-ray experiment on the Earth satellite *IMP-8*.	336
Figure 8.4	Average intensities (26-day) of 2.5-GV electrons and protons	337
Figure 8.5	Polar plot of the ratio of intensities measured at *Ulysses* to those measured at *IMP-8* at Earth as a function of *Ulysses* latitude for relativistic and low-energy galactic cosmic-ray protons and low-energy anomalous helium measured during *Ulysses*' FLS.	342
Figure 8.6	Compilation of latitude gradients measured for galactic cosmic rays and for anomalous components by *Ulysses* and, for comparison, by *Voyager 1.2* and *Pioneer 10/11* instruments, plotted as a function of particle rigidity	345
Figure 8.7	Daily average intensity of $> \sim 90$ MeV protons at *Ulysses*, detrended by subtracting running 27-day average intensities from the daily averages and average solar-wind velocities (6-hour) for the same period measured by the *Ulysses* SWOOPS instrument	347
Figure 8.8	Correlation between the amplitude of 26-day variation and the size of measured latitude gradients for a variety of energy intervals for protons, helium, carbon, and oxygen	348
Figure 8.9	Schematic of the COSPIN HET detector arrangement and orientation of the strips on the position-sensing detector to form a hodoscope	352
Figure 8.10	Mass histograms for C, N, O, Ne, Mg, and Si in the cosmic radiation as measured by the COSPIN HET	353
Figure 8.11	Mass histograms for Fe and Ni in the cosmic radiation as measured by the COSPIN HET (from Connell and Simpson, 1997a)	354
Figure 8.12	$^{49}V/^{51}V$ abundance ratio measured by the *Ulysses* HET compared with predictions for energy boosts via re-acceleration of 0 (top curve) through 300 MeV	361
Figure 9.1	Trajectories of the *Ulysses* and the *Galileo* spacecraft	375
Figure 9.2	Schematic configuration of the *Ulysses* dust detector (GRU)	377
Figure 9.3	Dust-impact rates observed by *Ulysses*	378
Figure 9.4	Dust-impact directions (rotation angle) observed by *Ulysses*	379
Figure 9.5	Dust-impact rate (impact charge $Q_I > 8 \times 10^{-14}$ C) observed by *Ulysses* during its south–north traverse around the time of ecliptic plane crossing (ECL)	380
Figure 9.6	Masses of dust particles observed by *Ulysses* during its south–north traverse. Ecliptic plane crossing coincided with perihelion passage	381

xxiv List of figures, tables and plates

Figure 9.7	Models of directional impact rates (of particles with masses $m > 10^{-17}$ kg) onto *Ulysses* from launch to completion of its first out-of-ecliptic orbit	384
Figure 9.8	*Ulysses* trajectory and geometry of dust detection – oblique view from above the ecliptic plane	386
Figure 9.9	Impact rate of dust particles observed by *Ulysses* around Jupiter fly-by	387
Figure 9.10	Dust impact directions observed around Jupier fly-by	388
Figure 9.11	Simulated arrival directions ROT of 10 nm dust particles from Jupiter to *Ulysses* compared with directions from which dust streams were observed by the *Ulysses* dust sensor	389
Figure 9.12	Jupiter's 'dusty ballerina skirt' formed by Jupiter stream particles moving away from the planet in a warped dust sheet	391
Figure 9.13	A Scargle-Lomb periodogram for two years of *Galileo* dust data	392
Figure 9.14	Histogram of the mass distribution of interstellar grains detected by the *Ulysses* dust instrument. The detection threshold for grains impacting with 20 km s^{-1} is $\sim 10^{-18}$ kg	397
Figure 9.15	Mass density per logarithmic mass interval	397
Figure 9.16	Ecliptic longitude and latitude of the upstream direction of interstellar dust grains measured by *Ulysses* after Jupiter fly-by	399
Table 1.1	Main trajectory characteristics	12
Table 1.2	Timeline of mission events	13
Table 1.3	The *Ulysses* hardware investigations	14
Table 1.4	Classical orbital elements for *Ulysses*	18
Table 2.1	Freezing-in temperatures of C, O, Si, and Fe observed in four sets of *Ulysses* data	60
Table 2.2	Overview of published elemental-fractionation models with their modes of ionization and separation of ions from neutrals	64
Table 2.3	Dependence of the ratio of alpha and proton temperatures T_α/T_p on the magnitude of heliographic latitude $\Theta = |\lambda|$ and on distance from the Sun R in AU	70
Table 2.4	Exponents α of the radial gradient of electron-core temperatures	83
Table 4.1	Criteria for mass flux and magnetic-field changes across idealized discontinuities	195
Table 5.1	Wave modes in the solar wind	230
Table 5.2	*Ulysses* URAP instrument parameters	232
Table 7.1	Densities of atoms at the heliospheric termination shock ($\sim$85–110 AU) and densities of atoms and ions in the local interstellar cloud (LIC) of elements and isotopes measured as pickup ions with SWICS on *Ulysses*	306
Table 7.2	Measured and derived average parameters in the LIC and in the very local neighbourhood of the solar system	308
Table 8.1	*Ulysses* radial gradient measurements (1–5 AU)	339
Table 8.2	*Ulysses* latitudinal gradient measurements (0°–±60°)	341
Table 8.3	Measurements of stable cosmic-ray isotopic ratios	355
Table 8.4	Measurements of radioactive clock isotopes	358
Table 8.5	Candidate electron-capture cosmic-ray secondary isotopes	363
Table 9.1	Parameters of the Jupiter dust streams	387
Plate 1	A model of the external field lines of the solar magnetic field at the time when *Ulysses* crossed the equator from south to north in early 1995	198–199

List of figures, tables and plates   xxv

Plate 2	A polar plot of solar-wind speed versus heliographic latitude observed by *Ulysses*	198–199
Plate 3	Contours of solar-wind speed as a function of latitude and Carrington longitude. Contours of the alpha-particle abundance N_α/N_p. Contours of the ionization temperature. Contours of density ratios of Mg to O	198–199
Plate 4	Four views of the Sun during the *Ulysses* fast latitude scan.	
Plate 5	Charge-state distributions for solar-wind carbon, oxygen, silicon, and iron ions accumulated during four 300-day intervals	198–199

1

The heliosphere and the *Ulysses* mission

André Balogh, Richard G. Marsden and Edward J. Smith

1.1 THE HELIOSPHERE

1.1.1 Introduction to the heliosphere

The subject of this book is the journey of discovery of the *Ulysses* spacecraft through the heliosphere, the vast volume of space that surrounds the Sun and reaches well beyond the orbits of the furthest planets in the solar system. The Sun dominates the properties of this volume of space through the solar wind, the constant flow of ionized particles emitted from the outer atmosphere of the Sun, the solar corona. Just as we are interested in the Sun as the star closest to us, we are also interested in studying the heliosphere and the solar wind as the only region of space of astrophysical significance that is accessible for direct observations. In addition, the importance of the Sun and the processes that link it to our neighbourhood on Earth make the study of the heliosphere a topic of special interest in space sciences.

In the past 40 years, space physics has become a recognized branch of astrophysics. Its topics focus on astrophysical phenomena in the general area of the solar system: the Sun itself, the heliosphere, the ionized and magnetic environments of the planets and their relationship with solar phenomena, and the propagation of cosmic rays in the heliosphere.

A unique aspect of space physics in the more general domain of astrophysical research is the possibility to make direct observations in the space medium itself. Although much can be learned by remote sensing of astrophysical phenomena (and there is no other means to observe most of the universe), accessing neighbouring regions of space with spacecraft that can return data from the medium to be studied is a great advantage. Given the current and foreseeable capabilities of space flight, our ability to make direct observations is restricted to regions of space in the solar system and its immediate vicinity. It is important to exploit this ability for understanding in depth and in detail at least these regions, to help us interpret indirect observations of more distant parts of the universe.

The development of our understanding of the heliosphere provides a good example for the importance of direct observations. Although the existence of the solar wind and the heliosphere was predicted in the 1950s, just before the dawn of the space age, there were competing theories that described the consequences of the very hot solar atmosphere and how it affected the properties of the medium in what was called interplanetary space. In particular, as late as 1960 there was considerable controversy between proponents of a fast, supersonic solar wind on the one hand, and scientists who suggested that the hot solar corona expands in the form of a gentle 'solar breeze'. Then, in the early 1960s the first space probes that left Earth orbit sent back data that unequivocally confirmed the existence of a continuous, fast stream of ionized particles from the Sun, the solar wind. After this, there was immediate and rapid progress in the discovery and theoretical interpretation of many phenomena associated with the Sun, the interplanetary medium and with what was called solar–terrestrial relations, the branch of geophysics that focused on solar effects on the Earth's ionosphere and magnetic field. These critically important first observations of the solar wind itself made the vital initial step in the rapid progress that has been made in understanding the Sun and how its emission of the solar wind affects a large volume of space around it. Many space probes have returned more data since then and heliospheric science is now built on the firm foundations of a wealth of observations, as well as on theoretical models that have been developed to explain the observations.

The progress made in physics of the heliosphere has of course led to new questions as theoretical models have attempted to extrapolate experimental observations to regions not yet visited by spacecraft. Given the complex processes that govern the medium, such extrapolation remains risky without the anchor of actual observations. As described in the following sections, space missions in the heliosphere before the 1990s were largely restricted to orbits in the ecliptic plane, the plane in which the Earth orbits the Sun. The reason for this is that the energy required to launch a spacecraft with a velocity sufficient to overcome the Earth's orbital velocity of $30\,\mathrm{km\,s^{-1}}$, not only to escape from Earth orbit but also out of the equatorial plane, is just too much for current rocket technology. As a result, much of what we knew of the heliosphere was based on data acquired in the proximity of this plane. The Sun rotates around an axis which is almost perpendicular to the ecliptic plane; the equatorial plane of the Sun makes an angle of only $7.25°$ with the ecliptic. This makes it difficult to observe the Sun's polar regions from the Earth, and spacecraft in the ecliptic plane can only return data from the equatorial region of the heliosphere. It is well known from solar observations, particularly from eclipse photographs and increasingly from sophisticated space-based solar observatories, that solar phenomena relevant to the heliosphere are usually very different at high solar latitudes.

The need for observations at high latitudes had been clear from early on, but such observations required a dedicated space mission which, in the end, took a long time to come. The mission, originally called the Out-Of-Ecliptic mission, was conceived in the early 1970s, with two spacecraft, but, as described in more detail in Section 1.2, eventually turned into the single spacecraft *Ulysses* mission, eventually launched in 1990.

The name 'interplanetary medium' has often been used to describe the region of space filled by the solar wind. This is an historical legacy of earlier space missions that were restricted to the orbital plane of the planets, the ecliptic plane. As is clearly demonstrated in Section 1.3, as well as throughout this book, *Ulysses* has revealed the true three-dimensional nature of the space dominated by the Sun. In the light of the knowledge gained through this highly successful mission, it is clear that the term 'heliospheric medium' is now more appropriate.

1.1.2 An outline of the origin and properties of the heliosphere

The heliosphere extends from the solar corona to an outer boundary where the solar wind encounters the interstellar medium. The outer corona of the Sun consists of a fully ionized gas threaded by magnetic fields rooted in the visible surface of the Sun, the photosphere. The coronal plasma is very hot, with a temperature in excess of a million degrees. It is still unclear just how the corona is heated to such temperatures; the most likely explanation is that waves from the lower layers of the solar atmosphere provide the necessary energy to heat the corona. The energy deposited in the coronal plasma appears also to be sufficient to accelerate it away from the Sun in the form of the solar wind. The speed of the solar wind varies between about 300 km s^{-1} to more than 800 km s^{-1}. This speed is well in excess of the speed of sound in the plasma.

The existence of the solar wind, as a continuous outflow of particles, was first surmised in the 1950s, based on the observation of the ion tail of comets. At first it was unclear what the speed and density of the solar wind would be; as already mentioned, one theory predicted a slow evaporation of the solar corona, in the form of the solar breeze. The main theoretical breakthrough came in 1958 when Eugene Parker of the University of Chicago (Parker, 1958, see also Parker, 1997) published his calculation of the expansion of the corona in the form of a continuous, supersonic stream of particles. This expansion provided the solution to an important puzzle: how is the pressure of the solar wind reduced at large distances from the Sun so that it can balance the pressure of the Local Interstellar Medium (LISM) estimated from astronomical observations? Even though Parker's solution was the only one that appeared to meet all the conditions both at the corona and far away from the Sun, it remained controversial for another 4 years, until measurements by the *Mariner 2* space probe (Neugebauer and Snyder, 1962) definitively confirmed the existence of Parker's solar wind. (Earlier measurements by Soviet probes to the Moon in 1959 also made measurements of the solar wind, but the results were not made public at the time.)

The variable speed of the solar wind is the result of different processes in the solar corona. Its other parameters (density, temperature and composition) were also found to be highly variable. There are two aspects of this general variability in the solar wind that are important to consider. First, what are the different processes and conditions in the solar corona that give rise to the variation? Second, what are the consequences of this variability for the dynamics and structure of the heliosphere?

Observations in the 1960s and 1970s made both in the vicinity of the Earth and on interplanetary probes showed that the solar wind is structured into fast ($>550\,\mathrm{km\,s^{-1}}$) and slow ($<450\,\mathrm{km\,s^{-1}}$) streams (e.g. see Schwenn, 1990). The characteristics of the wind also showed similar structuring in other parameters. The key to the origin of the different streams were provided by solar observations made on board *Skylab*. From these observations, the coronal regions responsible for the emission of fast solar-wind streams was successfully identified (Krieger *et al.*, 1973). Images of the Sun at wavelengths of low-energy X-rays not accessible to ground-based telescopes, showed that the solar corona was structured in the form of hot, bright loops and darker, cooler regions which, because of this dark appearance, were called coronal holes. The structuring was controlled by coronal magnetic fields: the loops were found to be magnetic structures with both ends anchored in the solar surface (so-called 'closed fields') whereas the magnetic field in the coronal holes was only anchored at one end, and the field lines from the coronal holes ('open fields') appeared to be dragged away from the Sun by the solar wind.

Once coronal holes were identified as the regions in the corona from which the fast solar wind originated, the origin of the slow solar wind still remained to be identified. This has proved to be a more difficult task. Even today this question has not been fully resolved, despite a wealth of more refined observations. It is clear that the slow wind is somehow associated with regions in the corona which are magnetically closed (the loop structures seen in the corona); however, in order to escape from the Sun, the constituents of the solar wind need to move (or be moved) onto magnetic field lines that are open. Significant progress in understanding the origin of the slow solar wind, with its much greater variability, was only made in the 1990s, not least by observations made by *Ulysses*, as described in Section 2 and, in more detail, in Chapter 2 by Marcia Neugebauer.

The second aspect, identified above, is the heliospheric consequence of variability in the solar wind. The solar wind is a plasma with an electrical conductivity close to infinity. One consequence is that the solar wind carries away the magnetic lines of force from the corona; another is that such plasma flows cannot mix, but can exercise a dynamic force on each other. This is the force that structures the heliospheric medium: solar wind streams of different speeds collide and compress each other to form a complex and evolving pattern as the solar wind flows away from the Sun into the distant reaches of the heliosphere.

The solar wind is variable not only as a function of the coronal region from which it originates, but, as these regions change in time, the resultant pattern of the solar wind also evolves. The major changes happen through the 11-year solar cycle: the solar corona undergoes considerable changes from one sunspot minimum to the next. At solar maximum much of the coronal material is firmly tied in very complex magnetic loops. Only small and transient coronal holes are present from which relatively short-lived, fast streams are emitted. Much of the solar wind emitted at solar maximum has slow speed and variable density. However, due to the complexity of the corona and frequent emergence of new magnetic flux from below the solar surface, the coronal structures often become unstable, resulting in large and frequent Coronal Mass Ejections (CMEs) which project considerable amounts of extra material and magnetic flux into the solar wind. From one solar rotation (about 26 days) to the next, the corona can change, due to restructuring resulting from the many instabilities that develop as the magnetic structures interact.

The heliospheric medium reflects fast-changing structures in the corona. The solar wind is generally slow and variable; faster streams are slowed down by their interaction with the dominating slow solar wind. Superimposed on the background variability, the counterparts of CMEs also bring a considerable amount of disturbance to the solar wind and the embedded Heliospheric Magnetic Field (HMF).

At solar minimum the configuration of the solar corona acquires a relative simplicity, compared with its state at maximum activity. Over the polar regions of the Sun, two large and stable coronal holes develop as the sunspot cycle wanes and remain there for a year or so after the minimum in the activity cycle. The region of confined plasma in the solar corona is restricted to the equatorial region. Fast solar wind from the coronal holes flows unimpeded into the heliosphere away from the equatorial region; slow streams are restricted to the vicinity of the solar equator, as described in detail in Chapter 2. The stability of coronal structures leads to stability in the flow patterns of the solar wind; at low to mid-latitudes fast and slow solar-wind streams interact to form a stable pattern of Corotating Interaction Regions (CIRs) which are not only characteristic of the epoch around solar minimum, but also represent the dominant structures in the inner and middle heliosphere (see Chapter 3 by Forsyth and Gosling).

Together with the solar wind, the HMF provides structure to the heliospheric medium. As mentioned above the HMF originates in the corona, but is carried away in the solar wind; given the high electrical conductivity of the plasma, the magnetic field can be regarded as 'frozen' into the solar wind. Measurements of the HMF allows us to link solar wind plasma to its solar origin. Parker's (1958) original theory of the solar wind also included a relatively simple magnetic field. This model assumed that the magnetic field is uniform around the Sun and is radially oriented on a surface in the outer corona where the solar wind originates.

Given the existence of the continuous flow of the solar wind, how is the outer boundary of the heliosphere determined? The simple sketch in Figure 1.1 provides an outline of the very complex answer to this question. In the first place, the solar wind eventually slows down; this occurs through a shock wave, the so-called termination shock, where the solar wind speed falls below the sound speed. No space probe has yet reached the termination shock, although *Voyager 1*, now at some 80 AU from the Sun (1 AU is the distance from the Sun to the Earth, about 150 million km) is thought to be getting close to it. Beyond the termination shock, a pressure balance exists between the Local Interstellar Medium (LISM) and the solar wind, through a surface called the heliopause. It is possible (although not proven as yet) that the interstellar wind (corresponding to the motion of the heliosphere through the LISM) may be fast enough to generate a shock wave, the heliospheric bow shock, upstream of the heliopause.

6　Introduction　[Ch. 1

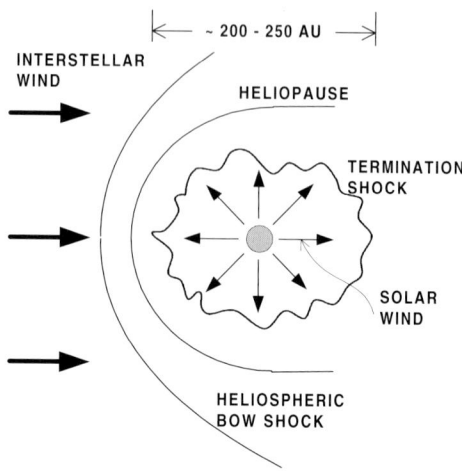

Figure 1.1. A schematic representation of the heliosphere.

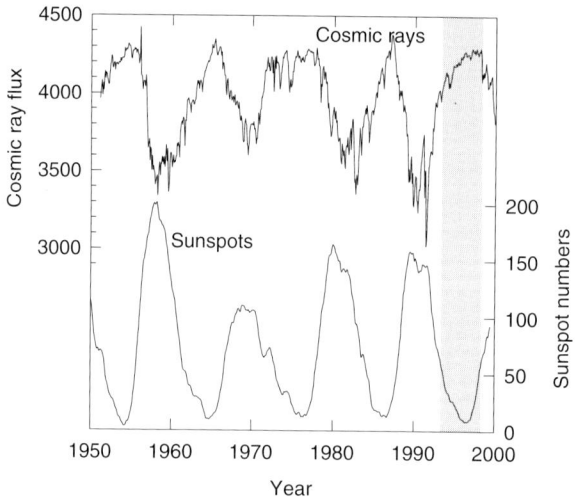

Figure 1.2. The anti-correlation between sunspot numbers and the intensity of cosmic rays detected at Earth that first led to the concept of the heliosphere as early as 1955 (Davis, 1955). The shaded area represents the period of the last solar minimum activity when *Ulysses* made the first three-dimensional observations in the heliosphere.

There is another argument for the existence of the heliosphere, as a finite volume in which the Sun dominates the properties of the medium. Originally, this argument predated the discovery of the solar wind and was related to the modulation of cosmic rays (Davis, 1955). It was noticed that the intensity of cosmic rays (very energetic particles that reach us from distant sources in the galaxy) was anti-correlated with the sunspot number, so that the cosmic ray intensity at Earth was at a maximum when there was a minimum in the number of sunspots. This is illustrated in Figure 1.2. The explanation offered was that, for reasons that were unknown at that time, solar activity influenced the access, or propagation of cosmic rays into the inner solar system; at high solar activity, cosmic rays encountered more difficult propagation conditions than when solar activity was low. The cause of this solar modulation of the cosmic-ray intensity became clear with the discovery of the solar wind and the embedded heliospheric magnetic field. The propagation of cosmic rays is controlled by the structure of the magnetic field; clearly, this structure is more complex at solar maximum than at solar minimum, leading to a reduction in cosmic ray intensity at sunspot maximum. Although this explanation became quite generally accepted, the theories and models that resulted were proved to be quite inadequate when *Ulysses* explored the polar regions of the heliosphere, as described in detail in Chapter 8 by Bruce McKibben.

1.1.3 Discovery missions to the heliosphere

Space probes have played a key role in exploration of the heliosphere. Since the early 1960s numerous missions have explored different regions in the vicinity of the Earth: in the distant, outer heliosphere; between the Sun and the Earth; and now, above the poles of the Sun. To end this section we describe a few key missions that contributed in a significant way to our knowledge of the properties and structures of the heliospheric medium, highlighting their contributions to our knowledge of the heliosphere.

The Soviet space missions to the Moon (*Lunik 1* and *2*) carried plasma detectors that first observed the solar wind. However, the first extensive set of data confirming the existence of the solar wind came from the *Mariner 2* mission to Venus in 1962, launched on 27 August 1962 (Davis *et al.*, 1964; Neugebauer and Snyder, 1962). Some of the properties of the embedded magnetic field were also observed on this mission (Coleman *et al.*, 1962). In the 1960s early Earth-orbiting spacecraft, with apogees reaching beyond the magnetosphere, discovered many of the basic properties of the solar wind and the magnetic field embedded in it. All the most important phenomena that shape and characterize the heliospheric medium were first noted by Earth-orbiting spacecraft. In particular, good agreement was found in general between the Archimedean spiral structure of magnetic field lines proposed by Parker (1958, 1963) and the observed orientation of magnetic field (Davis *et al.*, 1964; Ness and Wilcox, 1964) on the *IMP-1* mission. On the same mission the sector structure of the magnetic field, showing a recurring pattern of alternating polarities of the field as a function of solar longitude, was also identified (Wilcox and Ness, 1964). Other pioneering missions in Earth orbit were the *Vela* series in the 1960s and 1970s, operated primarily for monitoring nuclear tests on Earth, but able to measure solar-wind plasma as well.

The *Pioneer* series of missions, from the early 1960s, explored the region of space between Mars and Venus, as well as the distant heliosphere past Jupiter and Saturn. The first spacecraft to the outer planets were the *Pioneer 10* and *11* missions. *Pioneer 10* was launched on 2 March 1972; it reached Jupiter on 3 December 1973 to make the first close-up observations of the giant planet. After its encounter with Jupiter, the spacecraft followed a trajectory to the outer reaches of the heliosphere. *Pioneer 11* was launched on 5 April 1973 and reached Jupiter on 2 December 1974. At Jupiter the spacecraft was targeted in such a way that the fly-by would allow it to reach Saturn. This involved a trajectory which for the first time reached a heliolatitude of 16° above the ecliptic plane. The two missions were terminated in March 1997 and November 1995, respectively. At the time when the missions ended, *Pioneer 10* was at a heliocentric distance of 67 AU, while *Pioneer 11* was at 43 AU.

These two spacecraft were the first to make detailed observations of the heliospheric medium out to Saturn and beyond. The first phase of the *Pioneer* mission took place around the minimum-activity period in solar cycle 21 in the mid-1970s. They were the first to observe the recurring sequence of CIRs and their development with heliocentric distance (Smith and Wolfe, 1977); these large-scale structures, that consist of successively compressed and rarefied solar wind and magnetic field, are the most important features of the inner and middle heliosphere around solar minimum.

The inner heliosphere, between the Sun and the Earth's orbit, was the target of the joint German/NASA *Helios* mission. The two *Helios* spacecraft were launched on 10 December 1974 and 15 January 1976, respectively, into high-eccentricity heliocentric orbits with a perihelion of 0.3 AU and an aphelion of 1 AU. These two spacecraft collected valuable data on heliospheric processes in the region between the Sun and the Earth, in particular on the early evolution of heliospheric structures as a function of distance from the Sun (Schwenn and Marsch, 1990).

The *Voyager 1* and *2* spacecraft, launched in September 1977 and August 1977, respectively, still provide the only opportunity in the next decade or two to reach the outer boundary of the heliosphere. Since their launch the two spacecraft visited Jupiter and Saturn, and *Voyager 2* also flew by the two outer gas giants, Uranus and Neptune. Both spacecraft are now on trajectories that approach the outer boundaries of the heliosphere: *Voyager 1* at a speed of about $3.5 \, \text{AU/year}^{-1}$, *Voyager 2* at about $3.1 \, \text{AU/year}^{-1}$.

The first mission targeted specifically to explore the third dimension of the heliosphere is *Ulysses*, the subject of this book. This mission provided a significant step forward in our understanding of the heliosphere, by charting its structure as a function of heliolatitude.

1.2 *ULYSSES*: SCIENTIFIC OBJECTIVES AND MISSION CHARACTERISTICS

1.2.1 Introduction

Ulysses is a co-operative mission between the European Space Agency (ESA) and the National Aeronautics and Space Administration (NASA) aimed at increasing our knowledge of conditions and processes occurring in the inner heliosphere by means of *in-situ* observations covering the widest possible range of solar latitudes. Specific objectives of the scientific investigations carried out by *Ulysses* include the following:

- to determine the global, three-dimensional properties of the heliospheric magnetic field and the solar wind;

- to study the origin of the solar wind by measuring the composition of solar-wind plasma at different heliographic latitudes;
- to increase our knowledge of waves, shocks and other discontinuities in the solar wind by sampling plasma conditions that are different from those available near the ecliptic;
- to improve our understanding of solar modulation process affecting galactic and anomalous cosmic rays by measuring their latitudinal gradients and energy spectra;
- to improve our knowledge of the origin of galactic cosmic rays by measuring their isotopic composition;
- to study the acceleration and propagation of energetic particles of solar and interplanetary origin by observing their characteristics at low and high latitudes;
- to advance our knowledge of the local interstellar medium by measuring directly the neutral helium component that enters the heliosphere, as well as inferring its properties from measurements of interstellar pick-up ions;
- to improve our understanding of interplanetary and interstellar dust;
- to search for gamma-ray burst sources and, in conjunction with observations from other spacecraft, contribute to their identification with known celestial objects.

As is made clear in Section 1.3 and in the remaining chapters of this book, these objectives have been very successfully met in the context of the solar minimum heliosphere. In addition, *Ulysses* has made important contributions to our knowledge of the Jovian magnetosphere during the Jupiter fly-by phase, and has conducted a search for low-frequency gravitational waves using the spacecraft's radio communication link. Within the co-operative programme, ESA provided the spacecraft and is responsible for its operation. NASA provided the spacecraft power supply, the launch vehicles, and provides tracking via the Deep Space Network (DSN), and is responsible for data management.

1.2.2 The *Ulysses* orbit

Launched on 6 October 1990 *Ulysses* is the first, and so far only, spacecraft to venture into the unexplored regions above the solar poles. Direct injection into a high-inclination solar orbit is beyond the capabilities of even the most powerful launch vehicles, so *Ulysses* had to follow a circuitous path. The space probe, mounted atop a combined Inertial Upper Stage (IUS) and Payload Assist Module (PAM-S) rocket, was first brought to low-Earth orbit by the space shuttle *Discovery*. Following injection into an interplanetary transfer orbit, a gravity-assist manoeuvre at Jupiter was employed to change the orbital plane of the *Ulysses* space probe such that it reached high solar latitudes. It is interesting to note that the speed of *Ulysses* as it left the Earth's gravitational influence was 11.4 km s^{-1}, the greatest of any man-made object to date.

The prime scientific requirement for the *Ulysses* trajectory design was to maximize the time the spacecraft spends at high heliographic latitudes. As a convenient, but somewhat arbitrary, criterion the total observation time at latitudes greater than 70° was introduced. The period during which *Ulysses* was poleward of 70° heliographic latitude in a given hemisphere was designated a 'polar pass'. A minimum period of 150 days during the two polar passes (north and south) was originally specified. Maximum solar latitude was a secondary consideration. Additional constraints on the mission design were that:

- the perihelion distance should not be less than 1.28 AU for thermal reasons;
- the heliocentric radius at maximum latitude should not be larger than 2.3 AU for scientific reasons;
- the Sun-probe-Earth angle should not exceed 60° (excluding the first 30 days of the mission) to comply with the instrument design;
- Jupiter closest approach should be greater than 6.0 Jupiter radii to avoid radiation damage to instruments;
- Jupiter closest approach shall be at such longitude (time) to minimize radiation fluence.

The injection accuracy attained by the upper stage motors was extremely high. As a result only two small orbit adjustments were required to target the spacecraft to a position slightly north of Jupiter. It was possible to use some of the fuel budgeted for the first of these Trajectory Correction Manoeuvres (TCM-1) to improve the mission performance slightly. As a result, the final out-of-ecliptic orbit is such that *Ulysses* spends a total of 234 days above 70° solar latitude (typically 132 days at high southern latitudes and 102 days in the north) and achieves a maximum heliographic latitude of 80.2°. Since there was no a priori scientific reason to explore a given hemisphere first, the position of Jupiter (slightly south of the solar equator at the time of the fly-by) was exploited to optimize the flight path by sending *Ulysses* south.

Figure 1.3 shows a perspective view of the trajectory as seen from 15° above the ecliptic plane. In order to provide a more quantitative view, the orbit is shown in two alternative representations. In the first (Figure 1.4) heliographic latitude is plotted as a function of heliocentric distance. From this it can be seen that the polar passes in the two hemispheres are not symmetrical with respect to the Sun. In Figure 1.5, the ecliptic latitude and longitude of *Ulysses* in a fixed Sun-Earth reference system are projected onto the solar disc for the period March 1992 to February 1998. This representation is useful in that it allows us to correlate features and events on the surface of the Sun, as viewed from Earth, by *Ulysses* observations. As can be seen, the footpoint of *Ulysses* on the Sun is generally visible from the Earth between the months of December and June, crossing central meridian around March each year. Some key parameters of the *Ulysses* orbit are given in Tables 1.1 and 1.2, and a full description of the orbit in terms of its Classical Orbital Elements can be found in Section 1.2.7.

1.2.3 Scientific payload

The nine experiments that make up the *Ulysses* scientific payload were developed and provided by international teams of scientists, each led by one (or in some cases,

Sec. 1.2] *Ulysses*: scientific objectives and mission characteristics 11

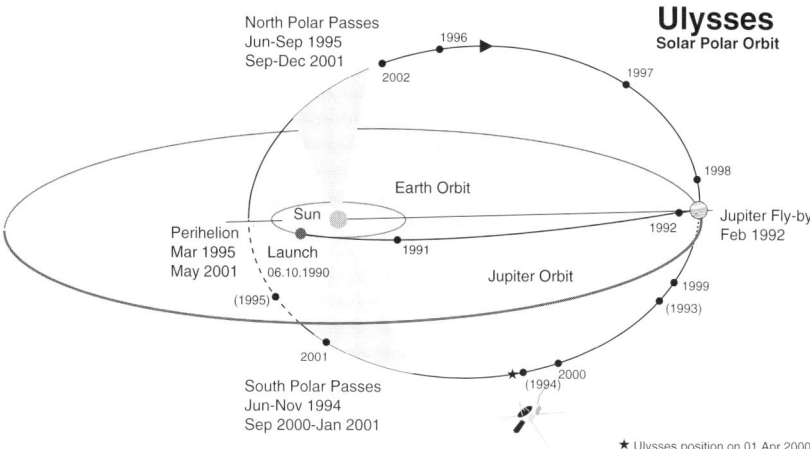

Figure 1.3. The *Ulysses* orbit viewed from 15° above the ecliptic plane (dots mark the start of each year).

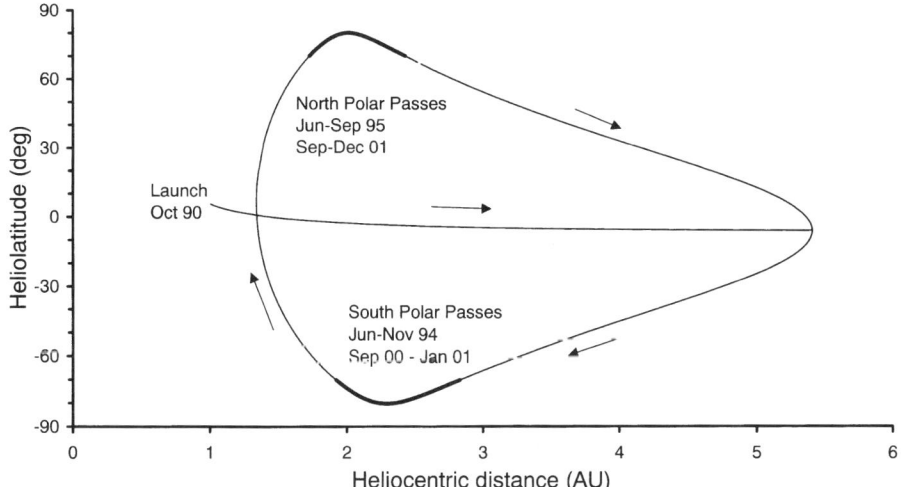

Figure 1.4. The heliographic latitude of the *Ulysses* spacecraft plotted as a function of heliocentric radial distance.

two) Principal Investigator(s). Countries involved in the *Ulysses* mission include Austria, Belgium, Canada, Finland, France, Germany, Greece, Hungary, Italy, Netherlands, Poland, Sweden, Switzerland, United Kingdom, and the USA. The payload, which is made up of two magnetometers, two solar-wind plasma instruments, a unified radio/plasma wave instrument, three energetic charged-particle instruments covering a wide range of energies and species, an interstellar

12 Introduction [Ch. 1

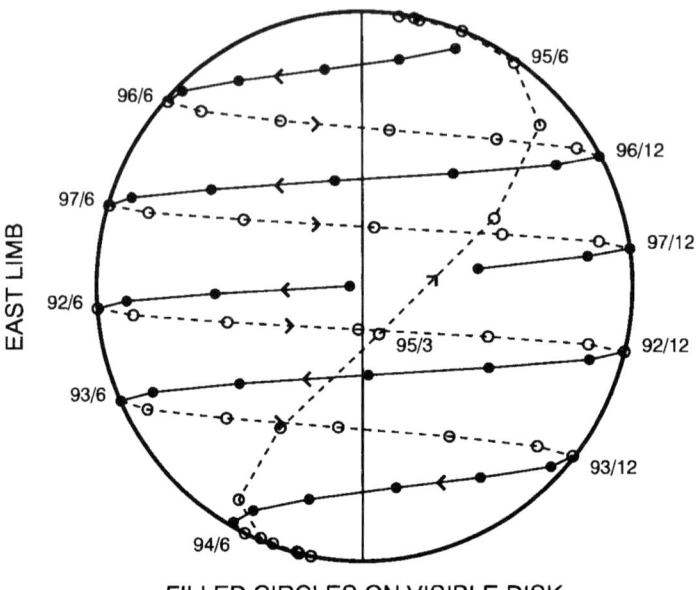

Figure 1.5. Projection of the *Ulysses* orbit (ecliptic latitude and longitude) on the solar disc as seen from Earth for the period March 1992–February 1998. Markers show the spacecraft position at 1-month intervals (filled circles and solid lines indicate footpoints on the visible disc).

Table 1.1. Main trajectory characteristics.

Parameter	
Orbital period (years)	6.195
Semi-major axis (AU)	3.373
Semi-minor axis (AU)	2.691
Inclination with respect to ecliptic (deg)	79.1

neutral-gas sensor, a solar X-ray/cosmic gamma-ray burst instrument, and a cosmic dust sensor, is summarized in Table 1.3. Many of the scientific instruments incorporate novel measurement techniques that were flown for the first time on *Ulysses*. Examples include the interstellar neutral-gas experiment that employs secondary electron or ion emission generated by the impact of a neutral atom on a LiF surface; the solar-wind ion-composition sensor that utilizes a combination of time-of-flight and energy signals to measure an ion's mass per charge and energy per charge simultaneously; and the high-resolution cosmic-ray isotope spectrometer that uses a new generation of position-sensing solid-state detectors. As a result of

Table 1.2. Timeline of mission events.

Event	Date (Yr Mo Dy)
Launch	1990 10 06
Jupiter fly-by	1992 02 08
First polar pass (S)	
start (70°S)	1994 06 26
maximum latitude (80.2°S)	1994 09 13
end (70°S)	1994 11 05
Perihelion (1.34 AU)	1995 03 12
Second polar pass (N)	
start (70°N)	1995 06 19
maximum latitude (80.2°N)	1995 07 31
end (70°N)	1995 09 29
Aphelion (5.41 AU)	1998 04 17
Third polar pass (S)	
start (70°S)	2000 09 06
maximum latitude (80.2°S)	2000 11 27
end (70°S)	2001 01 16
Perihelion (1.34 AU)	2001 05 23
Fourth polar pass (N)	
start (70°N)	2001 08 31
maximum latitude (80.2°N)	2001 10 13
end (70°N)	2001 12 10
Aphelion (5.41 AU)	2004 06 30

a comprehensive pre-launch programme, *Ulysses* is one of the magnetically 'cleanest' spacecraft ever launched.

The *Ulysses* on-board investigations are complemented by ",4 > Interdisciplinary and Guest Investigations, which combine data from several instruments to address specific problems of heliospheric science.

1.2.4 Spacecraft and mission operations

The *Ulysses* spacecraft is shown in its pre-launch configuration in Figure 1.6. Key design features of the spin-stabilized (5 rpm) spacecraft include the 1.65-m diameter, Earth-pointing High-Gain Antenna (HGA) that provides the communication link, and the Radioisotope Thermoelectric Generator (RTG) that supplies the spacecraft's electrical power. In order to satisfy the requirements imposed by the experiments with respect to electromagnetic cleanliness (EMC) and RTG radiation levels a 5.6-m radial boom, carrying several experiment sensors, is mounted on the opposite side of the spacecraft to the RTG. A 72.5-m tip-to-tip dipole wire boom and a 7.5-m axial boom serve as electrical antennas for the Unified Radio and Plasma Wave Experiment. Most of the scientific instruments are mounted on the main body of the

Table 1.3. The *Ulysses* hardware investigations.

Investigation	Acronym	Principal Investigator	Measurement
Magnetic field	VHM/FGM	A. Balogh Imperial College, London (UK)	Spatial and temporal variation in the heliospheric magnetic field: 0.01–44,000 nT
Solar-wind plasma	SWOOPS	D. J. McComas Los Alamos National Laboratory (USA)	Solar-wind ions: $225\,eV\,e^{-1}$ to $34.4\,keV\,e^{-1}$ Solar-wind electrons: 0.86–814 eV
Solar-wind ion composition	SWICS	G. Gloeckler University of Maryland (USA)	Elemental and ionic charge composition, temperature and mean speed of solar-wind ions: $140\,km\,s^{-1}(H^+)$ to $1,285\,km\,s^{-1}(Fe^{8+})$
Radio and plasma waves	URAP	R. J. MacDowall NASA/GSFC (USA)	Plasma waves, solar radio bursts, electron density and electric field: 0–60 kHz (plasma waves) 1–940 kHz (radio) 10–500 Hz (magnetic)
Energetic particles and interstellar neutral gas	EPAC/GAS	E. Keppler MPAe, Lindau (Germany)	Energetic ion composition: $300\,keV\,n^{-1}$ to $5\,MeV\,n^{-1}$ neutral He atoms
Low-energy ions and electrons	HI-SCALE	L. Lanzerotti Lucent Technologies, Bell Laboratories (USA)	Energetic ions: 50 keV to 5 MeV Energetic electrons: 30–300 keV
Cosmic rays and solar particles	COSPIN	R. B. McKibben University of Chicago (USA)	Energetic ions: $0.3\text{--}600\,MeV\,n^{-1}$ Electrons: 2.5–6,000 MeV
Solar X-rays and cosmic gamma-ray bursts	GRB	K. Hurley UC Berkeley (USA)	Solar flare X-rays and gamma bursts: 15–150 keV
Cosmic dust	DUST	E. Gruen MPK Heidelberg (Germany)	Dust particles: $10^{-16}\text{--}10^{-6}\,gm$

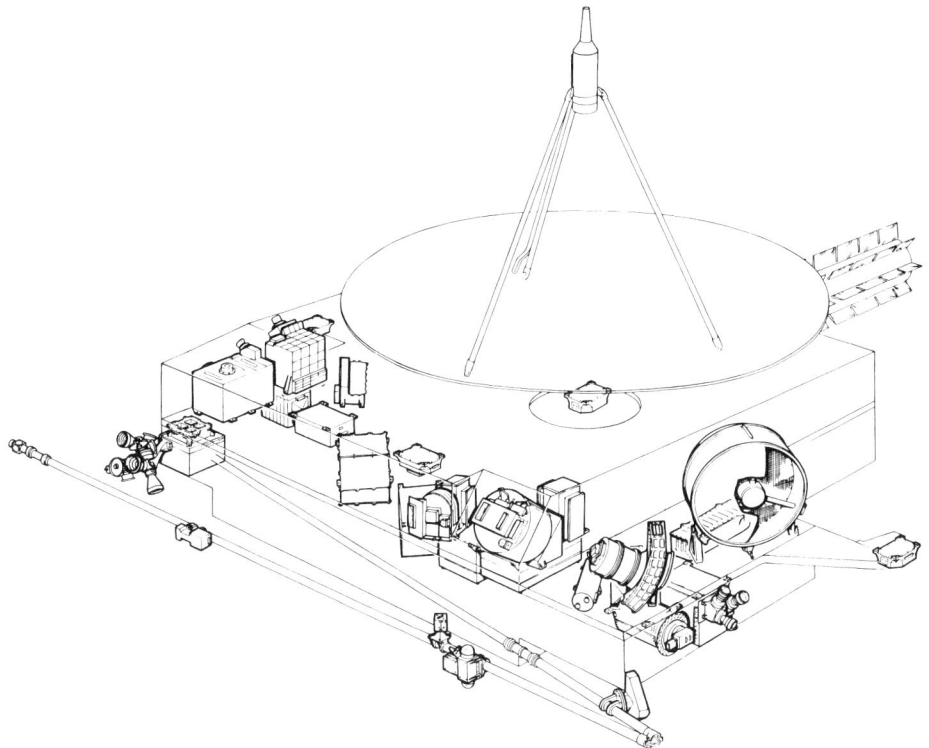

Figure 1.6. The *Ulysses* spacecraft in pre-launch configuration.

spacecraft as far as possible from the RTG (See Figure 1.6). The spacecraft mass at launch was 367 kg, including 55 kg for the scientific payload and 33.5 kg of hydrazine for orbit, attitude and spin-rate adjustments.

Near-continuous data throughout the mission is a prime scientific requirement. Since continuous coverage by ground stations is impossible for such a long-duration mission, data are stored on board and replayed, interleaved with real-time data, during periods of coverage (typically 8–10 hours in every 24 hours. The prime data rates are $1,024\,\mathrm{bits\,s^{-1}}$ for real-time data ('tracking mode') and $512\,\mathrm{bits\,s^{-1}}$ for stored data ('storage mode'). Mission operations for *Ulysses* are conducted from the Advanced Multi-Mission Operations System facilities, located at JPL (Jet Propulsion Laboratory), by a joint ESA/NASA mission operations team. JPL constitutes the focal point of the DSN, whose 34-m ground station subnet is used for communications with *Ulysses*.

1.2.5 Brief history

The idea of sending a probe to explore the regions of space far away from the ecliptic plane is by no means new. First mention of an out-of-ecliptic mission can be found in a meeting report from 1959 (Simpson et al., 1959), although at that time essential elements of such a mission, including precise interplanetary navigation, had yet to be proven. By the mid-1970s, however, a joint European Space Research Organisation (ESRO)/NASA Mission Definition Study had shown the feasibility of a dual-spacecraft concept that utilized a Jupiter gravity-assist manoeuvre to place two probes in oppositely directed, high-inclination orbits around the Sun. This mission, in which ESA (as ESRO has since become) and NASA were each to provide one probe, was given the name Out-Of-Ecliptic (OOE) Mission, with a planned launch in 1983.

The scientific investigations to be performed on the exploratory mission were selected by ESA and NASA jointly in 1978, with a total of seventeen flight experiments making up the payloads of the two spacecraft. Each probe carried a core set of instruments to measure the solar wind, the heliospheric magnetic field, energetic particles, cosmic rays, radio and plasma waves, and solar X-rays and cosmic gamma-ray bursts. In addition, the NASA probe was fitted with solar-imaging instruments, including a white-light coronagraph.

Having got off to a good start, by the beginning of the 1980s the programme's luck had started to change. In 1980 financial difficulties associated with the space shuttle forced NASA to delay the launch until the 1985 Jupiter opportunity. One year later, an even more serious blow was dealt when NASA cancelled its own spacecraft, and announced a launch delay of the remaining ESA probe by a further 13 months. Fortunately, commitment to the programme remained firm on the European side. In 1984, at the suggestion of science-team member Bruno Bertotti, the mission was renamed *Ulysses* in honour of the Greek hero of the Trojan war and King of Ithaca. Preparations for a launch aboard the space shuttle in May 1986 were well under way when, on 28 January of that year, the *Challenger* accident occurred. All shuttle launches were put on hold, and the *Ulysses* spacecraft went into storage. Finally on 6 October 1990 – nearly 8 years later than originally envisaged – *Ulysses* began its exploratory voyage aboard the shuttle *Discovery*.

The many disruptions and delays in the programme have clearly had an impact on the scientific goals of the mission. Most obvious is the lack of simultaneous observations above and below the ecliptic as originally foreseen with the two-spacecraft concept. Simultaneous *in-ecliptic* measurements, on the other hand, have been readily available from near-Earth platforms such as *IMP-8*, *WIND* and *ACE*. Cancellation of the proposed NASA probe, with its solar-imaging capability, also meant the loss of a unique chance to image the corona simultaneously from in- and out-of-ecliptic vantage points (e.g. together with *SOHO*).

Another effect of the various launch delays has been the shift in timing of the polar passes with respect to the solar cycle. If *Ulysses* had been launched in 1986, the first polar passes would have occurred near solar maximum, with the corona at its most complex. In reality, *Ulysses* arrived over the solar poles near solar minimum, and was therefore able to characterize the simpler, 'steady-state' solar-wind structure in great detail. These conditions matched closely those anticipated when the mission was first approved in the 1970s, albeit one solar cycle too late! Based on the excellent scientific output, and the good condition of the spacecraft and its payload, ESA and

NASA agreed to continue scientific operations beyond the original end-of-mission date (October 1995). In the first instance, this extension covered a period of 6.2 years, allowing observations to be made over the poles in 2000–2001, near solar maximum. A further extension, until September 2004, is now planned in order to explore the effects on the global heliosphere of the polarity reversal of the Sun's magnetic field.

1.2.6 Ulysses data archiving

The principal public archives for *Ulysses* data include the ESA Archive for Ulysses Data in ESA's Space Science Department at Estec, NSSDC, World Data Center A for Rockets and Satellites (co-located with NSSDC), and the Planetary Plasma Interactions Discipline Node of NASA's Planetary Data System (PDS/PPI) at the University of California at Los Angeles (Marsden *et al.*, 1996). Some *Ulysses* investigators distribute their data directly, in either quick-look or archive quality, via the Internet and the World Wide Web, while also using this method to submit the data to public archives. Most of the datasets are available in easy-to-use ASCII format and are accompanied by informative dataset descriptions. In several instances, comprehensive user guides have been provided for more detailed information on experimental hardware, science objectives, measured parameters, data processing, and data-quality issues. The PDS archive also consists of ASCII files, most of which contain one day of data. In 1999 a collection of eight CD-ROMs with archived data, covering the period from launch until September 1995, was issued as a special ESA publication (ESA SP-1230). This collection also includes the complete Jupiter fly-by dataset.

1.2.7 Ulysses orbital elements

The orbit of *Ulysses* can be described by a set of parameters called the *Classical Orbital Elements*, which can be specified for a given phase of the mission and are usually referenced to a particular time, or epoch, and frame of reference. The definitions of the quantities making up the classical orbital elements are as follows:

- mean distance (a) – the semi-major axis of the orbit measured in astronomical units (1 AU) = 149.59787 million km);
- the inclination (i) – the angle between the ecliptic plane and the plane of the orbit;
- eccentricity (e) – the eccentricity of the conic (0 = circle, <1 = eclipse, 1 = parabola, >1 = hyperbola) that describes the orbit;
- longitude of the ascending node (Ω) – the position in the orbit where the path of the spacecraft passes through the ecliptic plane, from below the plane to above the plane, measured from the vernal equinox (1st point of Aries);
- argument of perihelion (ω) – the angle in the plane of the orbit from the ascending node to the point where the spacecraft is closest to the Sun;

Table 1.4. Classical orbital elements for *Ulysses*.

Classical orbital element	Earth–Jupiter segment (valid from 16 October 1990 to 30 December 1991)	Jupiter fly-by (valid from 30 December 1991 to 19 March 1992)	Final out-of ecliptic orbit (valid from 19 March 1992 to 31 December 2001)
a (km)	134,510,6000	−676,299.74	504,594,094
e	0.889166	1.666109	0.60306
i (deg)	1.990719	142.19459	79.12801
Ω (deg)	12.85909	−43.055518	−22.51862
ω (deg)	7.72595	127.68470	−1.11377
M (deg)	$0.0365574 \times (J - 2,448,177.0755)$ where J is the Julian date of interest	–	$0.1591096 \times (J - 2,449,788.986)$
Perihelion date (Julian)	2,448,177.0755	2,448,661.003	2,449,788.986

- mean anomaly (M) – angle increasing uniformly with time by 360 degrees per orbital period from 0 at perihelion;
- true anomaly (TA) – the actual angle between the spacecraft position and the perihelion as seen from the Sun. This angle increases non-uniformly with time, changing most rapidly at perihelion.

In the case of *Ulysses* the reference frame used for the orbital elements in the 'Mean Ecliptic and Equinox of 1950.0'. The numerical values of the Classical Orbital Elements for the *Ulysses* trajectory are shown in Table 1.4.

In order to calculate the position of *Ulysses* in a given system of co-ordinates (e.g. heliocentric ecliptic co-ordinates) at a given date, we have to use the same procedure that whould be needed if we wanted to know the position of a planet or comet. In general terms this involves solving Kepler's equation:

$$E - e \sin E = M$$

using an iterative technique to obtain the intermediate quantity E called the eccentric anomaly from the mean anomaly M, both expressed in radians. The eccentric anomaly E is then used to compute the true anomaly TA, which tells us the actual position of the body (planet or spacecraft) in its orbit, where TA is given by:

$$TA = 2 \arctan \sqrt{\frac{1+e}{1-e}} \tan \frac{E}{2}$$

Armed with this information, and the appropriate orbital elements for the Earth if needed, we can then use the other classical orbital elements to find the co-ordinates of *Ulysses* in our reference system of choice (Figure 1.7). As an example, the position of *Ulysses* at 12:00 UT on 26 July 1999 (Julian date 2,451,386.0) has been computed

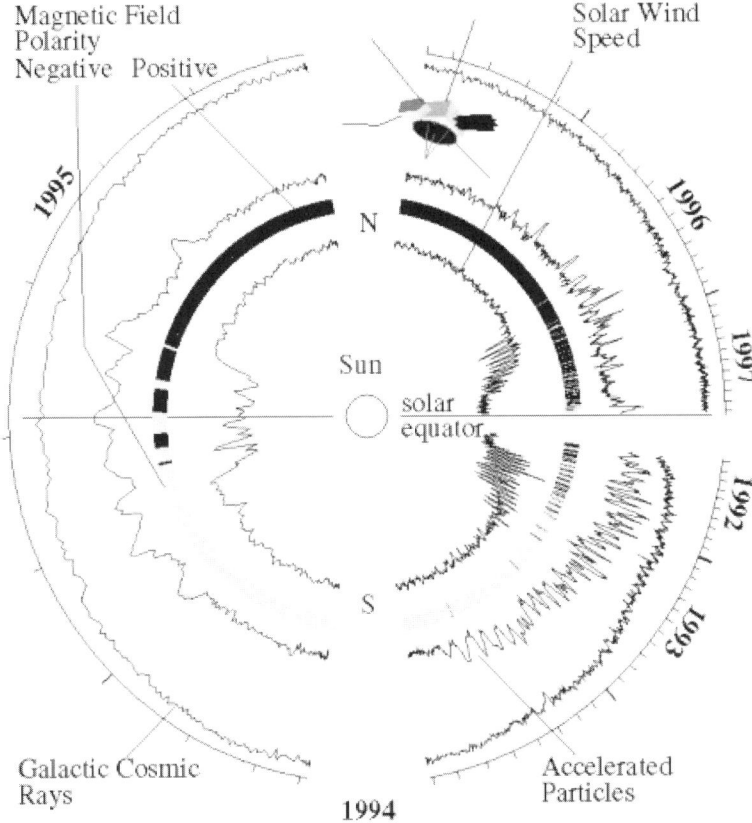

Figure 1.7. An overview of the *Ulysses* observations as a function of heliolatitude. In this polar plot, the speed of the solar wind, the polarity of the heliospheric magnetic field, and the intensities of energetic particles and cosmic rays measured by *Ulysses* are shown (from the centre, respectively). For a detailed discussion, see the text.

using the classical orbital elements listed above. The results are given in heliocentric ecliptic co-ordinates, and heliographic right ascension and latitude:

Date: 1999 07 26 12 00 00

Heliocentric range (AU)	4.726
Xh (AU)	−4.126
Yh (AU)	1.318
Zh (AU)	−1.891
Heliocentric ecliptic latitude (deg)	− 23.6
Heliocentric ecliptic longitude (deg)	162.3
Heliographic latitude (deg)	−30.8
Heliocentric right ascension (deg)	267.0

1.3 HIGHLIGHTS OF THE SCIENTIFIC DISCOVERIES OF THE *ULYSSES* MISSION

1.3.1 Introduction

Ulysses observations have established how the internal structure and properties of the heliosphere vary from the equator to the poles. The changes with latitude of four important constituents can be seen in Figure 1.7. This polar plot consists of concentric traces with an image of the solar corona at the centre. For the innermost trace, representing the solar wind, distance from the center is proportional to the speed at a given latitude. Each hemisphere contains two kinds of wind, variable slow wind at low latitudes and a steady high-speed wind at high latiudes reaching all the way to the pole. The two hemispheres appear to be symmetric.

In the trace outside the wind speed, distance from the centre is proportional to the intensities of energetic particles accelerated inside the heliosphere rather than at the Sun. They occur as a series of bursts at both low and high latitudes with intensities decreasing toward the poles. Although accelerated at low latitudes they are able to propagate into the region occupied by fast wind.

The third trace shows the intensities of high-energy galactic cosmic rays which enter the heliosphere from interstellar space. The intensities increase gradually from the equator to the poles again without being influenced by different solar-wind speeds.

The outermost trace represents the magnetic field which originates at the Sun and is transported into the heliosphere along with the solar wind. Although the field strength and direction vary systematically with latitude, only the polarity appears in this figure, opposite polarities being shown by the two shades. The field points outward in the north and inward or sunward in the south. The observed polarities are the same as those of the solar magnetic field at the respective poles.

Inspection of the figure shows that the boundary between fast and slow wind does not appear to be abrupt. Quasi-sinusoidal variation between high and low speeds is apparent in the right half of the figure. They are actually indicative of a coupling between the two regions. There are also many other aspects of this transition or coupling that are not represented in this figure. Thus, there are actually three distinct regions within which the scientific highlights of the mission can be characterized. A further distinction of the *Ulysses* results is whether the constituents originate at the Sun or enter the heliosphere from the local interstellar medium, specifically cosmic rays, dust and neutral and ionized gases (pickup ions).

The following description of scientific highlights begins with unique high-latitude observations in the polar heliosphere (i.e. the fast wind). Next, various aspects of the coupling between the high- and low-latitude heliosphere are presented. Finally, interstellar constituents are discussed along with conclusions drawn from observations about the medium outside the heliosphere. This overview of scientific results contains references to specific sections and figures in the chapters which follow and which provide detailed discussions.

1.3.2 High-latitude heliosphere

Polar solar wind

Ulysses has established the fundamental properties of the fast wind that originates at high solar latitudes in the polar coronal holes (Chapter 2). In so doing it has clarified the differences between fast and slow wind. It is safe to say that, as a result of *Ulysses*, we now know significantly more about the fast wind than about the low-latitude slow wind. One major reason for this state of affairs is the relative simplicity of the fast wind because of the absence of significant structure and hence the absence of troubling questions related to the dynamic evolution of the slow wind as it propagates outward. Another major reason is that the fast wind is known to be associated with coronal holes whereas the source of the slow wind is uncertain.

The basic properties of fast wind are known as a function of latitude (Figure 2.3). The velocity rises monotonically from the transition where the speed is about $500 \,\text{km}\,\text{s}^{-1}$ to approximately $750 \,\text{km}\,\text{s}^{-1}$ over the poles. The proton density falls off with distance, as expected, as r^{-2}. The product of the density n and speed v is constant on average, as in the ecliptic plane, an invariance apparently associated with the production of the solar wind. This invariance implies that the mass flux mnv is essentially independent of latitude. The momentum flux density or ram pressure of the wind, on the other hand, increases along with v and is larger at high latitudes than near the equator. The proton temperature T is correlated with speed, as in the ecliptic, so that the high-latitude wind is hot with $T \approx 6 \times 10^5 \,\text{K}$ in the fastest wind. Another characteristic feature is the relative constancy of the He/H ratio of 0.044.

The capability of *Ulysses* to measure heavy solar-wind ions and their charge states (degree of ionization) has revealed other basic differences between fast and slow wind. The composition of both types of wind exhibits the First Ionization Potential (FIP) effect, observed previously for energetic solar particles, with easily ionized atoms, such as metals, being more abundant than hard-to-ionize elements such as helium (Figure 2.11). Although this FIP effect is more obvious in the slow wind between fast streams, it persists in the fast wind as well. The ratio of the abundance of magnesium to oxygen, Mg/O, has been used as a measure of the strength of the FIP effect. The existence of the FIP effect in both types of wind shows that they are being influenced by physical processes characteristic of the lower lying chromosphere. Perhaps this coupling between the chromosphere and corona is related to the on-going replenishment of the corona which takes place several times a day as a consequence of the mass lost as solar wind.

The charge-state information can be used to derive the 'freezing-in' temperature of the constituent ions (i.e. the temperature in the corona at which electron collisions no longer influence ionization which then becomes invariant) (Figure 2.9). It has been the practice to derive the freezing-in temperature of oxygen T_O, as a typical measure of the freezing-in temperature of the wind.

These two parameters are highly correlated (Figure 2.12). They also reveal distinct differences between fast and slow winds with the Mg/O ratio and T_O being less in the former (Figure 2.13). Equally important is the finding that these parameters change abruptly thereby providing a means of identifying which type of wind is being observed. When both winds are observed at low latitudes, as when *Ulysses* first travelled toward the south solar pole, there are two transitions per solar rotation. The transition, although abrupt in these two parameters, is frequently not readily evident in the other plasma and magnetic-field measurements.

Non-radial solar wind expansion

Ulysses observations confirm that the solar wind expands non-radially near the Sun. In addition, a quantitative measure of the non-radial expansion and an understanding of the basic physical cause have been obtained. The divergence of polar plumes and other lines of evidence, principally from imaging, have long suggested such a departure from radial expansion. A measure of the extent of this divergence was obtained when the fast wind was found to extend downward in latitude to $\approx 45°$ while the coronal hole from which it originated extended to only $\approx 20°$ from the pole. An analysis of the spherical areas or solid angles occupied by both shows that the wind had expanded by a factor of 4.8.

The reason for the expansion is provided by the *Ulysses* magnetic-field measurements (Figure 2.31). The radial field component B_R was examined as a function of latitude since it is representative of the Sun's magnetic field in or near the corona. It is related to currents in the Sun and corona whereas the azimuthal component arises as a result of solar rotation and would be zero if the Sun were not rotating. A fundamental result is that, when corrected for distance from the Sun, the radial component is independent of latitude. On the other hand, particularly near sunspot minimum, the dipole component of the photospheric magnetic field is strong and nearly aligned with the Sun's rotation axis. Solar magnetographs show that, as expected, the magnetic field in the photosphere is stronger over the poles than near the equator. These observations can be reconciled with non-radial solar-wind expansion because the increased magnetic pressure at high latitude will drive the wind equatorward until magnetic equilibrium is established, at which distance the radial field will be independent of latitude as observed. Models developed prior to *Ulysses*, which presumed that the plasma pressure is dominant, have been modified to include the dominant magnetic force and show that this adjustment takes place rapidly and leads to an equilibrium 'surface' at about 5 solar radii.

Solar-wind observations permit testing of a hypothesis that solar-wind speed is inversely correlated with the expansion factor (Figure 2.2). The latter has been modelled by extrapolating the observed photospheric magnetic field to a source surface located at 2.5 solar radii, identifying the open flux which crosses that surface and comparing the areas occupied by the open flux at the source surface and in the photosphere. An inverse relation is, in fact, found in agreement with the hypothesis. The conclusion that the magnetic flux from the coronal hole spreads out by a factor of 4.8, combined with the measured flux at *Ulysses* ($B_R r^2$), leads to an estimate of the polar-cap field strength in the photosphere. Whether or not the field is uniform in the coronal hole or varies with latitude (as inferred from magnetograph observations) is relatively unimportant. The significant determinant is the expansion factor as inferred from *Ulysses*. The estimated field strength at the magnetic pole turns out to be approximately 7 Gauss (7×10^{-4} tesla). This estimate is consistent with magnetograph observations of the line-of-sight magnetic field. It refers to a time near sunspot minimum when the dipole field is close to its maximum strength.

Magnetic field and Alfvén waves

The magnetic field on average agrees qualitatively with the Parker model, decreasing in magnitude and becoming more radial as the latitude increases. Of particular interest is the spiral angle which is the most characteristic feature of the Parker solar-wind model (Figure 2.33). Small departures had been deduced from in-ecliptic measurements extending over many years raising the issue as to whether such departures occur at high latitude. A complicating factor is the ever-present fluctuations which not only spread the distribution of observed spiral angles over a large range, much larger than would be associated with changes in solar-wind velocity, but also introduce a marked asymmetry relative to the peak (most probable value). It is found that the most probable values agree best with the Parker model while the average values are consistently lower (more radial or less tightly wound) because of the asymmetry in the distribution of angles. Over the south pole, two peaks were actually observed, one less than and the other greater than the Parker angle. This observation turned out to be anomalous because subsequent comparisons in the north polar region produced a close correspondence with the Parker angle. Why the results differed in the two hemispheres is still unexplained.

The fast wind is characterized by large amplitude fluctuations in solar-wind magnetic field. These fluctuations are continuously present over a broad range of frequencies or periods which range from seconds to tens of hours. The fluctuations are typically as large or larger than the steady background magnetic field. They are well correlated with corresponding changes in solar-wind velocity while both the field magnitude and solar-wind speed remain constant (Figure 4.3). The dominant component has been identified as Alfvén waves which are propagating outward (away from the Sun) in both hemispheres. They share many characteristics of the Alfvén waves that have been studied for many years in the ecliptic. The long-period variation is consistent with being caused by random motion of the ends of the field lines in the photosphere (e.g. associated with random velocities of the plasma in solar granules and supergranules) (Figure 2.35).

The decrease in strength with distance is consistent with the waves being generated in the corona. A significant issue for the solar wind is the extent to which the waves contribute momentum since they represent an outward pressure in the corona. Calculations similar to those carried out previously for Alfvén waves in the ecliptic show that they make a contribution of at most a few percent (Smith *et al.*, 1995a). Another possible effect of the waves is that they may be a source of heat both in the corona and, as the waves are convected outward, in the wind. *Ulysses* measurements of the temperature or, equivalently, thermal speed of the heavy solar-wind ions supports this possibility. Generally, the temperature is proportional to ion mass, the proportionality depending on solar-wind speed. The same thermal speed for ions of different masses suggests that wave heating in the corona is responsible.

Attempts to understand how the Alfvén waves evolve with time and distance has led to analyses based on treating the superposed waves as turbulence (Chapter 4). The evolution is considered likely to involve non-linear effects, as in hydrodynamic or fluid turbulence, and the kinds of analysis that have been successful with the latter have been adapted to the study of Alfvénic turbulence. Techniques that have proven useful in analysing in-ecliptic data have been applied to high-latitude measurements. A weak component is propagating sunward in addition to the stronger outward-propagating component that dominates the correlation between the field and the plasma (Figure 4.4). The analyses reveal characteristic changes with distance similar to those seen at low latitudes but proceeding more slowly (Figure 4.9). Slower evolution may be caused by the absence of velocity differences/shears in the fast polar wind. Since the reduction in wave power is attributed to heating of the background plasma, it follows that turbulent heating is less at high latitude.

Cosmic rays in the polar regions

One of the eagerly awaited measurements from *Ulysses* was the cosmic-ray flux over the poles (Chapter 8). It was considered possible that the particles had much easier access to the inner heliosphere at high latitudes where the paths along the magnetic field were much shorter and diffusion, which opposes their entry, might be much less effective. However, no large-increase in cosmic-ray flux was observed. The explanation is that persistent large-amplitude fluctuations in the fast wind are keeping cosmic rays out of the polar regions. The possibility had been suggested beforehand that the weak radial field at high latitude would be dominated by transverse variation that would oppose entry of the cosmic rays. This prediction has now been confirmed.

The measured latitude gradients turned out to be much smaller than predicted by most cosmic-ray models prior to the *Ulysses* observations (Table 8.2). The small gradients were difficult to measure because of the gradual increase in cosmic-ray intensities as sunspot minimum approached. Another complication was the need to distinguish the radial gradient from the latitude gradient. Separation of these temporal and spatial dependences was achieved by taking advantage of simultaneous in-ecliptic measurements (made by *IMP-8*) and by concentrating on the fast latitude scan where the changes in radial distance were substantially reduced. A latitude gradient of only 0.3 per cent per degree was found for protons with energies of $\approx 1\,\text{GeV}$ and was less at lower and higher energies (Figure 8.5). The flux increase from the equator to the poles was only about 30 per cent whereas factors as large as 10 had been predicted.

An aspect of the solar magnetic field, whose significance was not appreciated prior to *Ulysses*, was the presence of a north–south asymmetry equivalent to displacement of the magnetic dipole by about $-10°$ (Figure 8.5). Evidence of this southward shift of the magnetic equator was first seen in measurements of galactic and anomalous cosmic rays during the rapid transit from pole to pole which showed a minimum near $-10°$ and higher fluxes of particles over the north pole. Retrospective examination of magnetic field data from *Ulysses* and the in-ecliptic spacecraft *WIND* revealed that the radial-field component was significantly different in the two hemispheres with the field being stronger in the south consistent with a southward displacement of the solar dipole (Smith *et al.*, 2000b). Magnetograph measurements also showed a difference in the field strengths at the poles with the south magnetic pole being stronger.

Both the *WIND* and magnetograph observations showed that the polar fields were varying in time and that the asymmetry disappeared while *Ulysses* travelled from the south into the north hemisphere. As a consequence the *Ulysses* measurements gave no indication of asymmetry and considerable confusion was caused initially in interpreting the magnetic-field evidence. A time variation is not inconsistent with cosmic-ray measurements because they sample a large volume of the heliosphere and it takes approximately a year for changes at the Sun to propagate to the outer boundary.

The existence of asymmetry in the *Ulysses* data, and evidence from magnetographs that asymmetries were seen in the last two sunspot cycles, indicates that this aspect of the three-dimensional heliosphere is not unusual although largely ignored prior to the *Ulysses* mission. It is likely to receive serious attention in the future as researchers contemplate the consequences for the solar wind, the magnetic field, and energetic particles.

Structure and variation in fast wind

Although the fast wind is essentially free of large variation in velocity and density, some relatively small structures were seen when *Ulysses* first entered and then remained inside plasma from the south-polar coronal hole. Several 'microstreams' were encountered which had some of the characteristics of the larger fast–slow stream effects typically found at low latitudes but which were uncorrelated with the streams simultaneously present at low latitudes (Figure 2.8). They were thought to be generated near the Sun, one hypothesis being that they were remnants of polar plumes, the high-density striations seen near the Sun's poles that outline magnetic lines of force. Their properties, however, failed to satisfy the expectations attributed to plumes and they were not observed subsequently either during the descent back to the equator or during latitude scans of both hemispheres. In spite of a general absence of large-scale structure in the high-latitude wind, a significant amount of microstructure (short-period variation which does not affect average properties) is present. The spectrum of Alfvén waves covers a broad range of frequencies from seconds to tens of hours. The correlation between the variation in field and solar-wind-velocity components (the 'Alfvénicity') gradually increases to a value of 0.9 at maximum latitude and is consistent with predominantly outward propagation (Figure 2.34). The ratio $\delta V/\delta B$ does not satisfy the usual relation for Alfvén waves but implies that the magnetic-energy density in the waves is about double their kinetic energy. This ratio is also found in low-latitude Alfvén waves although the reason is still not understood at present.

In addition to various statistical analyses of microstructure, specific structures were identified and analysed that have been studied for many years in the in-ecliptic solar wind (Chapter 4). Hydromagnetic discontinuities are a general feature of collisionless plasmas which correspond to abrupt changes in the magnetic field and plasma parameters (on timescales of seconds to minutes) (Figure 4.4). One class, the rotational discontinuity, is a large-amplitude Alfvén wave and represents a twist or kink propagating along the magnetic field. Tangential discontinuities, alternatively, are non-propagating structures that are convected along with the plasma. A different convected structure is the magnetic 'hole', a decrease in the magnetic-field strength lasting from seconds to minutes (Figure 4.29). At larger scales (varying from minutes to hours) there is the Pressure Balance Structure (PBS) in which the plasma pressure and magnetic field pressure vary simultaneously while their sum remains constant.

The rate of occurrence of directional discontinuities, a term that incorporates both rotational and tangential discontinuities, is a function of latitude (Figure 4.16). The occurrence rate is four to five times higher at high latitudes than it is in the ecliptic. Approximately two-thirds of the discontinuities can be classified as either rotational or tangential with the former being about ten times more frequent (than in the ecliptic) (Figure 4.23). The identity of the other one-third is ambiguous but could be a mixture of the two types of discontinuities that do not satisfy the criteria used to separate them. PBSs, extending from 3 to 24 hours in duration occupy about 5 per cent of the high-latitude data, are low contrast (involving changes in plasma and field of only ≈ 10 per cent) and are more frequently seen at low radial distances. Magnetic holes, restricted to those whose changes in direction are $< 10°$ which distinguishes them from other changes in field magnitude, occur about five times per day at high latitude or about ten times less frequently than at low latitudes where they are correlated with the large-scale solar-wind structure. In the fast wind their occurrence is independent of latitude and distance. The plasma within or near the holes is marginally stable with respect to the mirror mode in which the plasma displaces the field, supporting the view that they originate closer to the Sun as a result of this instability (Figure 2.22).

Radio and plasma waves at high latitude

Previous spacecraft measurements in the solar wind reveal the presence of a large variety of waves covering a broad range of frequencies from below 1 Hz to many MHz (Chapter 5). The reason for such diversity is the existence of a large number of modes within which waves can propagate in a magnetized plasma. The ion and electron motions that accompany the waves are strongly influenced by the presence of resonances at gyro and plasma frequencies. These resonances divide the parameter space defined by frequency and wavelength into domains within which distinctly different modes exist. For a given mode, waves can have multiple causes such as a streaming of electrons relative to the ions or temperature differences parallel and perpendicular to the magnetic field. Under proper conditions any or all of these modes can be excited. The study of the different modes has intrinsic scientific value. Moreover, the waves are diagnostic of the physical processes that govern how plasmas respond to departures from equilibrium.

Ulysses observations show a dominance of the same waves previously studied in/near the ecliptic (Gurnett 1991):

(1) line-of-sight radio waves, solar type-III bursts, propagating above the electron-plasma frequency (f_{pe});
(2) Langmuir waves (LWs) at or near f_{pe};
(3) Ion Acoustic Waves (IAWs) generated near the ion-plasma frequency (f_{pi}), but doppler-shifted by the faster solar wind to frequencies below f_{pe} but above the electron gyro frequency (f_{ge}), and
(4) Whistler Mode Waves (WMWs) typically found below $f_{ge}/2$.

These waves occupy four distinct frequency bands. The type-III bursts occur at the highest frequencies observed by *Ulysses* from several tens of kHz up to about 1 MHz. LW and electron-plasma oscillations are typically observed at a few tens of kHz. IAWs are observed at several kHz and WMWs lie in the lowest frequency band at and below hundreds of Hz. Another important distinction that permits identification of these modes is whether they are electromagnetic (with fluctuating electric and magnetic fields) or electrostatic (with fluctuating electric fields only). Type-III bursts and WMWs are electromagnetic. LWs and IAWs are electrostatic.

Type-III radio bursts are caused by electrons streaming from solar-flare sites along the spiralled magnetic field (Figures 5.1 and 5.7). A complex process is involved with electrostatic waves being generated initially at the local electron-plasma frequency (varying from 1 MHz to tens of kHz with radial distance), followed by conversion to electromagnetic radiation at and near the second harmonic. As the electrons travel away from the Sun, the decrease in these frequencies with distance produces the characteristic frequency-time profile.

Ulysses found a large decrease in the rate of occurrence of type IIIs with increasing latitude. This result was expected because few electron sources are located above the low-latitude sunspot belt and because the observations were made near solar minimum when activity is generally very low. However, the *Ulysses* measurements, combined with simultaneous observations by the *WIND* spacecraft, provide evidence of an intrinsic low-frequency cut-off unrelated to the local plasma frequency that might be related to stabilization of the electron beam so that it stops radiating waves and losing energy (Figure 5.6).

Other LWs are associated with characteristic variation in the HMF, notably magnetic holes and hydromagnetic discontinuities (Figure 5.12). The occurences are sporadic and increase with latitude probably because they are more common in the high-speed polar wind rather than as a result of a latitude dependence. The mechanism responsible for their generation is unclear.

LWs and electron-plasma oscillations are also continually observed at all latitudes as 'thermal radio noise' (Figure 5.15). They occur at and above f_{pe} and are influenced by the interaction of solar-wind plasma with the electrically conducting antenna. The characteristic dependence of amplitude on frequency can be modelled using six parameters to obtain values of the densities and temperatures of both the 'core' and 'halo' electrons and the speed and temperature of the protons. The inferred values agree closely with proton and electron measurements made by the solar-wind analyser and exhibit the same latitude dependence.

CMEs (Coronal Mass Ejections) and SEPs (Solar Energetic Particles) at high latitude

Although solar activity was low and decreasing during the first polar orbit, several CMEs were observed at high latitude (Section 3.3.3). A surprising feature of the observations was the speed of the CMEs which was the same as that of the surrounding fast solar wind. Apparently they were being subjected to the same acceleration processes near the Sun as the high-speed wind. Although there was no difference in the speed of the CMEs with respect to the wind, a pair of shocks were nevertheless observed. Simulations reveal that many of the shocks result from overpressure within the CME at the source which drives large-amplitude waves away from the interior which eventually steepen into shocks (Figure 3.27). This mechanism is reminiscent of, but different than, the shock pair that results from the interaction between fast and slow wind.

A major objective of the mission was to determine to what extent SEPs produced by solar flares could reach high latitudes. The low level of activity associated with sunspot minimum reduced the number of such events substantially. Nevertheless, a few SEP events were seen at high latitude, one at $-54°$ (during the ascent to the south pole) and the other at $-74°$ (during the return to the equator) (Section 6.3.1). The early event was associated with a CME accompanied by a pair of shocks. At least some of the particles were likely to have been produced locally in association with the passage of the CME. The later, higher latitude electron event had no such association with a CME but was accompanied by a type-III radio burst observed simultaneously by *Ulysses*. The observations favoured acceleration in the corona followed by propagation to high latitudes.

1.3.3 Coupling between the high- and low-latitude heliosphere

Introduction

The internal structure of the heliosphere originates at the Sun (Chapter 3). A relatively simple model has been used successfully to interpret the *Ulysses* observations as a function of latitude and longitude. The model includes the effect of the solar magnetic field which for simplicity is taken to be a magnetic dipole. The model also assumes three bands of solar wind symmetric about the magnetic equator with fast wind at high latitudes in both hemispheres and a band of slow wind at low latitudes (Figure 3.14). A thin boundary separates the two types of wind. The dipole is tilted relative to the Sun's rotation axis so that latitudes can be designated with respect to the dipole axis (heliomagnetic) or the rotation axis (heliographic). The magnetic fields in the two magnetic hemispheres have opposite signs (e.g. outward in the north, inward in the south). A Heliospheric Current Sheet (HCS), which serves as the magnetic equator, separates the oppositely directed fields and passes through the middle of the band of slow wind. Among the best-documented properties of the slow wind are those adjacent to the current sheet that separates the oppositely directed magnetic fields. The slow wind has low proton and electron temperatures and low density. The abundance of helium to hydrogen is reduced. These properties are representative of plasma in the Sun's equatorial 'streamer belt' representing electrons trapped on closed lines of force of the magnetic field (with both ends beginning and ending on the Sun) (Figure 3.3). The current sheet appears to be the outward extension of the streamer belt with the exception that the field lines in the solar wind, adjacent to the current sheet, are 'open' (with one end on the Sun and the other extending into space without returning to the Sun). One of the hypotheses for the origin of the slow wind involves magnetic reconnection of fields converting them from closed to open and allowing the wind to escape.

When evolution of the solar wind with distance is included, this 'tilted dipole model' accounts for the recurrent or corotating structures which dominate the internal structure of the heliosphere. The tilt angle brings fast wind down to low latitudes at certain longitudes where it can overtake slow wind emitted earlier in the same direction. The fast and slow winds interact without interpenetrating to form a region in which the plasma is compressed, a CIR (Figure 3.2). In this region the fast wind has been slowed and the slow wind speeded up. The sharp boundary between fast and slow winds at the Sun has widened into the interaction region. The distinction between fast and slow wind is preserved in the composition of the heavy ions (e.g. Mg/O) and the freezing-in temperature (T_O) which are unaffected by the interaction. On the other hand, the magnetic-field magnitude, the solar-wind density and the temperature increase as a result of the compression.

The compression causes a build-up of magnetic and plasma pressure at the interface between the fast and slow wind (Figure 3.4). The stresses exerted upstream and downstream of this region cause a characteristic deflection in the solar wind with the leading slow wind being deflected toward the west limb of the Sun and the trailing fast wind being deflected eastward. Beyond 1 AU large-amplitude waves travelling in both directions from the interface steepen into a pair of shocks. A forward shock travels ahead of the interface and a reverse shock travels sunward behind the interface while being convected outward by the faster moving solar wind. This pair of shocks is a characteristic feature of CIRs beyond the orbit of Earth.

At other longitudes the dipole tilt brings the fast–slow boundary and high-speed wind to low latitudes ahead of slow wind that is emitted later. The fast wind then simply outruns the slower trailing wind and a Corotating Rarefaction Region (CRR) is created (Figure 3.2). This unconstrained expansion produces a characteristic signature that is as distinct as for the CIR, but different. The speed declines monotonically from high to low values while the magnetic-field strength and solar-wind density are reduced to low values. One of the identifying features of CRRs is the time delay between the arrival of wind with different speeds. When the two portions of the wind are extrapolated back to the Sun using the measured speeds, they originate at the same, or nearly the same, solar longitude. A common origin implies that there is an abrupt change from fast to slow speeds near the Sun. Here again the boundary has broadened at large distances from the Sun although by a different physical mechanism than operates at CIRs. The distinction between fast and slow wind inside the CRR is preserved in the compositional and freezing-in differences.

Therefore, three global regions can be identified as a function of latitude. Near the magnetic equator and at low latitudes slow wind is present. At mid-latitudes both fast and slow wind interact to form CIRs and CRRs. At high latitudes only fast wind is present. These distinctions are basic to an understanding of the principal scientific results obtained by *Ulysses*.

Low-latitude solar wind

Ulysses spent 16 months in the low-latitude solar wind while en route to Jupiter. Solar activity was decreasing from the maximum in 1990. About 6 months after launch, in March–April 1991, the Sun suddenly produced a renewed burst of activity. A large number of strong flares and transients occurred in active regions located in the southern hemisphere. This outburst was reminiscent of those seen in August 1972, while *Pioneer 10* was en route to Jupiter, and in July 1982 when *Pioneer* and *Voyager* were well beyond 5 AU.

The *Ulysses* solar-wind measurements at ≈ 2.2 AU revealed a complex series of events involving several CMEs (Figure 3.19). Their duration was much longer than observed near Earth showing that they continued to expand beyond 1 AU. Several CME-associated shocks were identified during the 2 months. A major aspect of the activity was the large number of energetic particles emitted by the Sun and observed by *Ulysses* over a broad range of energies. Multiple energetic-particle injections led to a large continuous increase in intensity over a substantial volume of the inner heliosphere. The energetic particles interacted with complex magnetic fields travelling outward at the slower solar-wind speeds and were prevented from escaping directly into the outer heliosphere. Particle intensities in the inner heliosphere remained at a high level throughout the entire interval.

When solar activity subsided after approximately two solar rotations, the solar-wind structure underwent a distinct change and was dominated by a series of CIRs. *Ulysses* was sufficiently far from the Sun for the forward–reverse shocks to develop and become a characteristic feature of the CIRs. Energetic particles were observed at the CIRs accelerated by the shocks and the accompanying high level of upstream and downstream wave activity. Energetic particles and CIRs recurred at the solar rotation period and exhibited many of the properties seen on previous missions such as the tendency for particle intensities to be larger in the vicinity of the reverse shock.

Transition region

After the encounter with Jupiter, which redirected the spacecraft toward high solar latitudes, *Ulysses* remained in low-latitude slow wind until it reached $-13°$. Until *Ulysses* reached $-36°$, the wind speed alternated between low speeds of 400–450 km s^{-1} and speeds in excess of 700 km s^{-1} (Figure 3.7). At the same time, a major shift occurred in the magnetic-sector structure which began a gradual drift in longitude as the solar magnetic field was restructured (Balogh et al., 1993). The positive (northern hemisphere) sector gradually began to disappear until at $-30°$ the spacecraft passed completely above the HCS and only inward-directed polarities were seen at higher latitudes (Figure 3.8).

The periodic variation in speed showed that, however thin the transition between fast and slow wind might be near the Sun, at Ulysses the transition had a finite width in latitude. The transition to fast wind took place over a latitude band of tens of degrees. In solar latitude and longitude the tilted dipole model produces a sinusoidal variation in transition near the Sun. However, interaction between fast and slow wind deforms the boundary changing its shape and location relative to the evolved fast and slow wind. Thus the transition widens in latitude at some longitudes and becomes thinner at others. Analysis of the transition-region thickness is complicated; however, estimates of the spread in latitude range between $10°$ and $30°$.

Corotating shocks

As the latitude increased, fewer forward shocks were seen while reverse shocks continued to be present at higher latitudes. The explanation involves the three-dimensional character of the interaction regions. Because of the tilt of the magnetic axis and the velocity contours in heliographic coordinates, the interface that develops between high- and low-speed contours is also tilted (Figure 3.12). In both the north and south hemispheres the tilting causes the forward shocks to propagate equatorward and westward (in the direction of increasing solar longitude or direction of rotation of the Sun) while reverse shocks propagate poleward and eastward. The Ulysses observations confirmed this prediction of the tilted-dipole model (Figure 3.6).

Energetic particles associated with CIRs

As Ulysses rose in latitude, the intensities of CIR-accelerated particles systematically increased until a latitude $\approx -20°$, above which they began to decrease (Figures 6.2 and 6.3). The intensity maximum may be caused by the nature of the CIR shocks which were faster and stronger near that latitude as the pressure within the CIRs, the ultimate source of the shocks, grew to a maximum (Burton et al., 1996). At higher latitudes, despite generally declining fluxes, energetic particles continued to be present in the fast wind and after CIRS were no longer evident in the plasma and field (Figure 6.5). This unanticipated result spurred several attempts to explain how the particles, energized at lower latitudes in the vicinity of the CIRs, could reach such high latitudes.

An early suggestion is that the particles are arriving from further out in the heliosphere where corotating shocks can propagate to higher latitudes and populate the field lines leading back to the spacecraft with energetic particles (Roelof et al., 1996). Another possible explanation is diffusion of the particles across field lines to high latitudes from low-latitude CIRs (Kóta and Jokipii,

1998). This hypothesis is based on the random walk of open field lines driven by convective/turbulent motions in the photosphere.

A third possibility is a direct magnetic connection between high and low latitudes. A model has been developed that adds several features to the original Parker model of the heliospheric magnetic field that assumed radial fields corotating with the photosphere (Fisk, 1996). The features that were added were fast wind from polar coronal holes, tilting of the symmetry axis of the field (and the centroid of the coronal holes) relative to the Sun's rotation axis, and differential rotation of the photosphere. These changes produce global convective motions of the field superposed on corotation. The field lines depart from the spiralled-helix characteristic of the Parker field and allow a magnetic connection between high and low latitudes (Figure 2.36). Particles accelerated at CIRs can then follow field lines inward and upward to reach the spacecraft. Which of these models accounts best for the energetic particles at high latitude is still not clear.

Other properties of CIR-associated energetic particles were also established as *Ulysses* advanced toward the south (magnetic) pole. The ratio of hydrogen to helium was found to be an order of magnitude lower in CIRs than in solar events (Figure 6.5). The difference is attributed to the presence of singly ionized helium ions that originate as interstellar neutrals, become ionized inside the heliosphere, are picked up by the solar wind, and accelerated at CIR shocks.

Another observation of interest was the delay of up to several days between the arrival of recurrent energetic ions (of energy ≈ 0.5 MeV) and associated electrons (of energy (≈ 50 keV) (Figure 6.6). Both acceleration at distant CIR shocks and enhanced cross-field diffusion appear able to account for these observed delays.

The recurrent energetic ion fluxes decreased rapidly above $\approx 70°$ and were difficult to discern at higher latitudes and over the pole. After *Ulysses* passed under the south pole and began to descend to lower latitudes as part of the Fast Latitude Scan (FLS), energetic particles reappeared at a latitude of $\approx 50°$ (Figure 6.2). The phasing of these intensity peaks changed relative to the peaks seen during the ascent. This shift in longitude is attributed to a major change in configuration of the south-polar coronal hole (and therefore of the CIRs) during the transit across the pole.

Recurrent cosmic-ray decreases

Galactic cosmic rays are also affected by the internal structure of the heliosphere. Numerous time variations are superposed on the radial and latitude gradients. Recurrent decreases in cosmic-ray intensities are associated with CIRs (Figure 8.7). This effect is well known from previous observations in the ecliptic and began while *Ulysses* was at low latitudes. However, the decreases continued to higher latitudes (up to $70°$) where CIRs were no longer present. This behaviour is similar to that of lower energy recurrent ions and electrons which reach higher latitudes than CIRs. A satisfactory explanation of the cosmic-ray effect appears to involve enhanced diffusion of the cosmic rays across magnetic field lines. In this case the CIR causes a decrease at low latitudes whose 'shadow' is seen at high latitudes.

Plasma waves

Ion acoustic waves are typically seen at forward and reverse shocks accompanying fast solar-wind streams (Figure 5.9). They are well correlated with values of the electron to proton temperature ratio $T_e/T_p > 1$, as anticipated from theory. Another region in which T_e/T_p exceeds 1 is inside CMEs. IAWs have been found to be present for the several days it takes the CME to pass over *Ulysses* (Figure 5.11). Theory also indicates that IAWs should be strongly damped for $T_e/T_p < 1$. However, waves are observed sporadically at high latitudes in spite of $T_e/T_p < 1$, an observation which is unexplained.

At low latitudes, WMWs appear to be continually present at a low level. They may be associated with solar-wind heat flux carried by electrons in the halo of the distribution function. Sporadic increases are superposed on this background at shocks whether they accompany CIRs or CMEs. A possible mechanism is an electron cyclotron resonance activated by a temperature anisotropy in which $T_\perp/T_\parallel > 1$. WMWs are also present in the fast polar wind in spite of absence of significant fluctuations in B but in the presence of a significant heat flux (Figure 5.16).

Fast latitude scan

After reaching $-70°$, the spacecraft continued to gain speed as it travelled inward toward perihelion at 1.3 AU near the equator. The increased speed produced a rapid transit between 80°S and 80°N in only 8 months (FLS). As *Ulysses* returned to lower latitudes, recurrent energetic ions were absent over the south pole although the lower energy electrons continued to be present (Figure 6.2). Recurrent ions reappeared at $\approx -43°$, significantly lower in latitude than the $-70°$ at which they had been observed to 'fade out' during the preceding Slow Latitude Scan (SLS). Recurrent particles continued to be present as the spacecraft crossed the equator and entered the north heliosphere and were observed to $\approx 40°$N. A possible explanation for the difference in the latitudes at which the particles were observed is a decrease in seed particles, superthermal solar particles that are preferentially accelerated at CIRs, because of the continuing decline in solar activity.

The transition from fast to slower wind took place at $-23°$ and the transition to slow wind at $-19°$, a difference of only 4° (Gosling *et al.*, 1995). The transit through the low-latitude heliosphere then followed over an interval of two solar rotations during which five separate CIRs were observed twice in succession. Continuing northward the spacecraft left the slow wind at 19° and entered fast wind at 21°, a small difference of 2°. The transition region was significantly thinner than during the SLS, perhaps because the observations were made nearer the Sun (<2 AU as compared with ≈ 5 AU). The band of slow wind was 42° wide.

The first crossing of the heliospheric current sheet was at $-10°$ although the current sheet could have extended up to $-20°$ and been missed because of the rapid spacecraft motion in latitude while the Sun was rotating once (Smith *et al.*, 1995). Seven current-sheet crossings were identified in all and were related to a four-sector structure attributable to two polar coronal holes and a pair of lower latitude 'equatorial' coronal holes identified in solar data. The last crossing was seen at 18°N or close to the latitude of the last recorded slow wind. Throughout the *Ulysses* observations, the current sheet was found near the edge of the band of slow wind rather than being located in the middle as assumed in the tilted-dipole model near the Sun. This difference may be the result of fast–slow interactions which tend to displace the unperturbed slow wind relative to the current sheet, as is shown in the model when the configuration is extrapolated outward with such interactions taken into account (Figure 3.15).

Fewer CIR-associated shocks were found during the fast scan. Only seven tentative identifications were made of which four were identified in both the magnetic field and the plasma data. Reverse shocks were again seen at higher latitudes of $-36°$ and $+30°$. The presence of fewer shocks than seen during the slow scan is also probably a distance effect. CIR shocks typically form beyond 1 AU. A single CME was observed accompanied by a pair of shocks, as well as energetic particles, at $-23°$ near the transition from fast to slow wind. No CMEs were seen in the north.

Return to low latitudes

Ulysses passed over the north pole and slowly descended to lower latitudes and larger radial distances while solar activity was near a minimum. The north heliosphere was relatively quiet with the solar-wind speed and other properties repeating the gradual changes with latitude observed earlier in the mission (Figure 3.7). Recurrent energetic ions and electrons reappeared near 65°N after having been absent since before the north-polar passage. That these events were associated with CIRs occurring at lower latitudes became evident when *Ulysses* re-entered slower wind at 28° and a sequence of CIRs and CRRs began that persisted to 10°.

At lower latitudes only slow wind almost free of large-scale structure was seen. At this time the inclination of the Sun's magnetic axis was less than the latitudinal width of the band of slow wind, a condition that is characteristic of solar minimum. The fast wind ($>700\,\mathrm{km\,s^{-1}}$) last appeared at 18° while the slow wind persisted to 17°. These determinations allow the heliomagnetic latitude of the boundary between fast wind and the transition region to be estimated as 23° and the lower boundary between slow wind and the transition to be estimated as 14°. Thus, the finite width of the transition is 9°, significantly wider than seen nearer the Sun during the FLS and comparable with the width inferred during the south SLS.

The first crossing of the heliospheric current sheet took place at 25.5° (Forsyth *et al.*, 1997). The current sheet was not inclined by this angle which would be inconsistent with solar-wind observations. The high-latitude crossing was the result of a large 'bump' in the current sheet associated with an active region that appeared at the same longitude for several solar rotations.

An interesting aspect of magnetic-field observations was the direction of the magnetic spiral inside the CRRs. The most probable value of the differences between the observed and calculated spiral angle was $\approx 30°$, an unusually large discrepancy (Smith *et al.*, 2000a). In retrospect this discrepancy was also seen in CRRs during the prior SLS, as well as in much earlier *Pioneer 10, 11* data, and is now recognized to be a characteristic feature of CRRs. Although several hypotheses have been advanced to explain such a large departure, the explanation is still uncertain.

The sequence of about a dozen CIRs and their associated CRRs suggests that large numbers of corotating shocks would be seen. In fact, thirty corotating forward and reverse shocks were identified and, as during the south SLS, the reverse shocks tended to occur at higher latitudes (Figure 3.10). During the same interval of ≈ 18 months, three CMEs were observed accompanied by shocks. The low number of CMEs over such an extended time interval testifies to the ongoing minimum in the solar cycle.

1.3.4 Interstellar constituents and the medium outside the heliosphere

Introduction

The basic concept of the heliosphere is a region from which interstellar matter is excluded by the magnetized solar wind. The heliospheric magnetic field opposes the entry of interstellar plasma which has its own magnetic field. Very little is known for certain about the interstellar plasma and field or about the inward pressure that it exerts on the heliosphere. This lack of information is the prime reason for uncertainty about the size of the heliosphere (i.e. the distance from the Sun to the outer boundary, the heliopause, which separates solar and interstellar space). Unlike the solar wind, which is fully ionized, the interstellar medium is only partially ionized so that neutral atoms are an important constituent. These neutrals are unaffected by the distant solar wind and magnetic field and can penetrate deep inside the heliosphere. Inside, they travel under the influence of solar gravity before some become ionized by short-wavelength solar radiation or by transferring an electron to a solar-wind ion (charge exchange). After the neutral is transformed into an ion, it is picked up by the passing solar wind and transported back into the outer heliosphere.

In addition to the neutral gas, interstellar dust is present outside the heliosphere. Although electrically charged as a result of the presence of the warm interstellar plasma, the dust grains are generally massive enough that the weak solar wind cannot exclude them from entering the heliosphere where, as they approach the Sun, they are affected by the Sun's gravity and the increasingly strong magnetic field. There are also sources of dust within the heliosphere such as asteroids, dust left behind by comets and dust ejected from planetary magnetospheres such as Jupiter's.

An interstellar constituent that has been studied for many years is the cosmic ray gas. These fully ionized bare nuclei are accelerated to enormously high energies elsewhere in the Galaxy and are generally able to penetrate the heliosphere until the magnetic field and its variation becomes strong enough to influence their inward motion and drastically alter their spatial distribution and properties. Anomalous cosmic rays, on the other hand, are accelerated inside the heliosphere to very high energies. They begin life as interstellar neutrals with bulk speeds of a few tens of $km\,s^{-1}$ and temperatures of several thousand Kelvin (less than an electron volt).

They end up as singly ionized ions with energies of several hundred million electron volts after being transported outward to a possible shock inside the heliosphere at which the supersonic solar-wind flow is converted to subsonic flow (the termination shock) where they are presumably accelerated. Many of the energized particles are then able to penetrate back into the heliosphere while undergoing the same influences as galactic cosmic rays.

The observation of these constituents by instruments on *Ulysses* has produced significant scientific advances. Ionized interstellar neutrals (pickup ions) affect the evolution of the solar wind, their properties reveal much about physical conditions and processes inside the heliosphere including mechanisms that accelerate charged particles to high energies. Although they are observed within only 5 AU of the Sun, rather than in the outer heliosphere, they also contain information about the interstellar medium.

Dust

The first detection of interstellar dust was made by *Ulysses* while the spacecraft was near aphelion after leaving Jupiter (Section 9.5 and Figure 9.16). A principal discriminator related to their interstellar origin was the velocity of the dust which agreed in direction and speed ($26 \, \text{km} \, \text{s}^{-1}$) with the values inferred for interstellar gas from *Ulysses* measurements of neutral helium. The flux ($1.5 \times 10^{-4} \, \text{m}^{-2} \, \text{s}^{-1}$) was measured over an extended mass range and the mean mass was determined to be 3×10^{-13} gm (Figure 9.14). Such grains are thirty times more massive than those previously thought to be causing stellar extinction. The reason for the presence of such massive previously undetected dust in interstellar space is a puzzle. However, the spatial density of observed interstellar dust is consistent with zodiacal light observations as well as with the prior *Pioneer 10, 11* results. Smaller grains undetected by *Ulysses* are thought to be present in interstellar space but unable to reach 5 AU because of the combined effects of solar-radiation pressure and the heliospheric magnetic field.

Galactic Cosmic Rays (GCRs)

Cosmic-ray observations over the poles, measurements of the latitude gradient and evidence of a north–south asymmetry in the heliosphere are discussed in the subsection 'Cosmic rays in the polar region', above. *Ulysses* observations have also yielded estimates of average lifetimes of cosmic rays and the average density of the material through which they have travelled (Section 8.33). As heavy cosmic rays propagate through the Galaxy at nearly the speed of light, they occasionally strike a slow-moving interstellar particle. A 'splintering' of the cosmic-ray nucleus (spallation) occurs resulting in the production of individual nucleons (protons, neutrons, alpha particles) and a nucleus with a lower atomic mass. Thus, the original cosmic ray has been transformed into a lower mass isotope which can be radioactive and decay into a third stable isotope. Models account for various production and loss processes that influence the composition of the GCRs. Detailed calculations lead to predicted abundance ratios of various isotopes, such as $^{27}\text{Al}/^{26}\text{Al}$, as a function of cosmic-ray energies and the density of the medium through which they have travelled. Measured abundance ratios at specific energies can then be used to derive estimates of the average density of the galactic medium. *Ulysses* observations have dramatically increased the number of isotopes whose abundance ratios are known. The *Ulysses* measurements are made well inside the heliosphere at distances ≤ 5 AU, so energies must be corrected for losses between local interstellar space and the point of observation.

Isotope ratios lead to reasonably consistent values for the average density of 0.2–0.3 atoms cm^{-3}. This density is less than the average density of the Galaxy as a whole (≈ 1 cm^{-3}), so cosmic rays appear to spend much of their lifetime in a low-density region such as the galactic halo. Other analyses lead to an estimate of mass column density traversed by the cosmic rays of ≈ 10 gm cm^{-2}. This parameter is the product of interstellar mass (essentially the proton mass), average-number density and the path length of GCRs so that the latter can be determined as 1.6×10^{12} AU or 3×10^6 pc, approximately the distance to the galactic centre. At an average speed of $0.8c$ the corresponding lifetime is ≈ 20 Myr.

Neutral helium

The *Ulysses* GAS experiment detects and counts helium atoms arriving at the spacecraft. Over a few months, a map of the directions from which the particles arrive is obtained representing their distribution in latitude and longitude (Figure 9.16). To derive the helium parameters outside the heliosphere, a model is used which includes the effect of solar gravity and the rate of ionization (loss) of the helium. The particles arrive at *Ulysses* from two directions, either along a direct trajectory or indirectly, having already passed near the Sun and travelling outbound. The separate distributions can be modelled with inbound neutrals being lost due to photoionization and the outbound neutrals being lost due both to photoionization and impact ionization by solar-wind electrons while near the Sun.

The data obtained in the ecliptic at ≈ 5 AU and near perihelion at ≈ 1 AU lead to a direction for the influx (the velocity of the interstellar helium) at $74°$ longitude and $-5°$ latitude in ecliptic co-ordinates (based on the ecliptic pole and the direction to vernal equinox). The inferred speed, density and temperature of helium are 25 km s^{-1}, 0.014 cm^{-3} and $\approx 6{,}500$ K (Witte *et al.*, 1996). The density and temperature are consistent with inferences about interstellar hydrogen drawn from observations of resonantly scattered Lyman α. However, the inferred hydrogen speed is only ≈ 20 km s^{-1}. This difference is interpreted as the effect of charge exchange between interstellar hydrogen and interstellar protons outside the heliopause where the protons slow down as they approach the heliosphere. Thus, the helium measurements provide a quantitative estimate of this 'filtration' of hydrogen when it arrives at the heliopause.

Pickup ions

Ions are continuously created in the heliosphere as neutral atoms become ionized. The solar-wind is fully ionized so neutrals are absent but the cool interstellar gas can penetrate close to the Sun. The ions are then picked up by the solar-wind magnetic field. The *Ulysses* Solar Wind Ion Composition Spectrometer (SWICS) has produced significant new information about interstellar pickup ions (Chapter 7). A surprise has been the identification of a source of neutrals in the inner heliosphere (the inner source). *Ulysses* made the first measurements of interstellar pickup hydrogen even though it is the most abundant element in interstellar space. Singly ionized helium He^+ which is more difficult to ionize, had previously been detected at 1 AU. The distribution functions of the pickup ions have been fitted with models that incorporate ionization by short-wavelength solar radiation, charge exchange (of electrons) with solar-wind ions and electron impact (of neutrals that penetrate close to the Sun). The relative influence of these three processes varies from one constituent to another. The phase-space distribution function of H^+ is determined to be $\approx 10\,s^3\,km^{-6}$ in the fast polar wind at an average distance of 3.0 AU and a latitude of $-66°$ (Figure 7.2). This value is approximately the same as for $^4He^+$ at that distance. Doubly charged helium $^4He^{++}$ is also found at a flux level about one-thirtieth of that of H^+, the several sources being charge exchange of He with solar-wind alpha particles as well as photoionization and charge exchange of He^+ pickup ions (Figure 7.3). The $^3He^+$ isotope, which is rare in the solar wind, is present as a pickup ion with a flux $\cong 1/4,000$ that of $^4He^+$ (Figure 7.5).

Pickup ions have been studied in both the slow and fast wind. A characteristic feature in the unperturbed slow wind is the presence of high-speed tails extending above the usual cut-off of twice the solar-wind speed (Figure 7.7). Their presence indicates that some of the pickup ions are being accelerated locally. The high-speed tails are more pronounced in turbulent solar wind with the spectral shapes being similar behind both forward and reverse shocks. This similarity in spite of the differences in the shocks may imply acceleration by the downstream turbulence. Anomalous Cosmic Rays (ACRs) are considered to be pickup ions that have been carried out to the termination shock where they are accelerated. The *Ulysses* observations indicate that part of the acceleration is likely to occur inside the heliosphere as a result of pickup-ion interaction with solar-wind structures and turbulence.

Measurements of pickup ions have been combined with models to infer their abundance in interstellar space and to then estimate the inward pressure exerted by interstellar ions on the heliosphere (Sections 7.5 and 7.6). The measured flux of the pickup ion is first converted to a neutral density at the inner termination shock. The H fluxes outside the termination shock are adjusted for 'filtration' – the effect of charge exchange as the interstellar ions are slowed – because of their interaction with the heliosphere, relative to other neutrals or neutral helium. By modelling the region in which this occurs the interstellar densities of neutral H and He are obtained. The final step is allowance for the degree of ionization of the interstellar gas to arrive at a figure for interstellar pressure consisting of both the convected and thermal pressures. This result can be compared with the inward pressures corresponding to the location of the termination shock inferred from studies of the radial increase in flux of ACRs, which are believed to be accelerated at the shock. The difference is attributed to the additional inward pressure exerted by the interstellar magnetic-field (Figure 7.11). This leads to a value for the magnetic-field strength between 0.8 and

2.0 µG (80–200 pT). The inferred values of the interstellar field and ion densities lead to an Alfvén speed that is less than the convected speed of the interstellar medium. A detached bow shock upstream of the heliosphere is implied. When the total pressure derived in this manner is compared with astronomical estimates of the pressure in the local interstellar cloud a discrepancy is found. The pressure inferred from the *Ulysses* analysis is too small to balance the average pressure in the cloud. Several possible ways to reconcile this disagreement have been proposed. Perhaps the magnetic field is not uniform but varies from location to location or perhaps some additional component of interstellar pressure is being ignored.

A source of pickup ions inside the heliosphere

The successful identification of interstellar pickup ions has revealed the presence of another component that originates inside the heliosphere. The inner source ions identified by *Ulysses* are distinguished from interstellar ions because of differences in the abundance of heavy ions and the shape of their phase-space distribution functions. A distinguishing feature is the large flux of carbon ions C^+ which are rare as interstellar neutrals and have a rounded distribution function rather than the flat distributions associated with interstellar ions (Figure 7.13). This shape is attributed to adiabatic cooling of the ions as they are carried outward in the expanding solar wind and implies that they originate well inside the point at which they are observed.

Other ions from the inner source include singly ionized H, O, N and Ne (Figure 7.16). The abundance ratios of the inner source ions are similar to solar-wind abundances suggesting how they originate. The proposed explanation is that solar wind is absorbed by interplanetary dust and subsequently re-emitted as neutral atoms that are then re-ionized and picked up by the HMF. The presence of N^+ and Ne^+, which are depleted in dust, supports the solar wind as the source. The distribution functions of the inner source ions also exhibit high-energy tails indicating local acceleration is taking place. Therefore, the inner source ions are likely to be contributing to the ACR along with interstellar ions.

1.4 REFERENCES

Balogh, A., Erdös, G., Forsyth, R. J. and Smith, E. J. (1993) The evolution of the interplanetary sector structure in 1992. *Geophys. Res. Lett.* **20**, 2331.

Burton, M. E., Smith, E. J., Balogh, A., Forsyth, R. J., Bame, S. J., Phillips, J. L. and Goldstein, B. E. (1996) Ulysses out-of-ecliptic observations of interplanetary shocks. *Astron. Astrophys.* **316**, 313.

Coleman, P. J. Jr, Davis, L., Jr, Smith, E. J. and Sonett, C. P. (1962) *Science*, **138**, 1099.

Davis, L. E. Jr. (1955) Interplanetary magnetic fields and cosmic rays. *Phys. Rev.* **100**, 1440.

Davis, L. Jr. Smith, E. J., Coleman, P. J. Jr. and Sonett, C. P. (1964) Interplanetary magnetic measurements. In: *The Solar Wind*, Mackin, R. J., Jr and Neugebauer, M. (eds). Jet Propulsion Laboratory, Pasadena.

Duffett-Smith, P. (1988). *Practical Astronomy with Your Calculator*, 3rd edn. Cambridge University Press.

Fisk, L. A. (1996) Motion of the footpoints of heliospheric magnetic field lines at the Sun: Implications for recurrent energetic particle events at high heliographic latitudes. *J. Geophys. Res.* **101**, 15 547.

Forsyth, R. J., Balogh, A., Smith, E. J. and Gosling, J. T. (1997) Ulysses observations of the northward extension of the heliospheric current sheet. *Geophys. Res. Lett.* **24**, 3101.

Gosling, J. T., Bame, S. J., Feldman, W. C., McComas, D. J., Phillips, J. L., Goldstein, B. E., Neugebauer, M., Burkepile, J., Hundhausen, A. J. and Acton, L. (1995) The band of solar wind variability at low heliographic latitudes near solar activity minimum – plasma results from the Ulysses rapid latitude scan. *Geophys. Res. Lett.* **22**, 3329.

Gurnett, D. A. (1991) Waves and instabilities. In: *Physics of the Inner Heliosphere*, Vol. II, p.135, Schwenn, R. and Marsch, E. (eds). Springer-Verlag, Berlin.

Kota, J. and Jokipii, J. R. (1998) Modelling of 3-D corotating cosmic ray structures in the heliosphere. *Space Sci. Rev.* **83**, 137.

Krieger, A. S., Timothy, A. F. and Roelof, E. C. (1973) A coronal hole and its identification as the source of a high velocity solar wind stream. *Solar Phys.*, **23**, 123.

Marsden, R. G., Smith, E. J., Cooper, J. F. and Tranquille, C. (1996) Ulysses at high heliographic latitudes. *Astron. Astrophys.* **316**, 279.

Ness, N. F. and Wilcox, J. M. (1964) Solar origin of the interplanetary magnetic field. *Phys. Rev. Lett.* **13**, 461.

Neugebauer, M. and Snyder, C. W. (1962) The Mariner II preliminary observations, solar plasma experiment. *Science* **138**, 1095.

Parker, E. N. (1958) Dynamics of the interplanetary gas and magnetic fields. *Astrophys. J.* **128**, 664.

Parker, E. N. (1963) *Interplanetary Dynamical Processes*. Wiley-Interscience, New York.

Parker, E. N. (1977) Mass ejection and a brief history of the solar wind concept. In: Jokipii, J. R., Sonett C. P. and Giampapa, M. S. (eds), p. 3. The University of Arizona Press, Tucson.

Roelof, E. C., Simnett, G. M. and Tappin, S. J. (1996) The regular structure of shock-accelerated 40–100 keV electrons in the high latitude heliosphere. *Astron. Astrophys.* **316**, 481.

Schwenn, R. (1990) Large-scale structure of the interplanetary medium. In: *Physics of the Inner Heliosphere*, Marsch, E. and Schwenn, R. (eds) Springer-Verlag, Berlin.

Schwenn, R. and Marsch, E. (eds) (1990 and 1991) *Physics of the Inner Heliosphere*, Vols I and II Springer-Verlag, Berlin.

Simpson, J. A., Rossi, B., Hibbs, A. R., Jastrow, R., Whipple, F. L., Gold, T., Parker, E., Christofilos, N. and Van Allen, J. A. (1959) *J. Geophys. Res.* **64**, 1691.

Smith, E. J. and Wolfe, J. H. (1976) Observations of interaction regions and corotating shocks between one and five AU: Pioneers 10 and 11. *Geophys. Res. Lett.* **3**, 137.

Smith, E. J., Balogh, A., Neugebauer, M. and McComas, D. J. (1995a) Ulysses observations of Alfvén waves in the southern and northern solar hemispheres. *Geophys. Res. Lett.* **22**, 3381.

Smith, E. J., Balogh, A., Burton, M. E., Erdös, G. and Forsyth, R. J. (1995b) Results of the Ulysses fast latitude scan: Magnetic field observations. *Geophys. Res. Lett.* **22**, 3325.

Smith, E. J., Balogh, A., Forsyth, R. J., Tsurutani, B. T. and Lepping, R. P. (2000a) Recent observations of the heliospheric magnetic field at Ulysses: Return to low latitude. *Adv. Space Res.* **26**, 823.

Smith, E. J., Jokipii, J. R., Kota, J., Lepping, R. P. and Szabo, A. (2000b) Evidence of a North–South asymmetry in the heliosphere associated with a southward displacement of the Heliospheric Current Sheet. *Astophys. J.* **533**, 1084.

Wenzel, K.-P., Marsden, R. G., Page, D. E. and Smith, E. J. (1992) The Ulysses Mission. *Astron. & Astrophys. Suppl. Ser.* **92**, 207.

Wilcox, J. M. and Ness, N. F. (1964) Quasi-stationary corotating structure in the interplanetary medium. *J. Geophys. Res.* **70**, 5793.

Witte, M., Banaszkiewicz, M. and Rosenbauer, H. (1996) Recent results of the parameters of the interstellar medium from the Ulysses/GAS experiment. *Space Sci. Rev.* **78**, 289.

2

The solar-wind and heliospheric magnetic field in three dimensions

Marcia Neugebauer

2.1 OVERALL PATTERN OF HIGH- AND LOW-SPEED WINDS – THE BIG PICTURE

At the start of the *Ulysses* mission, the fundamental morphology of solar-wind plasma and the Heliospheric Magnetic Field (HMF) had already been established as a result of the *Helios 1, 2, Pioneer 10, 11, Voyager 1, 2* missions, as well as near-Earth missions. It was known that during periods of declining and low solar activity, there were large coronal holes in solar polar regions and that these coronal holes were the source of high-speed solar wind. This structure is illustrated by the model of solar and interplanetary magnetic fields shown in Plate 1. The red and blue features on the disc of the Sun are magnetograph data obtained in early 1995 (when *Ulysses* crossed the solar equator from south to north) coded such that outward-directed fields are red and inward are blue. The plate shows open field lines extending from the polar regions out into the heliosphere with one polarity in the north and the opposite polarity in the south. At lower latitudes, the field lines form closed loops which do not reach out into space, and the solar wind above those loops is slow and more highly variable than is the fast, high-latitude solar wind.

This simple picture was borne out by the *Ulysses* observations, which are summarized in Plate 2. In the centre of the figure, the corona, which is the hot ($\approx 10^6$ K) outer atmosphere of the Sun, is imaged against the disc in extreme ultraviolet wavelengths sensitive to temperatures in the lower corona. The less dense, cooler polar coronal holes appear dark at these EUV wavelengths. Outside the solar disc, Plate 2 shows the bright, dense coronal streamers extending outward from the equatorial regions of closed loops.

The outermost part of Plate 2 is a polar plot of daily averages of solar-wind speed measured by the SWOOPS instrument (Bame *et al.*, 1992) on *Ulysses*. The speed data have been colour coded with red and blue to denote HMFs pointing outward from and inward toward the Sun, respectively. The plot shows a dipolar structure, with an outward field in the north and an inward field in the south, in agreement with the polarity of the magnetic fields shown in Plate 1. There was fast wind over the poles and a rather narrow band of slow wind near the equator. Although the correspondence between solar and coronal images and the speed of the solar wind is quite striking, we should not read too much into the details of the correlations because each solar image was obtained at one particular time while more than 5 years were required to build up the *Ulysses* dataset. Nevertheless, the general correlation between fast, rather steady wind over the polar coronal holes, and slow, rather gusty wind over the equatorial streamers was observed throughout this period.

Because the solar magnetic field is tilted with respect to the Sun's rotation axis, the magnetic latitude of a spacecraft varies over the course of each solar rotation. Early in the *Ulysses mission* (bottom right quadrant of Plate 2), the tilt was quite large ($\sim 30°$) (Hoeksema, 1995), resulting in alternating high and low speeds as the spacecraft went from high to low magnetic latitudes and back again. The variation is less on the left side of Plate 2 than on the right because the tilt was less and because near its perihelion the spacecraft rapidly swept through the equatorial latitudes at a rate of $\sim 22°$ per solar rotation.

None of our previous understanding of the solar wind at solar activity minimum has been contradicted by the *Ulysses* observations. Now, however, we have better insight into the three-dimensional structure of the heliosphere and the HMF, the kinetic properties of the high-speed wind, and clues to coronal heating and solar-wind acceleration implicit in the elemental and charge-state abundances of minor ions.

This chapter focuses on the quasi-stationary state of the solar wind in the context of Plates 1 and 2. The many interesting physical processes observed in the interaction regions where the fast wind overtakes the slower wind in its path and the transient phenomena associated with coronal mass ejections are postponed to Chapter 3. Similarly, the topics of magneto-hydrodynamic (MHD) turbulence and fine-scale features in the IMF are postponed to Chapter 4.

2.2 SOLAR-WIND SPEED

As can be seen in Plate 2, poleward of $\sim 30°$ latitude the average solar-wind speed was consistently above $700 \, \text{km s}^{-1}$. For latitudes poleward of $\pm 36°$ the average speeds were $752 \pm 32.3 \, \text{km s}^{-1}$ in the south and $767 \pm 24 \, \text{km s}^{-1}$ in the north, where the quoted uncertainties are standard deviations of hourly averages (McComas *et al.*, 2000). The maximum 6-hour-averaged speeds observed outside transient events were 863 and $850 \, \text{km s}^{-1}$ in the southern and northern hemispheres, respectively (Phillips *et al.*, 1995c).

High speeds were expected at high latitudes for this period of declining solar activity. Using Interplanetary Scintillation (IPS) techniques, Rickett and Coles (1991) found that at latitudes $\pm 60°$, yearly average solar-wind speeds typically reached values between 550 and 650 km s^{-1} for periods of declining and low solar activity (specifically for 1973–1976 and 1983–1987). There are probably biases in these measurements that lead to the calculated speed being somewhat lower than the true peak solar-wind speed. The IPS technique measures the velocity component perpendicular to the line of sight integrated along the ray path between the Earth and a distant radio source with a weight proportional to the square of the electron density. Thus any line-of-sight component of velocity is not measured, and the contribution of any dense, low-speed region along the line of sight is exaggerated.

High speeds approaching those observed by *Ulysses* have been intermittently observed in the ecliptic near Earth during the descent from solar maximum when the tilt of the solar magnetic field is substantial and extensions of the high-speed flow from the polar coronal holes reach down to low latitudes, sometimes even crossing the equator. Feldman *et al.* (1976), for example, found average peak speeds of 741 km s^{-1} for nineteen encounters with high-speed streams by the *IMP-6*, *-7*, and *-8* Earth satellites in 1971–1974. For one of those streams the peak speed reached 820 km s^{-1}.

At intermediate latitudes the tilt of the solar magnetic dipole, together with the large-scale raggedness of the boundaries of polar coronal holes, produces considerable variation in solar-wind speed during the course of each solar rotation of ~25.5 days (Carrington rotation rate relative to *Ulysses*). As mentioned above, these effects account for the major variation between low- and high-speed wind seen, for example, from about 3:30 to 4:00 o'clock in Plate 2. For the initial southward leg of the trajectory some longitudinal variation in solar-wind speed could still be discerned at latitudes of 60 to 65°, poleward of which Phillips *et al.* (1995c) and Zurbuchen *et al.* (1996) found essentially no rotational modulation. Roberts and Goldstein (1998), however, found weak evidence for peaks in power spectra of solar-wind speed for 34-day periods and its harmonics for 208-day periods from *Ulysses* data spanning the entire latitude range −64 to −80.2 to −53°. They attribute the 34-day periodicity to the rotation rate of solar-surface features at a latitude of ±64°.

Two sets of contour diagrams of solar-wind speed as a function of solar latitude and longitude have been generated for different phases of the *Ulysses* mission. Figure 2.1 shows speed contours for the southern hemisphere in 1992–1993. The longitude in Figure 2.1 has been defined to correct for the deduced variable rotation rate of the high-speed stream. Specifically, the repetition rate of Jovian radio bursts, whose occurrence rate depends on the solar-wind speed at Jupiter, was used to determine a rotation period of 25 days through late 1992, followed by a rotation period of ~29 days in 1993 Macdowall *et al.*, 1995). The upper left panel of Plate 3 shows similar contours for the latitude range −47 to +42° obtained during the *Ulysses* Fast Latitude Scan (FLS) in 1994–1995. This figure was compiled from five solar rotations of SWOOPS data plus one rotation of data obtained by the *WIND* spacecraft at a latitude near 7°S (Neugebauer *et al.*, 1998). In this case, the two sets of data were mapped back to the Sun (assuming radial flow at constant speed) and binned and plotted as a function of the Carrington longitude of the calculated source region. Comparison of Figure 2.1 and Plate 3 reveals, as expected, that the latitude range or the tilt of the belt of low-speed wind was more highly inclined to the solar equator in

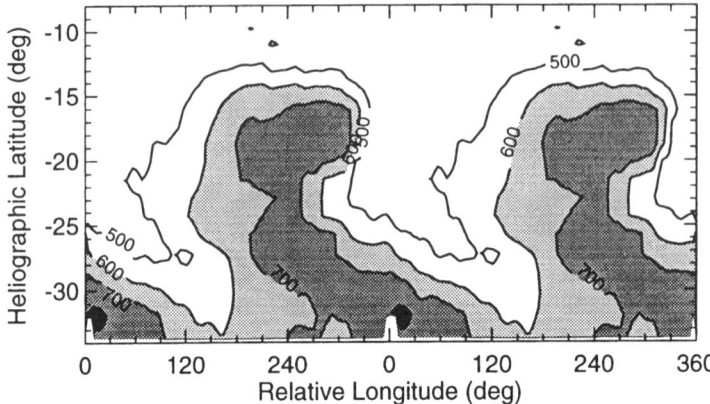

Figure 2.1. Contours of solar-wind speed versus heliographic latitude and a longitude defined by a rotation rate of 29 days relative to *Ulysses*. The speed data were acquired by SWOOPS from January 1992 through July 1993. Note that the data are repeated, showing two solar rotations (from Macdowall *et al.*, 1995).

1992–1993 than it was in 1994–1995, which was closer to the minimum of solar activity in mid-1996. Furthermore, the low-latitude extensions of the high-speed wind had simpler, less hooked shapes during the latter period.

The white asterisks in Plate 3 mark the location of crossings of the heliospheric current sheet (HCS) which separates the outward- from the inward-oriented interplanetary magnetic-field lines. While there is a general correlation of the HCS with the band of low-speed wind, the HCS crossings do not coincide with the very lowest speeds (Crooker *et al.*, 1997), and the speed was very non-uniform along the inferred location of the current sheet (Neugebauer *et al.*, 1998). This is not an entirely new finding. Variation in the minimum solar-wind speed with solar longitude can also be implied from figure 1 of Miyake *et al.* (1989) for the period March through July 1986 near the previous solar minimum.

Two different parameters have been used to characterize spatial variation in solar-wind speed. The first is based on the concept of heliomagnetic latitude in a co-ordinate system that matches the tilt of the solar-dipole field; in that co-ordinate system the heliomagnetic equator coincides with some average location of the heliospheric current sheet. Hakamada and Akasofu (1981) found that the solar-wind speed exhibited a positive latitudinal gradient in a co-ordinate system based on a wobbling solar dipole. Zhao and Hundhausen (1981) similarly found that the solar-wind speed observed in 1974 could be well organized in a heliomagnetic co-ordinate system tilted $30 \pm 10°$ from the solar rotation axis. The second approach to parameterizing solar-wind speed is based on an empirical anticorrelation between solar-wind speed and the magnetic expansion factor f which is a measure of the amount by which magnetic-field lines diverge between the surface of the Sun and interplanetary space (Wang and Sheeley, 1990).

Several different methods have been used to compute both the magnetic latitude and the expansion factor of any point in space from the distribution of magnetic fields on the solar surface. A common approach is based on the potential-field source-surface model first proposed in 1969 by Schatten *et al.* (1969). In that model the field is assumed to be curl free (i.e. no currents) between the solar surface and a source surface, commonly placed at 2.5 solar radii (R_s), where the field is forced to be radial to be consistent with the nearly radial flow of the solar wind. Alternatively, an MHD model can be used to determine the expansion factor and the magnetic latitude. Neugebauer *et al.* (1998) compared the results of an MHD calculation with several different source-surface models for the period of the *Ulysses* fast latitude scan. They found that the location of the HCS differed by as much as 21° from one model to another, but most of the intermodel differences were much smaller than that.

The top panel in Figure 2.2 shows solar-rotation averages of the solar-wind speed and the expansion factor during the *Ulysses* fast latitude scan as functions of magnetic latitude. The calculations were based on a source-surface model with an external current sheet; more detail can be found in the paper by Goldstein *et al.* (1996) from which the figure was taken. The speed appears to vary both with the expansion factor and magnetic latitude, but is a linear function of neither parameter. This result agrees with the conclusion of Wang *et al.* (1997) that the 20-year solar-wind data acquired near Earth cannot be adequately modelled on the basis of angular distance from the HCS alone.

The bottom panel of Figure 2.2 is a scatter plot of solar-rotation averages of speed versus expansion factor. Although a low value of f (less than ~7) appears to be a good predictor of high-speed wind, there is considerable scatter for the lower speeds and higher expansion factors. Part of the scatter arises from the extreme sensitivity of f to small changes in magnetic latitude very close to the heliospheric current sheet.

The magnetic models mentioned above can also be used to trace back the solar origin of any parcel of interplanetary plasma. Plate 4 shows the results of such a calculation based on MHD modelling for Carrington rotation 1893, when *Ulysses* crossed the equator during its FLS. In this plate, the boundaries of the coronal holes determined by ground-based observations in the He 10830 Å line are indicated by a dashed white line. The regions of model-determined open-field lines are shaded in grey. It has been noted that the He 10830 Å and the open-field line boundaries are generally related, but do not exactly coincide (e.g. Neugebauer *et al.*, 1998). In Plate 4 the footpoints of the field lines threading the *Ulysses* spacecraft have been colour coded according to speed, with red denoting fast wind and blue denoting slow. It is seen that the fastest wind originated deep within the coronal holes, with the speed decreasing toward the hole boundaries. The slow solar wind appears to have two distinct sources – the outer boundaries of the polar coronal holes and small, isolated coronal holes in equatorial regions. Similar calculations for the same period have been made using potential-field source-surface models, with very similar results (Neugebauer *et al.*, 1998).

There are two additional noteworthy features of the high-latitude speeds observed by *Ulysses*. First, even poleward of ±36°, where the spacecraft was

48 The solar-wind and interplanetary magnetic field in three dimensions [Ch. 2

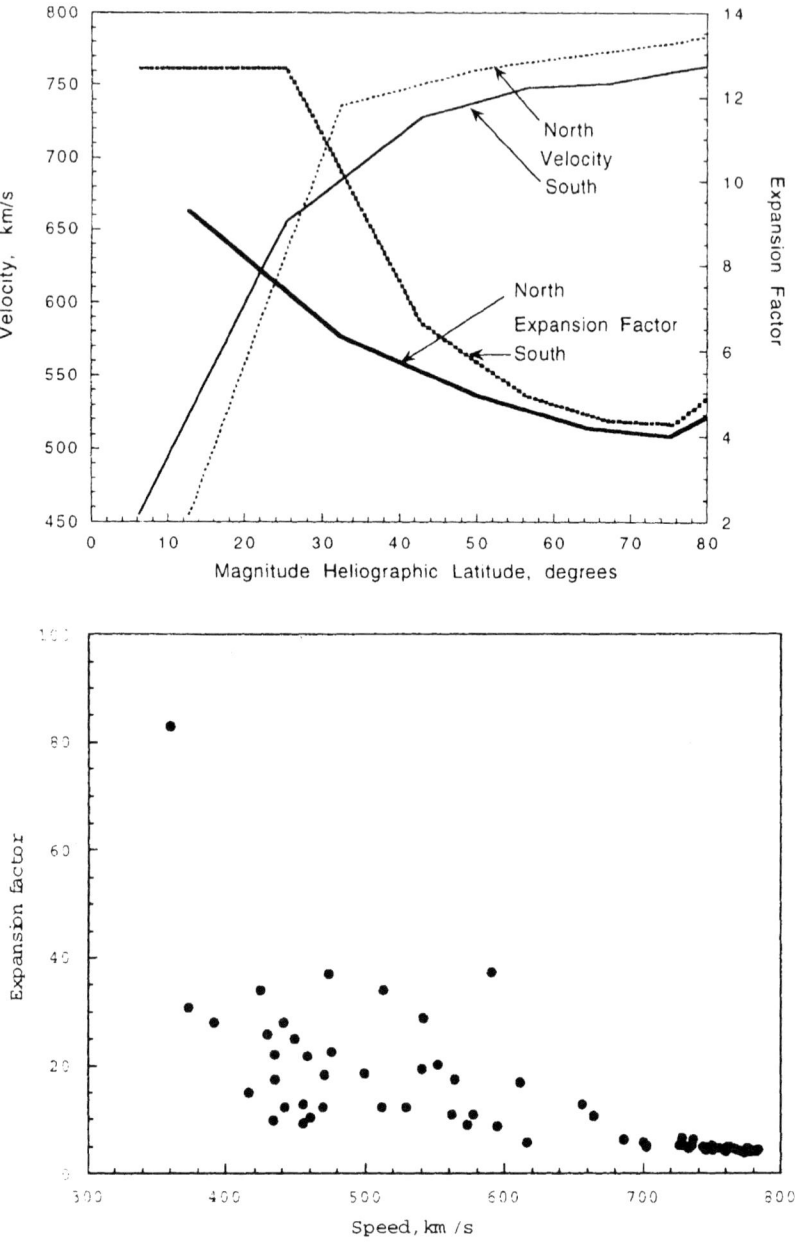

Figure 2.2. (top) Solar-rotation averages of solar-wind speed and of the expansion factor describing the divergence of the magnetic field between the surface of the Sun and the solar-wind plotted versus magnetic latitude for the period of the *Ulysses* FLS (bottom) Scatter plot of solar-rotation averages of expansion factor versus speed (from Goldstein *et al.*, 1996).

continuously embedded in the flow from polar coronal holes, the average speed continued to increase with increasing latitude at an average rate of $0.95\,\mathrm{km\,s^{-1}\,deg^{-1}}$ (McComas et al., 2000). Second, the speed was systematically higher in the north-polar region than in the south (see the top panel of Figure 2.2). This is only one of several modest north–south asymmetries resulting from the solar magnetic configuration at the time of the *Ulysses* FLS (Marsden et al., 1996). McComas et al. (2000) argue that these asymmetries in proton parameters are caused by temporal rather than spatial variation; they conclude that more energy goes into the polar solar wind during the declining phase of the solar cycle than during the minimum phase.

2.3 DENSITY AND MASS, MOMENTUM, AND ENERGY FLUXES

To a first approximation the density of the solar wind observed by *Ulysses* varied in response to the high- and low-speed streams, as expected; that is, there was a general inverse relation between the two parameters, in addition to which the density was enhanced in the interaction regions where fast wind ploughs into slower wind in its path and was depleted in rarefaction regions of decreasing speed.

At each latitude the density dropped off with an approximate inverse square of distance from the Sun, indicating no significant meridional flow. By fitting the entire dataset acquired at latitudes poleward of $\pm 36°$, McComas et al. (2000) derived the relation $N_\mathrm{p} = R^{-2.0}(3.36 - 0.0161\Theta)\,\mathrm{cm}^{-3}$, where N_p is the proton-number density, R is solar distance in AU, and Θ is the absolute magnitude of heliographic latitude. Close examination of their plate 3 shows, however, that the decrease in density with increasing latitude was due almost entirely to the inward pass from aphelion to the south pole, and that the rest of the first orbit showed essentially no latitudinal gradient in the proton density. The latitude gradient during the inward passage to the south pole was probably related to variation in the solar cycle rather than to true latitudinal gradients.

Figure 2.3 shows the latitude dependence of the magnitude of the mass-averaged fluid velocity

$$\mathbf{V}_\mathrm{f} = (N_\mathrm{p}\mathbf{V}_\mathrm{p} + 4N_\alpha\mathbf{V}_\alpha)/(N_\mathrm{p} + 4N_\alpha) \qquad (2.1)$$

for the FLS, where N_α is the alpha-particle number density, and $\mathbf{V}_\mathrm{p}$ and $\mathbf{V}_\alpha$ are ion velocities. Figure 2.3 also shows the mass density ($= N_\mathrm{p} + 4N_\alpha$) in $\mathrm{amu\,cm^{-3}}$, the mass flux ($N_\mathrm{p}V_\mathrm{p} + 4N_\alpha V_\alpha$) in units of $10^{12}\,\mathrm{amu\,m^{-2}\,s^{-1}}$, and the momentum flux ($N_\mathrm{p}V_\mathrm{p}^2 + 4N_\alpha V_\alpha^2$) in units of $10^{-9}\,\mathrm{Pa}$. Except for speed, all other parameters have been scaled by R^2. Each panel of Figure 2.3 shows the median, 5 per cent, and 95 per cent values of hourly average parameters binned by solar rotation and plotted versus heliographic latitude during the fast latitude scan. When features such as compression and rarefaction regions are washed out by averaging the data over a solar rotation, it is readily seen that the density is highest in the low-latitude, low-speed wind.

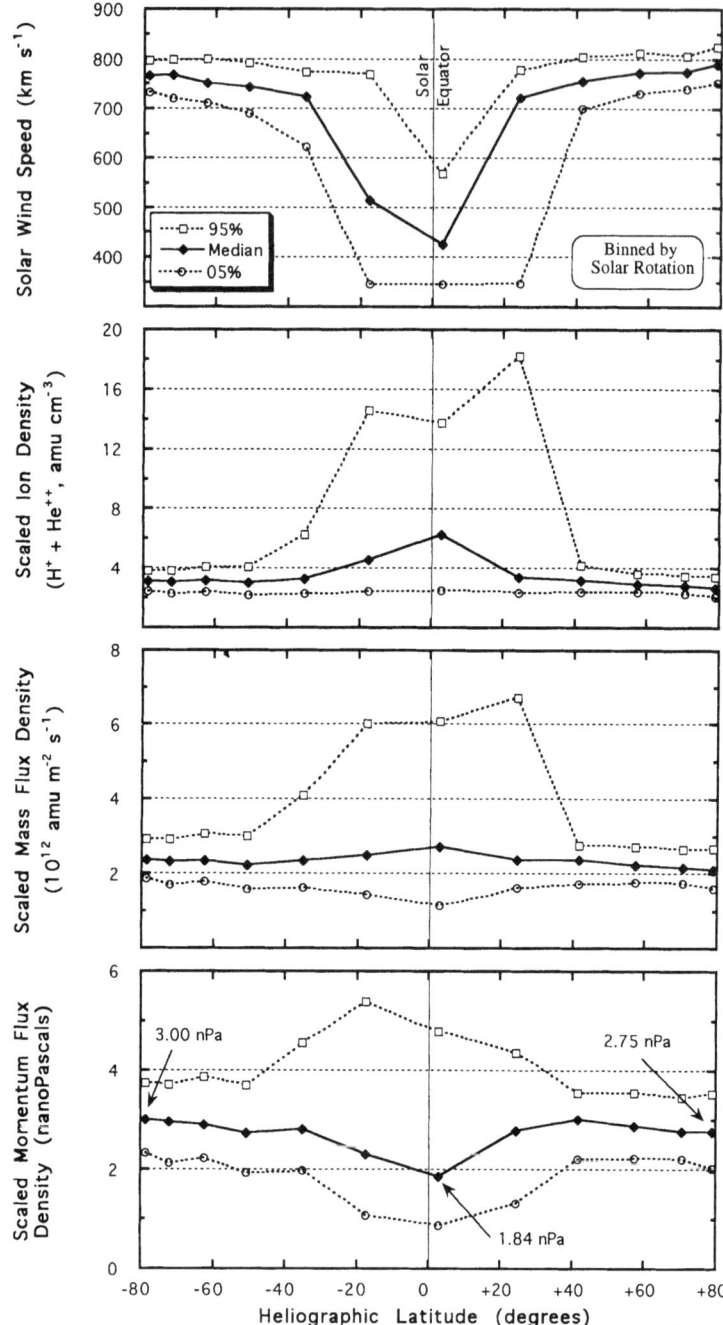

Figure 2.3. Median, 5%, and 95% values for selected fluid parameters over each solar rotation as a function of heliographic latitude during the *Ulysses* FLS (from Phillips *et al.*, 1995c).

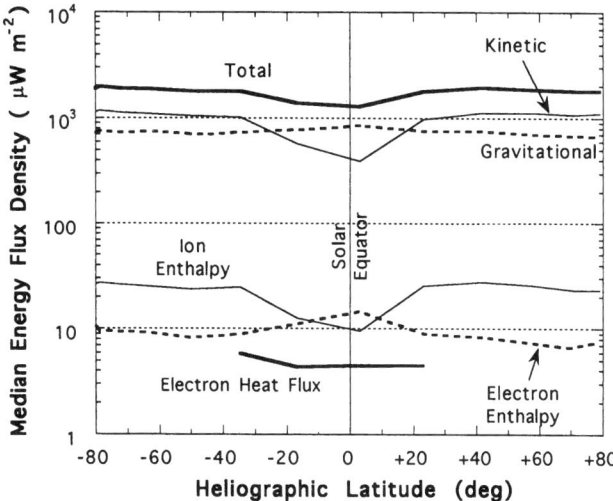

Figure 2.4. Median energy-flux densities, binned by solar rotation, versus heliographic latitude during the *Ulysses* FLS (from Phillips *et al.*, 1995a).

A long-standing question in space physics is the energy source for the acceleration of the solar wind (e.g. see Barnes, 1992). In early theories, the acceleration was assumed to arise from classical thermal conduction from the hot (over 10^6 K) corona. Additional inputs of energy or momentum are believed to be required for the fast wind at low latitudes, but Lallement *et al.* (1986) interpreted observations of backscattered solar Lyman-α radiation in the polar regions to suggest that perhaps the polar solar-wind flux was low enough that thermal conduction would suffice in polar regions. *Ulysses* measurements have shown that although the solar-wind flux in the polar regions normalized to 1 AU is only about two-thirds of the average equatorial flux (i.e. proton fluxes of $\sim 2 \times 10^8$ cm^{-2} s^{-1} compared with $\sim 3 \times 10^8$ cm^{-2} s^{-1}), it is still great enough to require energy input over and above that from classical electron-heat conduction (Barnes *et al.*, 1995).

Figure 2.4 shows solar-rotation-binned energy fluxes during the fast latitude scan, scaled to 1 AU by a factor R^2. The largest component measured at or beyond 1 AU is the kinetic energy of the ion flow (labeled 'kinetic' in Figure 2.4). The energy input near the Sun, however, must also include the work done against gravity between the solar surface and the point of observation; as shown in Figure 2.4, this gravitational energy flux is nearly the same as the energy flux associated with the bulk flow. How much energy must be supplied at the Sun? This figure can be approximated by taking the total energy shown in Figure 2.4 multiplying by 215^2 (because 1 AU = 215 R_s), and then multiplying by a factor of ~ 4.8 to account for the angular expansion between the coronal hole at the solar surface and the solid angle filled with high-speed flow from that coronal hole (Gosling *et al.*, 1995). The result is an average solar-energy input to the high-speed wind from the polar coronal holes of $\sim 4.4 \times 10^5$ erg cm^{-2} s^{-1}. A smaller value is required for the low-latitude solar wind.

Figure 2.4 also shows several other forms of energy flux in the solar wind, all of which are about two orders of magnitude less than the energy associated with the ion-bulk flow. In calculating these energy fluxes, the electron-heat flux was scaled to 1 AU as R^{-3} (Scime et al., 1994a), while the temperatures were scaled as $R^{-0.7}$, which are approximate empirical radial dependences discussed further in later sections. Not shown are the energy fluxes due to the magnetic field and the copious flux of hydromagnetic waves found at high latitudes; these energies are comparable with the enthalpy terms. Another missing factor is the ion-heat flux, which is much smaller than any of the factors shown.

Within the context of present models of the heliosphere, information about the shapes of the heliopause and the termination shock where the solar wind becomes subsonic can be surmised from the measured momentum flux. At least in its forward section, the location of the heliopause is expected to be determined by a pressure balance between the interstellar plasma (including cosmic rays) and the outward-flowing solar wind. Phillips et al. (1995c) pointed out that the decrease in momentum flux seen at low latitudes in the bottom panel of Figure 2.3 implies a possible peanut shape of the heliosphere. If the solar-wind-momentum flux decreases as R^{-2}, then the distance to the heliopause R_h should scale as the square root of the scaled momentum flux. Figure 2.5 is a polar plot of R_h versus heliographic latitude for all the data acquired during the first orbit of Ulysses about the Sun. The open circles are solar-rotation averages from the orbit of Jupiter into the south-polar region, the filled circles show data acquired during the fast latitude scan, and the triangles refer to the interval from the north-polar passage back out to >5 AU. There are two notable features in Figure 2.5. First, there does seem to be a pinched-in waist, or peanut shape, if only the FLS and northern-hemisphere data are considered. Second, the southern-hemisphere data (open circles) show the opposite effect, with a fat waist closer to solar-activity maximum in 1992. The peanut shape shown by Phillips et al. (1995) for the fast latitude scan was based on the median values of the momentum flux and shows a more pinched waist than is evident in Figure 2.5, which is based on average rather than median values. Because of the continuing merging of interaction regions with increasing distance and because of the finite response time of the heliopause and termination shock, the average value is probably the more meaningful one to use; Goldstein et al. (1996) present a figure comparing the momentum flux calculated both ways. In any case, the locations and shapes of the heliopause and termination shock are probably much more complex than can be characterized simply by the square root of the momentum flux. Some of the complexities in recent modelling of the outer heliosphere have been reviewed by Zank et al. (1996).

2.4 STRUCTURES IN THE HIGH-LATITUDE WIND

Although it is apparent from Plate 2 that the speed of the high-latitude solar wind was remarkably steady compared with speeds observed in equatorial regions, the jitter in the plot indicates that the polar solar wind was not without structure. Figure 2.6 provides power spectra of fluctuations in the three components (V_R, V_T, V_N) of

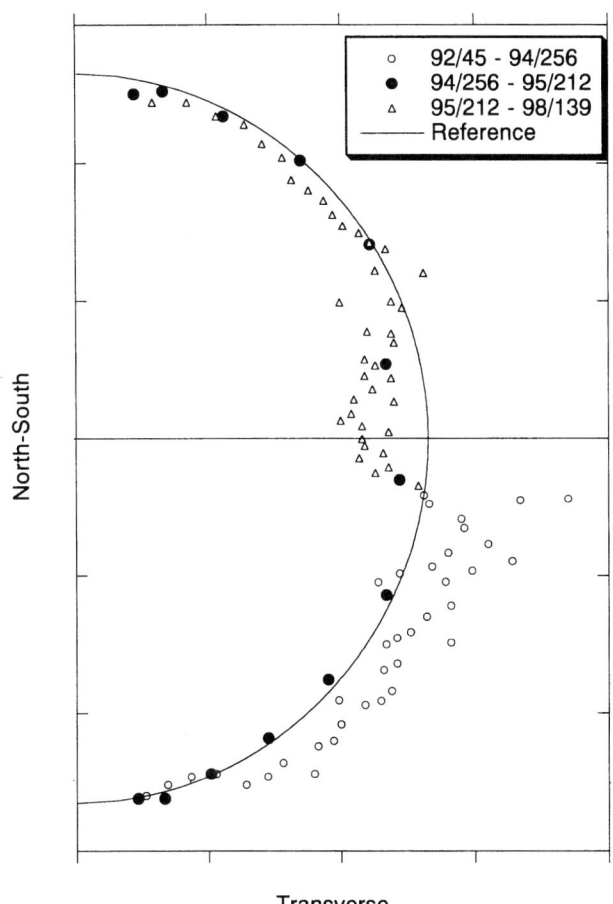

Figure 2.5. Polar plot of the square root of solar-rotation averages of the momentum flux of the solar wind versus heliographic latitude. FLS data are indicated by solid circles (from Phillips *et al.*, 1995c).

the proton-velocity vector in the high-speed wind from the south polar coronal hole. For periods shorter than about half a day, there is a similar amount of power in all three axes, whereas at longer periods the power in the radial component exceeds that in the two transverse components. The transverse fluctuations are due largely to outward propagating Alfvén waves. The power spectrum of the radial component V_R, however, reveals significant additional structure in the high-speed wind.

Examples of different types of fine structure in the high-latitude solar wind are shown in Figure 2.7. The top trace of this figure shows hourly average speed measured between 1 April and 24 April 1994, when *Ulysses* was at a solar latitude of $\sim 60°$S. At this time resolution, it is seen that the speed varied between ~ 700 and $800\,\text{km}\,\text{s}^{-1}$, with principal peaks or valleys separated by ~ 3 days. The middle trace in Figure 2.7 shows the plasma thermal pressure ($= \Sigma NkT$, where N is density in

54 The solar-wind and interplanetary magnetic field in three dimensions [Ch. 2

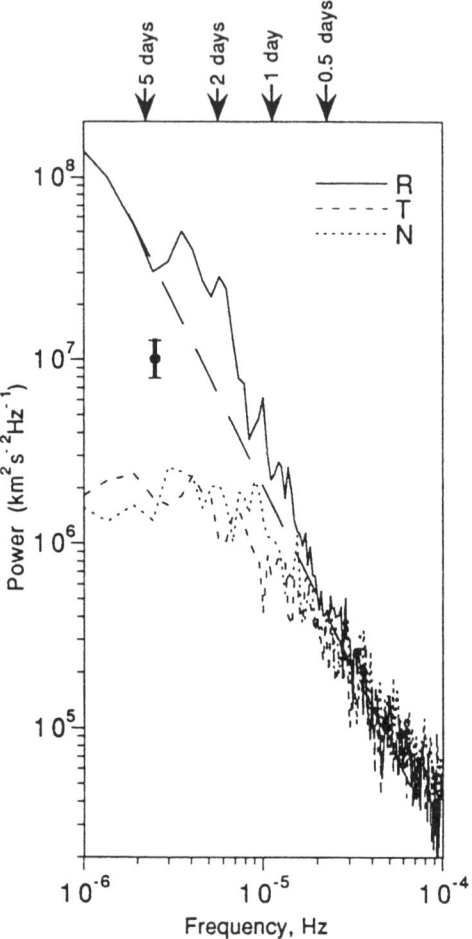

Figure 2.6. Power spectra of hourly averages of the radial R, tangential T, and normal N components the solar-wind proton velocity during seven solar rotations when *Ulysses* was continuously in the high-speed flow from the south-polar coronal hole (from Neugebauer *et al.* 1995).

particles cm^{-3}, k is the Boltzmann constant, T is temperature, and the sum is taken over protons, alpha particles, and electrons), and the bottom trace shows the magnetic pressure ($= B^2/8\pi$).

The most prominent structure in Figure 2.7 is associated with a coronal mass ejection (CME). CMEs are discussed further in Chapter 3; here it is only noted that there were large enhancements of both plasma and magnetic pressures following a forward shock (FS) and preceding a reverse shock (RS) caused by interplanetary expansion of the CME.

Plasma pressure enhancement marked as 'pressure balance' in Figure 2.7 is

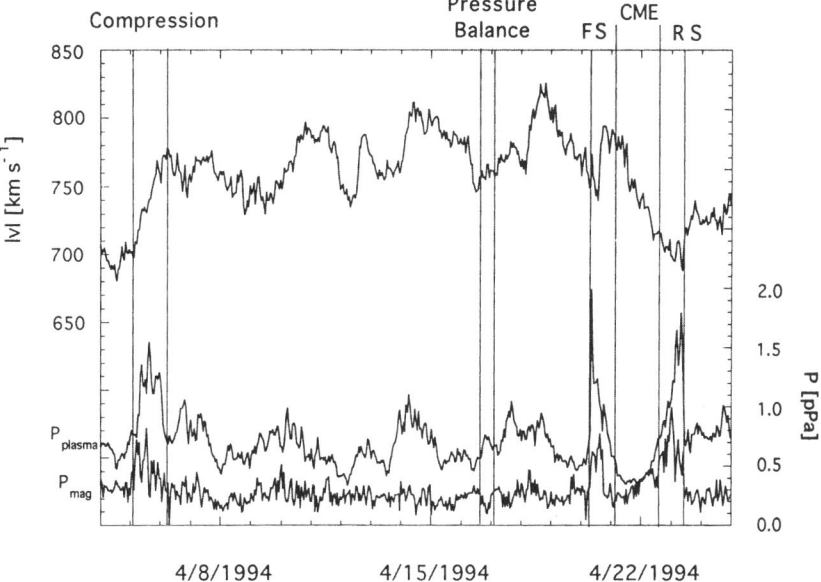

Figure 2.7. From top to bottom: hourly averages of solar-wind speed, plasma pressure (ΣNkT), and magnetic pressure ($B^2/8\pi$) for the 1–24 April 1994, observed by *Ulysses* at a heliolatitude of $\sim 60°$S (from McComas et al., 1995).

accompanied by a decrease in magnetic pressure such that the total pressure remained constant. In a survey of 230 days of high-latitude data, McComas *et al.* (1996) found many more Pressure Balance Structures (PBSs) with increases in $\beta = \Sigma NkT/(B^2/8\pi)$ than PBSs with decreases in β. Usually, both the density and the temperature were higher in the PBSs than in the surrounding plasma and than in the average high-latitude solar wind McComas *et al.*, 1995; 1996). Using a larger dataset 22 months in duration, Reisenfeld *et al.* (1999a) discovered a strong correlation between β and the helium abundance within the PBSs.

McComas *et al.* (1995, 1996) noted that the frequency of PBSs increased with decreasing distance from the Sun, reaching a rate of ~ 10 high-β structures per solar rotation inside 2 AU (McComas *et al.*, 1996). *Helios* data similarly showed the frequency of PBSs increasing between 1.0 and 0.3 AU in the ecliptic (Thieme *et al.*, 1990). This pattern, when combined with the variation in helium abundance and β, provides strong evidence that these PBSs are of solar origin. On the other hand, *Voyager* found an increasing frequency of PBSs as the spacecraft travelled outward to 10 AU (Vellante and Lazarus, 1987). Those more distant PBSs may have a dynamic origin, with streamlines or flux tubes with higher internal pressure expanding until they reach pressure equilibrium with their neighbours.

The final type of high-latitude structure illustrated in Figure 2.7 is the positive gradient of solar-wind speed with corresponding compressional increases of both plasma and magnetic pressures. These structures, called 'compressions' by McComas *et al.* (1995) and called 'microstreams' by Neugebauer *et al.* (1995), exhibit several distinctive properties, some of which are summarized in Figure 2.8. The figure is a superposed epoch analysis of 6-hour averages of solar-wind properties centered on twenty-nine well-resolved peaks and seventeen well-defined dips in the solar-wind speed detected during the seven solar rotations in 1994 when *Ulysses* was poleward of 60°S. The top panel shows the radial component of the proton velocity V_{pR}, which on average was $\sim 50\,\mathrm{km\,s^{-1}}$ higher at the peaks than at the dips. The second panel shows the normalized proton flux ($N_p V_{pB} R^2$), which was greatest on the leading edges of the peaks and lowest just ahead of the dips, as expected, by compression and rarefaction waves ahead of and behind the faster streams, respectively. Note that the flux was the same ($\approx 1.9 \times 10^8\,\mathrm{cm^{-2}\,s^{-1}}$) at the top of the peaks as at the bottom of the dips. The proton temperature, normalized by an $R^{-0.51}$ fit to all the data acquired in the flow from the southern-polar coronal hole, showed heating due to compression at the leading edges of the peaks and trailing edges of the dips and cooling due to rarefaction at the other sides of the extrema. At the peaks and dips themselves the faster plasma at the velocity peaks was hotter than the slower plasma at the bottoms of the velocity dips.

The bottom panel of Figure 2.8 shows that the faster plasma in the peaks had a slightly greater helium abundance than did the slower plasma in the dips. A more recent study by von Steiger *et al.* (1999) shows that microstreams had no statistically significant variation in relative abundances of ions heavier than helium. On average the carbon ions observed at microstream peaks had slightly (barely significant) lower charge states than did carbon ions observed near the dips. In summary, microstreams show a positive correlation between speed, proton temperature, and the ratio of alpha to proton fluxes, and perhaps a negative correlation with ion charge states.

PBSs and microstreams appear to be two separate populations of solar-wind structures, with not only different plasma properties, but also different timescales. Compressional microstreams were the dominant feature for periods longer than ~ 3 days, whereas PBSs dominated for periods less than ~ 1 day (McComas *et al.*, 1996).

What causes PBSs and microstreams? They are probably both of solar rather than interplanetary origin. Furthermore, it has been argued that if microstreams are interplanetary signatures of stationary structures with uniform size independent of latitude, then the time required for a structure to corotate past *Ulysses* would increase with latitude λ as $1/\cos\lambda$. No latitudinal dependence of microstream width was detected, however, which led to the suggestion that the solar counterparts of microstreams probably have a temporal component with a lifetime no longer than 2 or 3 days (Neugebauer *et al.*, 1995). To date, no similar analysis has been performed for PBSs.

Thieme *et al.* (1988, 1990) reported the observation by *Helios* of PBSs in the near-equatorial high-speed wind between 0.3 and 1 AU and postulated they were related to modulation of the solar wind by the solar supergranulation structure, which in turn is related to patterns of convection in the outer layers of the Sun. This remains a valid possibility.

Some of the structures may be related to polar plumes, which are bright, ray-like

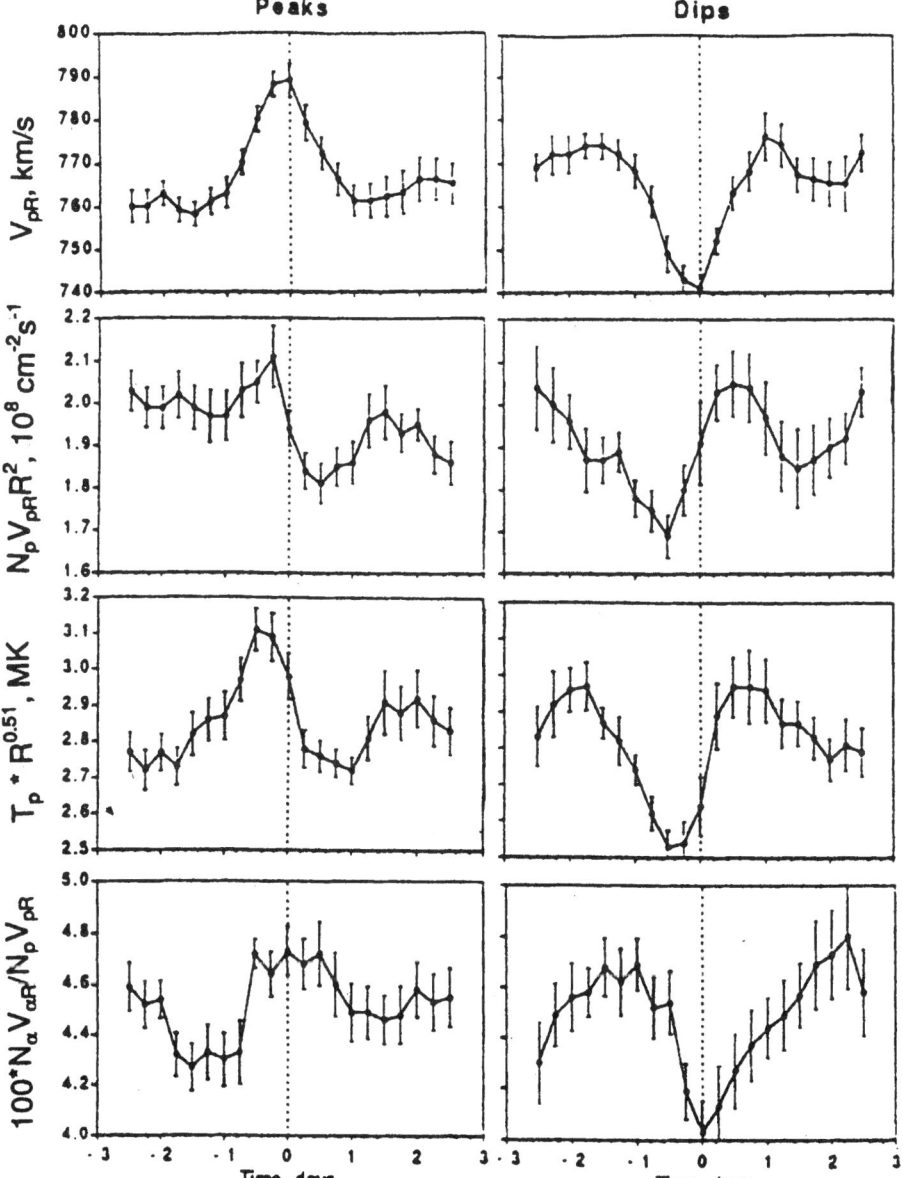

Figure 2.8. Superposed epoch analyses of variation of 6-hour averages in several plasma parameters with zero epochs set at the maxima (left) or minima (right) of the speed in the high-latitude solar-wind observed in the southern hemisphere. From top to bottom are plotted the radial component of the proton velocity, the proton flux normalized to 1 AU by a factor R^2, the proton temperature normalized to 1 AU by a factor $R^{0.51}$, and the ratio of the alpha particle to proton fluxes multiplied by 100 (from Neugebauer et al., 1995).

structures seen in images of the polar solar corona. Early observations suggested that individual plumes have lifetimes of several hours to several days, with an average of ~15 hours. A recent campaign to observe polar plumes with six separate instruments on *SOHO* revealed rapid (10-minute to hours) changes in detailed substructures and in overall brightness, but the large-scale locations and shapes of the plumes remained steady over the course of the 16-hour observation period and the brightest plumes could be seen on the previous and following days (Deforest, 1997).

Even though microstreams are most readily recognizable in the high-latitude solar wind near solar-activity minimum, they are not limited to high latitudes. They may be related to the ubiquitous irregular variation of the in-ecliptic solar wind which exhibits a fractal structure (Burlaga, 1975). It has also been noted that low-frequency peaks in the radial-component power spectrum shown in Figure 2.6 closely agree with some of the peaks of fluctuations in energetic-particle fluxes which Thomson *et al.* (1995) have suggested are caused by gravity-wave excitations (g-modes) of the Sun. Subsequent searches for solar g-modes in interplanetary plasma and field data from *Ulysses* and other spacecraft, however, found no evidence for solar g-modes in those data (Riley and Sonett, 1996; Hoogeveen and Riley, 1998, Denison and Walden, 1999). Finally, Roberts and Goldstein (1998) have suggested that the spectral peak near 3.3 days, which corresponds to a typical time between consecutive peak speeds of the microstreams, may be the tenth harmonic of the ~34-day rotation period of photospheric magnetic features near 70° latitude.

In summary, the origins of the PBSs and microstreams are still to be determined. If they are not related to polar plumes, a second open question is how the plumes affect the polar solar wind. It has been suggested that the outflow velocity of the material in the plumes may be much less than that in the interplume material and that at increasing distances from the Sun, where the Alfvén speed is lower than it is closer to the Sun, the shear flow between the plumes and the interplume wind becomes Kelvin–Helmholtz unstable and the plumes cease to maintain their identity (Parhi *et al.*, 1999). The question of the origins of PBSs and microstreams may not be answered until measurements are obtained much closer to the Sun, as with a solar-probe mission.

2.5 HEAVY IONS

Ulysses carries a new-generation instrument for the study of heavy ions in the solar wind. That instrument, the Solar Wind Ion Composition Spectrometer (SWICS), is described in detail by Gloeckler *et al.* (1992). Briefly, the measurement consists of four steps:

(1) the energy per charge of an ion is determined by a curved-plate electrostatic analyser;
(2) the ion is accelerated through a potential $\simeq 20\,\text{kV}$;

(3) its speed is determined by a time-of-flight measurement; and
(4) its total energy is measured in a solid-state detector.

The combined measurements of energy/charge, speed, and total energy allow calculation of an ion's mass and charge. The *Ulysses* SWICS has a great advantage over previous spaceflight ion-mass spectrometers whose measurements were limited to the ratio mass/charge with the consequent overlap or confusion of different ion species. The use of triple coincidence measurements (time-of-flight start time, time-of-flight stop time, and total energy) also virtually eliminates interference from background counts.

2.5.1 Ionization states of heavy ions

Except for the rare appearance of singly ionized helium from the Sun in coronal-mass ejections (see Chapter 3), and for non-solar ions picked up by the solar wind (Chapter 7), essentially all the solar-wind helium is in the doubly ionized state He^{++}, called alpha particles. The elements heavier than He have ranges of charge states that are diagnostic of conditions within a few solar radii of the Sun.

Plate 5 shows the distribution of charge states of carbon, oxygen, silicon, and iron ions in hourly averages observed by SWICS over four 300-day intervals. The intervals labelled 'South' and 'North' were acquired in the fast wind at high latitudes (day 1 1994 to day 304 1994 and day 182 1995 to day 125 1996, respectively), while those labelled 'Max' and 'Min' were acquired in the slow wind at low latitudes (day 244 1991 to day 182 1992, excluding the Jupiter encounter, and day 182 1997 to day 120 1998, respectively). The two fast-wind intervals have quite similar charge distributions as do the two slow intervals. There are, however, significant differences between the fast and slow winds.

These charge-state distributions can be converted to values of electron 'freezing-in' temperatures (also called ionization temperatures) at which the rate of ionization by ion-electron collisions is balanced by recombination to yield an equilibrium state. In the fast flow from coronal holes, the charge states of each of the four elements shown in Plate 5 can be well fitted by a single temperature. Values of these average freezing-in temperatures are given in Table 2.1. The charge state O^{5+} is not shown in Plate 5; its density in the high-speed wind is 0.54 per cent of the density of O^{6+} which corresponds to the same freezing-in temperature as that calculated from the $O^{7+}:O^{6+}$ ratio (Wimmer-Schweingruber *et al.*, 1998). The slow solar wind is different, however. The charge distributions for Si and Fe in Plate 5 look like mixtures of temperatures (von Steiger *et al.*, 2000); two-temperature fits to these data are included in Table 2.1.

Panel c of Plate 3 shows contours of the freezing-in temperature calculated from oxygen charge states T_O for five solar rotations during the *Ulysses* FLS. The inverse correlation between T_O and speed is readily apparent from comparison of Plates 3a and 3c. Ionization-state data confirm the idea that the high-latitude high-speed wind originates in polar coronal holes where the coronal temperature is lower than in other regions of the corona.

Table 2.1. Freezing-in temperatures of C, O, Si, and Fe observed in four sets of *Ulysses* data. The units are 10^6 K. Two values are given for the slow wind; the 'bulk' values are temperatures calculated from the distribution of the lower charge states and the 'max' values are calculated from the highest ionization states (see Plate 5) (data taken from von Steiger *et al.*, 2000).

Element	Fast, high latitude		Slow, low latitude	
	South	North	Max	Min
Carbon	0.98	0.93	1.37	1.34
Oxygen	1.12	1.05	1.64	1.52
Silicon – bulk	1.41	1.35	1.62	1.55
– max			1.66	1.59
Iron – bulk	1.26	1.20	1.14	1.10
– max			3.0	2.0

Another difference between ionization states in the fast and slow winds is their temporal variability. Figure 2.9 illustrates this point. It shows T_O and the solar-wind speed (here represented by the speed of alpha particles, rather than protons) versus time. The distinct difference between the ionization temperatures of the fast polar flow and the slower, equatorial flow extends down to the finest timescales resolved by SWICS (13 minutes). Wimmer-Schweingruber *et al.* (1997) found abrupt decreases in ionization temperature at the stream interfaces embedded in the corotating interaction regions (CIRs) (see Chapter 3). This discrete change in ionization temperature opens up the possibility of identifying the stream interfaces in the trailing edges of high-speed streams where the speed and kinetic temperature show no discontinuities (Burton *et al.*, 1999).

No systematic dependence of T_O on latitude has been reported for the periods when the spacecraft was completely within polar coronal-hole flow. Table 2.1 does, however, indicate that the fast solar wind had slightly higher freezing-in temperatures in the southern than in the northern hemisphere and that the freezing-in temperature in the slow wind was slightly higher in 1991–1992 ('Max') than in 1997–1998 ('Min'). Von Steiger *et al.* (2000) argue that the difference between the north and south data is a spatial rather than a temporal effect, which is opposite to the conclusion drawn by McComas *et al.* (2000) for various proton parameters. The difference between the Max and Min data can probably be ascribed to the decline in solar activity and in the frequency of CMEs between 1991 and 1998.

Although in the high-speed polar wind the charge-state distribution of each element can be represented by a single-ionization temperature as described above, each element has a different freezing-in temperature (see Table 2.1). The charge distribution of each element is characteristic of the height in the corona where ionization and recombination due to ion–electron collisions become negligible. The values of the frozen-in charge states depend on radial variation in electron density, electron temperature, and ion-flow speed as well as on the shape of the electron-distribution function and the ionization and recombination rates of each

Sec. 2.5] Heavy ions 61

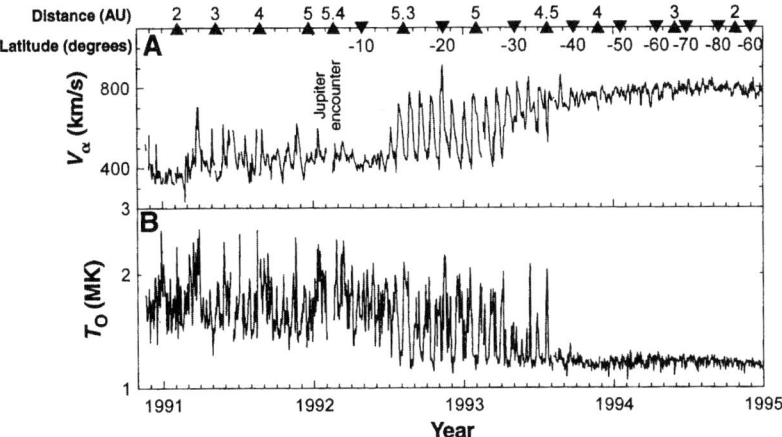

Figure 2.9. Time series of alpha-particle speed (top panel) and ionization temperature calculated from the charge states of oxygen ions (bottom panel) from the beginning of the *Ulysses* mission in 1991 through the end of 1993 when the spacecraft was in the high-speed flow from the south-polar coronal hole (from Geiss et al., 1995).

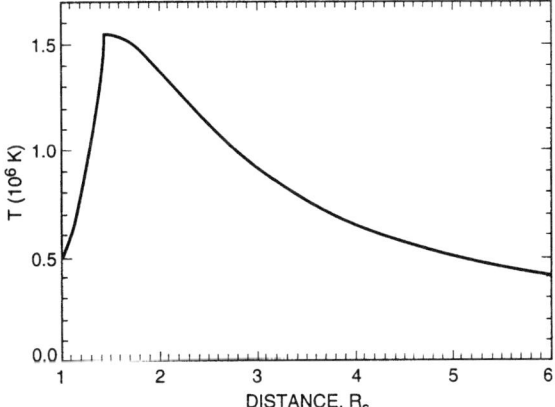

Figure 2.10. Coronal electron temperature versus distance from the Sun as calculated from the ionization charge-state distribution of heavy ions in the polar solar wind (after Ko et al., 1997)

ion species. The ionization and recombination coefficients are functions of electron energy. The electron-density profile can be obtained from coronagraph data from *Spartan 201* or *SOHO*. Most of the other electron parameters are poorly known, but can be modelled and adjusted to fit the distributions of ion charge states observed by *Ulysses*. The radial profile of electron temperature in the corona calculated by Ko et al. (1997) is shown in Figure 2.10. Ko et al. found they could not simultaneously fit the charge-state data for each of the five species studied (C, O, Mg, Si, and Fe) unless they allowed the radial gradient of the outflow speed of the heavy ions to decrease with increasing ion mass. The calculated electron temperature is reduced if it is further assumed that the more highly charged ions of each element have greater speeds than do ions with less charge (Ko *et al.*, 1998). Such a situation would be expected if the acceleration depended on the ion's mass/charge ratio or gyroradius. Within the parameter range they explored, Ko *et al.*, (1996) showed that the *Ulysses* charge-state data were consistent with an electron-distribution function that did not have a substantial suprathermal tail and that the existence of a modest tail did not have a large impact on the electron-temperature profile shown in Figure 2.10; the effect of non-thermal electron distributions is, however, mixed with the effects of charge-dependent ion speeds (Ko *et al.*, 1998). More recently, in response to the discovery based on data from the Ultra-Violet Coronagraph Spectrograph (UVCS) on *SOHO* (Kohl *et al.*, 1997) that the radial gradient of the flow speed of O^{5+} ions in the inner corona is much steeper than the speeds modelled by Ko *et al.* (1996, 1997, 1998), Esser *et al.* (1998) extended the range of coronal parameters used to match the *Ulysses* charge-state data, and found that steeper speed gradients require higher electron temperatures. Further observational constraints are required to obtain a unique solution for the electron-temperature profile in the corona. Recent near-Earth (*ACE* and *WIND* spacecraft) observations of a correlation of suprathermal electrons with the $O^{7+}:O^{6+}$ ratio (Wimmer-Schweingruber *et al.*, 1998) should help constrain the possibilities.

2.5.2 Abundance of heavy ions

Approximately 95 per cent of the solar wind is hydrogen and ~5 per cent is helium, while all the heavier elements together make up only ~0.1 per cent by number. The relative elemental abundances are not constant, but vary in ways that provide clues to the mechanisms responsible for the generation of the solar wind. Specifically, the abundance of an element in the solar wind depends on at least two factors – the time required for its neutral atom to become ionized (the First Ionization Time, or FIT) and the type of solar-wind flow. Plate 3d provides contours of the ratio of the density of low-FIT Mg ions to the density of high-FIT O ions for comparison with other aspects of the stream structure during the *Ulysses* FLS. The ratio Mg:O is generally greatest within the band of low-speed wind surrounding the HCS, but is quite non-uniform along that band; note the anomalously high Mg abundance at ~40° longitude near the equator.

The dependence of elemental abundance on FIT is summarized in Figure 2.11, which is a plot of the ratio of the abundance of an element (X) relative to oxygen to the abundance of that element, again normalized to oxygen, in the photosphere as a function of the time for the element to first become ionized. The data are given separately for the solar wind from polar coronal holes and for the interstream solar wind measured near the ecliptic. The data for Ar, Kr, and Xe were obtained from analyses of the *Apollo* lunar-foil experiment and lunar soils.

It is evident from Figure 2.11 that easily ionized elements, generally those with first-ionization potentials less than the energy of solar Lyman-α photons, are over-abundant by a factor of 1–2 in the coronal-hole flow and by a factor of 4–5 in the

Sec. 2.5]   Heavy ions   63

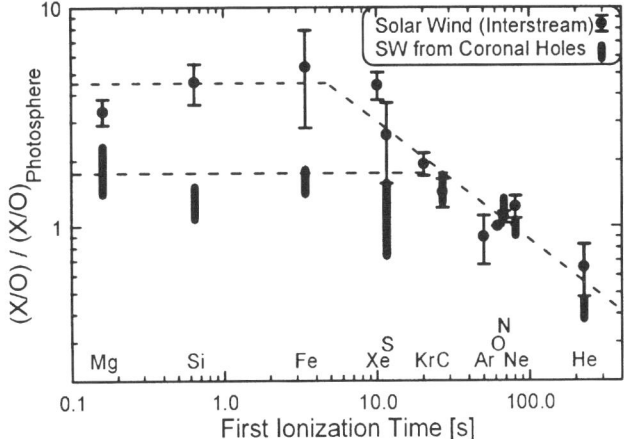

Figure 2.11. Elemental abundances in the solar wind as a function of the first ionization time. The dashed lines indicate patterns deduced for the solar wind from the slow interstream wind and the fast wind from polar coronal holes (from Geiss *et al.* 1995).

interstream wind. For elements with FIT $>\sim 30$ seconds, fractionation between the solar surface and the solar wind is approximately proportional to $(FIT)^{-1/2}$. Carbon is a key element whose ionization potential is close to the Lyman-α energy; its fractionation appears to be independent of the type of flow.

Von Steiger *et al.* (2000) have recently completed a re-analysis of the *Ulysses* SWICS data based on an improved, self-consistent analysis technique described by Schwadron *et al.* (2000). When their revised results are combined with revised values of elemental abundances in the photosphere (Grevesse and Sauval, 1998), they obtain results that are qualitatively similar to those shown in Figure 2.11, but which differ in detail. For low-FIT elements, Mg, Si, and Fe, they obtain an enhancement of photospheric abundances (relative to oxygen) of 2.6 for the slow wind and 1.9 for the fast wind. Both types of flow show a definite elemental fractionation, but the effect is stronger in the slow solar wind.

How can this fractionation of heavy elements in the solar wind be explained? Two sequential processes are assumed to be at work. At the 4,400-K minimum temperature just above the solar surface, the atmosphere consists mainly of neutral atoms, so the first step is initial ionization. The second step is removal of the ions to become part of the solar wind, which amounts to separation of the ions from the neutrals. Several different mechanisms have been proposed for accomplishing these two steps; they are summarized in Table 2.2 compiled by von Steiger (1996). None of these models fully and self-consistently explains the observed fractionation patterns. This is currently an area of active research.

One of the more remarkable findings of the *Ulysses* SWICS experiment was the striking correlation between the Mg/O abundance ratio and the ionization temperature of oxygen. Figure 2.12 shows the ratio of Mg to O ions (a low-FIT to a high-FIT element) versus the oxygen-ionization temperature. The abundance ratio tends

Table 2.2. Overview of published elemental-fractionation models with their modes of ionization and separation of ions from neutrals (after von Steiger, 1996).

Reference	Ionization mechanism	Separation mechanism
Vauclair and Meyer (1985)	Val-C model (Vernazza et al., 1981)	Gravitational settling
Geiss and Bochsler (1985)	UV photons, electrons	–
von Steiger and Geiss (1989)	UV photons, electrons	Density gradient across B, gravity
Antiochos (1994)	Heat flux from corona	Thermoelectric field
Ip and Axford (1991)	–	Rising B field
Hénoux and Semov (1992)	?	DC currents in B field
Marsch et al. (1995)	UV photons, electrons	Diffusion layer
Vauclair (1996)	Val-C model (Vernazza et al., 1981)	Rising B field
Tagger et al. (1995)	?	Waves in B field
Schwadron et al. (1999)	Photoionization	Waves in closed loops

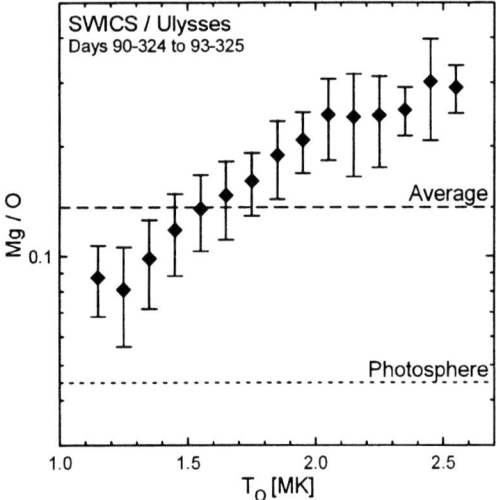

Figure 2.12. Scatter plot of the ratio of abundances of magnesium and oxygen ions detected by *Ulysses* through late 1993 versus the ionization temperature calculated from the ratio of the abundances O^{7+} and O^{6+} ions. The error bars denote standard deviations (from Geiss et al., 1995).

to level off at the lowest (coronal hole) and highest (interstream and CME) values of T_O. Figure 2.13 summarizes the elemental-abundance variation in a different format. It is a superposed epoch plot with the data for each of the entries into and exits from the high-speed streams observed in late 1992 and early 1993 aligned when the speed passed through $600\,\text{km}\,\text{s}^{-1}$. The data plotted are the alpha-particle speed, the

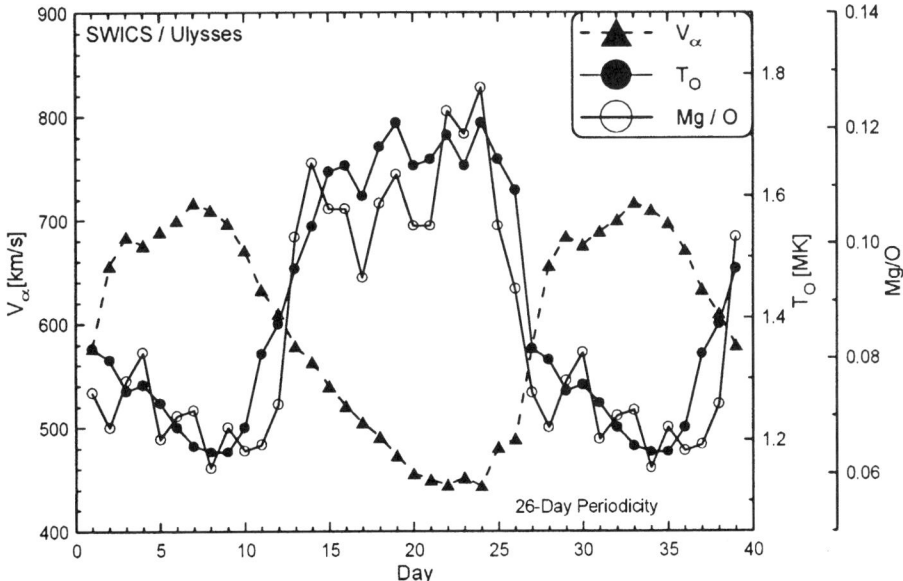

Figure 2.13. Superposed epoch plot showing the systematic variation in solar-wind speed, ionization temperature T_O, and the magnesium to oxygen abundance ratio as a function of the phase of solar rotation during the period of repetitive passage through high-speed streams from the polar coronal hole. The data are repeated to emphasize the periodicity (from Geiss *et al.*, 1995).

oxygen-ionization temperature T_O, and the Mg/O abundance ratio. The figure shows that the transition from one state to another is rather sharp, with few intermediate values. In their re-analysis of the *Ulysses* SWICS data, von Steiger *et al.* (2000) found a very similar result for T_O and Mg/O as well as for the freezing-in temperature of carbon and for the elemental ratios Si/O and Fe/O. These correlations suggest direct connections between the pickup of solar material as it first becomes ionized in the chromosphere to form the solar wind, the temperature higher up in the solar atmosphere in the corona where the final ionization state is determined, and the altitude where the acceleration occurs, suggesting that all three regions and processes are physically linked. These results force us to think of the whole chain of events as interrelated processes; it is no longer very fruitful to start theories about the acceleration of the solar wind with a boundary condition of a hot corona with a given chemical composition.

2.5.3 Helium

Helium is a special case among the elements in the solar wind because its relative abundance is the most variable and because it has been studied the longest under the greatest variety of conditions. Well before the *Ulysses* mission it was known that the

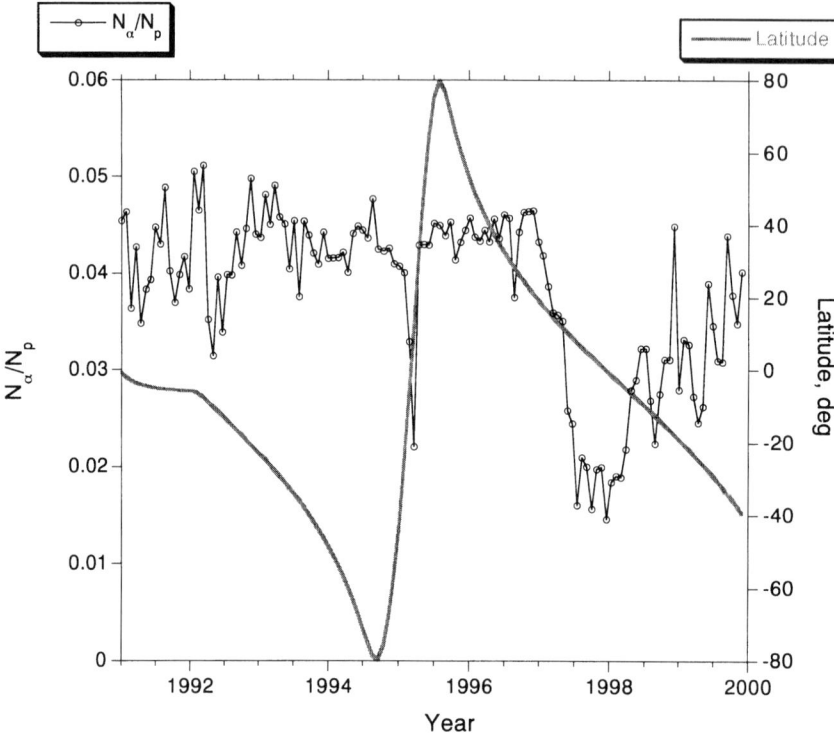

Figure 2.14. SWOOPS measurement of solar-rotation averages of the ratio of the densities of alpha particles and protons for the *Ulysses* mission through December 1999.

alpha-particle abundance relative to protons N_α/N_p is (1) lowest, sometimes vanishingly small, near the heliospheric current sheet (Borrini *et al.*, 1981), (2) often, but not always enhanced (sometimes reaching values as high as 0.4 (Robbins *et al.*, 1970)) in the plasma clouds from coronal-mass ejections, and (3) even in the absence of the HCS or CMEs, more variable and lower in the slow wind than in the fast wind associated with coronal holes. Schwenn (1990) gives average ratios of 0.038 and 0.048 for the slow and fast winds, respectively, for the period observed by *Helios*. Pre-*Ulysses* reviews of the properties of solar-wind helium are available in Schwenn (1990) and Neugebauer (1981). Panel b of Plate 3 shows a splotchy band of low helium abundance in the slow-wind region at the time of the *Ulysses* FLS.

Figure 2.14 displays solar-rotation averages of N_α/N_p observed by *Ulysses* SWOOPS for 1991 through 1999. The high variability in 1991–1992 can be attributed to a combination of HCS crossings and CMEs. The dip in early 1995 coincides with crossings of the HCS during the FLS. The extremely low solar-rotation averages in late 1997 and early 1998 are somewhat surprising; although the in-ecliptic helium abundance generally varies in concert with the solar-activity cycle (Feldman *et al.*, 1978a), long-term average values <0.02 are unusual. The low values of N_α/N_p observed in 1997–1998 when the spacecraft was at latitudes of 0 to $-20°$ may arise from the flatness of the HCS during the approach to solar-activity minimum, with *Ulysses* never being very far from the HCS for the entire period.

The helium abundance observed when *Ulysses* was in the fast high-latitude wind was less variable than at other times. This feature of coronal-hole flow had been previously noted in data acquired in the ecliptic by Bame *et al.* (1977), who suggested that this lack of variability might indicate that the coronal-hole helium abundance is the same as the photospheric abundance, with no fractionation. For radial flow, the conservation of particle fluxes requires that the ratio of the radial component of the alpha-particle flux $N_\alpha V_{\alpha R}$ to the radial component of the proton flux $N_p N_{pR}$ be independent of the radial distance R. In their analysis of solar-rotation averages over the entire first orbit of *Ulysses*, McComas *et al.* (2000) found a slow poleward rise in $N_\alpha V_{\alpha R}/N_p V_{pR}$ from 0.0438 at $\pm 36°$ latitude to 0.0452 at $\pm 80°$ latitude. This poleward gradient argues against Bame *et al.* (1977) suggestion that the abundances in the coronal-hole flow are the same as that in the Sun. From helioseismology it is now known that the helium abundance in the outer layers of the Sun is 0.084 (Pérez Hernández and Christensen-Dalsgaard, 1994), which implies that, even in polar coronal-hole flow, not all solar helium is picked up by the solar wind.

Why is the helium abundance so much more variable than the abundances of other elements? There are three possible reasons:

(1) Helium has the highest first-ionization potential and the longest first-ionization time of all the elements observed in the solar wind.
(2) In the corona the Coulomb drag between the outflowing proton plasma and the heavy minor ions is proportional to the drag coefficient $= Q^2/(2A - Q - 1)$, where Q and A are the ion charge and mass, respectively (Geiss *et al.*, 1970); according to this formula, the alpha-particle drag is smaller than for any other solar-wind ion.
(3) With ~ 20 per cent of the mass of the solar wind and an even greater fraction of the energy and momentum (because $V_\alpha > V_p$ as discussed in Section 2.6.3), the dynamics of alpha particles is an important factor in the acceleration of the solar wind.

The abundance of the isotope ^{3}He relative to ^{4}He in the Sun has important implications for the presolar deuterium abundance and for models of solar evolution and the Big Bang (Geiss and Reeves, 1972; Geiss and Gloeckler, 1998; Bochsler *et al.*, 1990). Measurements of the ratio ^{3}He : ^{4}He by the *Apollo* foil experiments, by *ISEE 3*, and by *Ulysses* have been summarized by Geiss and Gloeckler. The *Ulysses* mission provided an opportunity to obtain a whole-Sun value, as opposed to measurements limited to the ecliptic plane. Prior to *Ulysses*, in-ecliptic results ranged from a weighted average of five *Apollo* missions (all at times of relatively low solar-wind speed) of $(4.26 \pm 0.21) \times 10^{-4}$ to the long-term (1978–1982) average from *ISEE-3* of $(4.88 \pm 0.48) \times 10^{-4}$. Unfortunately, two separate analyses of the *Ulysses* SWICS data have yielded conflicting results. Bodmer *et al.* (1995) obtained an average value of $(4.08 \pm 0.77) \times 10^{-4}$ with a very small increase with increasing solar-wind speed. Gloeckler and Geiss (1998) and Bochsler et al. (1990), on the other hand, found the ^{3}He:^{4}He abundance ratio to be $(4.08 \pm 0.25) \times 10^{-4}$ in the slow, low-latitude wind and $(3.3 \pm 0.3) \times 10^{-4}$ in the high-speed coronal-hole flow. The cause of this discrepancy in interpretation of the *Ulysses* data remains to be determined. Recent near-Earth measurements by other spacecraft have shown large increases in ^{3}He:^{4}He within some CMEs (Gloeckler et al., 1999; Ho et al., 2000), but the question of isotopic variability in the quasi-stationary solar wind is still open.

2.6 ION-DISTRIBUTION FUNCTIONS

Feldman and Marsch (1997) recently published a review of kinetic phenomena in the solar wind which includes excellent summaries of the distribution functions of both ions and electrons and their variation as functions of position in the solar-wind stream structure and radial distance from the Sun. Rather than repeating that material, the present task is to consider those new features revealed by the *Ulysses* measurements.

2.6.1 Temperatures

If the expansion of the solar wind into the heliosphere were adiabatic and if the specific heat ratio γ were 5/3 as expected for a monatomic gas, the ratio $p/N^\gamma \propto T/N^{2/3}$ would be a constant of the motion, and for density N decreasing as R^{-2}, the temperature T would decline as $R^{-4/3}$. Under most circumstances, however, solar-wind ions are heated by interplanetary processes (such as wave–particle interactions or the passage of shocks) as they flow outward. Gazis et al. (1994) found that the near-ecliptic proton temperatures measured by *Pioneer 10* and *11* and by *Voyager 2* dropped off only as $\sim R^{-0.7}$ between 1 and 20 AU, and at greater distances appeared to level off toward a constant value below 10^4 K. An exception is the slow solar wind between 0.3 and 1.0 AU which was observed by *Helios* to expand approximately adiabatically (Marsch et al., 1982a). Liu et al. (1995) have used energy/charge spectra obtained separately for solar-wind protons and alphas by the *Ulysses* SWICS instrument to extend the *Helios* observations to 5.4 AU in the ecliptic. Liu et al. separated daily averages of the temperature data into three speed intervals: slow $<400 \text{ km s}^{-1}$; medium $= 400$ to 500 km s^{-1}; and fast $>500 \text{ km s}^{-1}$. For the slow wind their results agreed with the *Helios* results; the expansion was consistent with adiabatic flow with radial power-law exponents of 1.301 ± 0.129 and 1.337 ± 0.152 for protons and alphas respectively. At higher speeds interplanetary heating was clearly evident, even though data acquired within 30 min upstream and 3 hours downstream of interplanetary shocks were not included in the analysis.

Interplanetary heating was also evident in the fast high-latitude solar wind. Using 48-hour averages of the radial component of the proton-temperature tensor (which SWOOPS determines more accurately than the other components), Goldstein et al. (1996) found power-law dependencies of $R^{-1.03 \pm 0.05}$ in the north-polar region and $R^{-0.81\pm0.03}$ in the south-polar region for the distance range 1.55–3.03 AU. Feldman et al. (1998) obtained a similar power law of $R^{-1.02}$ for the northern hemisphere at latitudes above 50° (corresponding distance range = 1.5–3.2 AU) binned into 0.1-AU intervals of R. Finally, McComas et al. (2000) fitted solar-rotation averages of the radial proton temperature for all latitudes poleward of ±36° to an expression of the form:

$$T_p = R^a(T_o + b\Theta) \qquad (2.2)$$

to obtain $T_p = R^{-1.0}(258{,}000 + 223\,\Theta)$, where Θ is the absolute value of heliographic latitude. On average, the radial proton temperature was 3 per cent higher in the south-polar flow than in the north.

These power-law fits to the high-latitude proton temperature show less interplanetary heating than found for the average near-ecliptic wind sampled by the *Pioneers* and *Voyager 2*. Why should fast wind in the polar regions be heated less than the wind in the ecliptic? The explanation may be that there are more stream interactions, more shocks, and more turbulence-producing velocity shears at low latitudes than in the more uniform high-latitude flow.

The nature and form of interplanetary heating of solar-wind ions are not well characterized. In many theoretical models of solar-wind flow, a common expedient to get around that problem is to replace the energy equation by a polytrope relation of the form:

$$p/N^{\gamma^*} = \text{constant} \qquad (2.3)$$

where γ^* is called the polytrope index (Siscoe, 1983). For adiabatic flow of a monatomic gas, $\gamma^* = \gamma = 5/3$. Feldman et al. (1998) have derived values of γ^* for the fast high-latitude solar wind by examining the relation between temperature and density in compressions and rarefactions associated with six microstreams found in the polar solar wind and with multiple crossings into and out of the low-speed solar wind at solar distances of 4.0–5.1 AU. For the limited number of microstreams studied they found $\gamma^* = 1.70 \pm 0.14$, which is consistent with the adiabatic value of 1.67. For the study of corotating interaction regions, they filtered the data to remove shocked plasma, CME plasma, and plasma with speed $<600\,\text{km}\,\text{s}^{-1}$ and found $\gamma^* = 1.70$ and 1.66 for the southern and northern edges of the band of variable solar wind, respectively. Again, these values of the polytrope index are close to the adiabatic value. Feldman et al. concluded that the best overall polytrope relation for the high-speed wind between 1.5 and 4.8 AU is $T_p = (2.0 \pm 0.13 \times 10^5) N_p^{0.57}$, but that slightly different expressions apply in different locations or different times in the high-latitude solar wind. Specifically, localized regions of compression or rarefaction tend to be closer to adiabatic than is the average radially expanding wind. They suggest that Alfvén waves are a likely candidate for heating the high-latitude wind.

Ulysses also extended our knowledge of the temperatures of heavy ions in the solar wind to greater radial distances and latitudes. The different ion species in the solar wind are usually not in thermodynamic equilibrium. Two ion species with

Table 2.3. Dependence of the ratio of alpha and proton temperatures T_α/T_p on the magnitude of heliographic latitude $\Theta = |\lambda|$ and on distance from the Sun R in AU. The values are calculated from individual fits to T_α and T_p in the high-latitude solar wind by McComas et al. (2000).

R and Θ	35°	50°	65°	80°
1.5 AU	5.66	5.54	5.41	5.29
2.5 AU	6.27	6.13	6.00	5.86
3.5 AU	6.71	6.56	6.41	6.27
4.5 AU	7.05	6.90	6.74	6.59

masses m_i and m_j may have different temperatures T_i and T_j, which often satisfy the relation:

$$T_i/T_j \approx m_i/m_j \qquad (2.4)$$

When equation (2.4) is satisfied, the two ion species have the same thermal speed rather than the same temperature. Alfvén waves or other wave–particle interactions may be responsible for the equal-thermal-speed condition. Coulomb collisions tend to return the distributions toward equal temperatures if the Coulomb collision time τ_c is comparable with or less than the solar-wind expansion time τ_{ex}; the ratio τ_c/τ_{ex} is small in the cool, dense, slow solar-wind.

In their analysis of in-ecliptic SWICS data, Liu et al. (1995) found that the ratio of alpha to proton temperatures T_α/T_p did not change with radial distance. Average ratios were 3.7 ± 1.0, 3.9 ± 1.2, and 4.2 ± 0.8 for speed intervals of <400, 400–500, and >500 km s^{-1}, respectively, thus confirming the expectation that Coulomb collisions may play a role at the lower speed.

For the faster flow poleward of $\pm 36°$, McComas et al. (2000) used equation (2.2) to derive the relation $T_\alpha = R^{-0.80}(1.42 \times 10^6 - 871\Theta)$. The ratio of this fit to the alpha-particle temperature to the corresponding fit to the proton temperature yields the latitudinal and radial variation shown in Table 2.3. In the polar solar wind, T_α/T_p is consistently greater than the value of 4.0 expected from equation (2.4) and the alpha particles cool more slowly with distance from the Sun than do the protons. A likely explanation for these effects is discussed in Section 2.6.3.

Measurements by the SWICS instrument on Ulysses have extended the observations of thermal properties of ions heavier than helium to 5 AU and to high latitudes. The relation between the average kinetic temperatures of fourteen different ions from day 200 1992 to day 200 1993, a period when Ulysses was passing in and out of high-velocity streams once per solar rotation (latitude = 14–35°S; distance = 5.3–4.5 AU) is presented in Figure 2.15. The open circles indicate the temperature ratios predicted by equation (2.4), and the closed circles with error bars indicate the measurements together with statistical errors. The agreement between the observations and equation (2.4) is excellent except for O^{8+}, for which there are some difficulties with the measurements.

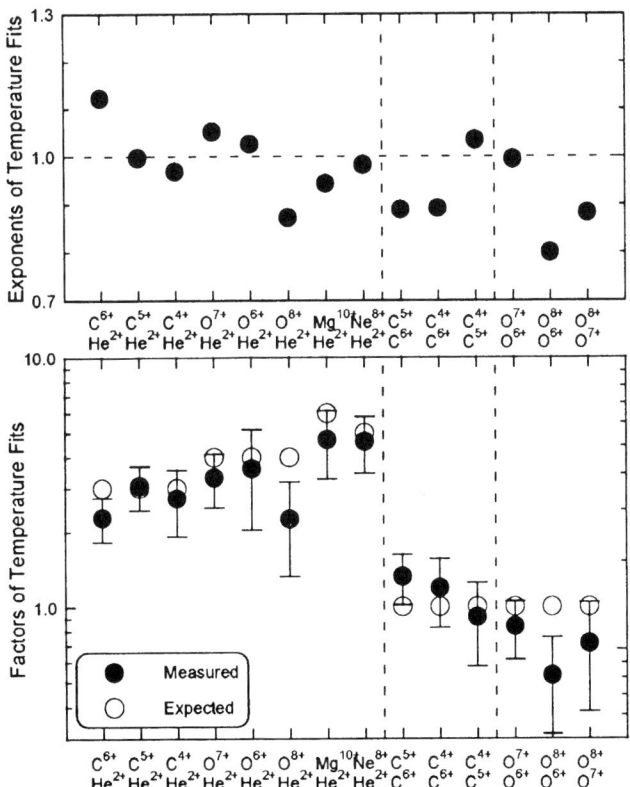

Figure 2.15. Kinetic temperature ratios for fourteen pairs of heavy ions observed by *Ulysses*/SWICS at mid-latitudes. The ratios expected for temperature proportional to mass are shown by open circles, and the filled circles indicate measured values (from von Steiger *et al.*, 1995).

There is, however, still some confusion concerning the circumstances under which ion temperature is proportional to ion mass in the fast solar wind. Another set of *Ulysses*/SWICS data is shown as filled circles in Figure 2.16, where the ratio T_O/T_{He} is plotted versus solar-wind speed. These data were acquired during the in-ecliptic phase of the *Ulysses* mission at radial distances of 2.5–5.3 AU. The ratio $T_O m_{He}/T_{He} m_O$ is seen to be <1 at low speeds, in agreement with most earlier measurements, but is significantly >1 at higher speeds. Figure 2.16 also shows data obtained by *WIND* near Earth between December 1994 and August 1995; these data agree that $T_{heavy} m_{light}/T_{light} m_{heavy}$ increases with speed. This pattern is similar to that found by analysis of 5 years of *ISEE-3* data which showed $T_O/T_{He} \approx 5$ over the speed range 400–600 km s^{-1} (Bochsler *et al.*, 1985). Other datasets, however, do not show such an effect. The *Ulysses* mid-latitude contours of $T_{C^{6+}}/T_{He^{2+}}$ on the right side of Figure 2.17 show no substantial rise at high temperatures which correlate with high speeds. Similarly, the temperatures of O, Fe, and Si ions observed near Earth by *SOHO*/Celias show no rise above

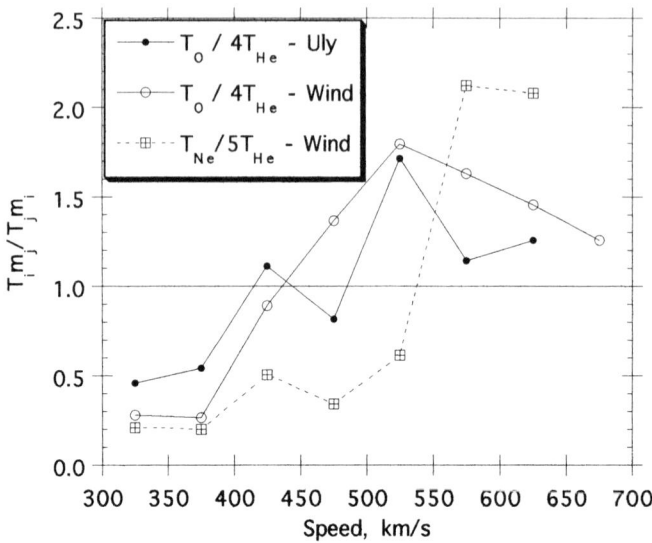

Figure 2.16. Mass-adjusted ratios of kinetic temperatures for oxygen and neon relative to helium as functions of solar-wind speed as measured in the ecliptic by *Ulysses and WIND* (data from Cohen *et al.*, 1996 and Collier *et al.*, 1996).

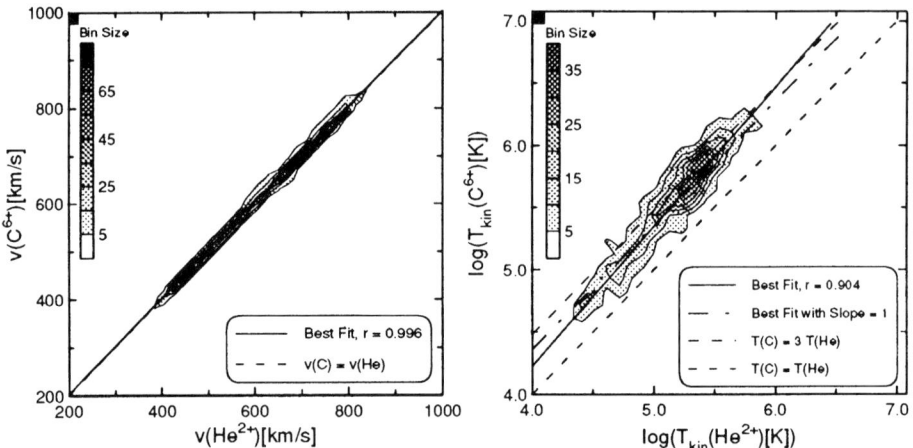

Figure 2.17. Correlograms of speeds (left) and kinetic temperatures (right) of C^{6+} versus He^{2+} ions observed by *Ulysses* at mid-latitudes (from von Steiger *et al.*, 1995).

$m_{\text{heavy}} T_p$ at high temperatures (Hefti *et al.*, 1998). Further work is required to link kinetic temperature ratios to specifics of stream structure and to dependence on distance from the Sun and on data-reduction techniques before we can understand the physical processes underlying these apparently conflicting results.

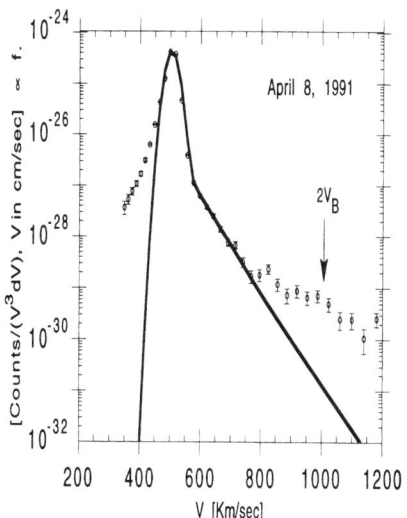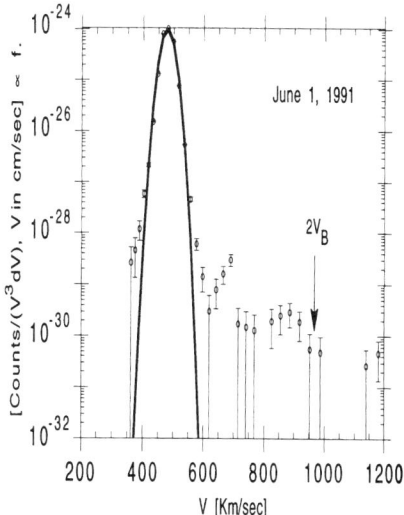

Figure 2.18. Two proton distributions observed by *Ulysses*. (left) A distribution with a pronounced high-energy tail observed on 8 April 1991. (right) A quiet-time distribution with very few (but above background) counts at high speeds (from Ogilvie *et al.*, 1993).

2.6.2 High-energy tails

The ability of SWICS to measure the energy/charge spectrum of an ion species without interference from ions with different values of mass or charge has enabled study of high-energy tails of ion-distribution functions. Using SWICS data, Ogilvie *et al.* (1993) found high-energy tails on proton spectra containing up to 1 per cent of the proton density in regions on the downstream (high-entropy) side of quasiperpendicular shocks between 1.33 and 3.14 AU. Examples of proton spectra with and without such tails are shown in Figure 2.18. High-energy tails were also observed on He^{2+} and O^{6+} distributions. The tails are approximately exponential in shape and persist to ~10 thermal speeds above the peak of the distribution and to speeds greater than twice the solar-wind speed. Individual tail events lasted between half a day and 5 days. Ogilvie *et al.* suggest that the tail particles are energized by shock-drift acceleration in quasiperpendicular shocks and differ from the accelerated particles found by Gosling *et al.* (1984) upstream of interplanetary shocks. This intermittent presence of high-energy tails on the proton distribution is also in contrast to the persistent presence of high-energy tails on the He, O, and Ne ions observed with the *WIND*/MASS experiment near Earth (Collier *et al.*, 1996).

2.6.3 Ion beams

Near the ecliptic it is well established that, despite their greater mass, alpha particles tend to flow away from the Sun faster than protons; see review by Feldman and Marsch (1997). This differential streaming: is greatest in the fast, low-density wind; is near zero in the slow, high-density wind; decreases with increasing distance from the Sun; and appears to have an upper limit set by the Alfvén speed. *Ulysses* has afforded the opportunity to study proton–alpha differential streaming under a broader range of conditions, with some interesting results.

First, a definition of terms is in order: $\Delta V = |\mathbf{V}_\alpha| - |\mathbf{V}_p|$ is the difference in alpha and proton speeds and $\mathbf{V}_{\alpha p} = \mathbf{V}_\alpha - \mathbf{V}_p$ is their vector-velocity difference. In their study of in-ecliptic *Ulysses* data, Neugebauer et al. (1994) found that the correlation of ΔV with proton speed V_p noted in *Helios* and other datasets disappeared at a solar distance of 2 AU. Beyond that distance, both ΔV and $V_{\alpha p}$ were largest in stream–stream interaction regions, especially downstream of interplanetary shocks.

Different results were found at higher latitudes where differential streaming could be followed beyond 5 AU. McComas et al. (2000) found that for latitudes poleward of $\pm 36°$ the speed difference ΔV declined as $R^{-1.39}$ and at a rate of -0.03 per cent per degree of latitude. The scaled value $\Delta V R^{1.39}$ was 41.1 ± 33.1 km s^{-1} in the south and 34.5 ± 27.4 km s^{-1} in the north.

The left side of Figure 2.19 shows that for speeds in the range 700–800 km s^{-1} the *Ulysses* measurements of $V_{\alpha p}$ are a reasonable extension of *Helios* data from 0.3 to 1 AU. A line with a slope R^{-1} is shown for comparison. Although the match-up of the two datasets is remarkably good, *Helios* data appear to fall off slightly more steeply than R^{-1}, while *Ulysses* data show a slightly shallower trend; the slight flattening of *Ulysses* data at the largest distances is probably an artefact arising from very small values of $V_{\alpha p}$ not being determined with sufficient accuracy to pass the quality-control criterion that $\mathbf{V}_{\alpha p}$ be aligned with the magnetic field within 20°.

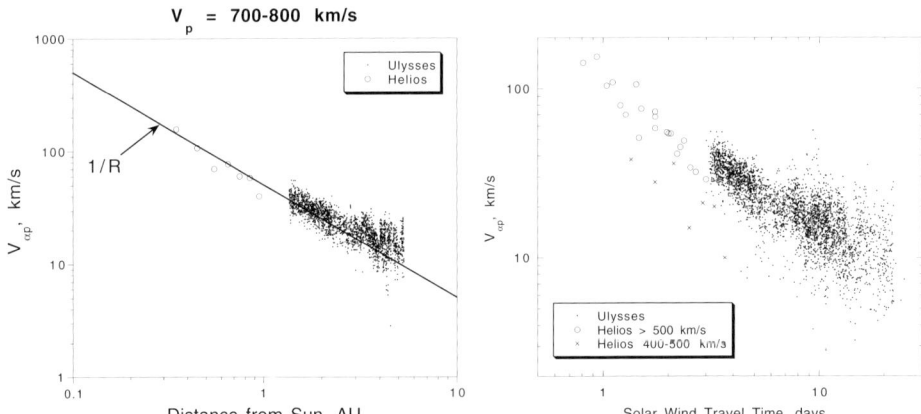

Figure 2.19. (left) Magnitude of the alpha–proton differential flow versus heliocentric distance for intervals with proton speed between 700 and 800 km s^{-1}. The dots denote 6-hour averages of *Ulysses* data and the open circles denote 0.1 AU averages of *Helios* data. (right) Averages (6-hour) of the magnitude of the alpha–proton differential flow as a function of travel time from the Sun. Dots denote 6-hour averages of *Ulysses* data; open circles are *Helios* data for proton speeds >500 km s^{-1}, and the crosses are *Helios* data for proton speeds in the range 400–500 km s^{-1} (from Neugebauer et al. 1996).

In Section 2.6.1 it was stated that the proton temperature fell off with increasing distance from the Sun more rapidly than did the alpha-particle temperature ($R^{-1.0}$ versus $R^{-0.8}$). Reisenfeld *et al.* (1999b) have found that the streaming energy lost by the alphas as $V_{\alpha p}$ declines goes into heating the alphas, thus explaining why alpha cooling is further from adiabatic than proton cooling.

As with *Helios* data, the magnitude of the differential streaming observed by *Ulysses* decreased with decreasing solar-wind speed. The distance and velocity dependencies could be combined into a single dependence on travel time τ from the Sun to the point of observation. The dependence of $V_{\alpha p}$ on τ is illustrated in the plot on the right side of Figure 2.19, which includes the entire range of speeds observed by *Ulysses* between December 1990 and September 1995. *Ulysses* data are a good extension of *Helios* data for *Helios* speeds $>500\,\mathrm{km\,s^{-1}}$. There is no obvious reason why all the data should fall on a single curve of the form $V_{\alpha p} \propto \tau^{-0.70\pm0.07}$, but such appears to be the case. Neugebauer *et al.* (1996) further found that the normalized parameter $V_{\alpha p}\tau^{0.70}$ was independent of solar latitude, thus ruling out any rotational effects on either acceleration or deceleration of the alphas relative to the protons. There was also no significant difference in the value of $V_{\alpha p}\tau^{0.70}$ between quasistationary and transient (CME) flows. Weak correlations were found between $V_{\alpha p}\tau^{0.70}$ and the amplitudes of fluctuations in both the magnitude and the direction of the interplanetary magnetic field, showing some effect of the wave field on differential streaming.

The *Helios* missions showed that out to distances of $\sim 1\,\mathrm{AU}$, $V_{\alpha p}$ is limited by and often equals the Alfvén speed $V_A = B/(4\pi\rho)$ (Feldman and Marsch, 1997). With increasing distance R from the Sun, the magnetic field strength B drops off approximately as $R^{-3/2}$ and plasma density ρ drops off as R^{-2}. The result is that poleward of $\pm 36°$ latitude, $V_A = R^{-0.49}(64.9 - 0.087\Theta)\,\mathrm{km\,s^{-1}}$ (McComas *et al.*, 2000). The differential alpha–proton streaming velocity $V_{\alpha p}$ decreases even more rapidly than does V_A. Figure 2.20 displays the drop in $V_{\alpha p}/V_A$ at long travel times (large R).

Near Earth and inside 1 AU, where $V_{\alpha p} \approx V_A$, the alphas appear to surf on the outward-propagating Alfvén waves, with the result that the alphas do not participate in the transverse fluctuations of the field and the proton velocity (Marsch *et al.*, 1982b). In the high-latitude solar wind, the alphas usually did participate in Alfvénic fluctuations, but their velocity fluctuations were often $180°$ out of phase with the velocity fluctuations of protons such that, for example, when the protons moved northward the alphas moved southward, and vice versa. This situation can occur when $V_{\alpha p}$ is greater than the speed of the waves in the reference frame of the fluid which has a velocity V_f given by equation (2.1). In such circumstances, in the reference frame of the alpha particles, the waves appear to be propagating back toward the Sun; such was in fact the case when out-of-phase alpha waves were seen. At low frequencies the properties of transverse MHD (Alfvén) waves are expected to satisfy the relation:

$$\delta \mathbf{V}_i = -(\omega/k - V_{i\|})\delta \mathbf{B}/B \qquad (2.5)$$

where $\mathbf{V}_i$ is the velocity of ion species i, ω is frequency, k is wave number, $V_{i\|}$ is the component of $\mathbf{V}_i$ parallel to the field $\mathbf{B}$, and B is the field magnitude. Operationally,

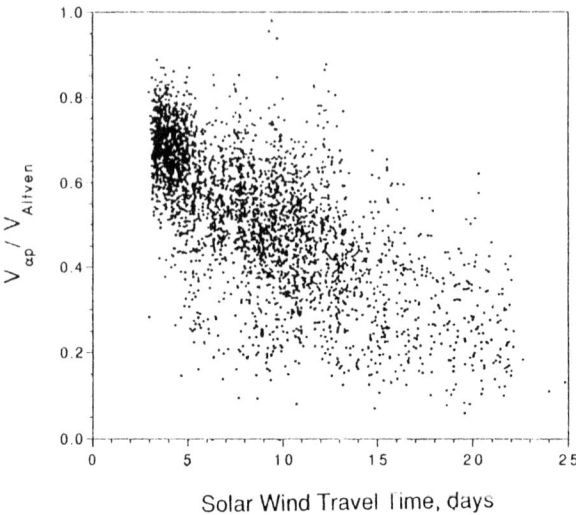

Figure 2.20. Averages (6-hour) of the ratio $V_{\alpha p}/V_A$ versus travel time from the Sun (from Neugebauer *et al.*, 1996).

the wave speed $V_{\text{wave}} = \omega/k$ can be determined from the slope of the correlation of the components of $\mathbf{V}_f$ with the components of $\mathbf{B}$; that is:

$$V_{\text{wave}} = \delta \mathbf{V}_f B / \delta \mathbf{B} \qquad (2.6)$$

As discussed later in Section 2.6.4, $V_{\text{wave}} \approx 0.6 V_A$ (Steinberg *et al.*, 1996). Figure 2.21 shows a scatter plot of the correlation between the normal (north–south) component of $\mathbf{V}_\alpha$ and the normal component of $\mathbf{B}$ as a function of $V_{\alpha p}/V_{\text{wave}}$. The sign of the correlation has been selected as positive (negative) if the variation in $V_{\alpha N}$ is in (out of) phase with the variation in V_{pN}. This figure shows that when $V_{\alpha p} > V_{\text{wave}}$ the proton and alpha oscillations are out of phase with each other, when $V_{\alpha p} < V_{\text{wave}}$ they are in phase, and when $V_{\alpha p} \approx V_{\text{wave}}$ the alphas surf and do not participate in the wave (i.e. $V_{\alpha p} \approx 0$). Brief intervals of such waves with the alphas oscillating out of phase with the protons have subsequently been detected near Earth with instruments on the *WIND* spacecraft (Steinberg *et al.*, 1996).

What about the velocities of ions heavier than helium? The von Steiger *et al.* (1995) study of heavy-ion kinetics included an analysis of the relative speeds of several heavy-ion species. An example of their results is shown on the left side of Figure 2.17, where the speed of the C^{6+} ion is compared with that of the He^{2+} ion, yielding a very high correlation with a slope of unity and a zero intercept. Their comparison of fourteen different pairs of nine different ion species (He^{2+} through Mg^{10+}) all showed equal speeds to a very high degree of accuracy over the range of speeds from 400–800 km s^{-1}. These results disagree with results obtained near Earth by *ISEE-3* and *SOHO*/Celias which indicate that Si and Fe ions lag behind the alphas (Schmid *et al.*, 1988; Bochsler, 1989; Hefti *et al.*, 1998). Perhaps the accelera-

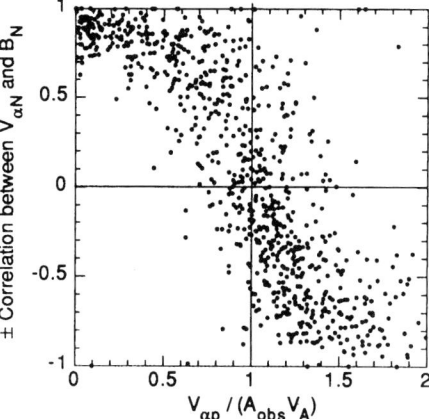

Figure 2.21. Correlation between $V_{\alpha N}$ and B_N as a function of $V_{\alpha p}$ (the difference between the field-aligned alpha particle and proton velocities) normalized to the observed wave speed. The sign of the ordinate is $+$ ($-$) if the alpha-particle motion is in phase (out of phase) with the proton motion (from Goldstein et al., 1995).

tion of ions as heavy as Si and Fe, which were not included in the study by von Steiger et al. (1995), is not yet complete at 1 AU.

In addition to alpha-particle beams and comoving heavy-ion beams, the solar wind also often exhibits more than one beam of protons (Feldman and Marsch, 1997 and references therein). The fast, polar solar wind often shows a secondary proton beam, with higher speed and lower density than the primary proton beam, together with an alpha-particle beam moving at the speed of the secondary proton beam. When only a single proton peak is present, it has the speed of the secondary proton peak in nearby spectra (Feldman et al., 1996). Feldman et al. (1996) suggest that the primary (lower speed) beam evolves from the ambient corona whereas the faster, secondary proton peak and the alphas are accelerated by intermittent jets resulting from magnetic reconnection.

A statistical study of the ratio of the field-aligned velocity difference between primary and secondary proton beams V_{pp} to the Alfvén speed shows that V_{pp}/V_A is everywhere less than an upper bound determined from the linear theory for electromagnetic proton–proton instabilities (Goldstein et al., 2000). This result provides good evidence that the relative streaming of the two proton beams is constrained by these particular microinstabilities. Although it is known that double-proton streaming becomes less pronounced at the largest heliocentric distances sampled by *Ulysses* (Feldman et al., 1996), determination of the other properties of double-proton beams in the high-latitude solar wind is still a work in progress.

In an analysis of *Ulysses* ion spectra near the heliospheric current sheet, Hammond et al. (1995) consistently found double-proton and alpha-particle beams on either side of the current sheet, but not within it. The faster beam has a lower proton density and a higher helium abundance (N_α/N_p) than the slower beam. It is conjectured that these double beams are formed by reconnection near the Sun in the boundary region between the open-field lines of coronal holes and the closed-field lines underlying coronal streamers (Hammond et al., 1995).

2.6.4 Anisotropies

In addition to high-energy tails and resolved double-ion beams, ion distributions can also be anisotropic, with different temperatures parallel and perpendicular to the magnetic field. It was well established by earlier missions that in the fast solar wind from coronal holes the core of the proton distribution has a temperature $T_{p\perp}$ perpendicular to the field which usually exceeds the temperature $T_{p\parallel}$ parallel to the field. This is taken as evidence for interplanetary heating by damping of ion-cyclotron waves. It is, however, difficult to study ion anisotropy at solar distances much beyond ~2 AU with *Ulysses* because of the relatively coarse angular resolution of the SWOOPS instrument (Bame et al., 1992). An effort is presently in progress to develop algorithms for obtaining improved estimates of ion anisotropy from SWOOPS. Despite this present lack of high precision, especially at larger solar distances, several studies that used estimates of plasma anisotropy measured by SWOOPS have provided interesting, if tentative, results:

(1) In a study of magnetic holes (localized regions of low field strength) between 1 and 5.4 AU and from the ecliptic to latitude 23°S, Winterhalter et al. (1994b) determined that series of holes tended to be found in plasma that was borderline stable against the mirror instability, which is driven by plasma anisotropy and for which the criterion is:

$$R_m = (\beta_\perp/\beta_\parallel)/(1 + 1/\beta_\perp) < 1 \quad \text{for stability} \tag{2.7}$$

where $\beta_\perp = 8\pi \Sigma NkT_\perp/B^2$ and $\beta_\parallel = 8\pi \Sigma NkT$. Figure 2.22 shows that most of the solar wind measured by *Ulysses* during 1991 and 1992 was quite stable with R_m usually <0.5, but R_m was between 0.5 and 1.0 during periods when multiple magnetic holes were observed. It should be noted that only proton and alpha-particle densities, temperatures, and differential streaming were used in the calculation of R_m in Figure 2.22; inclusion of the electron anisotropy and/or use of more accurate ion anisotropies might change the results slightly. These magnetic holes were found at all heliographic latitudes sampled by *Ulysses* (from Winterhalter et al., 2000).

(2) Figure 2.20 shows that over the course of the *Ulysses* mission $V_{\alpha p}$ was always less than the Alfvén speed, and Figure 2.21 shows that $V_{\alpha p}$ was sometimes greater than and sometimes less than V_{wave} as determined according to equation (2.6). Theoretically it is expected that:

$$V_{wave} = AV_A \tag{2.8}$$

where A is an anisotropy factor given by:

$$A^2 = 1 - 4\pi(p_\parallel - p_\perp)/B^2 \tag{2.9}$$

where $p_\parallel$ and $p_\perp$ are the pressures parallel and perpendicular to the magnetic

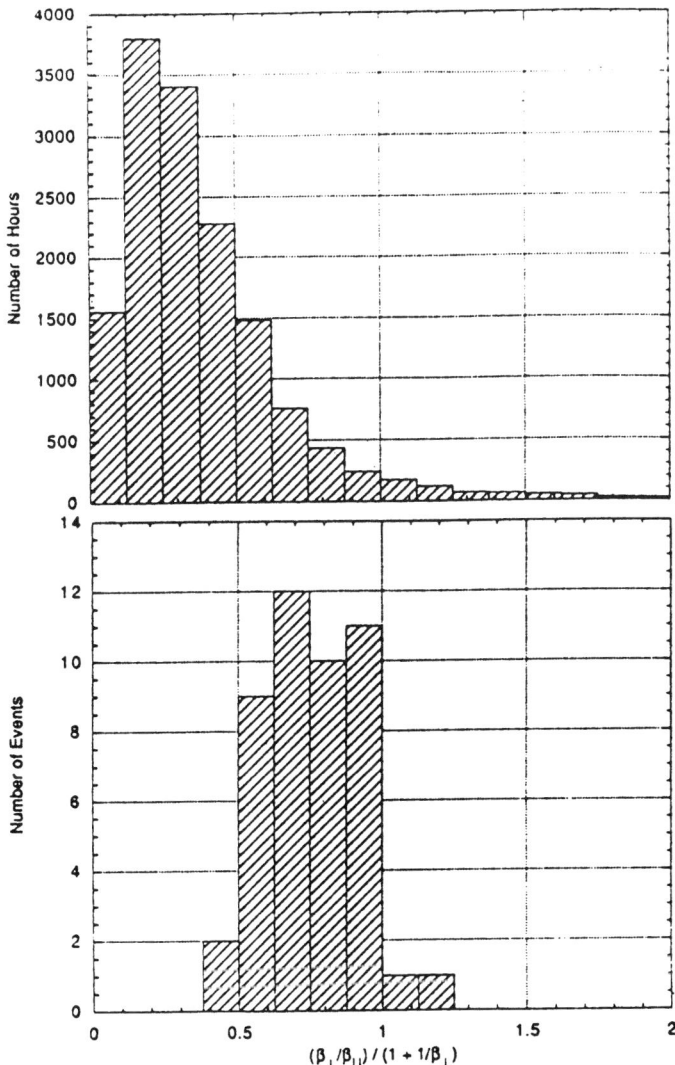

Figure 2.22. Distribution of the stability criterion for mirror-mode waves. $R_m = (\beta_\perp/\beta_\|)/(1 + 1/\beta_\perp) > 1$ for instability to occur. (top) Values of R_m calculated from hourly averages of proton and alpha distributions observed by *Ulysses* between 18 December 1990 and 31 December 1992. (bottom) Values of R_m adjacent to magnetic holes in the solar wind (from Winterhalter *et al.*, 1994a).

field. Goldstein *et al.* (1995) found that at high latitudes V_{wave} was consistently only about 70 per cent of the value expected from equations (2.8) and (2.9), even when the proton, alpha, and electron anisotropies and the alpha–proton differential streaming were accounted for in the calculation of pressures and the factor

A. This discrepancy was even larger than those consistently found in previous studies of in-ecliptic Alfvén waves (Belcher and Davis, 1971; Belcher and Solodyna, 1975) and rotational discontinuities (Belcher and Solodyna, 1975; Neugebauer *et al.*, 1984). Goldstein *et al.* (1995) suggest that perhaps the additional parallel pressure required to bring the two calculations of V_{wave} into agreement might be supplied by interstellar pickup ions (for which the angular distribution could not be measured by *Ulysses*) streaming along the interplanetary magnetic field with little pitch-angle scattering.

(3) Using the *Ulysses* URAP instrument, Hess *et al.* (1998) discovered enhancements of ion–acoustic-like waves in the vicinity of interplanetary shocks. Their examination of SWOOPS ion distributions and electron data indicate that beams and anisotropy in proton-distribution functions and enhanced electron-heat flux may contribute to wave instability, especially when $T_{\text{e}}/T_{\text{p}}$ is small.

2.7 ELECTRONS

2.7.1 Overview

Although previous sections have focused on the ion component of the solar wind, solar-wind electrons also present some interesting physics. The electric field required to maintain zero net charge in the solar wind and zero net current from the Sun and the electron heat conducted into the wind from the corona are important elements of the basic equations of solar-wind dynamics. Electron measurements on *Ulysses* have provided some insights into the processes responsible for electron interactions in the heliosphere, but have also raised some new questions.

The review article by Feldman and Marsch (1997) includes a good discussion of solar-wind electrons. Briefly, *in-situ* observations of interplanetary electrons are described in terms of four populations: photoelectrons, core, halo, and strahl.

Photoelectrons result from the interaction of sunlight with the spacecraft. Because solar photons eject electrons from sunlit surfaces, the electric potential of an interplanetary spacecraft is a few volts positive, which results in the spacecraft being surrounded by a cloud of trapped photoelectrons. Unless the spacecraft surface is electrically conducting, its dark side acquires a negative charge due to interplanetary electrons having much greater thermal velocities than the ions.

Away from a spacecraft most of the interplanetary electrons comprise the core population, which is often characterized as a bi-Maxwellian distribution which, near Earth, typically has a temperature near 10^5 K and an anisotropy $T_{\|}/T_{\perp}$ in the range 1.0–1.5.

Halo electrons form a superthermal tail of the core distribution; although anisotropic the halo is generally present at all pitch angles. The halo's velocity distribution is often fitted with a second bi-Maxwellian distribution with greater speed, higher temperature, and lower density than that describing the core. At 1 AU typical values are $N_{\text{halo}}/N_{\text{core}} \approx 0.05$ and $T_{\text{halo}}/T_{\text{core}} \approx 6$. Both populations drift with respect to the solar-wind frame with $N_{halo}V_{halo} + N_{core}V_{core} = 0$, in order to maintain zero current, and with the vector $\mathbf{V}_{halo} - \mathbf{V}_{core}$ aligned with the magnetic field.

The strahl, which is not always present, is a sharply field-aligned beam with sufficient energy to escape from the Sun's electrostatic-potential well. It is an essentially collisionless population which has been focused by the IMF due to the conservation of its magnetic moment between the Sun and the spacecraft. Near Earth the strahl is most intense and has the narrowest angular spread in the high speed wind (Feldman *et al.*, 1978c; Fitzenreiter *et al.*, 1998).

2.7.2 *Ulysses* instrumentation for electron measurements

Two of the instruments on *Ulysses* contributed to studies of solar-wind electrons: SWOOPS and URAP.

The electron detector in the SWOOPS package is a spherical-section electrostatic analyser; its design is described by Bame *et al.* (1992). Figure 2.23 shows a SWOOPS electron-energy spectrum, summed over all angles. At the lowest energies are the photoelectrons, which must be subtracted from the distribution before we can study interplanetary electrons. The simplest type of correction is a scalar adjustment for the additional acceleration of electrons toward the positively charged spacecraft, but Scime *et al.* (1994b) found that such a correction resulted in a physically unrealistic, non-gyrotropic distribution, and they developed a vector correction which gives satisfactory results. Above the photoelectron energy Figure 2.23 shows the core electrons, which in this case have been fitted to a Maxwellian distribution with $T = 1.2 \times 10^5$ K. At still higher energies, the distribution rises above the straight line resulting from the Maxwellian fit to the core; these are the halo and strahl electrons. The SWOOPS team has published a series of papers on the electron parameters measured by *Ulysses*, and some of those parameters have changed as

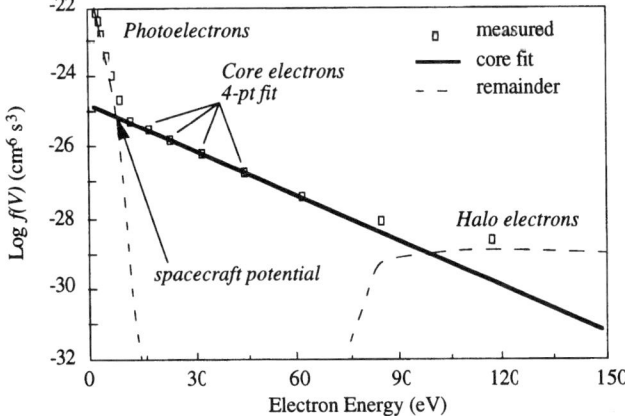

Figure 2.23. A typical energy spectrum, summed over all angles of incidence, measured by the SWOOPS electron spectrometer (from Scime *et al.*, 1999).

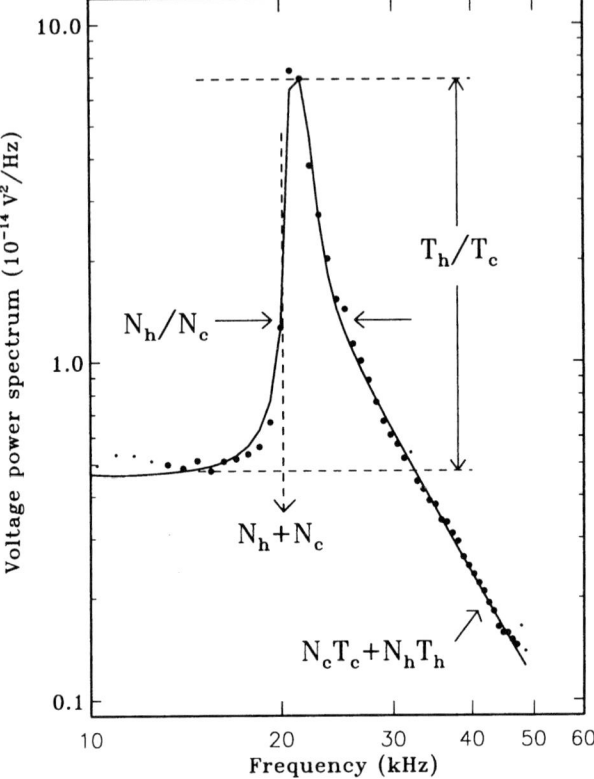

Figure 2.24. A typical quasi-thermal noise spectrum obtained by URAP. The dots are measurements and the solid line is the best fit to a two-Maxwellian distribution (from Maksimovic et al., 1995).

data reduction became more sophisticated. Only the most recent results, which include the vector correction for spacecraft charging, the use of the field direction measured by the *Ulysses* magnetometer to determine $T_\parallel$ and $T_\perp$, and the deletion of data acquired during URAP sounder operations, are presented here.

The URAP experiment is described by Stone et al. (1992). Its long-wire dipole electric antenna can detect voltage fluctuations induced by the random motion of ambient electrons which excite plasma waves near the plasma frequency f_p. Theoretical interpretation of the Quasithermal Noise (QTN) spectrum near f_p can provide information on the density and temperature of core and halo electrons (Meyer-Vernet and Perche, 1989). This technique is immune to the effects of photoelectrons and spacecraft potential (Meyer-Vernet et al., 1998). Figure 2.24 shows an example of a QTN spectrum with labels to show qualitatively how the different features of the spectrum relate to electron parameters. Any anisotropy or high-energy tails of the distributions lead to detailed fine structure near the peak of the spectrum which cannot be fitted by Maxwellian distributions. The QTN method is limited by the frequency range of the URAP experiment to electron densities $>\sim 0.4\,\mathrm{cm}^{-3}$.

2.7.3 Radial temperature gradient and polytrope index

Like the ions, solar-wind electrons cool as they travel away from the Sun, and, if the expansion is adiabatic, the polytrope index $\gamma_e^* = 5/3$ and the electron temperature $T_e \propto R^{-\alpha}$ where $\alpha = 4/3$. On the other hand, if Coulomb collisions are the dominant energy-exchange process for electrons, then theory predicts that $\gamma_e^* = 7/6$ and $\alpha = 1/3$ (Scudder and Olbert, 1979a).

Table 2.4 lists the most recent values of the exponent α determined from the radial gradients of the core temperature observed by *Ulysses* and compares them with each other and with the results from earlier near-Earth and interplanetary spacecraft. For near-Earth observations (*IMP 6–8*) the value of α is calculated from the measured value of γ_e^*. Most of the other entries were calculated by fitting a value of α to the core temperature as the spacecraft distance from the Sun changed. The exception is the entry for the *Ulysses–ICE* comparison. In those measurements Phillips *et al.* (1995a) compared the properties of the electrons observed by the *ICE* spacecraft at a heliocentric distance of 0.92–0.94 AU with the electron properties measured at *Ulysses* at 4.75–4.90 AU, with proper correction for

Table 2.4. Exponents α of the radial gradient of electron-core temperatures, where $T_{\text{core}} \propto R^{-\alpha}$.

R (AU)	Latitude	Spacecraft	α	Comments	Reference
1	Ecliptic	*IMP 6–8*	0.94	CIRs	Feldman *et al.* (1978b)
0.45–4.76	Ecliptic	*Voyager 2* and *Mariner 10*	0.35		Sittler and Scudder (1980)
0.3–1	Ecliptic	*Helios*	0.48–0.84	All data	Pilipp *et al.* (1990)
			0.49–0.74	Fast strams	
1.2–5	Ecliptic	*Ulysses* SWOOPS	0.85	All data	Scime *et al.* (1994a)
			0.91	Transients removed	
1.15–5.31	Ecliptic	*Ulysses* SWOOPS	0.82	All data	Phillips *et al.* (1995a)
			0.91	Without shocks and CMEs	
1–4.3	Ecliptic	*Ulysses* QTN	~Flat	$N_e > 0.4$/cc	Hoang *et al.* (1992)
0.92–4.9	Ecliptic	*Ulysses* SWOOPS and *ICE*	0.24–1.26	Highly variable from day to day	Phillips *et al.* (1995a)
1.3–2.3	-80 to $+80°$	*Ulysses* QTN	0.74		Issautier *et al.* (1998)
1.52–2.31	-80 to $-40°$	*Ulysses* QTN	0.64		Issautier *et al.* (1998)

the time lag between the two spacecraft, during a time when the two spacecraft were closely aligned, with the *Ulysses*–Sun–*ICE* angle ≈15°. They compared daily average properties at six different times in a high-speed stream, mostly in the rarefaction region on the trailing edge of the high-speed stream. As indicated by the entries in Table 2.4, there was a surprisingly large variation in the results, demonstrating the great inhomogeneity of the properties of interplanetary electrons.

All the values of α in Table 2.4 are smaller than the value 1.33 corresponding to adiabatic cooling of the core electrons. Some of the differences in results can be attributed to different methods of data reduction. For example, the *Ulysses*/SWOOPS gradients were fitted to median core temperatures for each solar rotation, whereas the in-ecliptic *Ulysses*/QTN gradients were calculated from average core temperatures over radial bins of 0.3 AU, and the high-latitude QTN gradients were fitted to all individual measurements. The lack of a measurable in-ecliptic gradient by the URAP/QTN method is probably caused by the frequency threshold of the URAP electric antenna; at increasing heliocentric distances the number of spectra with densities below the 0.4-cm^{-3} threshold, which almost certainly had correspondingly low core temperatures, would increase, so that the average temperature of spectra with $N_e > 0.4\,\mathrm{cm}^{-3}$ does not represent the true average temperature. It is also easy to understand the differences between radial temperature exponents for datasets with and without shock-associated intervals; we expect and find steeper spectra when pre- and post-shock intervals are deleted from the dataset. Neither estimate is more 'correct' than the other; if we are interested in the overall polytrope index for the purpose of modelling solar-wind expansion the total dataset is appropriate, whereas the gradient without the shock-associated data is more germane to studies of energy conversion over and above that provided by shocks. Figure 2.25 shows the solar-rotation median values of the core-electron temperature measured by SWOOPS as a function of heliocentric distance for the in-ecliptic phase of the *Ulysses* mission. Note that the spread of core temperatures is significantly smaller when pre- and post-shock and CME data are deleted. There does, however, seem to be a residual discrepancy

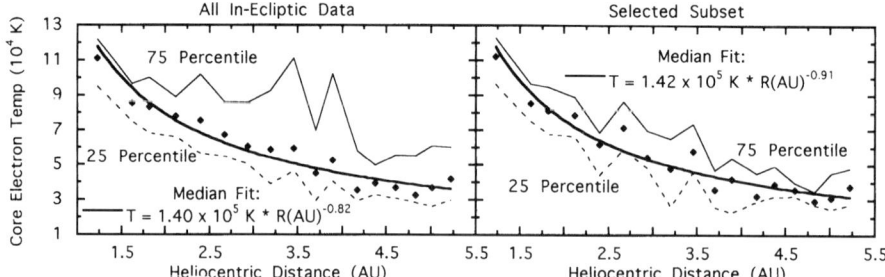

Figure 2.25. Median (diamonds) and 25 and 75 percentile values of core electron temperatures versus heliocentric distance for the *Ulysses* in-ecliptic trajectory, binned by solar rotation. All data are included in the left-hand figure, whereas pre- and post-shock and CME intervals have been deleted from the dataset on the right (from Phillips *et al.* 1995a).

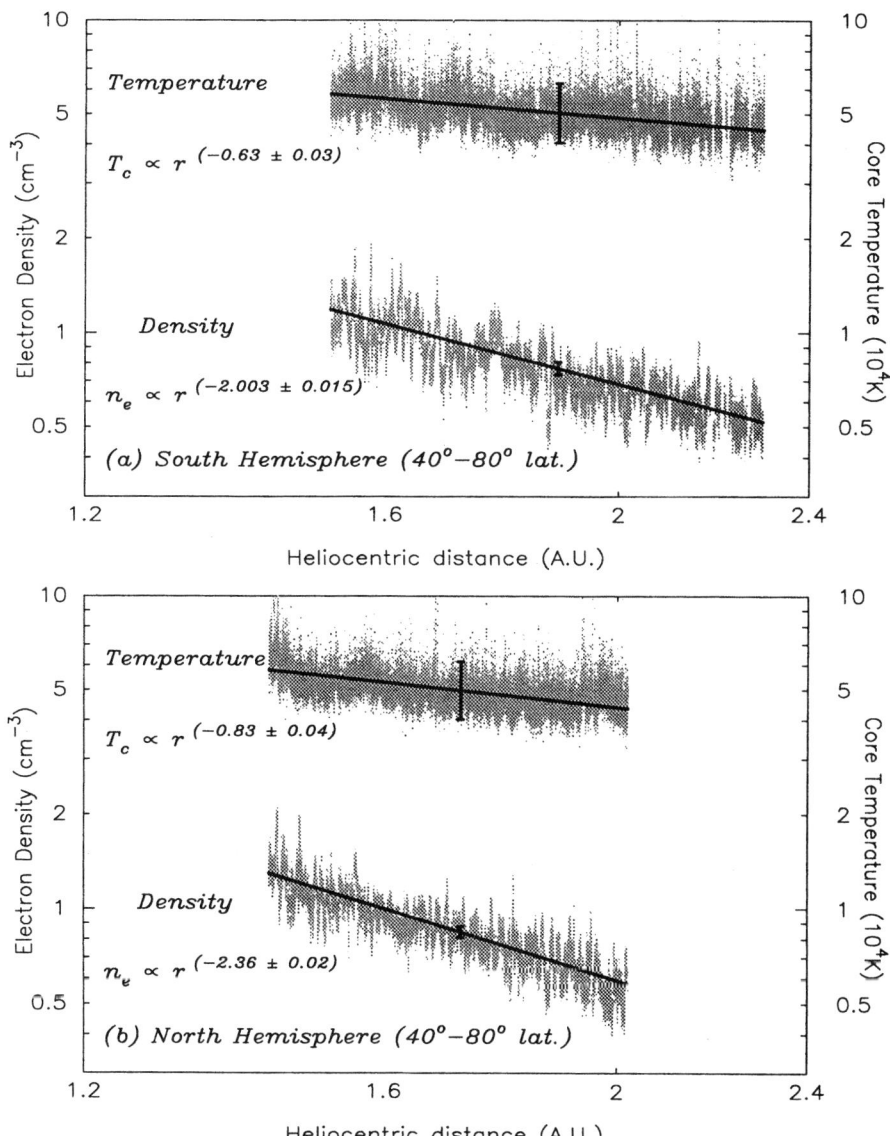

Figure 2.26. Radial variation in the core electron temperature and total electron density in the latitude ranges $\pm(40\text{--}80°)$ as determined by the URAP quasi-thermal noise technique. Southern hemisphere data are in the top panel, northern in the bottom (from Issautier et al., 1998).

between the *Ulysses* gradients and the *Voyager–Mariner* gradients, and to a lesser extent the *Helios* gradients.

The top panel of Figure 2.26 shows the radial variation in electron density and core-electron temperature calculated from the URAP/QTN during the *Ulysses* FLS from -80.2 to $-40°$. The inverse-square drop-off of the density suggests that neither a latitude gradient nor time variation were important and that the $R^{-0.63}$ fit to the temperature data may be the true radial gradient. The limitation of the QTN method to $N_e > 0.4\,\text{cm}^{-3}$, which was a problem for determining the in-ecliptic temperature gradients at large heliocentric distances, was not a problem for the densities encountered during the FLS.

The bottom panel of Figure 2.26 shows equivalent data for the northern hemisphere. There the density did not vary as R^{-2}, suggesting the presence of either temporal or latitudinal variation. The electron-core temperatures extrapolated to 1 AU were slightly higher in the north than in the south, suggesting a relation between core temperature and solar-wind speed.

Using SWOOPS data, Phillips *et al.* (1995a) approached the problem of latitude gradients somewhat differently than did the URAP investigators. For the data acquired at latitudes of -6 to $-48°$ before the south-polar passage, Phillips *et al.* normalized the observed core temperatures by the $R^{-0.91}$ radial variation observed in the ecliptic and found that $TR^{0.91}$ slowly declined poleward at a rate of 10^3 K per degree of latitude. Further analysis of all SWOOPS data for both polar passages should help sort out the apparent discrepancies between QTN and SWOOPS results.

2.7.4 Core anisotropy

With the *Ulysses* SWOOPS experiment it is possible, for the first time, to measure the anisotropy of electron distributions at heliocentric distances greater than 1 AU. Theoretically, we expect the combined effects of expansion in the spiral interplanetary magnetic field and Coulomb collisions to cause the anisotropy of core electrons, expressed as the ratio $T_\parallel/T_\perp$, to decrease with increasing distance from the Sun (Phillips and Gosling, 1990). In fact, however, the opposite behaviour was observed in the ecliptic; $T_\parallel/T_\perp$ for the dataset from which shock-heated and CME-associated intervals had been deleted slowly increased between 1.1 and 5.3 AU (Phillips *et al.*, 1995a). Phillips *et al.* (1995a) conclude that there may be some process that preferentially heats $T_\parallel$, but they do not suggest any specific mechanism. With increasing southern latitude the core $T_\parallel/T_\perp$ was approximately independent of latitude until $\sim -3.3°$, after which it showed a slow increase poleward to at least $-48°$; such an increase does agree with the expectations of Phillips and Gosling (1990). One unexplained peculiarity of the high-latitude data was a substantial number of spectra with $T_\parallel/T_\perp < 0.8$, a condition rarely encountered in the ecliptic.

Phillips and Gosling (1990) also predicted that the anisotropy would be controlled by the density, the speed, and the initial electron temperature at the Sun, with the density being the most important parameter. Denser plasma should undergo more collisions, cool more rapidly, and remain approximately isotropic. Working with a subset of data acquired between 5.0 and 5.4 AU at latitudes between -5.8 and $-24.5°$, Phillips *et al.* (1995a) did find the expected dependence of anisotropy on the density of the electron core population, with the distribution being nearly isotropic at higher densities. $T_\parallel$ was more variable than $T_\perp$.

2.7.5 Halo electrons and the interplanetary potential

Ulysses was the first spacecraft instrumented to study the halo component of interplanetary electrons at distances beyond 1 AU. Some of the results have been reported by McComas *et al.* (1992). They found that as the core population cools with increasing heliocentric distance, halo-electron distribution extends to lower energies so that no gap is found between the two populations.

There is an outward-directed polarization electric field in the interplanetary medium which accelerates ions and decelerates electrons by the amount required to keep the net charge of the Sun at zero. The interplanetary potential Φ is the integral of this field from a point in space to infinity; $e\Phi$ is the energy an electron must have to overcome this potential barrier and escape from the solar system. Two methods have been used to estimate the value of Φ. The first method, based on the idea that the halo electrons can escape whereas core electrons are trapped, is to set $e\Phi = E_B$, where E_B is the energy at which a breakpoint in the electron-energy spectrum is observed between the core and halo populations (see Figure 2.23). In the theory of Scudder and Olbert (1979), the combination of the velocity dependence of the Coulomb cross-section and collision dynamics yields:

$$E_B = 7kT_{core} \tag{2.10}$$

The second method of estimating the interplanetary potential is to set:

$$e\,d\Phi/dR = (1/N_{core})\,d(N_{core}kT_{core})/dR \tag{2.11}$$

which is an approximation of the electron-momentum-conservation equation in the limit that collisions, gravity, and inertial terms are ignorable. For $N_{core} \propto R^{-2}$ and $T_{core} \propto R^{-\alpha}$ the result is:

$$e\Phi = (1 + 2/\alpha)kT_{core} = [\gamma_e^*/(\gamma_e^* - 1)]kT_{core} \tag{2.12}$$

where γ_e^* is once again the polytrope index for electrons.

For high-speed streams observed near Earth by *IMP-6–8* Feldman *et al.* (1978c) found that the breakpoint energy E_B was persistently 15 eV greater than the value of $e\Phi \approx 30$ eV calculated by a method equivalent to that leading to equation (2.12) and that $e\Phi/E_B = 4.1$, rather than 7. Scudder and Olbert (1979) showed that the *Mariner 10* results were consistent with equation (2.10) and argued that Feldman *et al.*'s results could also be consistent if all experimental uncertainties are taken into account.

What does *Ulysses* contribute to knowledge of the interplanetary potential? For the *Ulysses* in-ecliptic SWOOPS data out to 4 AU, McComas *et al.* (1992) found that E_B decreased as $R^{-0.4}$, from $\sim$60 eV at 1 AU to $\sim$30 eV at 4 AU and that $E_B = 7.5kT_{core}$, in good agreement with equation (2.10). For high latitudes, Issautier *et al.* (1998) used their fit to the southern hemisphere URAP/QTN data shown in the top panel of Figure 2.26 together with equation (2.12) to calculate an interplanetary potential Φ at 1 AU of 27 V and a 1-AU core temperature of 7.5×10^4 K, to yield a ratio $e\Phi/kT_{core} = 4.2$ at high latitudes, in very close agreement with the results of Feldman *et al.* (1978c). What is needed to understand whether or not the break-point energy is a good measure of the interplanetary potential and whether or not the interplanetary potential is the same at high and low latitudes is to apply equation (2.12) to low-latitude *Ulysses* data and equation (2.10) to high-latitude data. If E_B is found to exceed $e\Phi$ from equation (2.12) in one or both latitude regimes, it would be an indication that halo electrons are scattered more efficiently than theoretically predicted by Scudder and Olbert.

From their comparison of *Ulysses* and *ICE* data described above, Phillips et al. (1995a) concluded that the density of the halo population N_{halo} has the steepest radial gradient when the total electron density is highest; such a result is consistent with scattering of electrons from the halo to the core, but is inconsistent with the conclusion of McComas et al. (1992) that the density ratio $N_{\text{halo}}/N_{\text{core}}$ remained constant ≈ 0.04 between 1.15 and 4.13 AU. It is clear that further analyses of *Ulysses* data are required to resolve some of these apparent discrepancies.

2.7.6 Electron-heat flux

The heat flux **q** is defined as:

$$\mathbf{q} = \int (m/2)\mathbf{U} U^2 f(V)\, d^3 V \tag{2.13}$$

where $\mathbf{U} = \mathbf{V} - \langle \mathbf{V} \rangle$ and $f(V)$ is the distribution function in velocity space. When a strahl is present, it carries most of the heat flux; otherwise, the heat flux is carried largely by the drift of halo electrons relative to the core. It had previously been determined from observations near Earth that, as expected theoretically, the heat flux vector **q** is aligned with the magnetic field, and at 1 AU it has a magnitude in the range 5–8 µW m^{-2} (Pilipp et al., 1990; Feldman et al., 1975). Both this 1-AU value and the electron-heat flux measured between 0.3 and 1.0 AU by *Helios 1* and *2* (Pilipp et al., 1990) are significantly smaller than the theoretical value expected on the basis of collisional interactions. The *Ulysses* mission allowed extension of observations out to 5 AU in the ecliptic and to high-latitude coronal-hole flow where modification of electron-distribution functions by interplanetary shocks is not important.

Ulysses in-ecliptic data provide the opportunity to examine any effects of the tighter winding of the interplanetary spiral and the consequent greater ratio of path length along the spiral to heliocentric distance as the distance from the Sun increases. For the data presented here, **q** was determined from the third moment of the total electron distribution (equation 2.13) without distinguishing the separate contributions of the core, halo, and strahl. As illustrated in Figure 2.27, *Ulysses* measurements of q match up quite well with measurements obtained inside 1 AU by *Helios*. The exponents of power-law fits to the in-ecliptic heat flux along the magnetic field depend somewhat on how the data are edited and averaged, but most values are close to $R^{-2.9}$ (Pilipp et al., 1990; Scime et al., 1996). As with the other *ICE–Ulysses* comparisons of longitudinally aligned observations described above, the fits of the heat-flux data also provided highly variable results, ranging from $R^{-2.54}$ to $R^{-3.81}$,

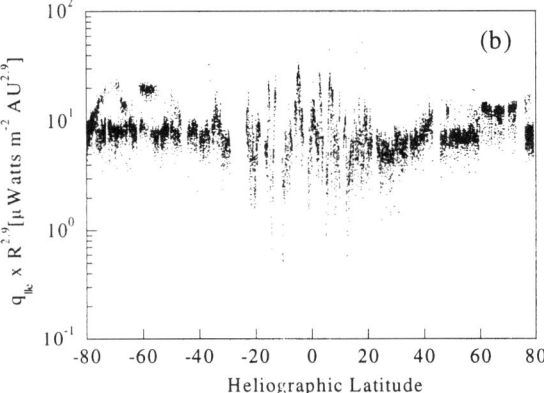

Figure 2.27. Decline in electron-heat flux with distance from the Sun as observed in the ecliptic by *Helios* and *Ulysses* (from Scime *et al.*, 1994).

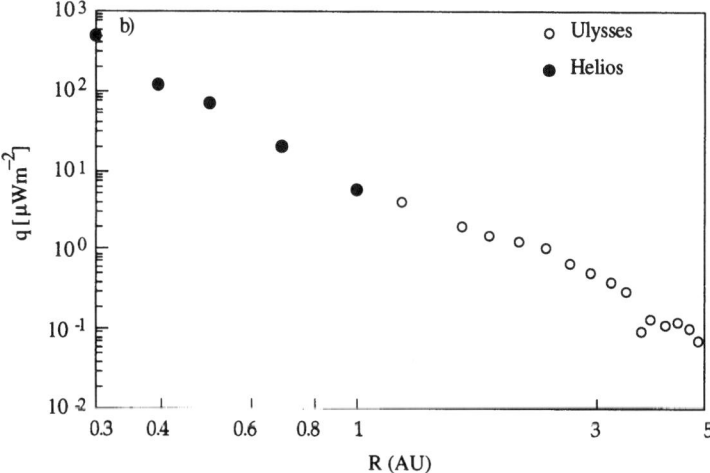

Figure 2.28. The electron heat flux multiplied by a factor $R^{2.9}$ to correct for dependence on distance from the Sun plotted versus latitude. The data were acquired by SWOOPS during the *Ulysses* FLS (from Scime *et al.*, 1999).

with all but one of the six comparisons giving heat-flux drop-offs steeper than $R^{-3.0}$ (Phillips *et al.*, 1995a).

At high latitudes, the interplanetary magnetic field was expected to be less tightly wound, and there was therefore an opportunity to compare the heat flux q at the same values of R but for particles that had travelled different distances along the field. After multiplying the high-latitude heat flux by $R^{2.9}$ to normalize for the radial gradient observed in the ecliptic, Scime *et al.* (1999) found little latitudinal dependence of the electron-heat flux; their results are shown in Figure 2.28.

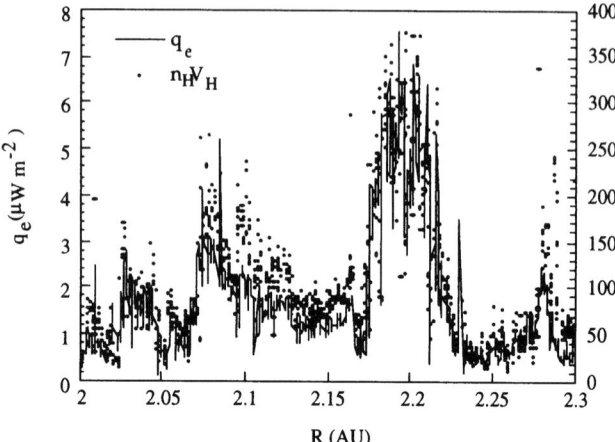

Figure 2.29. Electron-heat flux and magnetic-field magnitude measured in the ecliptic at heliocentric distances of 2.0–2.3 AU (from Scime *et al.*, 1994a).

At all latitudes and solar distances studied both near Earth and by *Helios* and *Ulysses*, the heat flux is less than that expected from collisional processes alone. There have been theoretical investigations of several instabilities that might limit the heat flux, and comparison with *Ulysses* data support the whistler-heat-flux instability proposed by Gary *et al.* (1994). The Gary *et al.* model successfully predicts the R^{-3} dependence of the upper bound for q. Scime *et al.* (1996) also found a rough anticorrelation of q with the plasma-wave amplitude detected by URAP at ∼9 Hz. The waves had more than enough power to pitch-angle scatter the halo electrons. One question that has not yet been answered, however, is to what extent the halo electrons scattered by the whistler waves contribute to the heating of the core at values greater than adiabatic.

Another aspect of *Ulysses* data was the close correlation between the heat flux q and the magnitude of the magnetic field, as illustrated in Figure 2.29. This correlation must be only a local effect, because the heat flux drops off much more steeply with heliocentric distance than does the field strength. Although we could argue that the correlation is an indication that the heat flux is locally conserved along a magnetic flux tube, it is perhaps more reasonable to assume that the data were obtained as the spacecraft cut across neighbouring flux tubes.

2.7.7 Strahl

In the previous section dealing with electron-heat flux, all electrons with energy above the break-point energy E_B were considered to be part of the halo component of the electron-distribution function. But some studies have concentrated on the very sharply focused beam of suprathermal electrons called the strahl. Because near Earth and in the 0.3–1.0-AU solar distances explored by *Helios* the

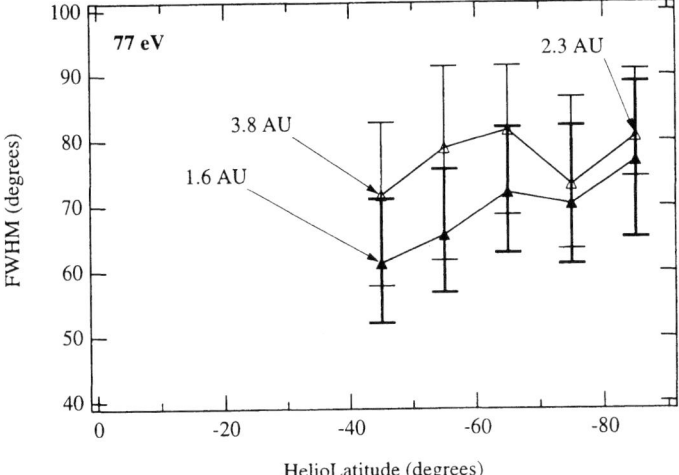

Figure 2.30. Median full width at half maximum of the strahl of 77-eV electrons versus heliolatitude. Error bars show the upper and lower quartiles of the measurements in each 10°-latitude bin. Unfilled triangles connected by thin lines were obtained before the south-polar passage and the filled triangles connected by the thick lines were obtained during the FLS (from Hammond et al., 1996).

strahl was observed exclusively in the nearly collisionless high-speed solar wind, the expectation was that the strahl would be a persistent feature of the high-speed flow from the polar coronal holes. Hammond et al. (1996) characterized the *Ulysses* observations of the strahl at various energies in terms of its pitch-angle width for a range of heliolatitudes and radial distances. Figure 2.30 displays their results in the form of the median full width at half maximum of the strahl at an energy of 77 eV. Both a latitudinal and a radial effect are apparent, with the radial effect appearing to be the stronger. The width of the strahl was found to increase with increasing radial distance, with the broadening becoming less at higher energies. Theoretically, we expect the opposite radial behaviour, as magnetic focusing in the expanding magnetic field should decrease the width of this beam of nearly collisionless electrons (Lemons and Feldman, 1983). We cannot appeal to the whistler-heat-flux instability, which was shown to be important in the ecliptic, because its growth requires $T_\parallel/T_\perp < 1$, which is contrary to observations in the fast high-latitude wind. Hammond et al. suggest that the extra angular scattering of the strahl may be caused by temporary magnetic mirrors formed by the large flux of Alfvén waves in the polar solar wind, but it has not yet been shown that the variation in magnetic-field strength is sufficiently large.

2.8 HELIOSPHERIC MAGNETIC FIELD

Magnetic fields play an important role in heliospheric physics. The solar magnetic field controls generation of the solar wind, determining whether it is fast or slow.

Once beyond the critical points where the solar wind becomes supersonic and super-Alfvénic, solar-wind dynamics together with solar rotation determine the properties of the HMF. The HMF in turn affects the trajectories of energetic particles of both solar and galactic origin, which are discussed in Chapters 4 and 8, respectively.

The simplest model for the direction of the HMF, with which data are often compared, is that of the Parker spiral (Parker, 1958), given by:

$$B_R/B_o = (R_s/R)^2 \qquad (2.14)$$

$$B_T/B_o = \Omega R_s^2 \cos \lambda / VR \qquad (2.15)$$

$$B_N/B_o = 0 \qquad (2.16)$$

where B_R, B_T, and B_N are the radial, tangential, and normal components of the magnetic field, R is distance from the Sun, R_s is the solar radius, V is the solar-wind speed, and λ is heliographic latitude. B_o is the field at the surface of the Sun, which is assumed to be radial. Ω is the (sidereal) rotation rate of the magnetic field at the solar surface. Snodgrass (1983) has derived an expression for the rotation rate of the photospheric magnetic field as a function of solar latitude:

$$\Omega = 3.101 - 0.464 \sin^2 \lambda - 0.328 \sin^4 \lambda \quad \mu\text{rad s}^{-1} \qquad (2.17)$$

The radial component B_R decreases as R^{-2} to conserve magnetic flux ($\nabla \cdot \mathbf{B} = 0$), while the rotation of the Sun drags around the footpoints of the field lines, which remain anchored in the Sun, to form a spiral pattern. The principal properties of the HMF revealed by *Ulysses* data are reviewed in this section, largely in the context of expectation based on this Parker spiral model.

2.8.1 Solar source and field polarity

As explained in Section 2.1, the polar plot of the solar-wind speed in Plate 2 is colour coded according to the direction of the HMF, with red denoting positive field directions, pointing out from the Sun, and blue denoting negative, inward fields. At high-latitudes the polarity of the HMF was almost entirely negative in the south and positive in the north, consistent with the magnetic polarity of the Sun during the declining phase of solar cycle 22. There are, however, small patches of the 'wrong' polarity. Balogh *et al.* (1999) have demonstrated that these inversions are caused by large-scale folds in the field rather than by flux tubes with reversed polarity originating at the Sun.

At low latitudes the polarity of the HMF switched back and forth between outward and inward fields. Changes in polarity mark crossings of the which is a (probably warped) surface tilted with respect to the solar equator. As the Sun and the HCS rotated, *Ulysses* was alternately above and below the HCS, observing the sector structure of the low-latitude HMF. The latitudinal extent of the region of alternating polarities was narrower during the FLS (left side of Plate 2) than during the first latitude climb of *Ulysses* (lower right quadrant of Figure 2.2). This is almost

2.8.2 Field strength

Figure 2.31 shows a different representation of the HMF observed by *Ulysses*, with the daily-average radial component of the HMF (B_R), scaled by the square of the distance from the Sun, plotted versus time for the FLS from 80°S to 80°N in 1994 and 1995; the heliographic latitude of *Ulysses* is shown at the top of the figure.

For all solar distances for which the direction of the solar-wind velocity vector has been measured (i.e. distances >0.3 AU), the flow is nearly radial with typical deflections of only a few degrees resulting from waves and from the interactions of fast and slow solar-wind streams. If the flow were radial all the way out from the solar surface, the field at high latitudes would be stronger than the equatorial field because one of the properties of a dipole field is that its strength increases poleward. Figure 2.31 shows, however, that the strength of the normalized radial component (i.e. $B_R R^2$) of the HMF was ~3 nT, independent of latitude. The implication of this observation is that there are latitudinal flows, from pole to equator, in the corona. The polar fields are sufficiently strong that their pressure exceeds the plasma pressure by at least an order of magnitude. Thus, as the open polar magnetic fields spread out over surrounding closed magnetic loops, the plasma moves equatorward along the expanding field lines. This process is implicitly included in MHD models of the solar wind, but the model of Pneuman and Kopp (1971), for example, still retains more latitudinal variation in $B_R R^2$ than was observed by *Ulysses*.

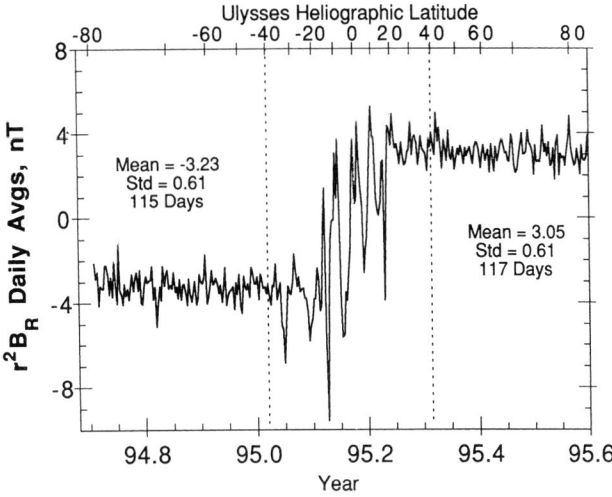

Figure 2.31. *Ulysses* observations of daily averages of the radial component B_R of the HMF, normalized to a solar distance 1 AU, versus time (bottom scale) and versus heliographic latitude (top scale) (from Smith *et al.*, 1997).

Another noteworthy feature of Figure 2.31 is that the value of $B_R R^2$ was ~6 per cent greater in the southern hemisphere than in the north. A larger difference between the two hemispheres is, however, implied by the *Ulysses* measurements of galactic and anomalous cosmic rays during the FLS which revealed 10–20 per cent higher particle fluxes in the northern hemisphere than in the south and a flux minimum near 10°S (Simpson *et al.*, 1996; Heber *et al.*, 1996; Mckibben *et al.*, 1996). Models of the HMF based on extrapolations of the field, observed in the solar photosphere, into the heliosphere also suggested that at the time of the FLS the HCS was displaced as much as 10° southward of the heliographic equator (see Neugebauer *et al.* (1998) for plots of the locations of the HCS based on different models). For the total magnetic flux emerging from the north-polar regions to be balanced by the flux entering the south-polar regions, a southward displacement of the HCS implies weaker fields and less modulation of cosmic rays in the north than in the south. The seven reversals of the HMF observed by the *Ulysses* magnetometer during the FLS did not, however, appear to be consistent with a southward displacement of the HCS (Smith and Balogh, 1995; Erdös and Balogh, 1998). A recent comparison of *Ulysses* magnetometer data with simultaneous data acquired in the ecliptic by the *WIND* spacecraft has resolved this apparent disagreement; there was significant temporal variation in the solar magnetic field and the HMF during the time of the FLS, which, when combined with the *Ulysses* trajectory, led to misinterpretation of the average latitude of the HCS (Smith *et al.*, 2000).

2.8.3 Field direction

Figure 2.32 shows the value of the Parker spiral angle ($\phi_P = \tan^{-1} B_T/B_R$) expected along the *Ulysses* trajectory according to equations (2.14)–(2.17) and in response to

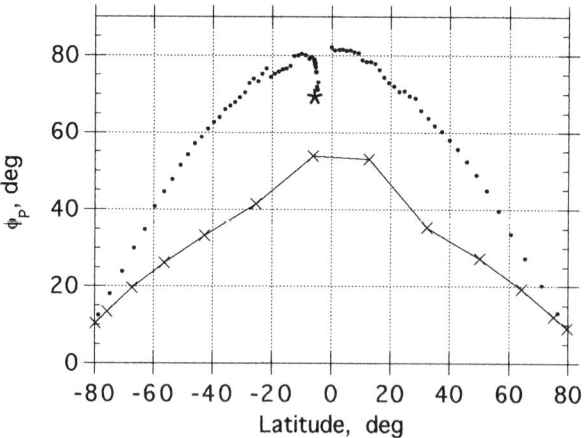

Figure 2.32. Solar-rotation averages of the azimuthal angle of the HMF expected on the basis of the Parker spiral, the *Ulysses* trajectory parameters, and the observed solar-wind speed. The FLS is indicated by connected × symbols, the other intervals by dots.

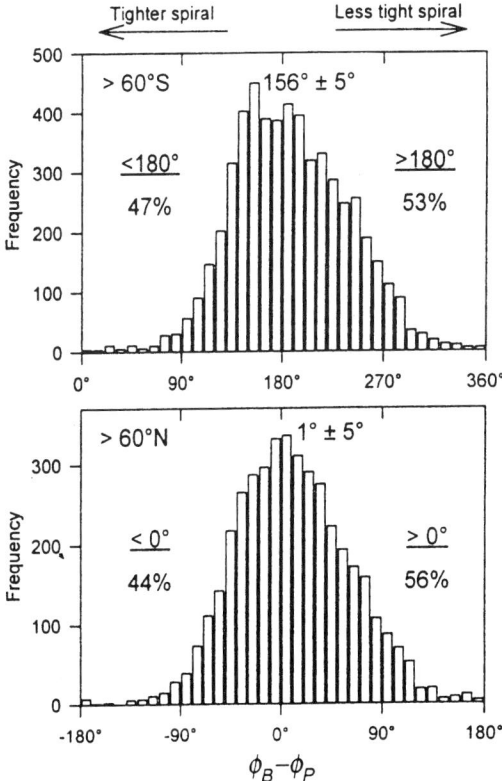

Figure 2.33. Histograms of the direction of the high-latitude HMF (ϕ_B) relative to the expected Parker spiral angle (ϕ_P) (from Forsyth et al., 1996).

changes in R, λ, and V (with the inward/outward polarity of the field ignored). The calculation is based on solar-rotation averages of hourly averages of spacecraft location and solar-wind speed. Starting near the centre of the diagram marked by *, when *Ulysses* was outbound in the ecliptic at 3 AU, ϕ_P became more tangential, reaching a value near 80°, as the spacecraft moved away from the Sun. Then after the Jupiter fly-by, the spacecraft moved simultaneously inward and southward until reaching 80°S, where $\phi_P \approx 10°$. The FLS took the spacecraft inward to perihelion at 1.3 AU and then slightly outward again to 80°N. The fine-scale bumps in the curve are caused by variation in solar-wind speed encountered at low latitudes.

Forsyth et al. (1996) compared the measured spiral angle ϕ_B with ϕ_P; their high-latitude results are shown in Figure 2.33 as histograms of hourly-averaged field directions divided into 10° bins of $\phi_B - \phi_P$. Although the most probable direction of the field in the southern hemisphere was more tightly wound by 24° than the Parker spiral, the most probable direction of the field in the northern hemisphere agreed well with the Parker model. In both hemispheres, there was a broad spread of directions about the Parker spiral angle.

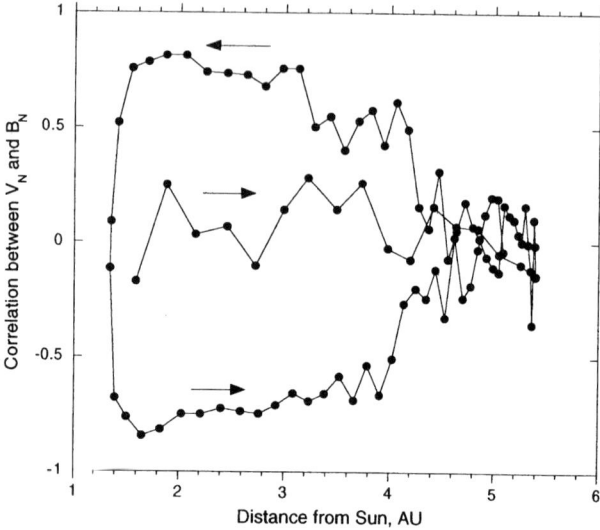

Figure 2.34. Correlation between normal components of the proton velocity V_N and the magnetic field B_N along the *Ulysses* trajectory. Each point represents a solar-rotation average calculated from hourly averages of V_N and B_N. The arrows indicate increasing time along the *Ulysses* trajectory. As expected for Alfvén waves propagating away from the Sun, the correlation is positive in the southern hemisphere where the interplanetary field is inward, and negative in the north where the field is outward (from Neugebauer, 1999).

Large-amplitude Alfvén waves contributed most to the the angular spread observed in the polar regions; these waves, and their evolution with distance from the Sun, are discussed in greater depth in Chapter 4. As *Ulysses* passed to higher latitudes, hourly variances in each of the components of the magnetic field and the velocity steadily increased. The fluctuations in **B** and **V** had all the properties expected of Alfvén waves propagating away from the Sun: the transverse components of **B** and **V** were highly correlated over a broad frequency range corresponding to periods <1 to >10 hours while magnitudes B and V were nearly constant (Smith *et al.*, 1995). Figure 2.34 shows solar-rotation averages of the hourly correlation between the normal components B_N and V_N over the course of the *Ulysses* mission through 1997. For pure, outwardly-propagating Alfvén waves, this correlation would be +1 in the southern hemisphere and −1 in the north. Following along the *Ulysses* trajectory, the figure shows very little correlation out to Jupiter, at distances >4 AU, or at the perihelion equatorial crossing, but strong correlations in the solar wind from polar coronal holes. The amplitude of these waves was great enough to account for much of the spread of observed field directions about the most probable value shown in Figure 2.33. At times the wave amplitude was so large that the field appeared to reverse direction, which accounts for the occasional blue (red) intervals near the north (south) poles in Plate 2.

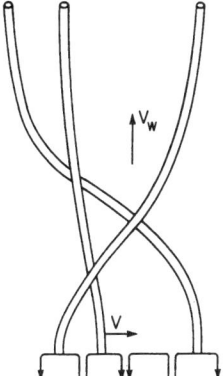

Figure 2.35. A diagram demonstrating how the dragging of field lines by solar convection generates transverse fields at the solar surface (from Jokipii and Kóta, 1989).

What generates these waves? One idea is that they are unabsorbed remnants of the wave field that heats the corona and accelerates the solar wind to high speeds in coronal holes. Recent observations suggest that higher frequency ion–acoustic waves may play a more important role in heating the corona. Another source of transverse waves is summarized by the drawing in Figure 2.35 from a paper by Jokipii and Kóta (1989). Long before there were any high-latitude data from *Ulysses*, they suggested that the motions of the footpoints of interplanetary magnetic-field lines caused by convection in the outer layers of the Sun would generate a transverse component of the field near the Sun. Those disturbances would travel out along the field lines as Alfvén waves. Jokipii and Kóta also pointed out that it follows from equations (2.14) and (2.15) that transverse components at the solar source fall off as $1/R$, whereas radial components fall off at $1/R^2$. Thus substantial transverse fields at the Sun would yield a predominance of transverse fields at large distances from the Sun. The Jokipii and Kóta model is a modification of the Parker model in which the underlying pattern is still the Parker spiral but the instantaneous field varies widely about the average spiral leading to an average field that is mostly transverse at large distances, even in polar regions.

Two other extensions of the basic Parker model that include the effects of processes in the lower solar atmosphere have been proposed by Smith and Bieber and by Fisk. From their study of solar-cycle variation in the HMF in the ecliptic plane, Smith and Bieber (1991) suggested (1) that there was a transverse component of the field at the Sun arising from the differential rotation of the solar magnetic field and (2) that the dependence of solar rotation on latitude (equation 2.17) should be modified to account for the latitudinal transport of field lines equatorward from the polar regions as discussed above in connection with the latitudinal independence of the radial component of the HMF. Heber *et al.* (1996) found that the Smith and Bieber model gave a better fit to the latitude variation in cosmic-ray protons during the FLS than did a simple Parker model. Cosmic-ray modulation is covered in much greater detail in Chapter 8.

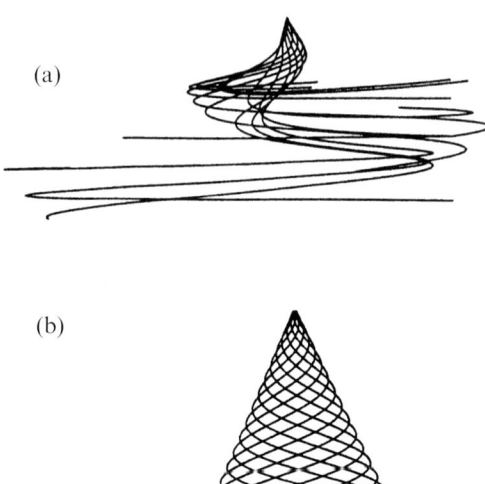

Figure 2.36. Projection of tracings of magnetic-field lines originating at a latitude of 70°S according to (a) the Fisk model and (b) the Parker spiral field (from Fisk, 1996).

The most recent model of the HMF is by Fisk (1996) and Fisk *et al.* (1999). That model takes into account (1) the angular offset or tilt between the Sun's rotation axis and the axis of its magnetic dipole, (2) the latitudinal flow, discussed above, from the regions of high magnetic pressure near the magnetic pole to lower magnetic latitudes, (3) the differential or latitude-dependent rotation of the solar magnetic field as given by equation (2.17), and (4) the rigid, latitude-independent rotation of the boundaries of the coronal holes from which the high-speed wind comes. Their picture is that closed magnetic loops open as they differentially rotate into a coronal hole and then reconnect to close again when they reach the boundary at the other side of the hole. The result is a complicated geometry in which an interplanetary field line may be observed at a latitude more than 40° away from the latitude at which its footpoint is tied to the Sun. Figure 2.36 contrasts the complex traces of field lines emanating at a solar latitude of 70(S for the Fisk model with the simple field lines corresponding to the basic Parker model.

In summary, *Ulysses* data have shown that the interplanetary magnetic field at polar latitudes is more complex than can be explained on the basis of the simple Parker model. The field lines probably connect to the Sun at latitudes different from those at which they are observed in the solar wind; the field direction is dominated by waves; and the direction of the field at the solar boundary is perhaps not purely radial.

2.9 ACKNOWLEDGMENTS

I thank Peter Bochsler for his careful review of this work and his suggestions for improvement. This review was performed at the Jet Propulsion Laboratory under a

2.10 REFERENCES

Antiochos, A. K. (1994) The physics of coronal closed-field structures. *Adv. Space Res.* **14**(4) 139.

Balogh, A., Forsyth, R. J., Lucek, E. A., Horbury, T. S. and Smith, E. J. (1999) Heliospheric magnetic field polarity inversions at high heliographic latitudes. *Geophys. Res. Lett.* **26**, 631.

Bame, S. J. (1992), The Ulysses solar wind plasma experiment. *Astron. and Astrophys. Suppl.* **92**, 237.

Bame, S. J., Asbridge, J. R., Feldman, W. C. and Gosling, J. T. (1977) Evidence for a structure-free state at high solar wind speeds. *J. Geophys. Res.* **82**, 1487.

Barnes, A. (1992) Acceleration of the solar wind. *Rev. Geophys.* **30**, 43.

Barnes, A., Gazis, P. R. and Phillips, J. L. (1995) Constraints on solar wind acceleration mechanisms from Ulysses plasma observations: The first polar pass. *Geophys. Res. Lett.* **22**, 3309.

Belcher, J. W. and Davis, L. Jr. (1971) Large-amplitude Alfvén waves in the interplanetary medium. *J. Geophys. Res.* **76**, 3534.

Belcher, J. W. and Solodyna, C. V. (1975) Alfvén waves and directional discontinuities in the interplanetary medium. *J. Geophys. Res.* **80**, 181.

Bochsler, P. (1989) Velocity and abundance of silicon ions in the solar wind. *J. Geophys. Res.* **94**, 2365.

Bochsler, P., Geiss, J. and Joos, R. (1985) Kinetic temperatures of heavy ions in the solar wind. *J. Geophys. Res.* **90**, 10779.

Bochsler, P., Geiss, J. and Maeder, A. (1990) The abundance of 3He in the solar wind – a constraint for models of solar evolution. *Solar Phys.* **128**, 203.

Bodmer, R., Bochsler, P., Geiss, J., von Steiger, R. and Gloeckler, G. (1995) Solar wind helium isotopic composition from SWICS/Ulysses. *Space Sci. Rev.* **72**, 61.

Borrini, G., Gosling, J. T., Bame, S. J., Feldman, W. C. and Wilcox, J. M. (1981) Solar wind helium and hydrogen structure near the heliospheric current sheet: a signal of coronal streamers at 1 AU. *J. Geophys. Res.* **86**, 4565.

Burlaga, L. F. (1975) Interplanetary streams and their interaction with the *Earth Space Sci. Rev.* **17**, 327.

Burton, M. E., Neugebauer, M., Crooker, N. U. von Steiger, R. and Smith, E. J. (1999) Identification of trailing edge solar wind stream interfaces: A comparison of Ulysses plasma and composition measurements. *J. Geophys. Res.* **104**, 9925.

Cohen, C. M. S., Collier, M. R., Hamilton, D. C., Gloeckler, G., Sheldon, R. B., von Steiger, R. and Wilken, B. (1996) Kinetic temperature ratios of O^{6+} and He^{2+}: Observations from WIND/MASS and Ulysses/SWICS. *Geophys. Res. Lett.* **23**, 1187.

Collier, M.R., Hamilton, D. C., Gloeckler, G., Bochsler, P. and Sheldon, R. B. (1996) Neon 20, oxygen 16, and helium 4 densities, temperatures, and suprathermal tails in the solar wind determined with WIND/MASS. *Geophys. Res. Lett.* **23**, 1191.

Crooker, N. U., Lazarus, A. J., Phillips, J. L., Steinberg, J. T., Szabo, A., Lepping, R. P. and Smith, E. J. (1997) Coronal streamer belt asymmetries and seasonal solar wind variations deduced from Wind and Ulysses data. *J. Geophys. Res.* **102**, 4673.

DeForest, C. E. (1997) Polar plume anatomy: Results of a coordinated observation. *Solar Phys.* **175**, 393.

Denison, D. G. T. and Walden, A. T. (1999) The search for solar gravity-mode oscillations: An analysis using Ulysses magnetic field data. *Astrophys. J.* **514**, 972.

Erdös, G. and Balogh, A. (1998) The symmetry of the heliospheric current sheet as observed by Ulysses during the fast latitude scan. *Geophys. Res. Lett.* **25**, 245.

Esser, R., Edgar, R. J. and Brickhouse, N. S. (1998) High minor ion outflow speeds in the inner corona and observed ion charge states in interplanetary space. *Astrophys. J.* **498**, 448.

Feldman, W. C. and Marsch, E. (1997) Kinetic phenomena in the solar wind. In: Jokipii, J. R., Sonett, C. P. and Giampapa, M. S. (eds) *Cosmic Winds and the Heliosphere*, p. 617, University of Arizona Press, Tucson, AZ.

Feldman, W. C., Asbridge, J. R., Bame, S. J., Montgomery, M. D. and Gary, S. P. (1975) Solar wind electrons. *J. Geophys. Res.* **80**, 4181.

Feldman, W. C., Asbridge, J. R., Bame, S. J. and Gosling, J. T. (1976) High-speed solar wind flow parameters at 1 AU, *J. Geophys. Res.* **81**, 5054.

Feldman, W. C., Asbridge, J. R., Bame, S. J. and Gosling, J. T. (1978a) Long-term variations of selected solar wind properties: Imp 6, 7, and 8 results. *J. Geophys. Res.* **83**, 2177.

Feldman, W. C., Asbridge, J. R., Bame, S. J., Gosling, J. T. and Lemons, D. S. (1978b) Characteristic electron variations across simple high-speed solar wind streams. *J. Geophys. Res.* **83**, 5285.

Feldman, W. C., Asbridge, J. R., Bame, S. J., Gosling, J. T. and Lemons, D. S. (1978c) Electron heating within interaction zones of simple high-speed solar wind streams. *J. Geophys. Res.* **83**, 5297.

Feldman, W. C., Barraclough, B. L., Phillips, J. L. and Wang, Y.-M. (1996) Constraints on high-speed solar wind structure near its coronal base: A Ulysses perspective. *Astron. Astrophys.* **316**, 355.

Feldman, W. C., Barraclough, B. L., Gosling, J. T., McComas, D. J., Riley, P., Goldstein, B. E. and Balogh, A. (1998) Ion energy equation for the high-speed solar wind: Ulysses observations. *J. Geophys. Res.* **103**, 14.

Fisk, L. A. (1996) Motion of the footpoints of heliospheric magnetic field lines at the Sun: Implications for recurrent energetic particle events at high heliographic latitudes. *J. Geophys. Res.* **101**, 15547.

Fisk, L. A., Zurbuchen, T. H. and Schwadron, N. A. (1999) On the coronal magnetic field: Consequences of large-scale motions. *Astrophys. J.* **521**, 868.

Fitzenreiter, R. J., Ogilvie, K. W., Chornay, D. J. and Keller, J. (1998) Observations of electron velocity distribution functions in the solar wind by the WIND spacecraft: High angular resolution strahl measurements. *Geophys. Res. Lett.* **25**, 249.

Forsyth, R. J., Balogh, A., Horbury, T. S., Erdös, G., Smith, E. J. and Burton, M. E. (1996) The heliospheric magnetic field at solar minimum: Ulysses observations from pole to pole. *Astron. Astrophys.* **316**, 287.

Gary, S. P., Scime, E. E., Phillips, J. L. and Feldman, W. C. (1994) The whistler heat flux instability: Threshold conditions in the solar wind. *J. Geophys. Res.* **99**, 23391.

Gazis, P. R., Barnes, A., Mihalov, J. D. and Lazarus, A. J. (1994) Solar wind velocity and temperature in the outer heliosphere. *J. Geophys. Res.* **99**, 6561.

Geiss, J. and Reeves, H. (1972) Cosmic and solar system abundances of deuterium and helium-3. *Astron. Astrophys.* **18**, 126.

Geiss, J. *et al.* (1995) The southern high-speed stream: Results from the SWICS instrument on Ulysses. *Science*, **268**, 1033.

Geiss, J. and Bochsler, P. (1985) Ion composition in the solar wind in relation to solar abundances, in *Rapports Isotopiques dans le Système Solaire*, p. 213. Cepadues-Editions, Toulouse.

Geiss, J. and Gloeckler, G. (1998) Abundances of deuterium and helium-3 in the protosolar cloud. *Space Sci. Rev.* **84**, 239.

Geiss, J., Hirt, P. and Leutwyler, H. (1970) On acceleration and motion of ions in corona and solar wind. *Solar Phys.* **12**, 458.

Geiss, J., Gloeckler, G. and von Steiger, R. (1995) Origin of the solar wind from composition data. *Space Sci. Rev.* **72**, 49.

Gloeckler, G. and Geiss, J. (1998) Measurement of the abundance of helium-3 in the Sun and in the local interstellar cloud with SWICS on Ulysses. *Space Sci. Rev.* **84**, 275.

Gloeckler, G., et al. (1999) Unusual composition of the solar wind in the 2–3 May 1998 CME observed with SWICS on ACE. *Geophys. Res. Lett.* **26**, 157.

Gloeckler, G. L. et al. (1992) The solar wind ion composition spectrometer. *Astron. Astrophys. Suppl.* **92**, 267.

Goldstein, B. E., Neugebauer, M. and Smith, E. J. (1995) Alfvén waves, alpha particles, and pickup ions in the solar wind. *Geophys. Res. Lett.* **22**, 3389.

Goldstein, B. E. et al. (1996) Ulysses plasma parameters: Latitudinal, radial, and temporal variations. *Astron. Astrophys.* **316**, 296.

Goldstein, B. E., Neugebauer, M., Zhang, L. D. and Gary, S. P. (2000) Observed constraint on proton-proton relative velocities in the solar wind. *Geophys. Res. Lett.* **27**, 53.

Gosling, J. T., Bame, S. J., Feldman, W. C., Paschmann, G., Sckopke, N. and Russell, C. T. (1984) Suprathermal ions upstream from interplanetary shocks. *J. Geophys. Res.* **89**, 5409.

Gosling, J. T. et al. (1995) The band of solar wind variability at low heliographic latitudes near solar activity minimum: Plasma results from the Ulysses rapid latitude scan. *Geophys. Res. Lett.* **22**, 3329.

Grevesse, N. and Sauval, A. J. (1998) Standard solar composition. *Space Sci. Rev.*, **85**, 161.

Hakamada, K. and Akasofu, S.-I. (1981) A cause of solar wind speed variations observed at 1 AU. *J. Geophys. Res.* **86**, 1290.

Hammond, C. M., Feldman, W. C., Phillips, J. L., Goldstein, B. E. and Balogh, A. (1995) Double ion beams associated with the heliospheric current sheet. *J. Geophys. Res.* **100**, 7881

Hammond, C. M., Feldman, W. C., McComas, D. J., Phillips, J. L. and Forsyth, R. J. (1996) Variation of electron-strahl width in the high-speed solar wind: Ulysses observations. *Astron. Astrophys.* **316**, 350.

Heber, B., et al. (1996) Spatial variation of >40 MeV/n nuclei fluxes observed during the Ulysses rapid latitude scan. *Astron. Astrophys.* **316**, 538.

Hefti, S. (1998) Kinetic properties of solar wind minor ions and protons measured with SOHO/CELIAS. *J. Geophys. Res.* **103**, 29697.

Hess, R. A., MacDowall, R. J., Goldstein, B., Neugebauer, M. and Forsyth, R. J. (1998) Ion acoustic-like waves observed by Ulysses near interplanetary shock waves in the three-dimensional heliosphere. *J. Geophys. Res.* **103**, 6531.

Hénoux, J. C. and Somov, B. V. (1992) First ionization potential fractionation. In: C. Mattok (ed.) *Coronal Streamers, Coronal Loops, and Coronal and Solar Wind Composition, ESA/SP-348*. p. 325.

Ho, G. C., Hamilton, D. C., Gloeckler, G. and Bochsler, P. (2000) Enhanced solar wind $^3He^{2+}$ associated with coronal mass ejections. *Geophys. Res. Lett.*, **27**, 309.

Hoang, S., et al. (1992) Solar wind thermal electrons in the ecliptic plane between 1 and 4 AU: Preliminary results from the Ulysses radio receiver. *Geophys. Res. Lett.* **19**, 1295.

Hoeksema, J. T. (1995) The large-scale structure of the heliospheric current sheet during the Ulysses epoch. *Space Sci. Rev.* **72**, 137.

Hoogeveen, G. W. and Riley, P. (1998) The search for solar gravity-mode oscillations in the solar wind using Ulysses plasma data. *Solar Phys.* **179**, 167.

Ip, W.-H. and Axford, W. I. (1991) On the first ionization potential effect in the solar corona. *Adv. Space Res.* **11**(1), 247.

Issautier, K., Meyer-Vernet, N., Moncuquet, M. and Hoang, S. (1998) Solar wind radial and latitudinal structure: Electron density and core temperature from Ulysses thermal noise spectroscopy. *J. Geophys. Res.* **103**, 1969.

Jokipii, J. R. and Kóta, J. (1989) The polar heliospheric magnetic field. *Geophys. Res. Lett.* **16**, 1.

Ko, Y.-K., Fisk, L. A., Gloeckler, G. and Geiss, J. (1996) Limitation on suprathermal tails of electrons in the lower solar corona. *Geophys. Res. Lett.* **23**, 2785.

Ko, Y.-K., Fisk, L. A., Geiss, J., Gloeckler, G. and Guhathakurta, M. (1997) An empirical study of the electron temperature and heavy ion velocities in the south polar coronal hole. *Solar Phys.* **171**, 345.

Ko, Y.-K., Geiss, J. and Gloeckler, G. (1998) On the differential ion velocity in the inner solar corona and the observed solar wind ionic charge states. *J. Geophys. Res.* **103**, 14539.

Kohl, J. L. (1997) First results from the SOHO ultraviolet coronagraph spectrometer. *Solar Phys.* **175**, 613.

Lallement, R., Holzer, R. E. and Munro, R. H. (1986) Solar wind expansion in a polar coronal hole: Inferences from coronal white light and interplanetary Lyman alpha observations. *J. Geophys. Res.* **91**, 6751.

Lemons, D. S. and Feldman, W. C. (1983) Collisional modification to the exospheric theory of solar wind halo electron pitch angle distributions. *J. Geophys. Res.* **88**, 6881.

Liu, S., Marsch, E., Livi, S., Woch, J., Wilken, B., von Steiger, R. and Gloeckler, G. (1995) Radial gradients of ion densities and temperatures derived from SWICS/Ulysses observations. *Geophys. Res. Lett.* **22**, 2445.

Macdowall, R. J. (1995) The three-dimensional extent of a high speed solar wind stream. *Space Sci. Rev.*, **72**, 125.

Maksimovic, M., Hoang, S., Meyer-Vernet, N., Moncuquet, M., Bougeret, J.-L., Phillips, J. L. and Canu, P. (1995) Solar wind electron parameters from quasi-thermal noise spectroscopy and comparison with other measurements on Ulysses. *J. Geophys. Res.* **100**, 19881.

Marsch, E., Mühlhaüser, K.-H., Rosenbauer, H., Schwenn, R. and Neubauer, F. M. (1982a) Solar wind helium ions: observations of the Helios solar probes between 0.3 and 1 AU. *J. Geophys. Res.* **87**, 35.

Marsch, E., Mühlhäuser, K.-H., Schwenn, R., Rosenbauer, H., Pilipp, W. and Neubauer, F. M. (1982b) Solar wind protons: three-dimensional velocity distributions and derived plasma parameters measured between 0.3 and 1 AU. *J. Geophys. Res.* **87**, 52.

Marsch, E., von Steiger, R. and Bochsler, P. (1995) Element fractionation by diffusion in the solar chromosphere. *Astron. Astrophys.* **301**, 261.

Marsden, R. G., Smith, E. J., Cooper, J. F. and Tranquille, C. (1996) Ulysses at high heliographic latitudes: An introduction. *Astron. Astrophys.* **316**, 27.

McComas, D. J. (2000) Solar wind observations over Ulysses first full polar orbit. *J. Geophys. Res.* 105.

McComas, D. J., Bame, S. J., Feldman, W. C., Gosling, J. T. and Phillips, J. L. (1992) Solar wind halo electrons from 1–4 AU. *Geophys. Res. Lett.* **19**, 1291.

McComas, D. J., Barraclough, B. L., Gosling, J. T., Hammond, C. M., Neugebauer, M., Balogh, A. and Forsyth, R. (1995) Structures in the polar solar wind: plasma and field observations from Ulysses. *J. Geophys. Res.* **100**, 19893.

McComas, D. J., Hoogeveen, G. W., Gosling, J. T., Phillips, J. L., Neugebauer, M., Balogh, A. and Forsyth, R. (1996) Ulysses observations of pressure-balance structures in the polar solar wind. *Astron. Astrophys.* **316**, 368.

McComas, D. J. *et al.* (1998) Ulysses' return to the slow solar wind. *Geophys. Res. Lett.* **25**, 1.

McKibben, R. B., Connell, J. J., Lopate, C., Simpson, J. A. and Zhang, M. (1996) Observations of galactic cosmic rays and anomalous helium during Ulysses passage from the south to the north solar pole. *Astron. Astrophys.* **316**, 547.

Meyer-Vernet, N. and Perche, C. (1989) Tool kit for antennae and thermal noise near the plasma frequency. *J. Geophys. Res.* **94**, 2405.

Meyer-Vernet, N., Hoang, S., Issautier, K., Maksimovic, M., Manning, R., Moncuquet, M. and Stone, R. G. (1998) Measuring plasma parameters with thermal noise spectroscopy. In: Pfaff, R. R., Borovsky J. and Young, D. T. (eds) *Measurement Techniques in Space Plasmas: Fields, Geophysical Monograph Ser. Vol. 103*, p. 205. American Geophysics Union.

Miyake, W., Mukai, T. Oyama, K.-I., Terasawa, T., Hirao, K. and Lazarus, A. J. (1989) Thin equatorial low-speed region in the solar wind observed during the recent solar minimum. *J. Geophys. Res.* **94**, 15359.

Neugebauer, M. (1981) Observations of solar-wind helium. *Fund. of Cosmic Physics.* **7**, 131.

Neugebauer, M. (1999) The three-dimensional solar wind at solar activity minimum. *Rev. Geophys.*, **37**, 107.

Neugebauer, M., Clay, D. R., Goldstein, B. E., Tsurutani, B. T. and Zwickl, R. D. (1984) A reexamination of rotational and tangential discontinuities in the solar wind. *J. Geophys. Res.* **89**, 5395.

Neugebauer, M., Goldstein, B. E., Bame, S. J. and Feldman, W. C. (1994) Ulysses near-ecliptic observations of differential flow between protons and alphas in the solar wind. *J. Geophys. Res.* **99**, 2505.

Neugebauer, M., Goldstein, B. E., McComas, D. J., Suess, S. T. and Balogh, A. (1995) Ulysses observations of microstreams in the solar wind from coronal holes. *J. Geophys. Res.* **100**, 23389.

Neugebauer, M., Goldstein, B. E., Smith, E. J. and Feldman, W. C. (1996) Ulysses observations of differential alpha-proton streaming in the solar wind. *J. Geophys. Res.* **101**, 17047.

Neugebauer, M. *et al.* (1998) The spatial structure of the solar wind and comparisons with solar data and models. *J. Geophys. Res.* **103**, 14587.

Ogilvie, K. W., Geiss, J., Gloeckler, G., Berdichevsky, D. and Wilken, B. (1993) High-velocity tails on the velocity distribution of solar wind ions. *J. Geophys. Res.* **98**, 3605.

Parhi, S., Suess, S. T. and Sulkanen, M. (1999) The generation of smooth high speed solar wind from plume-interplume mixing. Paper presented at *Solar Wind Nine, CP471*, edited by Habbal, S. R. p. 433. American Institute Physics, Woodbury, NY.

Parker, E. N. (1958) Dynamics of the interplanetary gas and magnetic fields. *Astrophys. J.* **128**, 664.

Pérez Hernández, F. and Christensen-Dalsgaard, J. (1994) *Mon. Not. Roy. Astron. Soc.* **269**, 475.

Phillips, J. L. and Gosling, J. T. (1990a) Radial evolution of solar wind thermal electron distributions due to expansion and collisions. *J. Geophys. Res.* **95**, 4217.

Phillips, J. L., Feldman, W. C., Gosling, J. T. and Scime, E. E. (1995a) Solar wind plasma electron parameters based on aligned observations by ICE and Ulysses. *Adv. Space Res.* **16**(9)95.

Phillips, J. L., Bame, S. J., Gary, S. P., Gosling, J. T., Scime, E. E. and Forsyth, R. J. (1995b) Radial and meridional trends in solar wind thermal electron temperature and anisotropy: Ulysses. *Space Sci. Rev.* **72**, 109.

Phillips, J. L. *et al.* (1995c) Ulysses solar wind plasma observations from pole to pole. *Geophys. Res. Lett.* **22**, 3301.

Phillips, J. L., Goldstein, B. E., Gosling, J. T., Hammond, C. M., Hoeksema, J. T. and McComas, D. J. (1995d) Sources of shocks and compressions in the high-latitude solar wind: Ulysses. *Geophys. Res. Lett.* **22**, 3305.

Pilipp, W. G., Miggenrieder, H., Mühlhäuser, K.-H., Rosenbauer, H. and Schwenn, R. (1990) Large-scale variations of thermal electron parameters in the solar wind between 0.3 and 1 AU. *J. Geophys. Res.* **95**, 6305.

Pneuman, G. W. and Kopp, R. A. (1971) Gas-magnetic field interactions in the solar corona. *Solar Phys.* **18**, 258.

Reisenfeld, D. B., McComas, D. J. and Steinberg, J. T. (1999a) Evidence of a solar origin for pressure balance structures in the high-latitude solar wind. *Geophys. Res. Lett.* **26**, 1805.

Reisenfeld, D. B., Gary, S. P., Gosling, J. T., McComas, D. J., Steinberg, J. T., Goldstein, B. E. and Neugebauer, M. (1999b) Energetics of alpha-proton streaming in the high latitude solar wind (abstract). *Eos. Trans. Amer. Geophys. Un. Suppl.* **46**, F799.

Rickett, B. J., and Coles, W. A. (1991) Evolution of the solar wind structure over a solar cycle: Interplanetary scintillation velocity measurements compared with coronal observations. *J. Geophys. Res.* **96**, 1717.

Riley, P. and Sonett, C. P. (1996) Interplanetary observations of solar g-mode oscillations? *Geophys. Res. Lett.* **22**, 1541.

Robbins, D. E., Hundhausen, A. J. and Bame, S. J. (1970) Helium in the solar wind. *J. Geophys. Res.* **75**, 1178.

Roberts, D. A. and Goldstein, M. L. (1998) Evidence for a high-latitude origin of lower latitude high-speed wind. *Geophys. Res. Lett.* **25**, 595.

Schatten, K., Wilcox, J. W. and Ness, N. F. (1969) A model of the interplanetary and coronal magnetic fields. *Solar Phys.* **6**, 442.

Schmid, J., Bochsler, P. and Geiss, J. (1988) Abundance of iron ions in the solar wind. *Astrophys. J.* **329**, 956.

Schwadron, N. A., Fisk, L. A. and Zurbuchen, T. H. (1999) Elemental fractionation in the slow solar wind. *Astrophys. J.* **521**, 859.

Schwadron, N. A., *et al.* (2000) Techniques for analysis of data from time-of-flight instruments. *J. Geophys. Res.* **105**.

Schwenn, R. (1990) Large-scale structure of the interplanetary medium. In: Schwenn and Marsch, E. (eds) *Physics of the Inner Heliosphere. 1. Large-Scale Phenomena.* p. 99. Springer-Verlag, Berlin.

Scime, E. E., Bame, S. J., Feldman, W. C., Gary, S. P., Phillips, J. L. and Balogh, A. (1994a) Regulation of the solar wind electron heat flux from 1 to 5 AU: Ulysses observations. *J. Geophys. Res.* **99**, 23 401.

Scime, E. E., Phillips, J. L. and Bame, S. J. (1994a) Effects of spacecraft potential on three-dimensional electron measurements in the solar wind. *J. Geophys. Res.* **99**, 14776.

Scime, E. E., Gary, S. P., Phillips, J. L., Balogh, A. and Lengyel-Frey, D. (1996) Electron energy transport in the solar wind: Ulysses observations. *Solar Wind 8*, Winterhalter, D. *et al.*, p. 210. American Institute of Physics, Woodbury, NY.

Scime, E. E., Badeau, A. E. Jr. and Littleton, J. E. (1999) The electron heat flux in the polar solar wind: Ulysses observations. *Geophys. Res. Lett.* **26**, 2192.

Scudder, J. D. and Olbert, S. (1979a) A theory of local and global processes which affect solar wind electrons. 1. The origin of typical 1 AU velocity distribution functions – steady state theory, *J. Geophys. Res.* **84**, 2755.

Scudder, J. D. and Olbert, S. (1979b) A theory of local and global processes which affect solar wind electrons. 2. Experimental support. *J. Geophys. Res.* **84**, 6603.

Simpson, J. A., Zhang, M. and Bame, S. (1996) A solar polar north-south asymmetry for cosmic ray propagation in the heliosphere: The Ulysses pole-to-pole rapid transit. *Astrophys. J.* **465**, L69.

Siscoe, G. L. (1983) Solar system magnetohydrodynamics. In: Carovillano, R. L. and Forbes, J. M. (eds) *Solar-Terrestrial Physics*, p. 11, D. Reidel, Norwell, MA.

Sittler, E. C. and Scudder, J. D. (1980) An empirical polytrope law for solar wind thermal electrons between 0.45 and 4.76 AU: Voyager 2 and Mariner 10. *J. Geophys. Res.* **85**, 5131.

Smith, C. W. and Bieber, J. W. (1991) Solar cycle variation of the interplanetary magnetic field spiral. *Astrophys. J.* **370**, 435.

Smith, E. J. and Balogh, A. (1995) Ulysses observations of the radial magnetic field. *Geophys. Res. Lett.* **23**, 3317.

Smith, E. J., Balogh, A., Neugebauer, M. and McComas, D. (1995) Ulysses observations of Alfvén waves in the southern and northern solar hemispheres. *Geophys. Res. Lett.* **22**, 3381.

Smith, E. J., Balogh, A., Burton, M. E., Forsyth, R. and Lepping, R. P. (1997) Radial and azimuthal components of the heliospheric magnetic field: Ulysses observations. *Adv. Space Res.* **20**(1), 47.

Smith, E. J., Jokipii, J. R., Kóta, J., Lepping, R. P. and Szabo, A. (2000) Evidence of a north-south asymmetry in the heliosphere associated with a southward displacement of the heliospheric current sheet. *Astrophys. J.*

Snodgrass, H. B. (1983) Magnetic rotation of the solar photosphere. *Astrophys. J.* **270**, 288.

Steinberg, J. T., Lazarus, A. J., Ogilvie, K. W., Lepping, R. and Byrnes, J. (1996) Differential flow between solar wind protons and alpha particles: First WIND observations. *Geophys. Res. Lett.* **23**, 1183.

Stone, R. G. (1992) The unified radio and plasma wave investigation on Ulysses. *Astron. Astrophys. Suppl.* **92**, 291.

Tagger, M., E. Falgarone and Shukurov, A. M. (1995) Ambipolar filamentation of turbulent magnetic fields. *Astron. Astrophys.* **299**, 940.

Thieme, K. M., Marsch, E. and Schwenn, R. (1988) Relationship between structures in the solar wind and their source regions in the corona. Paper presented at *Proceedings of the Sixth International Solar Wind Conference, NCAR/TN-306 + Proc*, edited by Pizzo, V. J. Holzer, T. E. and Sime, D. G. (1988) p. 317. National Center for Atmospheric Research, Boulder, CO.

Thieme, K. M., Marsch, E. and Schwenn, R. (1990) Spatial structures in high-speed streams as signatures of fine structures in coronal holes. *Ann. Geophys.* **8**, 713.

Thomson, D. J. Maclennan, C. G. and Lanzerotti, L. J. (1995) Evidence for solar g-mode modulation of interplanetary charged-particles and magnetic fields. *Nature*, **376**, 139.

Vauclair, S. (1996) Element segregation in the solar chromosphere and the FIP bias: The 'skimmer' model. *Astron Astrophys.* **308**, 228.

Vauclair, S. and Meyer, J.-P. (1985) Diffusion in the chromosphere, and the composition of the solar corona. Paper presented at *Proceedings 19th International Cosmic Ray Conference*, **4**, 233.

Vellante, M. and Lazarus, A. J. (1987) An analysis of solar wind fluctuations between 1 and 10 AU. *J. Geophys. Res.* **92**, 9893.

Vernazza, J. E., Avrett, E. H. and Loeser, R. (1981) Structure of the solar chromosphere III. *Astrophys. J. Suppl.* **45**, 635.

von Steiger, R. (1996) Solar wind composition and charge states. Paper presented at *Solar Wind 8*, edited by Winterhalter, D., Gosling, J. T., Habbal, S. R., Kurth, W. S. and Neugebauer, M. (1985) pp. 193. American Institute of Physics, Woodbury, NY.

von Steiger, R. and Geiss, J. (1989) Supply of fractionated gases to the corona. *Astron. Astrophys.* **225**, 222.

von Steiger, R., Geiss, J., Gloeckler, G. and Galvin, A. B. (1995) Kinetic properties of heavy ions in the solar wind from SWICS/Ulysses. *Space Sci. Rev.* **72**, 71.

von Steiger, R., Fisk, L. A., Gloeckler, G., Schwadron, N. A. and Zurbuchen, T. H. (1999) Composition variations in fast solar wind streams. Paper presented at *Solar Wind Nine, CP471*, edited by Habbal, S. R. p. 143. American Institute Physics, Woodbury, NY.

von Steiger, R., Schwadron, N. A., Geiss, J., Gloeckler, G., Fisk, L. A., Hefti, S., Wilken, B., Wimmer-Schweingruber, R. F. and Zurbuchen, T. H. (2000) Composition of quasi-stationary solar wind flows from SWICS/Ulysses. *J. Geophys. Res.* **105**, 27217.

Wang, Y.-M. and Sheeley, N. R. Jr. (1990) Solar wind speed and coronal flux-tube expansion. *Astrophys. J.* **355**, 726.

Wang, Y.-M., Sheeley, N. R. Jr., Phillips, J. L. and Goldstein, B. E. (1997) Solar wind stream interactions and the wind speed-expansion factor relationship. *Astrophys. J.* **488**, L51.

Winterhalter, D., Neugebauer, M., Goldstein, B. E., Smith, E. J., Bame, S. J. and Balogh, A. (1994a) Ulysses field and plasma observations of magnetic holes in the solar wind and their relation to mirror-mode structures. *J. Geophys. Res.* **99**, 23371.

Winterhalter, D., Neugebauer, M., Goldstein, B. E., Smith, E. J., Tsurutani, B. T., Bame, S. J. and Balogh, A. (1994b) Magnetic holes in the solar wind and their relation to mirror-mode structures. *Space Sci. Rev.* **72**, 201.

Winterhalter, D., Smith, E. J., Neugebauer, M., Goldstein, B. E. and Tsurutani, B. T. (2000) The latitudinal distribution of solar wind magnetic holes. *Geophys. Res. Lett.* **27**.

Wimmer-Schweingruber, R. F., von Steiger, R. and Paerli, R. (1997) Solar wind stream interfaces in corotating interaction regions. *J. Geophys. Res.* **102**, 17407.

Wimmer-Schweingruber, R. F., von Steiger, R., Geiss, J., Gloeckler, G., Ipavich, F. M. and Wilken, B. (1998) O^{5+} in high speed solar wind streams: SWICS/Ulysses results. *Space Sci. Rev.* **85**, 387.

Zank, G. P., Pauls, H. L. and Williams, L. L. (1996) Modelling the outer heliosphere. Paper presented at *Solar Wind Eight*, edited by Winterhalter, D., p. 599. American Institute of Physics, Woodbury, NY.

Zhao, X.-P. and Hundhausen, A. J. (1981) Organization of solar wind plasma properties in a tilted, heliomagnetic coordinate system. *J. Geophys. Res.* **86**, 5423.

Zurbuchen, T., Bochsler, P. and von Steiger, R. (1996) Coronal hole differential rotation rate observed with SWICS/Ulysses. Paper presented at *Solar Wind Eight, Proceedings of the Eighth International Solar Wind Conference, AIP Conference Proceedings 382*, edited by Winterhalter D. *et al.*, p. 273. American Institute of Physics, Woodbury, NY.

3

Corotating and transient structures in the heliosphere

R. J. Forsyth and J. T. Gosling

3.1 INTRODUCTION

In this chapter we examine the large-scale structures that develop in the heliosphere due to interactions between solar-wind flows of different speed and origin. Chapter 2 has already discussed the large-scale organization of fast and slow-speed solar wind in the heliosphere near solar minimum. *Ulysses* confirmed that relatively uniform high-speed wind (~ 750 km s^{-1}) originates from the polar coronal holes, and that the more variable low-speed wind has its origin in the vicinity of the streamer belt. This picture is nicely illustrated by Figure 2.2. When a spacecraft, such as *Ulysses*, is at low or mid-latitudes, both fast and slow solar-wind streams are encountered as the solar source regions rotate beneath the spacecraft. In addition to these relatively persistent sources, mass ejections from the Sun produce transient solar-wind streams in the heliosphere of a wide range of speeds up to and exceeding 1,500 km s^{-1}. Since the solar wind is nearly an infinitely conducting plasma, the different flows cannot interpenetrate each other. Thus as one type of flow catches up with or leaves behind a flow of different speed, large-scale regions of compression and rarefaction develop.

In Section 3.2 we discuss the corotating structures that develop as a result of the quasi-stationary interaction between the fast and slow-speed solar-wind streams. The compression regions that develop as a result of this interaction form a spiral pattern in the heliosphere which corotates with the Sun as long as the source regions remain relatively unchanged. These are referred to as Corotating Interaction Regions, (CIRs).

Section 3.3 goes on to discuss the transient structures that are produced in the heliosphere as a result of Coronal Mass Ejections (CMEs).

The *Ulysses* mission has provided a wealth of new information on both these phenomena, primarily as a result of its orbit covering the full range of heliographic latitudes up to 80°. This orbit revealed the full three-dimensional structure of CIRs for the first time. For transient flows, the major new result has been the observation of a different type of interaction between the mass ejection flow and the surrounding solar wind at high latitudes.

The discussion in this chapter will be primarily based on the solar-wind plasma and magnetic field observations from *Ulysses*. The extensive *Ulysses* energetic-particle observations related to CIRs and transient structures are dealt with separately in Chapter 6. We also emphasize that the observations presented here were made during the declining and minimum phases of the solar-activity cycle. It is expected that the observations presently being made at solar maximum will have a somewhat different character.

3.2 COROTATING STRUCTURES

3.2.1 CIR formation and structure

We begin by summarizing the underlying physics involved in the formation of CIRs and by presenting a brief review of what was learned about CIRs prior to *Ulysses*. This enables us to put the new *Ulysses* results which follow into proper context. For further background we refer the reader to two recent tutorial reviews including material on CIRs (Gosling, 1996; Gosling and Pizzo, 1999) and a volume dedicated to CIRs (Balogh *et al.*, 1999) resulting from a series of workshops held on the topic between 1996 and 1998, which contains detailed chapters dealing with all aspects of the subject.

Formation and evolution of CIRs

For CIRs to form it is necessary that the rotation of the Sun leads to sources of fast solar wind moving into the same invariant longitude from which slower wind was previously emitted. The underlying physics can be best understood by considering the interaction of the fast and slow wind along a particular radial line extending out from the Sun in one dimension. Suppose that slow wind initially flows from the corona along this line. At some time later, the rotation of the Sun supplies a coronal-hole source of fast solar wind at the footpoint of our radial line. Because it is travelling faster, plasma in the fast wind will catch up with the slow wind ahead of it. Due to the fact that the fast and slow wind originate on different field lines, the two plasmas cannot interpenetrate; in other words the faster plasma cannot overtake the slower plasma ahead. Thus, even though the flows are collisionless, a compression region builds up at the leading edge of the fast-wind stream. Eventually solar rotation causes slow flow from a different coronal region to become aligned with the footpoint of the radial line. Since the fast wind is running away from this second slow-wind region, a rarefaction develops on the trailing edge of the fast-wind stream.

Figure 3.1 shows the results of a simple spherically symmetric one-dimensional gas dynamic simulation (Hundhausen, 1973) to illustrate the above discussion. Initially ($t = 0$) the simulation region is filled with a steady-state slow wind of speed $325 \, \text{km s}^{-1}$ whose pressure decreases with distance from the Sun as in an

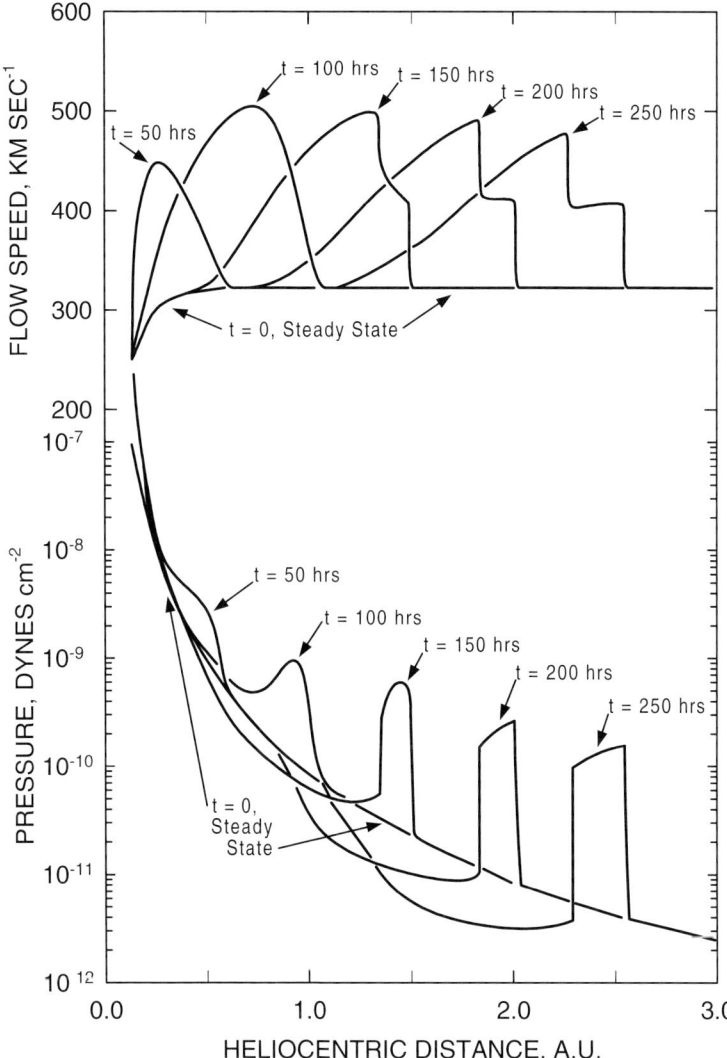

Figure 3.1. A plot of flow speed and pressure from a simple one-dimensional simulation showing the formation and evolution of interaction regions with distance from the Sun (From Hundhausen, 1973).

adiabatic expansion. The figure shows snapshots in time of the flow speed and pressure plotted against distance following the introduction of a high-speed stream simulated by increasing and then decreasing the pressure at the inner boundary. The pressure profiles at $t = 50$ hours and $t = 100$ hours illustrate the compression region which develops on the leading edge of the high-speed stream. The rarefaction on the trailing edge of the stream becomes apparent at later times.

A forward-propagating pressure wave develops at the leading edge of the compression which acts to accelerate the slow wind ahead, while a backward-propagating wave, known as a reverse wave, acts to decelerate fast wind behind the compression region. The region of high pressure bounded by the forward and reverse waves is referred to as the interaction region. The net effect of the interaction is to transfer momentum and energy from the fast wind to the slow wind, thus limiting the steepening of the fast stream, and damping out the speed difference between the fast and slow streams. If, however, the speed difference is greater than about twice the fast-mode speed (the characteristic speed with which small-amplitude pressure waves propagate in a plasma), the leading edge of the fast stream steepens faster than the compression region can expand, so that the compression region initially becomes narrower. The resulting non-linear increases in pressure cause the forward and reverse waves to steepen into shocks. Once the shocks have formed, the interaction region can expand since the shock waves can propagate much faster than the fast-mode speed. Most of the acceleration of the slow wind and deceleration of the fast wind then takes place discontinuously at the forward and reverse shocks as shown by the step-like velocity profile at later times in Figure 3.1. With increasing heliocentric distance the shocks gradually damp out the speed differences between the fast and slow wind. We note that because both the forward and reverse shocks are convected away from the Sun by the highly supersonic flow of the solar wind the downstream region of the reverse shock is sampled before the upstream region by a stationary observer.

In two dimensions in the solar equatorial plane, the process described above occurs at all longitudes. The state of evolution of the flows is, however, a function of heliographic longitude. Thus, assuming that the pattern of fast and slow-wind sources is time stationary at the Sun, the interaction region assumes a spiral configuration in the equatorial plane as shown in Figure 3.2. As the sources rotate, the whole spiral pattern corotates with the Sun, hence the term *Corotating* Interaction Regions (Smith and Wolfe, 1976). Since the pressure ridge within the interaction region is inclined relative to the radial direction, the pressure gradients provide forces also in the azimuthal direction (Siscoe *et al.*, 1969). In the leading part of the interaction region the slow solar-wind flow is deflected in a westward sense, and in the trailing part of the interaction region the fast wind is deflected eastwards. Such westward and eastward deflections in the solar-wind flow are common in 1 AU observations (e.g. Siscoe *et al.*, 1969; Gosling *et al.*, 1972, 1978). Thus the forward and reverse waves have a propagation direction with both azimuthal and radial components. When the waves steepen into shocks the flow deflections take place discontinuously at the shock fronts.

Observations of CIR formation and evolution

The two *Helios* spacecraft, which orbited the Sun in the ecliptic between 0.3 and 1 AU, provide the most comprehensive set of observations of stream-interaction regions in the early stages of evolution, thoroughly discussed by Schwenn (1990). By 1 AU the forward and reverse waves typically have not yet steepened into shocks

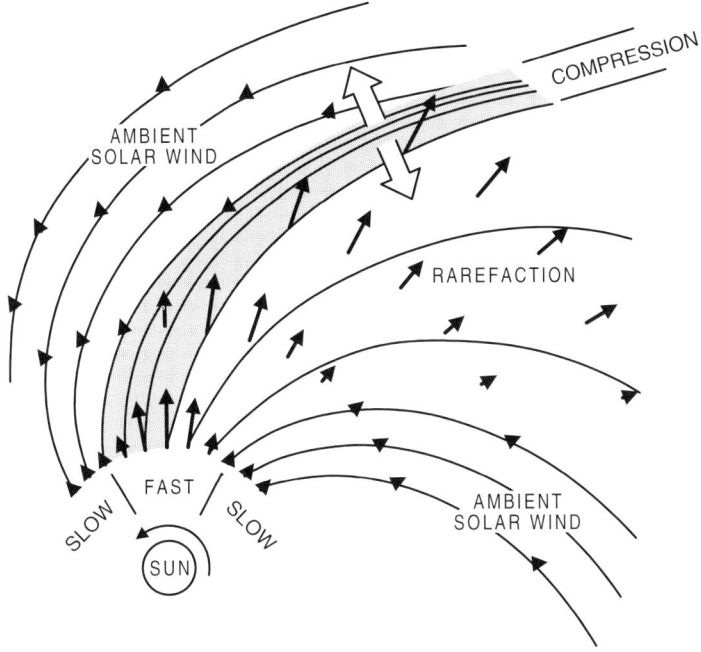

Figure 3.2. A schematic diagram of a corotating interaction region in two dimensions in the equatorial plane. The solid lines represent magnetic-field lines while the length of the arrows are a measure of the flow speed. (From Pizzo, 1978).

(Gosling et al., 1972), but the shocks usually do form by about 2–3 AU (e.g. Smith and Wolfe, 1976; Hundhausen and Gosling 1976; Gosling et al., 1976a). Indeed, data from the *Pioneer* spacecraft in the mid-1970s provided the first observations of a long-lived sequence of CIRs with fully developed forward–reverse shock pairs (Smith and Wolfe, 1976). With increasing distance, the speed differences are further reduced by the propagating shocks, and the compression regions expand and begin to merge with other compression regions, producing merged interaction regions that are the dominant structures in the low-latitude outer heliosphere beyond about 10 AU (Burlaga, 1984; Burlaga et al., 1997).

The patterns of fast and slow wind leading to CIR development

CIR development depends on the organization of the sources of fast and slow solar wind in the solar corona. In the descending and minimum phases of the solar cycle the corona commonly contains two large polar coronal holes that are the source of most of the fast solar wind. These are centred on the magnetic poles of the Sun and contain magnetic-field lines open to the outer heliosphere. Lower latitudes are dominated by closed magnetic structures close to the Sun underlying the streamers that can be seen in coronagraph images, typically organized in a band around the

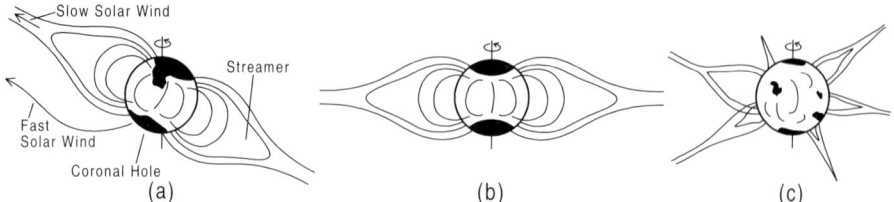

Figure 3.3. Schematic diagrams of the variation of the coronal magnetic field with the solar cycle: (a) declining phase; (b) solar minimum; (c) solar maximum (adapted from Suess et al., 1998).

magnetic equator (Chapter 2). This band is the source of the slower and more variable speed solar wind and is often referred to as the streamer belt (Gosling et al., 1981).

Figure 3.3 schematically illustrates how the structure of the solar corona varies through the solar cycle. Part (a) shows the situation in the declining phase where the magnetic dipole is strongly tilted with respect to the rotation axis. Thus the band of slow solar wind is also tilted at the same angle. A spacecraft situated at a low fixed latitude will alternately see fast solar wind from a coronal hole followed by slow solar wind from the streamer belt as the Sun rotates. This is the situation required for CIRs to develop, in particular since the large-scale structure of the corona does not vary very much on the timescale of the solar rotation at this phase of the solar cycle. As solar activity continues to decline the tilt of the magnetic axis to the rotation axis decreases until the two are nominally aligned at solar minimum as shown in part (b). Localized warps in the streamer belt, however, still allow CIRs to develop.

For comparison, part (c) shows the situation at solar maximum. The polar coronal holes are smaller and streamers are present over a wide range of latitudes at this time. Thus stable streams of fast solar wind are less likely at solar maximum and CIRs which do form are likely to also be less time stationary. Smaller coronal holes may still produce fast solar-wind streams which do not persist for as long as a solar rotation. Such streams will lead to interaction with slower wind in the same way as described above, but the resulting interaction regions will not form a large-scale, long-lived corotating pattern.

CIR morphology

The above discussion provides the framework for understanding CIR observations in the heliosphere in the radial-distance range 0.3–5.4 AU prior to *Ulysses*. To illustrate the basic plasma and magnetic-field morphology of CIRs we have chosen an example of a CIR observed by *Ulysses* near 5 AU in November 1992. Although it was observed at a latitude of 23°S it shows many of the characteristics that had previously been documented from near-ecliptic data. A number of the plasma and field parameters from this event are plotted in Figure 3.4. The top panel shows the solar-wind speed profile where a Forward Shock (FS) early on day 334 can be seen as the step-like increase in the speed marking the leading boundary of the CIR. This particular event was terminated by a Reverse Wave (RW) on day 336 rather than a reverse shock. Subsequent panels show the azimuthal and normal components of the velocity, proton density, proton temperature, magnetic-field strength, magnetic-field azimuthal and meridional angles, and the total pressure (plasma + magnetic field, calculated as $2n_p k T_p + B^2/2\mu$, where electron pressure is assumed equal to ion pressure). In agreement with our discussion of the radial evolution above, the density, temperature, magnetic-field strength, and pressure are significantly elevated within the interaction region on the leading edge of the high-speed stream.

Apart from the forward and reverse shocks (or waves), there are two other significant boundaries within the interaction region. The first of these is the *Stream Interface* shown on Figure 3.4 by the vertical line labelled SI. An SI is defined as a discontinuous drop in density and increase in temperature within an interaction region. It should be noted that a well-defined discontinuity matching this definition is not discernible in every interaction region. The existence of such discontinuities was first noted by Belcher and Davis (1971) while the term *stream interface* was introduced by Burlaga (1974). Belcher and Davis proposed the existence of a sharp transition between the slow and fast solar-wind flows near the Sun, and that the discontinuities which they observed in interplanetary space were the result of the evolution with distance of this transition within the interacting flows. Burlaga (1974) proposed an alternative interpretation that the boundaries became sharp due to the non-linear evolution with distance of what might initially be a gradual transition between low and high temperatures near the Sun. The presently accepted concept of an SI dates from the work of Gosling *et al.* (1978) who, in agreement with Belcher and Davis, described the interface as the boundary separating what was originally dense slow gas near the Sun from what was originally rare fast gas near the Sun. They noted the defining characteristics of an SI at 1 AU as a sharp drop in proton density, a sharp rise in proton temperature, a sharp discontinuity in the flow direction consistent with the flow deflections first observed by Siscoe *et al.* (1969), a jump in flow speed (all these are seen in the example in Figure 3.4), and a steepening in the slope of the time profile of the flow speed. Also, at 1 AU the interface was usually close to the maximum of the total pressure within the interaction region, the time profile of the pressure often being symmetrical about the interface. This is to be expected since the forward and reverse pressure gradients should be centred on the boundary between what was originally the slow and fast solar wind. The CIR in Figure 3.4 does not exactly fit this pattern probably because it was observed near 5 AU. As illustrated in Figure 3.1, the pressure ridge at the centre of a CIR weakens with distance from the Sun, and by 5 AU the major pressure changes have shifted to the shocks. It was also noted that the alpha particle to proton ratio and the alpha particle flow speed relative to the protons changed abruptly at the interface (Gosling *et al.*, 1978), consistent with the picture that the plasmas on either side of the interface have different origins. A further signature was that the magnetic-field variability increased at SI (Belcher and Davis, 1971), consistent with the presence of Alfvén waves, now known to be a feature of the fast solar wind (Chapter 4).

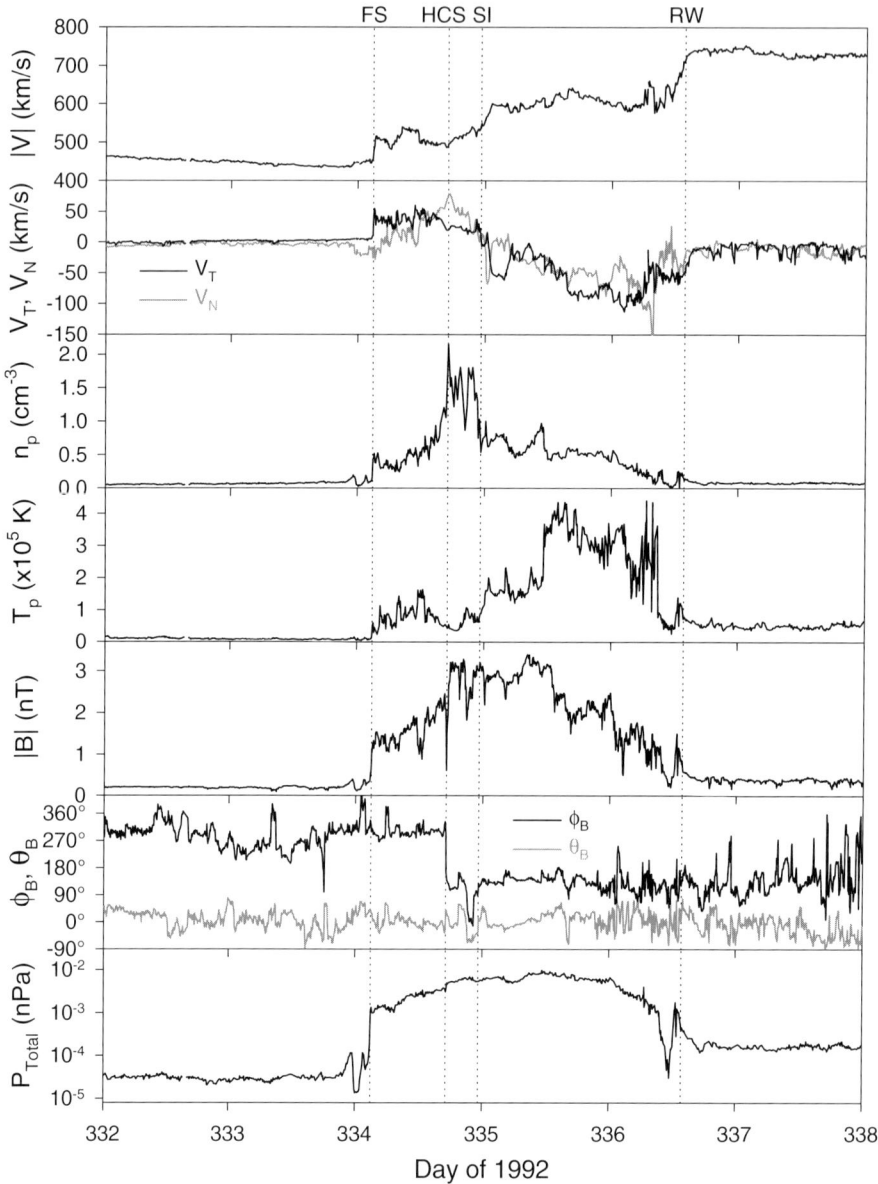

Figure 3.4. An example of the variation with time of selected plasma and magnetic field parameters through a CIR observed by *Ulysses* in November 1992. From the top, the parameters plotted are the solar wind speed, the tangential and normal components of the velocity, proton number density, proton temperature, magnetic field strength, magnetic field azimuthal and meridional angles, and the total pressure. The vertical lines indicate the forward shock (FS), heliospheric current sheet (HCS), stream interface (SI) and reverse wave (RW). The measurements were made by the SWOOPS and magnetometer instruments onboard *Ulysses* (Bame *et al.*, 1992; Balogh *et al.*, 1992).

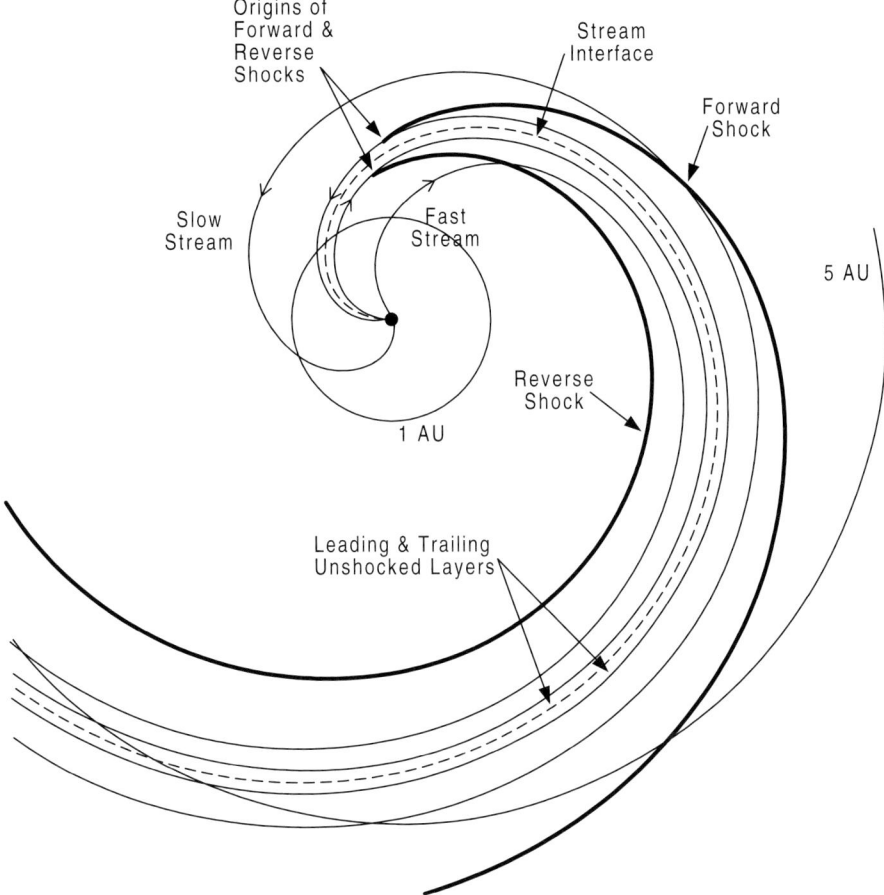

Figure 3.5. An illustration of the two-dimensional geometry of CIRs showing the shocks, stream interface and unshocked regions. (From Crooker *et al.*, 1999).

In summary, then, the SI is a fundamental boundary within a CIR separating what was originally slow solar wind near the Sun, from what was originally fast solar wind. Since magnetic-field lines at the interface should map back to the coronal-hole boundaries at the Sun, magnetic-field lines should not cross the SI which implies that it should be a tangential discontinuity in the plasma and magnetic-field data.

We note that not all the plasma observed between the shocks and the SI has been processed by the shocks. Since the shocks typically do not develop until well beyond 1 AU, there is a region of plasma on either side of the SI which remains unshocked (e.g. Gosling *et al.*, 1976a; Palmer and Gosling, 1978). Figure 3.5 illustrates the idealized geometry of CIRs showing the forward and reverse shocks, the SI and the unshocked regions. This is of particular importance for understanding magnetic-field line connectivity in CIRs when interpreting energetic-particle observations (Chapter 6).

A further boundary, usually found within CIRs at 5 AU, is the Heliospheric Current Sheet (HCS), also known as a magnetic-field sector boundary. An example is seen in Figure 3.4, identified as the discontinuous 180° change in the azimuthal angle of the magnetic field. The HCS is a large-scale structure in the heliosphere in its own right and represents the boundary between the northern and southern magnetic hemispheres (Chapter 2) (i.e. it is the outward extension of the Sun s magnetic equator). Near the Sun the HCS is embedded in the band of slow solar wind associated with the streamer belt. Unlike the SI, the HCS does not have to lie within a CIR. As the forward wave propagates into the slow solar wind ahead it eventually overtakes the HCS. Thus, with increasing distance from the Sun it becomes more likely to find that the HCS has been swept into a CIR. About two-thirds of all clean HCS crossings at 1 AU occur within CIRs (Borrini et al., 1981), and the majority of CIRs observed by *Pioneer 10* beyond 5 AU contained HCS crossings (Thomas and Smith, 1981). As in the example in Figure 3.4 we would always expect to find the HCS between the forward shock and SI. This pattern has been confirmed by observations (Gosling et al., 1978). We note that the polarity of the magnetic field behind the HCS can be used to infer the magnetic field polarity of the coronal hole from which the fast wind driving the CIR originates. Thus the example in Figure 3.4 is consistent with a fast stream originating in a coronal hole with inward pointing magnetic field.

3.2.2 *Ulysses* CIR observations

Overview

During its first polar orbit of the Sun, there were two intervals when the *Ulysses* observations were dominated by quasistable sequences of CIRs. Figure 3.6 shows these two sequences in the context of the orbital latitude. The solar-cycle influence is indicated in the top panel which shows the average tilt of the heliospheric current sheet, given by the average of the predicted maximum northern and southern latitudes of the HCS as inferred from the Stanford source surface model (e.g. Hoeksema, 1995). The plot thus includes the effect of warps in the HCS as well as the overall tilt. This can be used as an approximation for the overall tilt of the band of low-speed solar wind which, at least near solar minimum, extends over a relatively narrow latitude range either side of the HCS. Also plotted on this panel is the absolute value of the latitude of *Ulysses*. The second panel shows daily averages of the solar-wind speed. The two sequences of CIRs occur where the solar-wind speed alternates between high and low values at approximately the solar rotation period.

The first of the CIR sequences highlighted in Figure 3.6 was driven by a high-speed solar wind flow which recurred at *Ulysses* for 15 solar rotations between July 1992 and August 1993 (e.g. Philips et al., 1995). The hourly averaged solar wind speed, magnetic field strength and proton number density through this period are

Sec. 3.2] Corotating structures 117

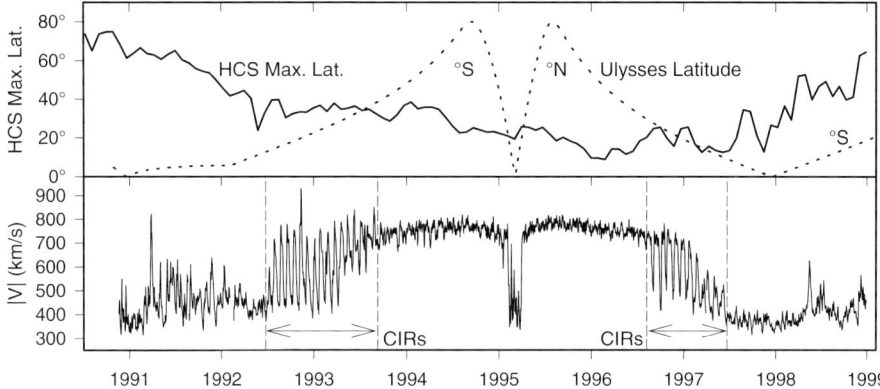

Figure 3.6. Solar wind speed throughout the *Ulysses* mission (lower panel) highlighting the two periods where CIRs dominated the observations. The upper panel shows the maximum latitude of the heliospheric current sheet inferred from the Stanford 'classic' source surface model, and the absolute value of the *Ulysses* latitude.

plotted in the upper half of Figure 3.7. During this time *Ulysses* moved inwards in heliospheric distance from 5.3 to 4.5 AU and towards higher southern latitudes from 13°S to 36°S. The second sequence occurred between August 1996 and May 1997 as *Ulysses* returned from high northern latitudes (McComas et al., 1998) from 30°N down to 11°N. The data from this period are plotted in the lower half of Figure 3.7 on exactly the same scale to allow comparison. In this case the distance range traversed was from 4.1 out 5.0 AU.

A shorter interval of less well defined CIRs was encountered in the second half of 1991 while *Ulysses* was still in the ecliptic plane, which appeared primarily as a solar rotation recurrence of increased magnetic-field strength (Balogh et al., 1993a; Gonzalez-Esparza et al., 1996). In addition, a small number of CIRs were encountered while *Ulysses* was briefly transiting low latitudes in 1995 (Gosling et al., 1995a; Smith et al., 1995). This period, apparent in the middle of Figure 3.6, is usually called the 'fast latitude scan'.

Southern hemisphere CIRs

The first of the two CIR sequences in Figure 3.6 has been the most extensively studied. The 15 recurrences of the high-speed solar wind stream have been numbered in the first panel of Figure 3.7 following (Bame et al., 1993). A number of papers in the literature have subsequently used this scheme to identify the corresponding CIRs. Figure 3.6 shows that this sequence of CIRs became established at the end of a period of decreasing tilt of the HCS and that the HCS tilt then remained stable at about 30° for a long time thereafter. The appearance of the high-speed stream was associated with the development of an equatorward extension of the southern polar coronal hole (Bame et al., 1993), consistent with the changing HCS

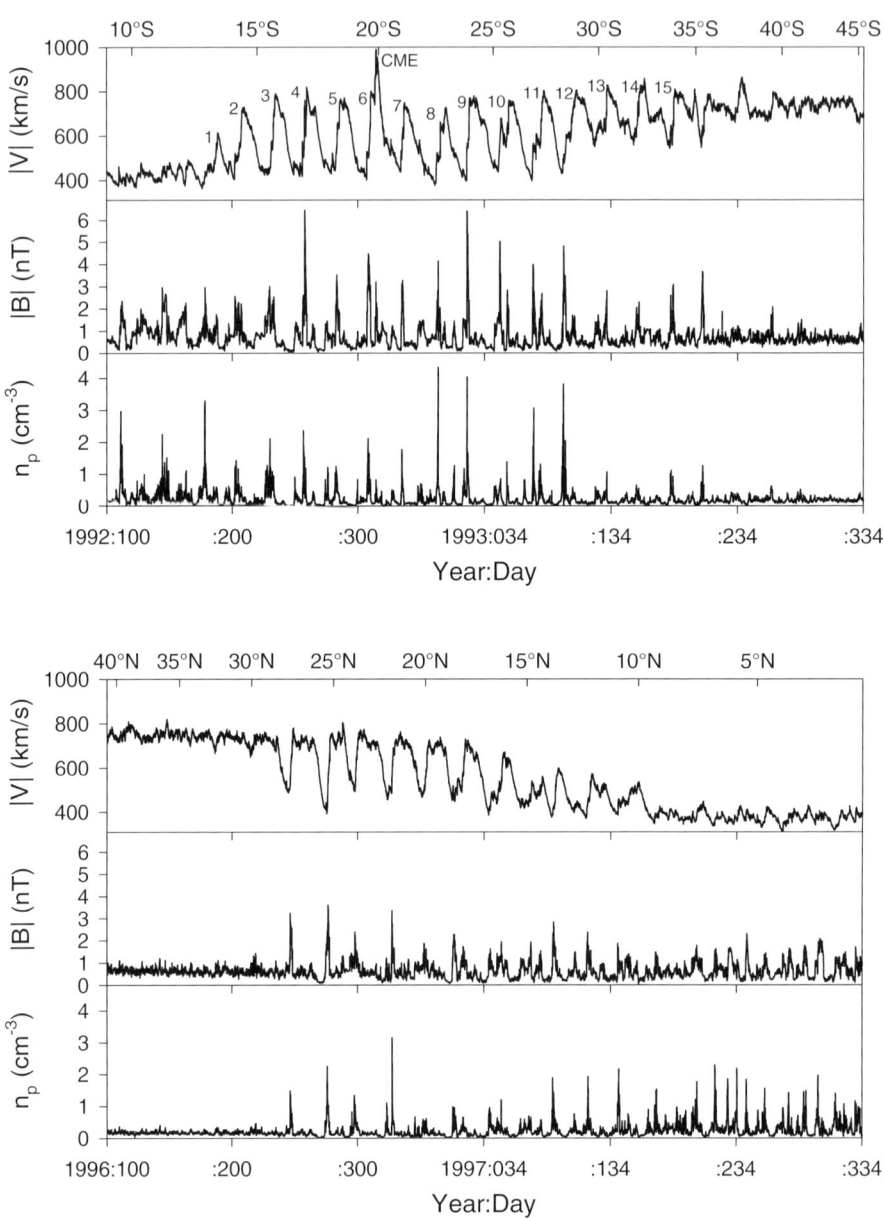

Figure 3.7. Solar wind speed, magnetic field strength and proton number density plotted through the southern hemisphere CIRs (top three panels) and the northern hemisphere CIRs (lower three panels). The heliographic latitude is labelled along the top axis. The two sets of plots are on the same scales. The numbering of the high-speed streams in the southern hemisphere follows Bame et al. (1993).

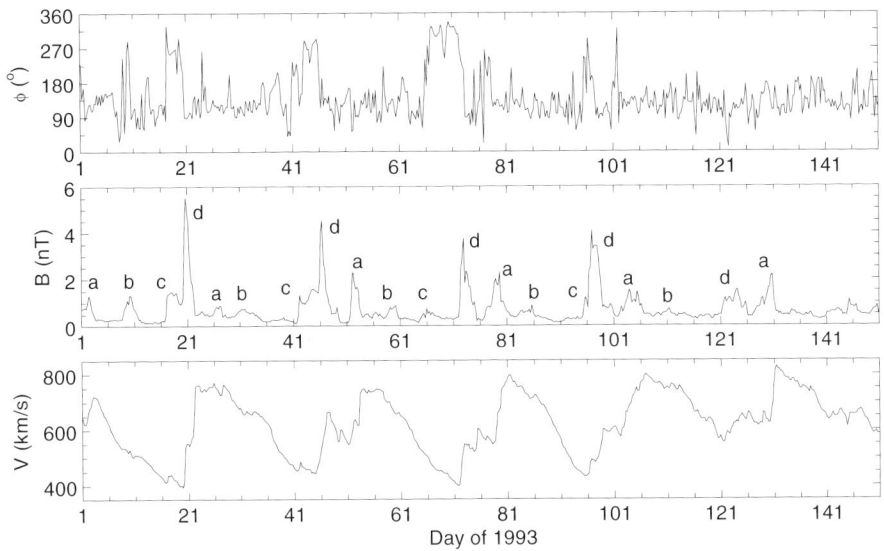

Figure 3.8. Magnetic-field azimuth angle, magnetic-field strength, and solar wind speed for a subset of the southern hemisphere CIRs. The letters a, b, c and d represent different recurrent features in the magnetic field strength. (From Smith *et al.*, 1993).

tilt, bringing the fast wind down to lower latitudes in a narrow longitude range. At the same time the magnetic sector structure changed in phase from being synchronous with the sidereal solar rotation period (25.4 days), to having an eastward drift corresponding to a rotation period of about 27 days (Balogh *et al.*, 1993b). Both the development of the coronal-hole extension and this change in the magnetic polarity pattern are consistent with a major reordering of the large-scale coronal magnetic field. A recurrent high-speed stream also appeared at 1 AU in the ecliptic plane at approximately the same time, consistent with origin from the same coronal hole extension (Bame *et al.*, 1993). This finding supports the view that the appearance of the high-speed stream at *Ulysses* was not due simply to *Ulysses* travelling to higher latitudes. By contrast, the 1993 transition from the alternating streams to continuous high-speed solar wind was only a latitude effect (Philips *et al.*, 1994).

Figure 3.8, adapted from Smith *et al.*, (1993) shows a subset of the southern hemisphere CIRs during the first 5 months of 1993 on an expanded scale to bring out detail in the magnetic-field structuring. The magnetic-field azimuth angle is shown in the top panel, the magnetic-field magnitude is shown in the second panel, and the flow speed is shown in the lower panel. An azimuth angle of the order of 100° indicates inward polarity fields of the southern hemisphere, while that of about 280° indicates outward polarity fields of the northern hemisphere. This figure confirms that the high-speed streams had the polarity appropriate to the southern hemisphere. It also indicates that beyond about day 100 of 1993, when *Ulysses* passed to latitudes higher than 28°S, the spacecraft no longer encountered the HCS. Beyond this time the minimum solar-wind speed increased to ~550 km s^{-1}, consistent with the spacecraft no longer fully penetrating into the streamer belt. As a result, the magnetic-field enhancements in the CIRs (Figure 3.8, panel 2) decreased owing to the reduced difference in speed between the flows. The disappearance of HCS crossings and of the lowest-speed wind occurred earlier than predicted by the tilt values shown in the top panel of Figure 3.6. This is most likely due to limitations in the potential field model used to compute the coronal magnetic field, from which the tilt angle has been derived.

Multiple enhancements of the magnetic field evident in the second panel of Figure 3.8 indicate a relatively complex speed structure closer to the Sun. Early in the CIR series the dominant magnetic-field compression is that labelled d which is bounded by the forward and reverse shocks of the CIR associated with the dominant high-speed stream. Three other apparently recurrent magnetic-field compressions are, however, evident in the figure, labelled a, b, and c. Of these, based on the magnetic polarity information in the top panel, it appears that compression c is caused by a weak stream with origin north of the heliomagnetic equator. This compression duly disappears once *Ulysses* no longer encounters the HCS. Beginning with the second cycle in Figure 3.8, the main reverse shock moves over to compression a which then grows and eventually becomes dominant as compression d declines. It is possible that these changes are a result of a gradual evolution in the structure of the coronal-hole boundary or of the position of *Ulysses* relative to a ripple in the boundary. The compression b does not coincide with any speed increases, but does coincide with changes in the gradient as the speed declines. This compression may have been created by a separate weaker stream which has already merged with the dominant high-speed stream.

Northern hemisphere CIRs

Ulysses observations of CIRs as the spacecraft returned from high northern latitudes in 1996/1997 are summarized in the lower half of Figure 3.7, again showing, from top to bottom, solar wind speed, magnetic field strength and proton number density. The speed profile of these streams are quite different from those of the southern hemisphere. In this period the underlying tilt of the heliospheric current sheet was much lower ($\sim10°$) as expected close to the solar minimum. There was, however, a localized warp in the current sheet due to a single large solar active region that deflected the current sheet up to higher northern latitudes over a relatively narrow range of longitudes (Forsyth *et al.*, 1997; Riley *et al.*, 1999). It is this warp that caused the increase in the maximum latitude of the HCS apparent in Figure 3.6 beginning near the middle of 1996. Thus the whole band of low-speed solar wind extended northwards in this localized longitude range. As a consequence, *Ulysses* first encountered the northern edge of the band at $\sim30°$N (Gosling *et al.*, 1997), a higher latitude than would otherwise have been expected at this phase of the solar cycle. During the first few solar rotations *Ulysses* sampled primarily high-speed wind, only dipping briefly into the low-speed wind once per rotation.

The magnitude of the compressions in the magnetic field and density were considerably less in the northern hemisphere than in the southern hemisphere (Figure 3.7). This is likely a result of the smaller difference in speeds on the leading edges of the streams, which in turn may be a consequence of the lower streamer belt tilt. *Ulysses* did not penetrate very far into the band of low-speed wind on each rotation and so did not reach the lowest solar wind speeds.

Composition signatures

The *Ulysses* 1992/1993 observations confirmed that there are clear composition differences between the fast and slow solar wind (Geiss *et al.*, 1995 and Section 2.5). Of particular interest here is the composition change as the spacecraft crossed the various boundaries within a CIR. Analyses of both the southern (Wimmer-Schweingruber *et al.*, 1997) and northern hemisphere CIRs (Wimmer-Schweingruber *et al.*, 1999) confirm that at the stream interface the oxygen and carbon freezing-in temperatures, and the magnesium/oxygen abundance ratio, change from values characteristic of the low-speed wind to those of the high-speed wind. Figure 3.9, adapted from Wimmer-Schweingruber *et al.* (1997), provides a representative example of the freezing-in temperature behaviour through a CIR. Density and temperature changes on day 363 clearly indicate the SI boundary while those on days 361 and 365, as well as the corresponding speed changes, identify the forward and reverse shocks bounding the CIR. The previously mentioned rise in the proton to alpha ratio at the SI (Gosling *et al.*, 1978) is also apparent. Both the oxygen and carbon freezing-in temperatures drop at the interface from the higher variable values of the slow wind to the lower, more constant, values of the high-speed wind. As expected, the ionization temperatures are unaffected by the shocks. Thus composition signatures are an additional useful marker within a CIR that help to identify solar wind that was originally fast or slow when it left the Sun. Composition evidence for multiple SI boundaries, of which there must be an odd number, was found in some CIRs (Wimmer-Schweingruber *et al.*, 1997, 1999). These were interpreted as being due to structure in coronal hole boundaries close to the Sun. A comparison between CIRs observed in the northern and southern hemispheres (Wimmer-Schweingruber *et al.*, 1999), showed that the SI of the northern set had less well defined kinetic signatures than in the south. This difference, as well as the fact that fewer recurrences of the CIRs were observed in the north, is thought to be related to the tilt of the streamer belt at the time. The lower tilt led to interface orientations being less inclined to the direction of solar-wind flow in the north and thus the interaction was less severe (see also Gosling and Pizzo, 1999).

Ulysses plasma and composition data have been used to search for the equivalent of SI on the trailing edges of high-speed solar wind streams (Burton *et al.*, 1999). In principle, there should be a boundary in this region marking the change back from originally fast solar wind to originally slow solar wind. Such boundaries are harder to identify in the rarefaction regions because the speed profiles stretch out the signatures. Abrupt drops were identified in specific entropy ($T/n^{\gamma-1}$, where γ is the polytropic index) in nearly every rarefaction region associated with the

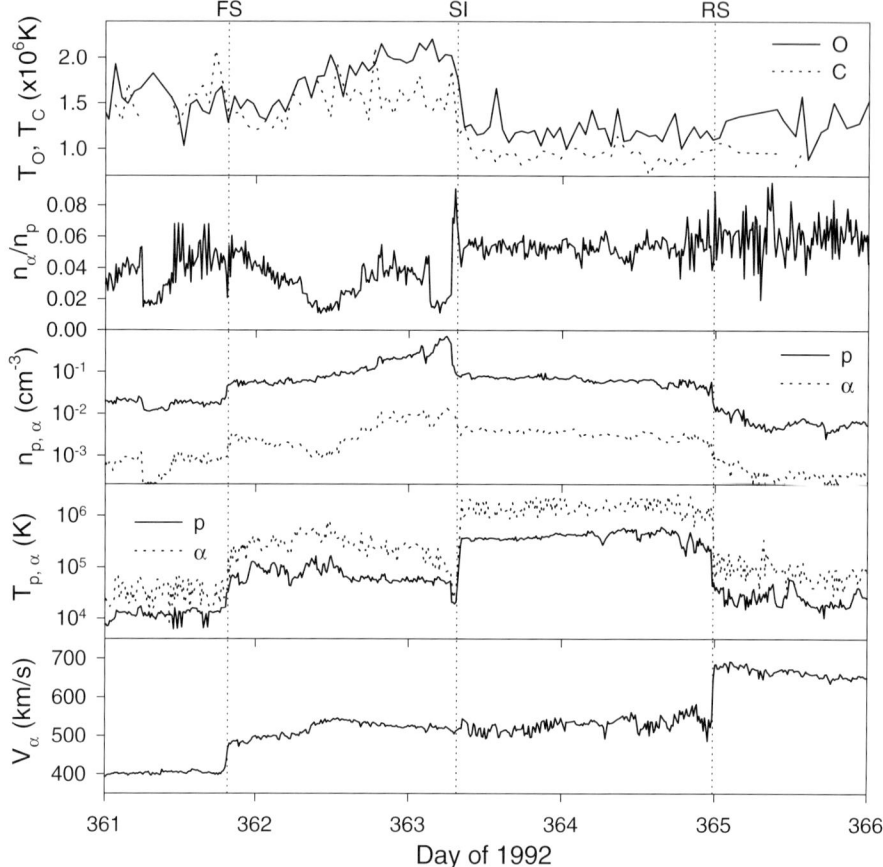

Figure 3.9. Variation of oxygen and carbon freezing-in temperatures (top panel) through a CIR observed by Ulysses in December 1992. Plotted in subsequent panels are the alpha to proton ratio, proton and alpha number densities, proton and alpha temperatures, and the alpha particle velocity, all measured by the SWICS instrument on Ulysses (Gloeckler et al., 1992). (From Wimmer-Schweingruber et al., 1997).

southern high-speed streams (Burton et al., 1999) The composition data showed a change from that of the fast wind back to that of the slow wind at the same boundary, but on a less abrupt timescale, confirming these as the boundaries between the originally fast and slow wind. The average duration of the specific entropy change was 51 minutes, which corresponds to a boundary thickness of $\sim 10^6$ km, while the shortest was 5 minutes, equivalent to a thickness of $\sim 10^5$ km.

CIR shock characteristics

Figure 3.10 provides a summary of the corotating forward and reverse shocks observed during the two main southern and northern hemisphere CIR periods.

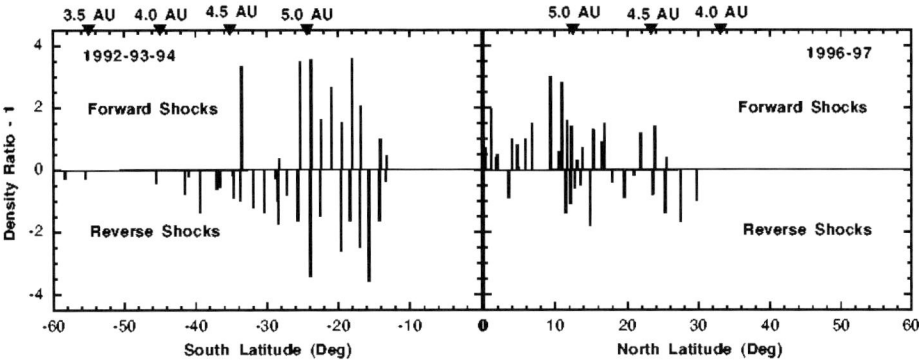

Figure 3.10. A summary, from Gosling and Pizzo (1999), of the corotating forward and reverse shocks observed by *Ulysses* in both the southern (left panel) and northern (right panel) hemispheres. Forward shocks are plotted upwards and reverse shocks downwards. The height of the lines is a measure of the shock strength.

See Balogh *et al.* (1995) and McComas *et al.* (1998) for details of all the *Ulysses* shock timings during these periods and Gonzalez-Esparza *et al.* (1996) for a study of their association with corotating or transient flows. Figure 3.10 shows the southern hemisphere shocks on the left and the northern hemisphere shocks on the right, with latitude on the lower horizontal axis and distance on the upper axis. The parameter plotted on the vertical axis, the 'density ratio – 1' of the shocks, is a measure of their strength. (The density ratio is the ratio of the downstream density to the upstream density.) The ratios for the forward shocks are plotted upwards from the centre line, while those of the reverse shocks are plotted downwards.

In the southern hemisphere all the strongest forward shocks except one are confined to latitudes within 26°S of the heliographic equator (Gosling *et al.*, 1993a, 1995b). Note that this latitude is only slightly less than the tilt of the heliospheric current sheet at this time (Figure 3.6). The majority of the shocks below this latitude occurred as forward-reverse shock pairs as expected at these heliocentric distances and are of comparable strengths. Unlike the forward shocks, CIR associated reverse shocks continued to occur regularly at latitudes poleward of 26°S, until the spacecraft reached 42°S, with still an occasional, but no longer regular, occurrence thereafter. The strengths of these reverse shocks gradually decreased as *Ulysses* moved further poleward from the band of variable-speed wind. A study of the latitude dependence of a number of the shock parameters in the southern hemisphere (Burton *et al.*, 1996) confirmed, in particular, that the strongest shocks were observed at mid-latitudes where the velocity gradients were greatest.

In the northern hemisphere a similar pattern of shock occurrence was observed, although the pattern is somewhat less clearly organized than in the south. Only reverse shocks were observed poleward of 26°N, but in this case none were observed poleward of 30°N (Gosling *et al.*, 1997). Forward shocks began below 26°N, again similar to the highest latitude at which the heliospheric current sheet was observed, even though the underlying tilt of the solar dipole was lower at this time. In general, the strengths of all the shocks during this period were weaker than those at comparable latitudes in the southern hemisphere. Below about 10°S, mostly forward shocks were observed. It is likely that the majority of the differences in the pattern of shock occurrence between the southern and northern hemispheres can be attributed to the lower tilt of the solar dipole during the latter interval and hence the differing geometry of the heliospheric current sheet and band of low-speed solar wind. During the fast latitude scan (Gosling *et al.*, 1995a), reverse waves that had not yet steepened into shocks were observed on either side of the band of variable-speed solar wind, consistent with the observations in Figure 3.10.

Flow deflections at CIR shocks

We have already discussed the east-west flow deflections commonly associated with CIRs in the ecliptic plane. *Ulysses* observations provided the first clear evidence of a systematic pattern of north-south flow deflections associated with CIRs, thus extending our understanding into three dimensions (Gosling *et al.*, 1993a; Gosling *et al.*, 1997) Figure 3.11 shows two good examples of the flow deflections observed by *Ulysses* at CIRs, the first from the southern hemisphere and the second from the northern hemisphere. From top to bottom, the solar-wind speed, azimuthal flow angle, meridional flow angle and proton pressure are plotted. The CIR forward

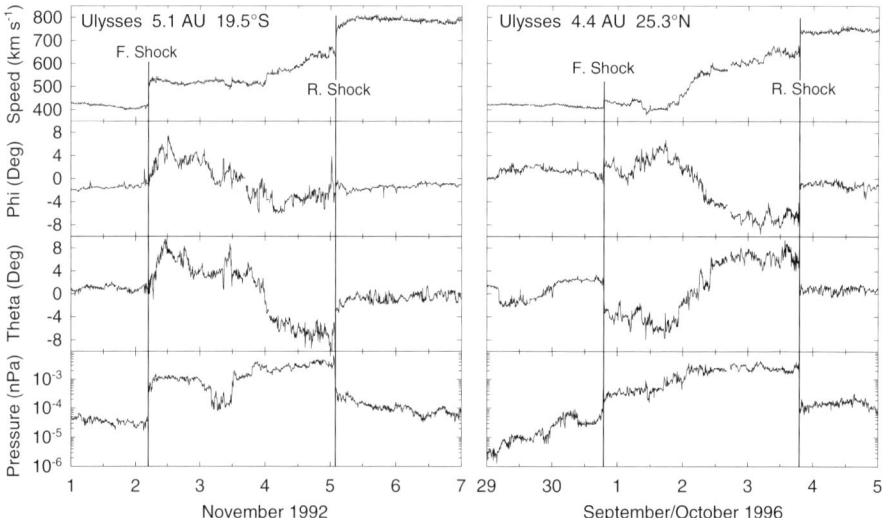

Figure 3.11. Examples of flow deflections from two CIRs, one southern and one northern hemisphere. Parameters plotted are the solar wind speed, azimuthal and meridional flow angles and the proton pressure. (From Gosling *et al.*, 1995b; 1997).

and reverse shocks are identified by vertical lines, where the step-like changes in the speed and pressure occur. Azimuthal and meridional deflections away from the radial are clear in the regions downstream of both the forward and reverse shocks. (Remember that a spacecraft observes the downstream region of the reverse shock before it observes the upstream region.) In the southern hemisphere, following the forward shock there is a positive (or westward) deflection in the azimuthal flow and a positive (i.e. northward or equatorward) deflection in the meridional flow. Downstream of the reverse shock there is a negative (or eastward) deflection in the azimuthal flow and a negative (or poleward) deflection in the meridional flow. In the northern hemisphere example, the deflections in the azimuthal flow are as before, consistent with the spiral geometry of the interaction regions in the azimuthal plane. The deflection of the meridional flow behind the forward shock is, however, now negative (i.e. southward, or equatorward again since we are now in the northern hemisphere) while downstream of the reverse shock the deflection is positive (or poleward). In the same way as the azimuthal deflections were interpreted as showing that the forward shocks have a westward component in their propagation direction (and eastward for reverse shocks), the meridional deflections indicate that forward shocks have an equatorward component in their propagation direction, and that the reverse shocks have a poleward component. This poleward propagation of the reverse shocks explains why *Ulysses* generally observed only reverse shocks at latitudes poleward of where the fast and slow flows actually interacted in both the southern and northern hemispheres. It is also consistent with the dominance of forward shocks at low latitudes that was noted in the northern hemisphere. Quantitative analyses of the orientation of the shock fronts from the velocity deflections (Riley *et al.*, 1996) and the magnetic field deflections (Burton *et al.*, 1996) at the shocks confirm that the orientations are, in general, consistent with the propagation directions just described.

The pattern of flow deflections and shock propagation directions observed in the southern hemisphere are consistent with the predictions of a global model of corotating solar-wind flows developed by Pizzo (Gosling *et al.*, 1993a; Pizzo, 1991). This model assumes a pattern of slow and fast solar-wind near the Sun that is organized around a tilted dipole geometry. Flow deflections observed during the fast latitude scan (Gosling *et al.*, 1995c) and later as the spacecraft returned from high northern latitudes (Gosling *et al.*, 1997) continued to be consistent with this model. The flow deflections and shock propagation directions arise from the three-dimensional orientation of the surface separating the fast and slow solar-wind flows close to the Sun. In the declining and near minimum phase of the solar cycle, when the streamer belt has a simple tilt angle with respect to the rotation axis as in Figure 3.12, the boundary between the fast and slow wind is also inclined at this angle. Since the forward and reverse waves arise from pressure gradients transverse to the interface, in the northern hemisphere the forward waves acquire equatorward and westward components of propagation and the reverse waves acquire poleward and eastward components. In section 3.2.3 we will discuss the modelling work and its comparison with the *Ulysses* observations in greater detail.

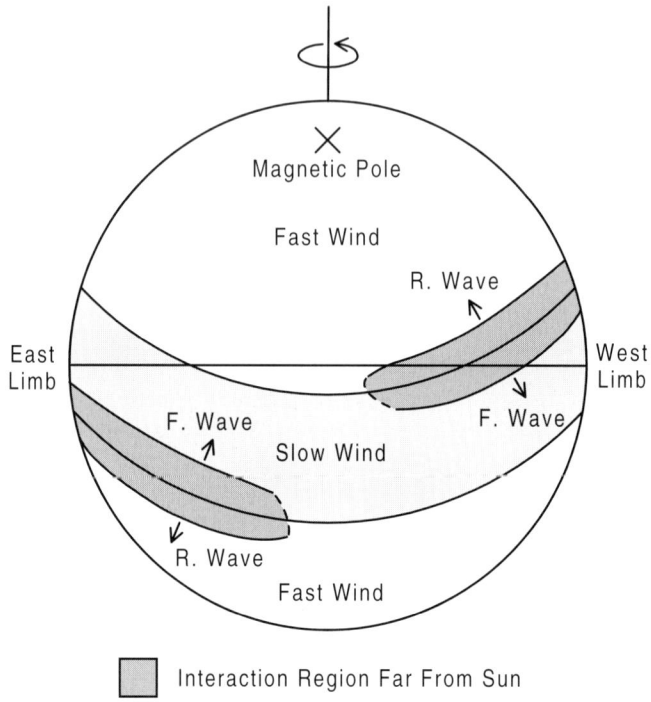

Figure 3.12. Schematic diagram illustrating how the tilted streamer belt leads to equatorward propagating forward shocks and poleward propagating reverse shocks. The magnetic pole is tilted forward from the plane of the paper (from Gosling and Pizzo, 1999).

Recent analysis of the magnetic-field structure of CIRs observed in both the southern and northern hemispheres (Clack *et al.*, 2000) lends support to the above view. The magnetic field vectors follow a planar ordering throughout the CIRs with the orientation of the planes being consistent with the orientation of the interaction regions predicted by the above model, oppositely inclined in the southern and northern hemispheres.

SI orientations in near-ecliptic interaction regions observed between 1 and 5 AU by *Ulysses* were different from those observed by the *Pioneer* and *Voyager* spacecraft (Gonzalez-Esparza *et al.*, 1997) in the same distance range. The majority of the *Ulysses* interfaces were nearly perpendicular to the equatorial plane, while those at *Pioneer* and *Voyager* had significant latitudinal tilts similar to the *Ulysses* mid-latitude results discussed above. The suggested interpretation (Gonzalez-Esparza *et al.*, 1997) is that the CIRs observed by *Ulysses* just after solar maximum, might be associated with fast solar wind from equatorial coronal holes and hence did not acquire the tilts characteristic of the simplified coronal and heliospheric geometry nearer solar minimum. In addition, during the *Ulysses* fast latitude scan (FLS), the latitudinal tilts of CIRs were less well organized near the equator (Gosling *et al.*, 1995c).

Counterstreaming electron events at CIR shocks

In Section 3.3.1 we will discuss the use of the suprathermal electron heat flux for inferring the topology of magnetic field lines. Suprathermal electrons in the solar wind form a population separate from the core or thermal electrons (Section 2.7). These electrons typically are strongly beamed along the magnetic field. Under normal circumstances the suprathermal electrons are heated in the solar corona and thus stream outwards from the Sun (Feldman *et al.*, 1975). Prior to *Ulysses*, counterstreaming suprathermal electrons (i.e. travelling in both directions along the magnetic field) had been observed in the solar wind only on the presumably closed magnetic field lines associated with ICMEs and on field lines connected to planetary and cometary bow shocks.

A new result from *Ulysses* was that counterstreaming suprathermal electron events were observed upstream of virtually all CIR-associated shocks beyond about 2 AU from the Sun (Gosling *et al.*, 1993b). These counterstreaming suprathermal electrons were a result of shock-heated electrons leaking out into the upstream regions, a process earlier associated only with planetary shocks. Events of this type had not previously been identified at 1 AU because corotating shocks have typically not formed at that distance and, if they have, are still relatively weak. ICME-driven shocks, which are more common at 1 AU, typically have a component of their shock normal directed along the background magnetic field away from the Sun with the result that any shock heated electrons escaping upstream travel in the same direction as the normal solar wind suprathermal heat flux.

The counterstreaming events observed by *Ulysses* upstream of both the forward and reverse shocks lasted on average 2.4 days. The intensity of the events was strongest close to the shock front and faded with increasing distance upstream. Figure 3.13, which shows the magnetic field geometry expected for a CIR, helps to illustrate why these suprathermal electron events are found to be counterstreaming. The dashed curve on this figure shows the effective trajectory of *Ulysses* through the CIR, and the shaded regions indicate schematically where the counterstreaming electrons were observed. For both the forward and reverse shocks the field lines that intersect the shock fronts, if followed away from the shock front on the upstream side, eventually connect back to the Sun. Thus a flux of shock heated electrons streaming away from the shock front on the upstream side will have a propagation direction opposed to that of the normal solar wind suprathermal electron flux.

Commonly the reduction of intensity of the counterstreaming electron beam with distance from the shock was accompanied by a spreading of the beam in pitch angle (Gosling *et al.*, 1993b). This was interpreted as a consequence of magnetic mirroring and scattering. The average duration of 2.4 days corresponds to a distance along a field line to the shocks of the order of 15 AU. An alternative suggestion for the beam fading with distance was simply that a smaller number of electrons were heated by the shocks further from the Sun because of the radial fall off of the solar-wind density.

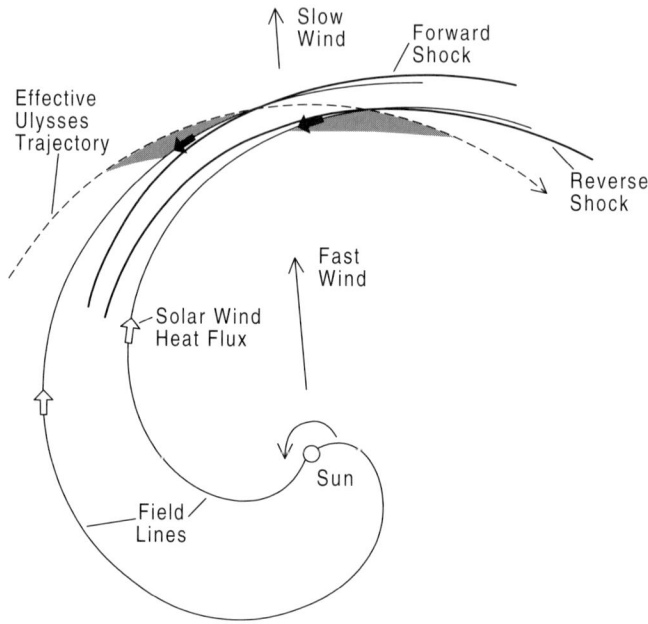

Figure 3.13. The magnetic field geometry associated with the counterstreaming electron beams produced by CIR shocks. The shaded regions indicate where the counterstreaming electrons were observed. (From Gosling *et al.*, 1993b).

3.2.3 Three-dimensional models of CIRs

Prior to *Ulysses*, the development and evolution of CIRs could be reasonably well understood on the basis of one and two-dimensional models. As we have described, a three-dimensional model is necessary to describe fully the dynamical processes involved in CIR formation and evolution. The model which has been specifically compared to the *Ulysses* observations is that developed by Pizzo (1991, 1994; Pizzo and Gosling, 1994). A tutorial introduction to the assumptions, application and results of modelling CIRs is given by Gosling and Pizzo (1999). In addition, an analytical model of CIR morphology has recently been presented by Lee (2000).

Pizzo's model (Pizzo, 1991, 1994) prescribes a pattern of fast and slow solar-wind sources on a spherical surface r_0 near the Sun and solves the ideal magnetohydrodynamic (MHD) equations to follow the evolution and interaction of the resulting flows outward through the heliosphere. It is assumed that in the frame of reference corotating with the Sun, this large-scale pattern of fast and slow wind sources does not change with time. Observations that coronal hole boundaries are stable for many solar rotations indicate that this assumption is reasonably justified. No attempt is made to describe the flows within the inner solar corona where the solar wind is still accelerating. Rather, r_0 is taken to be far enough from the Sun, typically ~18–30 solar radii, that the flows are already supersonic and radial. Thus the non-radial flows closer to the Sun are ignored.

The MHD equations describe conservation of mass, momentum, energy and magnetic field in a plasma and thus determine the behaviour of the velocity, magnetic field, mass density and gas pressure. The magnetic field is assumed to be perfectly frozen-in to the plasma. Because the dynamics in the solar wind are dominated by the energy in the bulk flow, the internal energy equation for the solar wind is taken to be a polytropic law relating the gas pressure and density. The numerical techniques used to solve the equations and step the solution out in radial distance through the heliosphere have been described by Pizzo (1982).

Some limitations of this model are that all non-steady effects are neglected. Thus complications arising from the gradual time evolution of the coronal structure and hence the pattern of fast and slow solar-wind flows cannot be described. Small scale effects such as waves and instabilities are also not included and effects of differential rotation are neglected. The advantage of the model is that it gives results in good agreement with observations for the large-scale flow patterns in the heliosphere; but it is less useful for describing smaller scale temporal and local effects that arise in CIRs.

An idealized pattern of fast and slow solar wind organized by the tilted dipole geometry, shown in Figure 3.14, describes the flows at the inner boundary of the model. The solar magnetic dipole axis is tilted at an angle α to the rotation axis, and a band of uniformly slow, dense, cold, solar wind flow encircles the dipole equator. Outside of this band, a uniformly fast, tenuous, hot solar wind originates from the polar regions. The transition region between these two flow regimes is assumed to be relatively narrow, of the order of a few degrees. All the flows are taken to be purely radial at the inner boundary r_0. This flow pattern is undoubtedly much simpler than the true pattern of fast and slow solar-wind flows near the Sun but serves to demonstrate how CIRs arise.

Figures 3.15–3.17 illustrate the results of this model applied to the time period when *Ulysses* was observing CIRs in the southern hemisphere in 1992–1993 (Pizzo and Gosling, 1994). The tilt angle α was set at 30° to match the approximate tilt of the heliospheric current sheet and streamer belt at that time. Figure 3.15 shows latitude versus longitude maps of the resulting flow pattern at 1 AU (upper panel) and 5 AU (lower panel). These grey-scale maps range from white, representing fast solar wind flow, through to black, representing slow solar wind flow. The white line embedded in the slow flow indicates the location of the HCS. The nearly sinusoidal pattern in the band of slow flow at 1 AU simply arises from projecting the band, symmetric about the dipole equator, onto the latitude-longitude map. The 5 AU panel has been shifted 90° in longitude to allow for corotation. A spacecraft at a fixed latitude and heliocentric distance samples the flow patterns shown by moving across the figure horizontally from right to left, taking ~25.4 days (the solar rotation period) to cross the figure. A compressive interaction between the fast and slow flow will take place at longitudes where the spacecraft crosses a boundary from slow to fast wind, that is at ~90° longitude in the northern hemisphere and at ~270° longitude in the southern hemisphere, as indicated by the annotations on the

TILTED-DIPOLE FLOW GEOMETRY

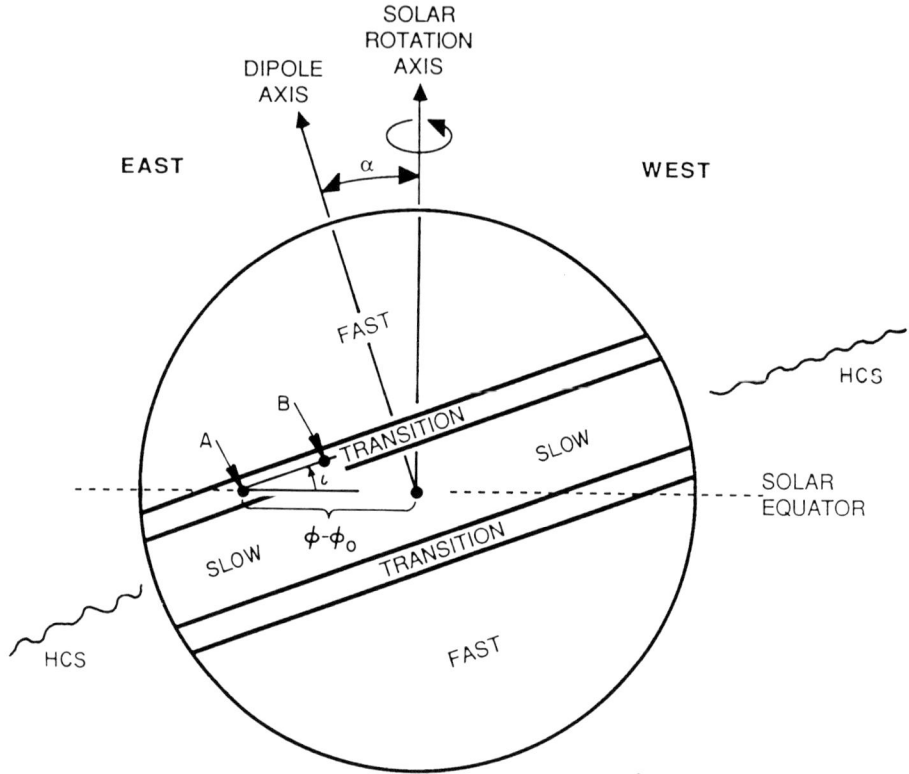

Figure 3.14. The tilted dipole flow geometry used at the inner boundary of the three-dimensional simulations of CIRs (from Pizzo, 1991).

figure. At longitudes where a transition from fast to slow wind occurs rarefactions are present.

At 1 AU the pattern of fast and slow wind has not changed very much from the input pattern near the Sun, although CIRs have begun to develop in both hemispheres. At the two longitudes referred to above, the fast/slow wind boundary is closer to the heliospheric current sheet than elsewhere as a result of the interaction that has taken place. By 5 AU (lower panel of Figure 3.15) forward-reverse shock pairs have developed bounding the interaction regions near 90° and 270° longitude. The locations of the shocks and their approximate directions of propagation have been annotated on the figure. The forward shocks are represented by the sharp transitions from black (slow flow) to grey on the right edges of the interaction regions, while the reverse shocks are represented by the sharp transitions from grey to white to the left edges of the interaction regions. Note that the HCS has been overtaken by the forward wave propagating through the slow wind ahead of

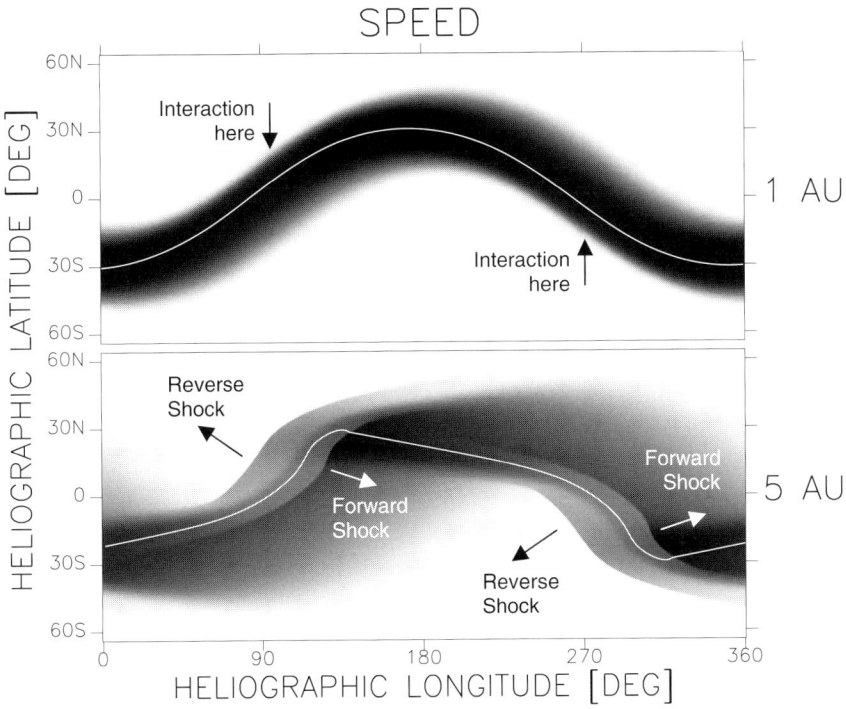

Figure 3.15. A latitude–longitude grey scale map of solar wind speed from the three dimensional CIR simulation of Pizzo and Gosling (1994) at 1 AU (upper panel) and 5 AU (lower panel). White represents the fastest wind and black the slowest wind.

the interaction region in agreement with observations. The SI (representing the boundary between what was originally slow and fast wind in the initial conditions at r_0) is just visible as a change from slightly darker to slightly lighter grey to the left of the HCS boundary within the interaction regions. It is clear from the figure that the forward shocks propagate westward and equatorward into the slow wind, while the reverse shocks propagate eastward and poleward into the fast wind. By comparison with the panel above it can be seen that the orientation of the shock fronts is related to the orientation of the boundaries between the fast and slow flow at 1 AU, which in turn arises from the tilted dipole geometry of the initial conditions. Figure 3.15 also shows that the southern hemisphere interaction region at ∼270° longitude has an inclination opposite to that of the northern hemisphere interaction region at ∼90° longitude. This agrees well with the opposite orientations of the CIRs which *Ulysses* encountered in the southern and northern hemispheres (Gosling et al., 1997). One final feature to note in Figure 3.15 is the gradual darkening of the shading from right to left across the lower panel at say 30°S after passing through the reverse shock near 270° longitude. This indicates a gradual reduction in flow speed which is a signature of the rarefaction region behind the CIR.

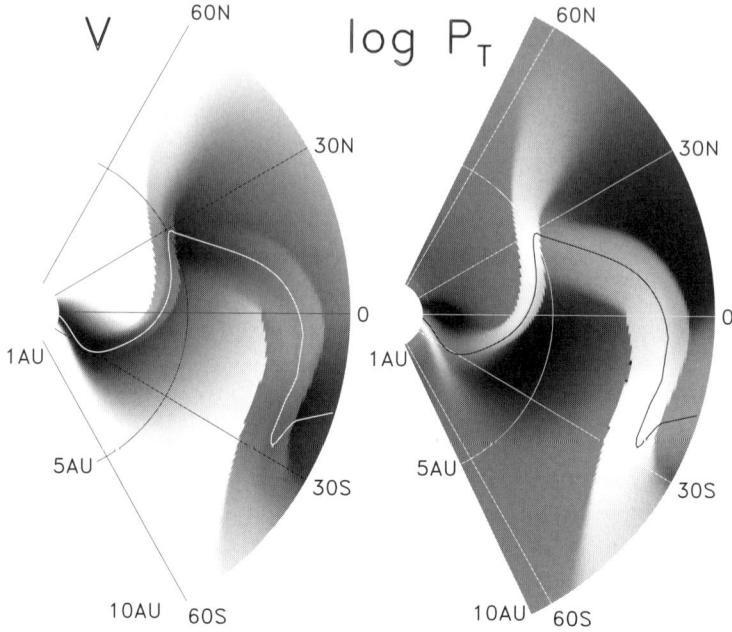

Figure 3.16. A meridional plane map of the solar wind speed (left) and pressure (right) from the CIR simulation of Pizzo and Gosling (1994).

Figure 3.16 shows a slice in a meridional (north–south) plane through the spiral structure of the CIRs, extending from 1 to 10 AU. The panel on the left shows the speed structure in the same grey scale as above, while the panel on the right shows the pressure structure with high pressure regions indicated in white. For this selected case, the northern hemisphere CIR is approaching 5 AU and thus is useful for comparison to the *Ulysses* observations, while the southern hemisphere CIR is approaching 10 AU. The southern hemisphere CIR has expanded more than the northern one, having evolved more. The figure also illustrates how the interactions between the fast and slow wind distort the HCS away from the simple ballerina skirt picture illustrated, for example, by Thomas and Smith (1981). At about 4 AU, between about 0° and 30°N latitude, the northern hemisphere CIR has pushed the HCS further out from where it would be located if the flow speed had been uniform and has caused the HCS to become sharply kinked at its maximum latitude. In addition, the maximum latitude of the HCS has been reduced slightly.

Figure 3.16 also shows how the high pressure region associated with the southern hemisphere CIR, with the HCS embedded within it, extends northward and sunward from the equator at about 8 AU to intersect the northern hemisphere CIR near 5 AU and 30°N. Pizzo and Gosling (1994) point out that this extension into the northern hemisphere is a consequence of the initial conditions. In a radial

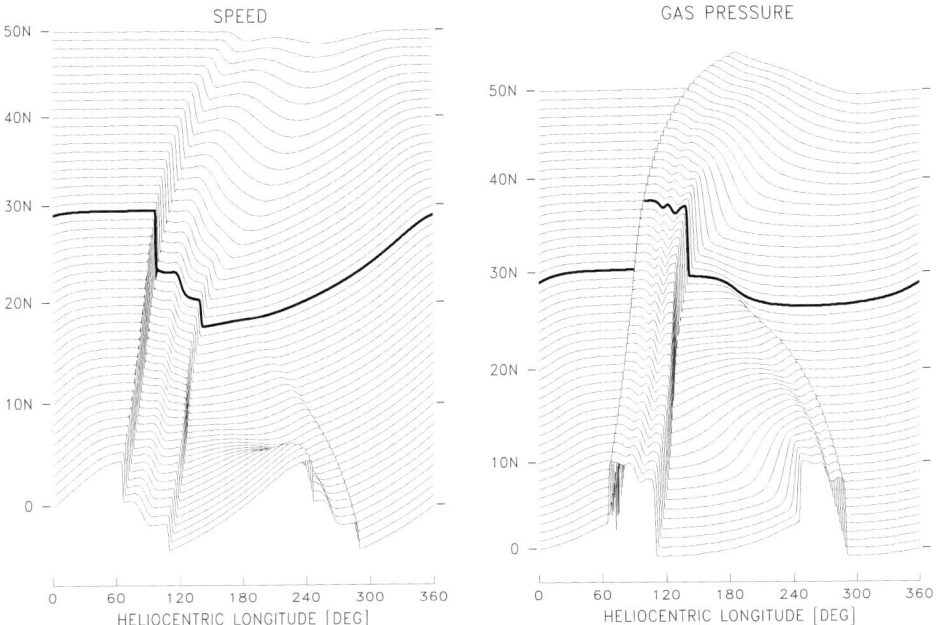

Figure 3.17. Stack plots of the latitude and longitude variation of speed (left) and pressure (right) from the CIR simulation of Pizzo and Gosling (1994).

cut through this structure at, for example, about 25°N, a spacecraft would observe successively the relatively weak northern extension of the southern CIR, a crossing of the HCS into southern hemisphere magnetic-field polarity, the strong forward shock of the northern hemisphere CIR, a return crossing of the HCS into the northern hemisphere polarity, and, finally, the reverse shock. Apart from interchanging the hemispheres, this scenario bears a close similarity to what *Ulysses* observed in 1993 while in the southern hemisphere, as already illustrated in Figure 3.8. *Ulysses* entered the high field strength region labelled c early on 17 January, identified as associated with the northern hemisphere by the crossing of the HCS which follows close behind. The forward shock leading the southern hemisphere CIR labelled d, and then the HCS again, are both encountered early on 20 January. At this time *Ulysses* was located at a latitude of about 24°S.

Finally, to illustrate the relationship between the variation of CIR structure with latitude observed by *Ulysses* and that predicted by the model, we show in Figure 3.17 two stack plots from the simulation at 5 AU. The first shows the solar wind speed as a function of longitude, and the second the gas pressure as a function of longitude, both stacked at 1° intervals of latitude between 0° and 50°N. The darker trace is at a latitude of 29°N, the maximum latitude of the HCS at 5 AU. Again, a spacecraft would sample these structures moving from right to left across the figure at the appropriate latitude.

At the equator, a symmetrical two stream pattern and corresponding interaction regions with forward-reverse shock pairs are seen. The southern hemisphere CIR on the right rapidly weakens with latitude, but the northern hemisphere CIR on the left continues to have the shock pair structure up to the maximum HCS latitude in agreement with the *Ulysses* observations. Above this latitude the forward shock rapidly weakens into a more extended forward wave but the reverse shock persists for at least another $10°$ in latitude. By $50°$N only weak forward and reverse waves remain as the structures gradually merge into the uniform high-speed solar wind flow. The forward shock weakens and disappears above the HCS maximum latitude partly because the angle of attack of the fast wind against the slow wind of the streamer belt becomes much more oblique there as can be seen in Figure 3.15. In addition, the northward extension of the southern hemisphere forward wave eventually intersects the forward shock of the northern hemisphere CIR because of their opposite orientations; the resulting interaction contributes to the weakening of the forward shock.

In summary, this type of model reveals how the large-scale three-dimensional structure of CIRs is intimately related to the geometry of fast and slow solar-wind sources in the corona and hence to the structure of the coronal magnetic field. The model succeeds remarkably well in reproducing many aspects of the CIRs observed by *Ulysses*.

3.3 TRANSIENT FLOWS

3.3.1 The origin of CMEs and their signatures in the heliosphere

Solar origin of CMEs

The existence within the solar-wind of transient flow disturbances driving shock waves was first proved observationally (e.g. Gosling *et al.*, 1968) in early 1 AU spacecraft data from the late 1960s although their existence had been postulated much earlier. A brief historical summary of some of the early ideas on transient solar wind disturbances is given by Gosling (1997). It was recognized that transient shock disturbances were driven by plasma of markedly different character from the normal solar wind (e.g. Hirshberg *et al.*, 1972) while the observation of counter-streaming suprathermal electrons within some of the disturbances suggested that this plasma often has a closed magnetic field geometry (Montgomery *et al.*, 1974). In this chapter we will use the acronym CME specifically to describe the ejection of plasma from the solar corona, and the acronym ICME to describe the ejected material observed *in situ* far from the Sun.

The specific association of transient disturbances in the heliosphere with the solar eruptions now known as CMEs came from the first space-borne coronagraphs, especially that of *Skylab* (MacQueen *et al.*, 1974). More recent coronagraph observations of CMEs include those from the Solar Maximum Mission (SMM) (e.g. Hundhausen, 1993) and from *SOHO* (e.g. Howard *et al.*, 1997). Figure 3.18 shows an

Figure 3.18. An example of a coronal mass ejection observed by the LASCO coronagraph on SOHO on 5th October 1996. (The SOHO/LASCO data are produced by a consortium of the Naval Research Laboratory (USA), Max-Planck-Institut fuer Aeronomie (Germany), Laboratoire d'Astronomie (France), and the University of Birmingham (UK). SOHO is a project of international co-operation between ESA and NASA.)

example of a CME observed by the Large Angle Spectroscopic Coronagraph (LASCO) on *SOHO* in October 1996, an event which eventually propagated out to the position of *Ulysses*, as will be discussed in Section 3.3.2. As in the example in the figure, CMEs often show up in the coronagraph observations as vast 'plasma bubbles' erupting from the solar atmosphere. They usually originate from closed magnetic field regions in the corona. Indeed, the images suggest that the mass ejections themselves contain closed magnetic loops that connect back to the Sun. Thus CMEs can be thought of as eruptions of solar material from regions of closed magnetic field in the corona where the magnetic forces had previously been strong enough to prevent the plasma from expanding outwards.

The details of the initiation of CMEs are not well understood. CMEs usually occur near magnetic field neutral lines in the solar atmosphere, often associated with prominence or filament structures, beneath closed magnetic field structures in the corona (e.g. Hundhausen, 1998). It is generally thought that an instability causes them to erupt. At solar minimum neutral lines are more common in the solar equatorial regions; thus at solar minimum mass ejections usually only occur at low latitudes (Hundhausen, 1993). At solar maximum neutral lines are, however, common at all latitudes and hence also are CMEs. The rate of occurrence of CMEs varies with the solar cycle. One study estimated the rate from the whole sun as ~ 0.2 per day at solar minimum, and ~ 3.5 per day at solar maximum

(Webb and Howard, 1994). In coronagraph images, CMEs are often observed to have a three part structure (Hundhausen, 1988), consisting of a bright curved leading part, followed by a dark cavity and an embedded prominence, mirroring the structure of the coronal streamers from which such CMEs originate. From coronagraph observations, the mass ejected in a single CME ranges from 10^{15} to 10^{16} g (e.g. Hundhausen, 1993) while ejection speeds in the field of view range from less than 100 km s^{-1} to greater than 1500 km s^{-1} (Gosling et al., 1976b; Howard et al., 1985; Hundhausen et al., 1994). Soft X-ray observations, e.g., from the *Yohkoh* spacecraft, often show new coronal loops forming in the aftermath of CMEs (Webb et al., 1976; Kahler, 1977; Sheeley et al., 1983; Hiei et al., 1993; Hudson and Webb, 1997). In addition, solar flares (intense brightenings at optical and X-ray wavelengths) often occur in association with CMEs (Gosling et al., 1974; Munro et al., 1979), but there is substantial evidence that solar flares are not the cause of CMEs (Gosling, 1993).

Interplanetary propagation and evolution of CMEs

It has not yet been possible to follow a CME all the way from the corona out into interplanetary space. The closest *in-situ* observations have been those of *Helios* at 0.3 AU. Thus the details of how features observed in the coronagraph images evolve into the signatures found in the solar wind are not entirely understood. The speed of the ICME compared with that of the solar wind leading and trailing the ejection is, however, an important factor in determining the evolution. Near the ecliptic, where most of the in-situ observations prior to *Ulysses* have been made, ICMEs can have speeds which greatly exceed that of the ambient low-speed solar wind and thus can produce compression regions and drive shock waves ahead of them (Sheeley et al., 1985; Gosling et al, 1991). These are referred to as transient shocks, as opposed to the corotating shocks discussed in the first part of this chapter, although the basic physics leading to their formation is essentially the same.

The slower ICMEs observed in the solar wind near 1 AU have speeds comparable to that of the ambient solar wind and thus do not drive shock waves. ICMEs often expand as they propagate out into the solar wind (Burlaga et al., 1981; Gosling et al., 1987) and thus often produce a gradual decline in speed as they pass a spacecraft. If the solar wind flow behind an ICME is faster than that of the ICME itself, then an interaction occurs on the trailing edge of the ICME in the same way as between the fast and ordinary slow solar wind.

Identification of CMEs in interplanetary space

There are a number of plasma and magnetic field signatures that qualify as unusual compared to the normal solar wind but which are commonly observed within ICMEs (Gosling, 1990; 1993; 1996b; Neugebauer and Goldstein, 1997). Very few of these, however, occur consistently for every ICME, and some can arise from other sources. Additionally, the start and stop times of the different signatures often do not coincide. Thus one commonly relies on the presence of several of these signatures for the identification of an ICME.

To illustrate some of the typical signatures at low latitudes, we present in Figure 3.19 an example of the magnetic field and plasma parameters of an ICME observed at *Ulysses* between March 15 and March 18 of 1991. At this time *Ulysses* was still in its in-ecliptic cruise to Jupiter and was at a heliocentric distance of 2.4 AU and a latitude of $3.5°$S. From top to bottom the parameters plotted are the magnetic-field strength, azimuthal angle and meridional angles of the magnetic-field, the solar-wind speed, proton density, alpha particle to proton number density ratio, and proton temperature.

Two dashed vertical lines drawn across all the panels of the plot indicate the start and end times of the ICME plasma as identified (Phillips, 1997) by bi-directional (or counterstreaming) suprathermal electrons (BDE). Since the suprathermal electron population in the solar wind has its origin in the hot solar corona and is constrained to flow along the interplanetary magnetic field (Feldman *et al.*, 1975), in the normal solar wind a unidirectional suprathermal electron heat flux propagates outward from the Sun along the magnetic field. This is a consequence of the fact that the field lines in the normal solar wind are effectively connected to the corona only at one end. Within ICMEs, however, suprathermal electrons often beam in both directions along the magnetic-field lines, suggesting that the field lines within ICMEs commonly are connected back to the Sun at both ends. The topology could be either a simple closed loop or a more complex helical flux rope-type structure that is still magnetically connected to the Sun at both ends. These two possible field configurations are illustrated schematically in the upper and lower panels of Figure 3.20.

The event shown in Figure 3.19 is an example where a fast ICME produced a forward shock wave. The shock wave is evident as the instantaneous increase in field strength, velocity, and temperature some 17 hours ahead of the BDE signature. The $180°$ change in the azimuthal angle just downstream of the shock suggests that the heliospheric current sheet has been swept up by the shock driven ahead of the ICME. The magnetic field is elevated between the shock and the ICME as a result of the interaction between the ICME and the slower ambient solar wind ahead. The interaction commonly also increases the density in this region, although in Figure 3.19 this increased density is masked by a greater increase associated with the heliospheric current sheet crossing. The end of this ICME event, as defined by the BDE signature, is marked by another forward shock wave propagating into the back of the ICME structure.

Since corotating shock waves generally do not form until well beyond 1 AU, most shock waves observed inside 1 AU are driven by fast ICMEs. Since *Ulysses* spends much of its time near 5 AU, however, shock waves alone do not serve as a good ICME identifier in the *Ulysses* data. Most of the ICMEs in the *Ulysses* data were originally identified either from a bi-directional suprathermal electron signature or, less often, from a magnetic cloud signature, which we describe next.

In magnetic field data a common signature of an ICME is a characteristic coherent rotation of the magnetic-field direction consistent with passage through a magnetic flux rope-like structure (e.g. Burlaga *et al.*, 1981; Gosling, 1990). An example can be seen in the two magnetic-field angles in Figure 3.19, with the

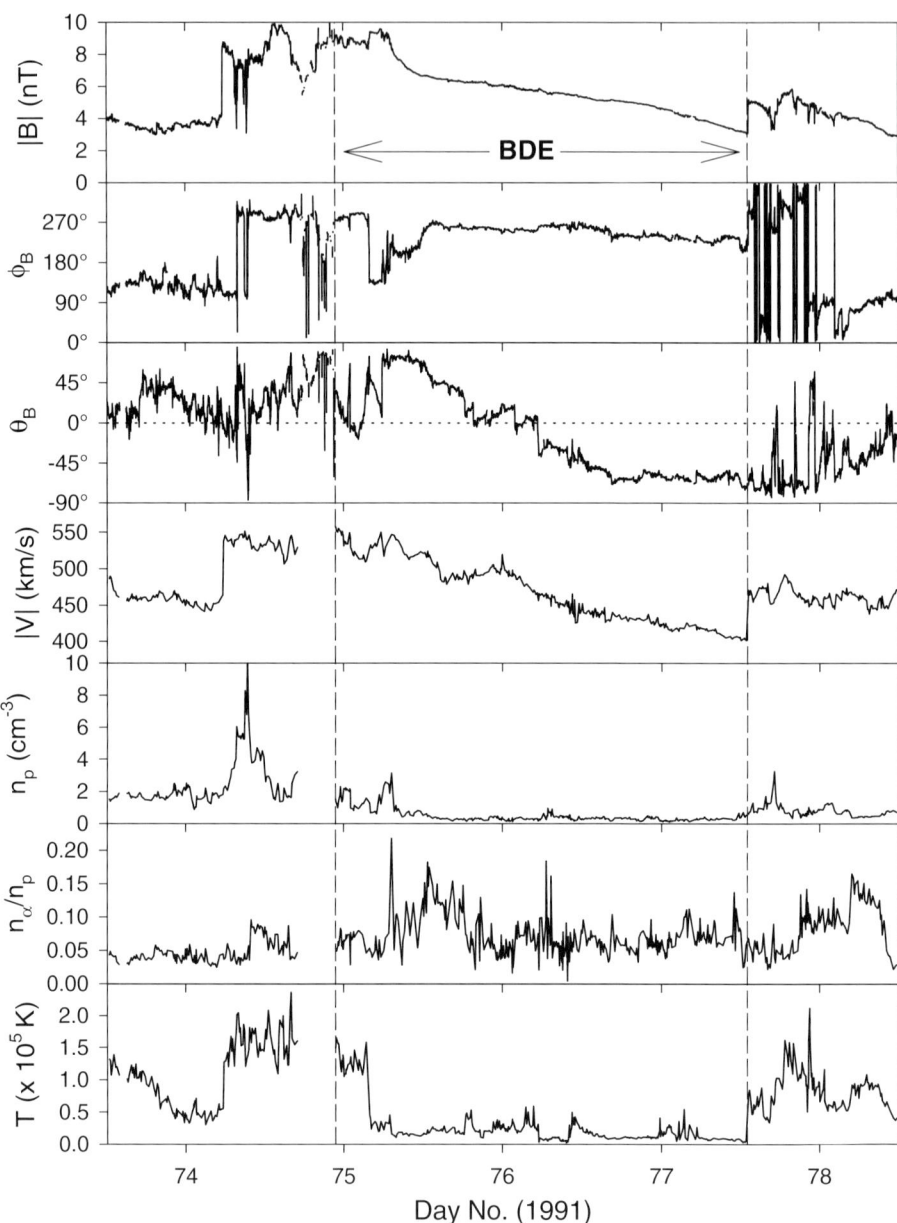

Figure 3.19. Plasma and magnetic field parameters for an ICME observed by *Ulysses* during March 15–18, 1991. From the top, the parameters plotted are the magnetic field strength, the azimuthal and meridional angles of the magnetic field, the solar wind speed, proton density, proton to alpha ratio and proton temperature. The vertical dashed lines show the boundaries of the ICME plasma as identified (Phillips, 1997) by the bi-directional electron (BDE) signature.

Sec. 3.3] Transient flows 139

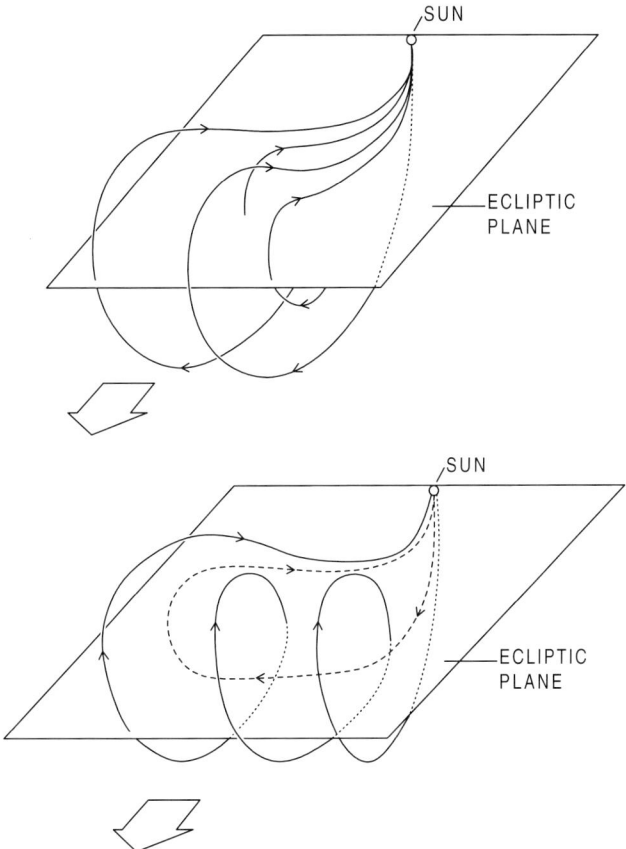

Figure 3.20. Sketch showing two possible configurations of magnetic field lines closed back to the Sun within an ICME. The top panel shows simple loops connecting back to the Sun, whereas the lower panel shows a flux rope magnetic field configuration. (From Gosling, 1990).

rotation in this case being mostly in the meridional angle. In this example the rotation begins some time after the start of the BDE signature. These flux rope-like structures consist of helical field lines wound around a central axis, as illustrated in the lower panel of the sketch in Figure 3.20. Such field rotations, however, are observed in only about a third of all ICMEs (near the Earth) (Gosling, 1990). A subset of those ICMEs with flux rope signatures also have strong magnetic field and low plasma beta and are referred to as magnetic clouds (e.g. Burlaga, 1991). In magnetic clouds, because of the low beta, the magnetic field has relaxed to a nearly force-free configuration. Thus magnetic clouds often have low variance magnetic fields (Burlaga, 1991). Both high magnetic-field strength (e.g. Burlaga and King, 1979) and low-variance magnetic fields commonly occur for non-cloud

ICMEs as well. In the example of Figure 3.19 the magnetic field strength does remain high but gradually falls as the event proceeds. Given the low proton density and temperature and the relatively high magnetic field strength within the flux rope portion of the ICME, this region is clearly filled with a low beta plasma and matches the criteria for identification as a magnetic cloud. The low variability of the field strength is particularly obvious although it does not begin until a few hours after the BDE signature. The field direction is also less variable inside the ICME than it is outside.

The four lower panels of Figure 3.19 show plasma parameters during this ICME. The declining speed profile within the event indicates that the ICME was expanding and producing a rarefaction as it propagated out through the heliosphere. As already described, the proton density is relatively high ahead of the ICME, while within the ICME itself, coinciding with the flux rope signature, the density is notably low. This low pressure is probably an effect of the expansion of the ICME. The alpha to proton abundance ratio is elevated within the ICME, particularly in comparison to that upstream of the shock. This is typical of ICMEs (Hirshberg et al., 1972; Borrini et al., 1982), although the reasons are not well understood, and unusual abundances and charge states of heavy ions are also common (Galvin, 1997). The proton temperature is elevated ahead of the ICME due to heating by the shock, and is markedly low within the flux rope portion of the ICME. Such low proton and also electron temperatures are often observed (e.g. Gosling et al., 1973; Montgomery et al., 1974) and are believed to be the result of the fact that ICMEs commonly expand. In Figure 3.19 many of the plasma parameters have a closer correspondence in time to the flux rope signature in the magnetic field than to the times of the BDE signature. As already stated, this mismatching in time between the various signatures is not uncommon within ICMEs. The presence of the BDE signature ahead of the magnetic cloud nonetheless indicates that the spacecraft must also have been sampling closed magnetic field lines in this region.

It is uncertain whether the flux ropes observed in interplanetary space pre-exist in the corona or are formed by reconnection as part of the aftermath of the eruption process or both. Some authors (e.g. Burlaga, 1991; Rust, 1994) suggest that erupting prominences and the surrounding cavity (Low, 1997) are pre-existing flux ropes that are carried out into the solar wind as CME ejecta. A correlation exists between the preferred sense of helicity (left handed in the northern, and right handed in the southern hemisphere) of the flux ropes inferred for prominences and the sense of helicity of the corresponding magnetic clouds (Bothmer and Schwenn, 1998; Rust, 1994). It seems unlikely that the region containing the bright prominence material itself is the direct counterpart of the interplanetary flux rope since this region commonly forms only a small part of the erupting CME as seen in coronagraph images (e.g. Gosling, 1990). A number of authors (e.g. Low, 1997) have, on the other hand, interpreted the low density cavity seen ahead of the prominence material in an erupting CME as a region containing a flux rope magnetic field structure, and argue that this is the region which propagates out to become the flux rope or magnetic cloud observed by in-situ spacecraft. Recently there have been a small number of observations at 1 AU of magnetic clouds containing a short plug of cold, high

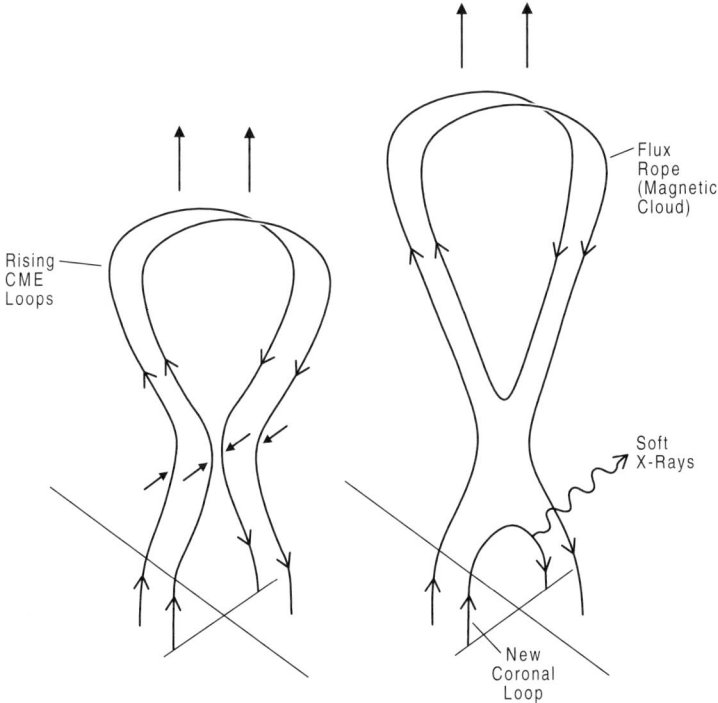

Figure 3.21. Sketch illustrating how reconnection between the legs of neighbouring magnetic field loops which are sheared relative to one another leads to the formation of a flux rope within a CME. (From Gosling, 1993).

density material at the rear of the cloud (Burlaga *et al.*, 1998, Gopalswamy *et al.*, 1998). These high density regions have a number of characteristics consistent with originally being prominence material.

An alternative model (Gosling, 1990; Gosling *et al.*, 1995d; Gosling, 1999) is based on the observation that new magnetic loops commonly form in the corona in the regions where CMEs have recently erupted (e.g. Sheeley *et al.*, 1983). This suggests that magnetic reconnection is occurring beneath the departing CMEs (e.g. Kopp and Pneuman, 1976). In a two dimensional picture this process would produce a detached loop of magnetic field in the solar wind and a new magnetic field loop in the corona, attached to the Sun at both ends. Gosling (1990) pointed out that in reality it is unlikely that reconnection would occur between the two legs of the same field line as any shear would mean that these legs would be well separated in space. Thus where reconnection does occur, it is likely to be between the legs of two different field loops as illustrated in Figure 3.21. This process forms a helical structure which is still connected to the Sun at both ends, as well as a new coronal loop beneath the CME. If this process were to occur multiple times in coronal loops which were expanding outwards in a CME then a flux rope structure in the solar wind would be the result. Note that this explanation of how the flux rope forms is not inconsistent with the association of the central dark cavities seen in coronagraph images of CMEs with the flux ropes observed in interplanetary space. The eventual formation of a field line structure completely disconnected from the Sun is also possible. Some evidence for the disconnection of magnetic field loops has been found in dropouts of the suprathermal electron heat flux within ICMEs observed by *ISEE-3* (Gosling, 1999) and *WIND* (Larson *et al.*, 1997). The topic of reconnection will be explored further in section 3.3.2 in the context of *Ulysses* observations.

Fate of CMEs in the outer heliosphere

The fate of ICMEs as they continue to propagate outward into the distant heliosphere (beyond 5 AU, out of the region on which this chapter is focused) depends on their speed relative to the ambient solar wind. Some ICMEs may merge with CIRs to form localized merged interaction regions (e.g. Burlaga *et al.*, 1986). At solar maximum when there is a relatively high number of fast ICMEs at all latitudes, the pressure waves driven by many of these may merge in the outer heliosphere to form a so-called global merged interaction region (e.g. Burlaga *et al.*, 1993).

3.3.2 Summary of *Ulysses* ICME observations

The *Ulysses* exploration of the heliosphere began in October 1990, not long after the maximum of solar cycle 22. Thus many transient flows associated with CMEs occurred in the early months of the mission. In particular, a large number of ICMEs were encountered during March 1991, culminating in a sequence of events which included a very fast (1,000 km s^{-1}) ICME (Phillips *et al.*, 1992). Of 25 ICMEs documented during the 15 month in-ecliptic phase of the mission (Phillips, 1997), 6 occurred during March 1991. The example shown in Figure 3.19 is the second of these.

Once *Ulysses* had flown past Jupiter in February 1992 and had begun travelling southward in its polar orbit, the first few months were dominated by slow solar wind including approximately nine CME-related events (Gonzalez-Esparza *et al.*, 1998). The strong recurrent high-speed stream became established in July 1992 after which time the frequency of encounters with CME ejecta became much lower (Phillips *et al.*, 1995; Gonzalez-Esparza *et al.*, 1996). Yet, one ICME event in November 1992 produced solar wind speeds reaching 1,000 km s^{-1}. Once *Ulysses* had passed southwards into latitudes where the slowest solar wind was no longer observed, only six definite ICMEs were observed (Gosling *et al.*, 1994a). A further three events showed counterstreaming suprathermal electrons without any corroborating signatures to eliminate connection with corotating reverse shocks. The highest latitude ICME event in this first southerly phase of the mission was at 60.5°S (Phillips *et al.*, 1995). A number of these high-latitude events had unusual characteristics which will be discussed more fully in Section 3.3.3. The decline in the number of ICMEs encountered as *Ulysses* travelled southwards is due both to subsiding solar activity and to the fact that *Ulysses* was climbing to higher latitudes away from the streamer belt from which the majority of CMEs originate.

When *Ulysses* next crossed the streamer belt region, during the fast latitude scan, only one definite ICME was encountered, exactly as the spacecraft made its main entry into the band of slow solar wind in February 1995 (Gosling *et al.*, 1995a). No ICMEs were then detected by *Ulysses* for the whole duration of the north polar pass until October 1996, just after *Ulysses* had again begun to encounter the band of low-speed solar wind at ~4.5 AU (Gosling *et al.*, 1997). During the subsequent four months three definite ICME signatures were observed (Gosling *et al*, 1997; Forsyth *et al.*, 1997; Funsten *et al.*, 1999), all originating from the same band of active longitudes.

Identification of the solar origin of ICMEs observed by **Ulysses**

Given the time, location, and speed of an ICME observed by any interplanetary spacecraft, it is possible to estimate the travel time of the event from the Sun by assuming constant travel speed. Hence, the approximate time that the mass ejection left the Sun can be determined and, also, the approximate Carrington longitude and latitude of the ejection source. It is then possible to search through ground or space-based solar observations for evidence of the occurrence or aftermath of a CME which corresponds to the event detected by the spacecraft. An immediate limitation for *Ulysses* is that both the space-based solar observatories that have been operating in parallel with *Ulysses*, *Yohkoh* in Earth orbit, and *SOHO* at the L1 point, always view the Sun from the vicinity of Earth, as of course do the ground-based observatories. For a substantial part of each year *Ulysses* is on the far side of the Sun as viewed from the Earth and hence the CME regions of origin cannot be observed. A further limitation is that ICME speeds need not be constant. The faster CMEs decelerate as they interact with slower ambient wind ahead and behind, and the slower CMEs accelerate as they interact with faster wind.

Despite these limitations, two studies (Weiss *et al.*, 1996; Funsten *et al.*, 1999) have made comparisons between ICMEs seen by *Ulysses* and their corresponding signatures in solar images. The first of these (Weiss *et al.*, 1996) surveyed observations from the *Yohkoh* Soft X-ray Telescope (SXT) in association with *Ulysses* ICMEs observed between November 1991 and June 1994. For the reason discussed above, only ICMEs observed by *Ulysses* between November and June of each year could be included. The study was also limited to those ICMEs that could be mapped back with confidence using the constant speed assumption. CMEs themselves are not usually observed directly in the *Yohkoh* images; rather the soft X-ray images show evidence of the restructuring of the solar corona associated with the departure of the CMEs. For five *Ulysses* ICMEs it was possible to identify soft X-ray events sufficiently isolated in time and location in the SXT images to reduce the possibility of an incorrect association.

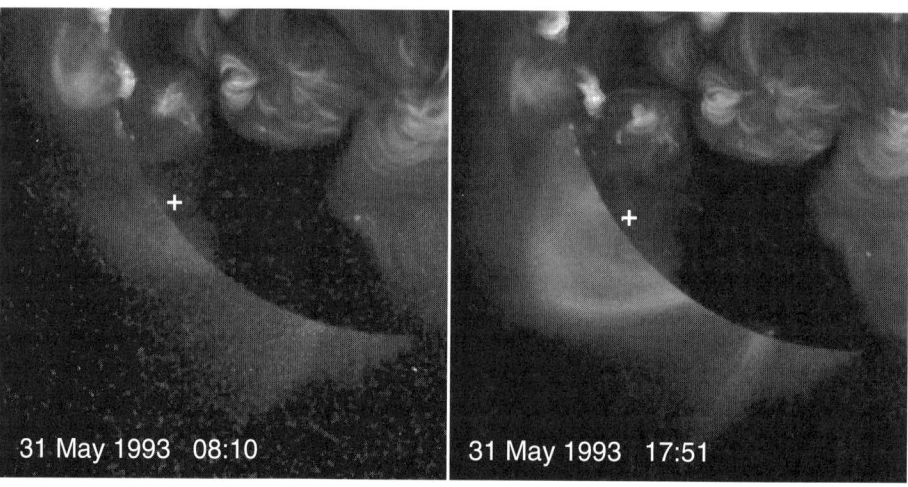

Figure 3.22. X-ray images from the *Yohkoh* spacecraft showing the coronal loops that formed behind the June 1993 CME observed by Ulysses (Weiss *et al.*, 1996).

For three events the associated soft X-ray signature was the appearance of a newly formed arcade of loops connected to the solar surface at both ends. This is consistent with what we would expect if reconnection takes place within the legs of CMEs that have erupted (Hiei *et al.*, 1993; Hundhausen, 1997). Figure 3.22 from Weiss *et al.* (1996) shows one example of the formation of new coronal loops which correlated with a *Ulysses* encounter with an ICME (Gosling *et al.*, 1994b). At *Ulysses* the ICME was observed on 9–13 June 1993 when *Ulysses* was at 32.5°S and at 4.6 AU. Mapping the event back at the observed central speed of 750 km s^{-1} yielded a launch time that fell between the times of the two *Yohkoh* images shown. Note the formation of the large coronal loops at about 30°S on the east limb of the Sun, which developed between 0810 and 1751 UT on 31 May 1993. At this time *Ulysses* was located just behind the east limb. The other two events were associated with an active region flare, an event which may take place following the eruption of a CME, although not necessarily directly below it (e.g. Hundhausen, 1997). All five events included long-duration X-ray events (e.g. Hundhausen, 1997), suggesting that they were all associated with the formation of new loops by reconnection of the magnetic-field lines behind departing CMEs.

Three of the ICMEs studied contained magnetic flux rope signatures and were low beta structures. A different three showed a signature of overexpansion (Section 3.3.3). No one-to-one correspondence was found between any of these signatures and the two types of soft X-ray signature, thus suggesting that the interplanetary characteristics of CMEs cannot be predicted well on the basis of soft X-ray observations of their post ejection features in the corona.

The second study (Funsten *et al.*, 1999) of the solar origin of ICMEs focused on the three ICMEs encountered between October 1996 and January 1997 as *Ulysses* returned from high northern latitudes and was situated roughly above the west limb of the Sun as viewed from Earth and *SOHO*. In this case the *Ulysses* events were compared to white light images from the LASCO on *SOHO*. The relatively low frequency of CME occurrence at this time made identification of associated events comparatively easy. The first two ICMEs were clearly associated with single CMEs on the west limb of the Sun, while the third was related to a series of ejections producing a complex interplanetary signature at *Ulysses*, apparently due to the interaction of the multiple ejecta. Successive *SOHO* images were used to infer the speed at which the CMEs left the Sun. Knowledge of this speed allows deductions to be made about how the ICMEs evolved while travelling from the Sun out to *Ulysses*.

One event studied arrived at *Ulysses* on 14 October, 1996 (Funsten *et al.*, 1999). This was associated with a CME observed by *SOHO* leaving the Sun on October 5. Figure 3.18 shows a LASCO image of the event. The speed of the CME in the plane of the sky at the *Ulysses* latitude was $650 \, \text{km s}^{-1}$. The leading edge of the ICME as it passed over *Ulysses*, however, was travelling at $850 \, \text{km s}^{-1}$ despite evidence from the travel time that it must have decelerated in transit. Assuming a constant deceleration with distance from the Sun gave an estimated ejection speed at the Sun of $935 \, \text{km s}^{-1}$, thus allowing the longitude of the ejection to be calculated as $49°$ behind the west limb. The *Ulysses* data show that this ICME was expanding as it passed over the spacecraft. Thus the leading edge of the ICME was decelerated as it expanded into slower solar wind ahead of it.

Magnetic field topology of ICMEs

It has been suggested (Gosling, 1990; Gosling *et al.*, 1995d) that the magnetic field rotations observed by spacecraft in interplanetary space may be signatures of flux ropes formed by three-dimensional reconnection occurring within the magnetic fields beneath erupting CMEs. The June 1993 ICME observed at *Ulysses* was a flux rope ICME and the *Yohkoh* observations (Figure 3.22) showed that new magnetic loops formed in the corona after the CME departed from the Sun (Weiss *et al.*, 1996). This lends support to the argument that reconnection did occur behind that particular CME. The ICME took over 3 days to pass *Ulysses* (Gosling *et al.*, 1994b). Counter-streaming suprathermal electrons, however, disappeared during a 10-hour period in the second half of this ICME (Gosling *et al.*, 1995d), suggesting that *Ulysses* sampled open magnetic-field lines within the body of the ICME flux rope. A mixture of open and closed field lines within an ICME can be produced by three-dimensional reconnection, as simulations of the reconnection process in the Earth's magnetic tail reveal.

The successive stages of this process are illustrated in the left hand panel of Figure 3.23. The first stage (a) of the process forms a flux rope within the CME which is connected to the Sun at both ends, as already discussed in Section 3.3.1. In (b) a further reconnection takes place between the flux rope and an open field line of the normal solar wind to produce a flux rope which is connected to the Sun at one end and is open to the outer heliosphere at the other end. This portion of the flux

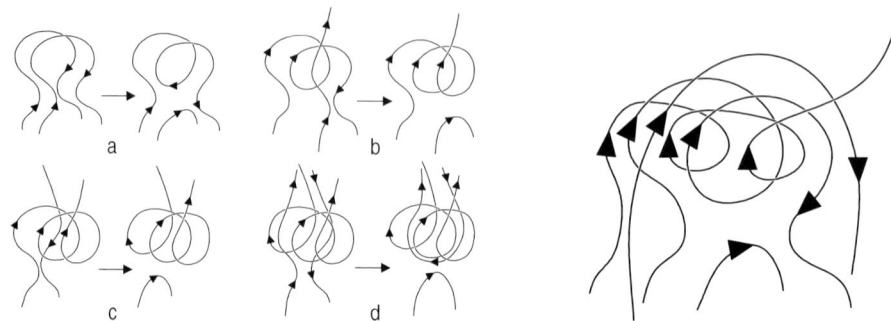

Figure 3.23. A diagram illustrating the successive stages of 3D reconnection, eventually leading to disconnection (from Gosling *et al.*, 1995d).

rope would be consistent with the interval in the June 1993 ICME which showed only unidirectional suprathermal electrons. Panel (c) shows a further reconnection with another open field line, producing a flux rope now completely disconnected from the corona, while in panel (d) two open solar wind field lines reconnect behind the CME to produce a single disconnected field loop. Both types of disconnected field line (c and d) are connected to the outer heliosphere at both ends and would produce electron heat flux dropouts. The schematic in the right hand panel shows a mixture of field line topologies within an ICME resulting from these kinds of reconnection that is qualitatively consistent with the June 1993 *Ulysses* example.

Evidence of open magnetic field lines embedded within ICMEs has also come from *Ulysses* energetic particle data both from an ICME observed by *Ulysses* in February 1995 (Bothmer *et al.*, 1996) and the same June 1993 event described above (Armstrong *et al.*, 1994). In the latter case, however, the interval in which it is reported does not match that of the unidirectional suprathermal electrons (Gosling *et al.*, 1995d).

A separate observation related to ICME topology is that the minor ion composition, based on analysis of *Ulysses* data of the O^{+7}/O^{+6} ratio, appears to differ from that of the ambient solar wind only for those ICMEs which exhibit magnetic cloud properties (Henke *et al.*, 1998). The origin of this effect is not understood.

Relationship of ICMEs with the heliospheric current sheet

CMEs often originate from within helmet streamers in the solar corona (e.g. Hundhausen, 1988). These helmet streamers are also the regions above which solar magnetic field lines of opposite polarity come together to form the HCS (Chapter 2). A study of fourteen magnetic clouds observed by the *ISEE-3* spacecraft revealed that eight were encountered at sector boundaries (Crooker *et al.*, 1998). A number of the magnetic clouds observed by *Ulysses* similarly were encountered at the heliospheric current sheet. Figure 3.24 shows magnetic field data from an ICME

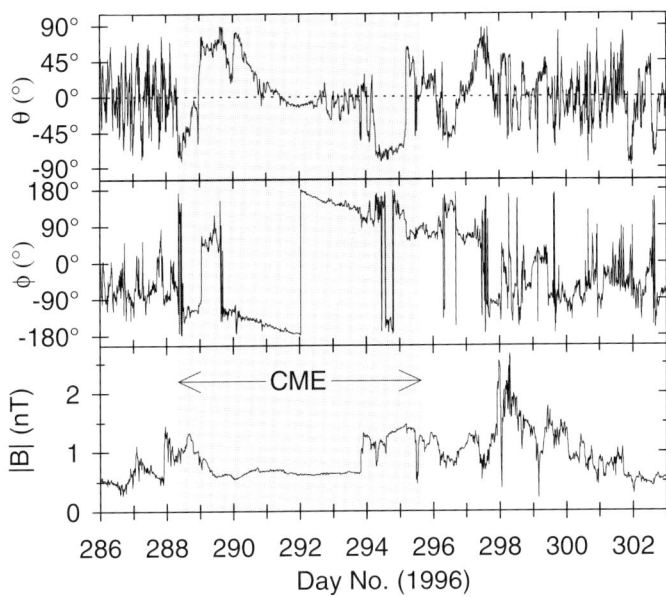

Figure 3.24. *Ulysses* magnetic field data from days 286 to 302 (12 to 28 October) of 1996. The figure shows a magnetic cloud separating opposite polarity magnetic field lines. (From Forsyth et al., 1997).

containing a magnetic cloud observed by *Ulysses* at 4.4 AU and 24°N in October 1996. The three panels show the meridional and azimuthal angles describing the magnetic field direction and the magnetic field magnitude. The azimuthal angle was close to −90° before the ICME arrived, consistent with a spiral field line of northern hemisphere polarity. Once the ICME had passed over *Ulysses*, the azimuth angle was close to +90°, consistent with a spiral field line of the opposite southern hemisphere polarity. Thus the magnetic cloud carried the field reversal associated with the sector boundary separating the opposite polarity magnetic fields. The schematic diagram in Figure 3.25 illustrates the solar origin of a flux rope in three-dimensional reconnection and the interplanetary relationship between the ICME and the heliospheric current sheet. Flux ropes formed by three-dimensional reconnection thus form an occlusion in the heliospheric current sheet. Rather than draping around or pushing apart the heliospheric current sheet, the ICME locally replaces it (Crooker and Intriligator, 1996; Crooker et al., 1998).

A further *Ulysses* result relating ICMEs and the heliospheric current sheet is the observation (McComas et al., 1994) of reconnection across a current sheet driven by the compression wave propagating ahead of a fast ICME.

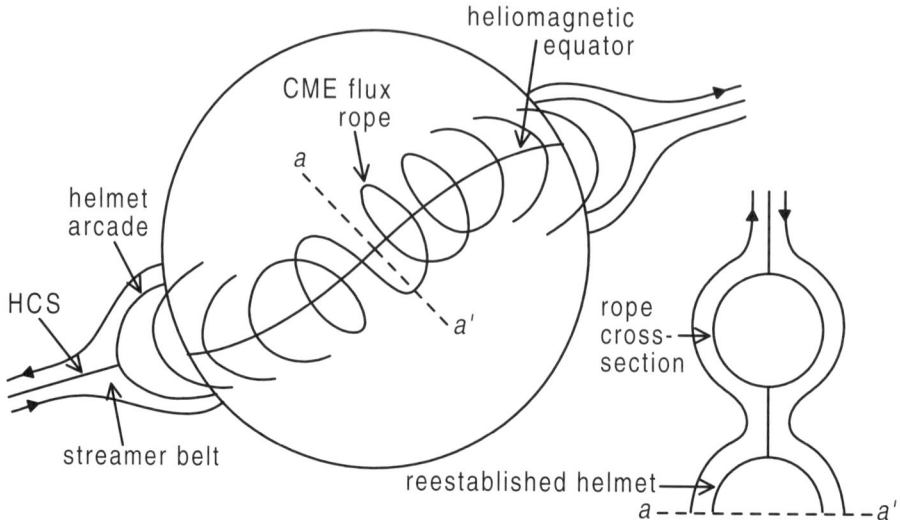

Figure 3.25. A schematic diagram, reproduced from (Crooker et al., 1998), illustrating how a magnetic flux rope originating from 3-D reconnection in a helmet streamer propagates outwards as an occlusion in the heliospheric current sheet.

3.3.3 ICMEs observed at high latitudes

Speeds of ICMEs at high latitudes

Relatively high speed appears to be a particular characteristic of the ICMEs observed by *Ulysses* at high latitudes. Of the six definite ICMEs observed at southerly latitudes greater than 30°S in 1993 and 1994, one had an average speed of 650 km s^{-1} while all the others had speeds greater than 700 km s^{-1} (Gosling et al., 1994c; Phillips et al., 1995), the average over all events being 730 km s^{-1} (Gosling, 1996). These speeds are faster than 95 per cent of all ICMEs observed near the ecliptic plane at 1 AU (Gosling, 1996). At these latitudes *Ulysses* was sampling high-speed coronal hole solar wind for the majority of the time (Phillips et al., 1994). None of these ICMEs were travelling fast enough to drive forward shock waves into the ambient wind as a result of their relative motion. In addition, except in two events, there was no evidence that the ICMEs were accelerated by faster solar-wind pushing from behind or pulling from in front via a rarefaction. This suggests that the high speeds of the ICMEs at high latitudes were primarily a result of their initial acceleration near the Sun and were not due to interaction with the leading and trailing ambient wind as they travelled out from the Sun. Because coronagraph observations (e.g. Hundhausen et al., 1994) have revealed that the speeds of ICMEs within 5 solar radii of the Sun vary from less than 100 km s^{-1} to greater than 1,200 km s^{-1} and that there is no systematic variation of this distribution with latitude, and because slower interplanetary ICMEs observed near the ecliptic plane at 1 AU have a minimum speed similar to that of the ambient slow wind characteristic of low latitudes (e.g. Gosling, 1990), it can be concluded that all ICMEs reach at least a speed equal to that of the minimum speed of the ambient solar wind in which they are propagating (Gosling et al., 1994c). The observation that ICMEs at high latitudes were faster than near the ecliptic then may simply be due to the fact that they tend to accelerate up to the minimum speed of the fast solar wind. Another possibility is that the *Ulysses* result is a statistical anomaly associated with a small number of events.

Why slow ICMEs have a minimum speed equal to that of the surrounding ambient wind is not well understood. It is possible that once a slow CME has been triggered, the plasma released then becomes subject to the same basic acceleration processes as that of the normal solar wind (Gosling et al., 1994c). Alternatively, the CME might be accelerated up to near the ambient solar wind by a dynamic interaction with the surrounding wind, a possibility which has been explored in both gas dynamic (Gosling and Riley, 1996) and MHD simulations (Cargill et al., 1996). Comparison with the *Ulysses* observations suggests that this type of interaction could have played a part in the acceleration of two out of the six ICMEs discussed above (Gosling and Riley, 1996).

Composition of high-latitude ICMEs

Low-latitude ICMEs, reviewed earlier, often have significantly higher helium abundance than the ambient solar wind. On the other hand, all of the ICMEs observed at high latitudes by *Ulysses* during its first orbit had helium abundances similar to that of the ambient solar wind (Phillips et al., 1995; Barraclough et al., 1996). This is unexpected if we regard the helium abundance as a characteristic related to the region from which the solar wind plasma originates and may be a statistical anomaly since not all portions of all ICMEs exhibit increased helium abundance. Further insight into the problem comes from analysis of an ICME observed by *Ulysses* at high latitude and *IMP-8* near the ecliptic, discussed in greater detail below. The helium abundance in the near-ecliptic portion of the ICME was highly elevated compared to the high-latitude portion (Gosling et al., 1995e) (see Figure 3.29 below). It could be argued that in this case the high helium abundance was characteristic of only a central portion of the ICME which was not encountered by *Ulysses* at high latitudes (Barraclough et al., 1996).

The heavy ion composition measurements for the *Ulysses* high-latitude ICMEs show a similar behaviour to those for helium (Galvin, 1997). Both the ion composition relative to that of protons and the relative abundance of the charge states of individual ions are similar to those of the ambient fast solar wind at high latitudes. The disappearance of the characteristic ICME composition signatures appeared to be gradual as *Ulysses* moved to ever increasing latitude.

Overexpansion of high-latitude ICMEs

A new discovery by *Ulysses* was that many of the ICMEs encountered at mid to high latitudes were overexpanding (i.e. were expanding due to a high internal pressure)

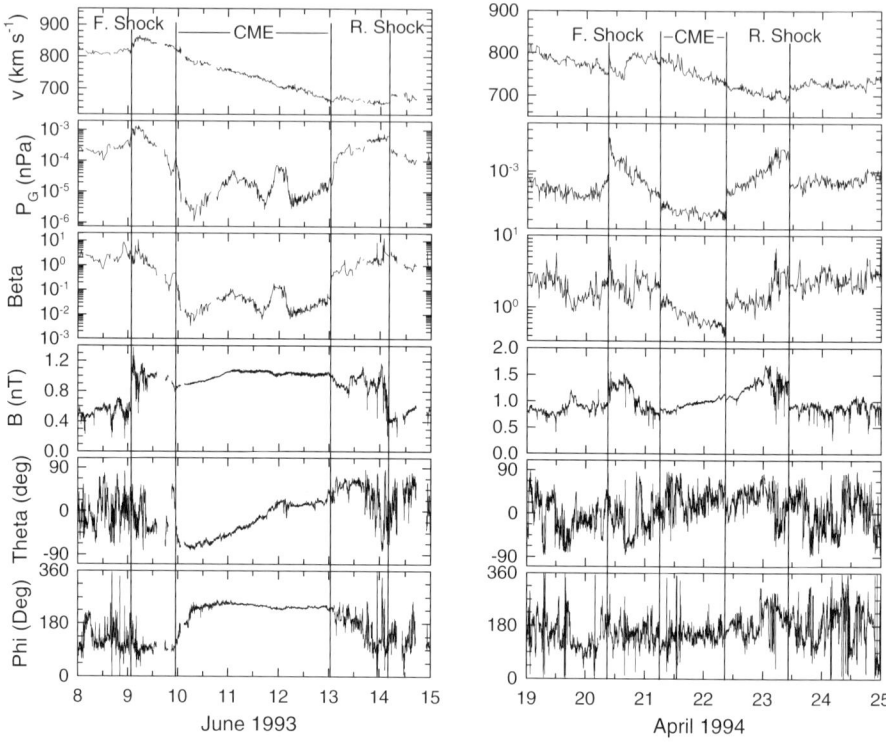

Figure 3.26. Plasma and field parameters from the June 1993 and April 1994 ICMEs observed by *Ulysses*. From the top, parameters plotted are the solar wind speed, gas pressure, plasma beta, magnetic field strength, and magnetic field meridional and azimuthal angles (from Gosling *et al.*, 1994a; 1994b).

with respect to the ambient solar wind (Gosling *et al.*, 1994a,b). The first event of this type was observed in June 1993 when *Ulysses* was at 4.6 AU and 32°S (Gosling *et al.*, 1994b), which was bracketed by a forward–reverse shock pair. Selected plasma and magnetic field parameters for this event are plotted in the left hand panel of Figure 3.26. The forward and reverse shocks and the interval of counterstreaming suprathermal electrons are marked by vertical lines. Other signatures confirming this event as an ICME are the low plasma beta and the smoothly rotating strong magnetic field with low variance, which indicate that this event was also a magnetic cloud. The speed of the centre of the ICME was slower than the solar wind ahead and faster than the solar wind behind. Therefore, the ICME was neither running into the plasma ahead nor being overtaken by the plasma behind, and the shocks could not have been produced by relative motion between the ICME and the surrounding ambient plasma. The declining speed profile from the front to the rear of the event indicates that the ICME was expanding as it passed *Ulysses*. The symmetrical

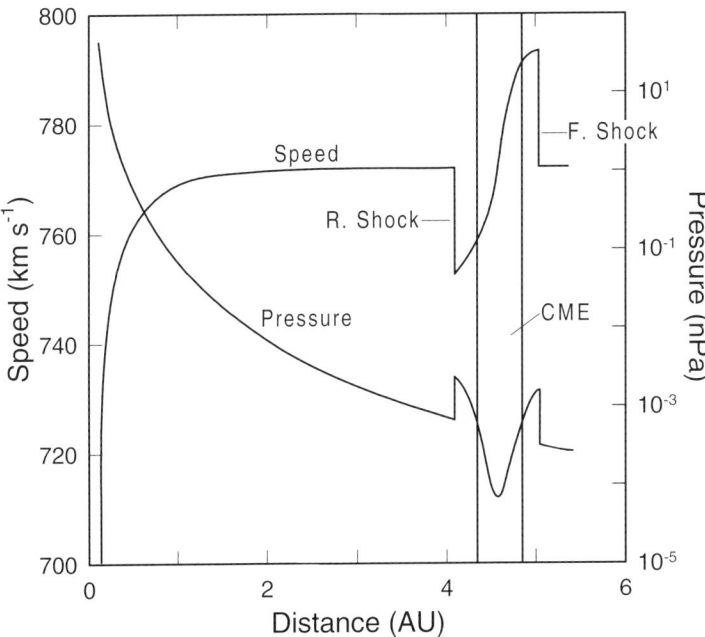

Figure 3.27. Solar wind speed and pressure plotted against heliocentric distance for a simulated ICME that has propagated out to 5 AU. The disturbance was initiated at 0.14 AU as a factor 4 bell shaped increase in density, 10 hours wide. (From Gosling et al., 1994a).

structure of the pressure profile with two high pressure pulses on either side of the ejecta suggest that the forward–reverse shock pair was the result of an the overexpansion of the ICME due to a high initial internal pressure (Gosling et al., 1994b). Indeed, the total pressure in the centre of the ICME shown in Figure 3.26 was still higher inside the ICME than outside. The high initial pressure near the Sun may have been due to high density, high temperature, or high field strength, or some combination of these.

Following the identification of this first event, two additional ICMEs with forward–reverse shock pairs driven by overexpansion were identified during the first *Ulysses* cruise to high southerly latitudes (Gosling et al., 1994a). The righthand panel of Figure 3.26 shows plasma and field parameters for one of these, from April 1994. In contrast to the June 1993 event, this event was not a magnetic cloud. It is a particularly nice example of the symmetrical pressure profile bounding the ICME created by the over-expansion. Of the six definite ICMEs encountered southward of 30°S during the initial transit to high southerly latitudes, five showed signatures of overexpansion (Gosling et al., 1998). Subsequently two other relatively high latitude ICMEs showing overexpansion signatures were identified during the first *Ulysses* orbit, one in February 1995 at 22°S during the

FLS and the other in October 1996 at 24°N. The timings, positions and other characteristics of all these events are tabulated in Gosling et al. (1998).

The physics behind the development and propagation of these overexpanded ICMEs has been explored using a simple one-dimensional gas dynamic simulation (Gosling et al., 1994a). Figure 3.27 shows the result of such a simulation for a disturbance which was initiated at 0.14 AU as a bell shaped density pulse (while temperature and speed were held constant) in an otherwise time-stationary high-speed solar-wind expansion. This mimics the ejection of a CME with a high internal pressure, due to the high density, into the fast solar wind. The figure shows radial profiles of speed and pressure through the disturbance as its leading edge reaches 5 AU. Due to the high initial internal pressure, the supersonic CME expands with distance both forward and backward into the surrounding solar wind, increasing from an initial radial width of 0.17 AU to a width of 0.5 AU by the time it reaches 4.6 AU. Forward and reverse pressure waves, driven by the expansion, form ahead of and behind the ICME and steepen into shocks well before the disturbance reaches 4.6 AU. At this distance the overall radial separation of the shocks has increased to ~0.95 AU. A rarefaction develops inside the ICME, producing a pressure minimum at its centre. Despite the limitations of the simulation due to the one-dimensional nature and the absence of magnetic-field effects, the similarities between the simulation in Figure 3.27 and the observations in Figure 3.26 are clear. In particular the simulation reproduces the symmetrical nature of the pressure profile about the centre of the ICME and the anti-symmetric shape of the speed profile caused by the propagation of the forward and reverse waves into the surrounding solar wind. Figure 3.28 schematically illustrates a meridional slice through an over-expanding CME. This diagram has been shaded so that black represents compression and white represents rarefaction and helps to illustrate that the expansion takes place in all directions, not just on the leading and trailing edges.

The *Ulysses* high latitude observations reveal that not all overexpanding ICMEs produce forward-reverse shock pairs. Sometimes only a reverse shock is detected, while in others cases only a forward shock is detected. One-dimensional gas dynamic simulations (Gosling et al., 1998) reveal that such effects are probably related to speed gradients within the ambient wind and within the ICMEs themselves. For example, the simulations demonstrate that when the ambient wind ahead of a high-pressure ICME runs away from the ICME faster than the ICME can expand into it, no forward shock forms ahead of the ICME. Similarly, when a high-pressure ICME runs away from slower trailing wind faster than it can expand into the trailing wind, no reverse shock forms behind the ICME. The rarefaction associated with such relative motion produces forces that contribute substantially to the overall expansion of the ICME.

Another set of one-dimensional simulations (Riley and Gosling, 1998) have addressed the question of whether an over-expanding ICME should eventually implode due to the rarefaction that develops within the ICME interior. Similar implosions do eventually occur when explosions are set off in the Earth's atmosphere. These simulations reveal that, for the heliospheric case, implosions do not occur at any distance from the Sun so long as the ambient wind pressure falls off

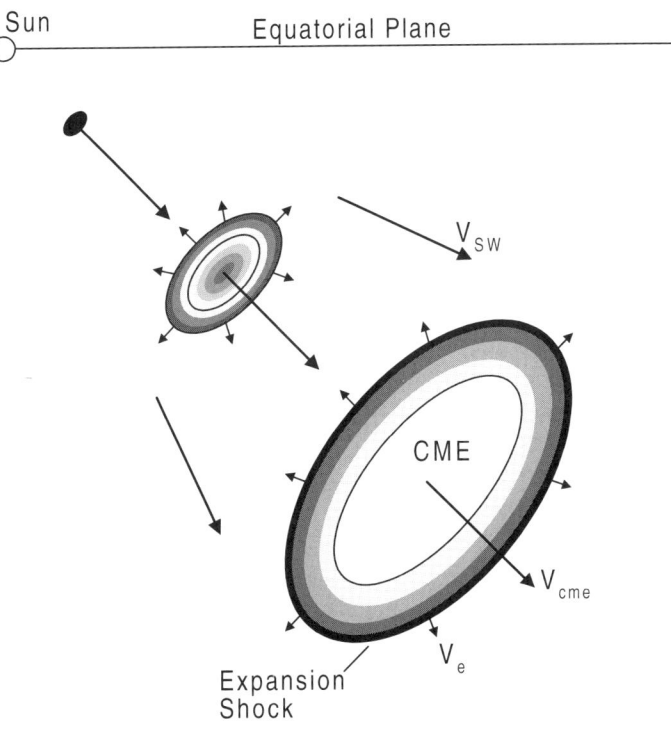

Figure 3.28. Schematic diagram illustrating an overexpanding ICME propagating in the solar wind. The shading represents the distribution of pressure. The large radial variation in the ambient pressure has been removed (from Gosling *et al.*, 1994a).

adiabatically with increasing heliocentric distance. The presence of pickup ions, which increase the plasma pressure in the distant heliosphere and which were not included in the simulations, might alter this conclusion.

Whereas the simulations described above use a density increase at the inner boundary to generate the high internal pressure of the CME and ignore effects associated with the magnetic field, a recent study utilising an MHD simulation has explored the propagation of a disturbance containing a magnetic flux rope, which gives also an elevated magnetic field pressure within the ICME (Cargill *et al.*, 2000). This simulation showed that the flux rope could maintain its coherent structure as the ICME expanded outwards into the solar wind, and thus be observed by a spacecraft at 5 AU as a magnetic cloud. This is consistent with the observation that about half of the high-latitude events observed by *Ulysses* contained magnetic clouds.

None of the simulations have addressed the detailed processes by which the high internal pressure is generated in the corona, since the disturbances are introduced into the simulations outside the critical point where the solar wind goes supersonic. If the internal pressure of the CME is already high before the disturbance has propagated beyond the critical point, then one wonders how the reverse wave manages to escape into interplanetary space at all rather than running back into the Sun (Gosling et al., 1994a). For the reverse wave to escape, the initial outward speed of the CME close to the Sun would have to exceed the fast mode speed there so that the reverse wave can propagate sunward at a speed slower than that at which the CME is travelling (Gosling et al., 1995e). Since reverse waves associated with overexpanding ICMEs to date have only been observed in the fast wind, this suggests that the critical point for the fast wind may often lie very close to the solar surface.

Multispacecraft observations of ICMEs at both high and low latitudes

Two studies have compared ICMEs observed by both *Ulysses* and spacecraft near the Earth (Hammond et al., 1995; Gosling et al., 1995e). An ICME observed by *Ulysses* in February 1994, when the spacecraft was at 3.5 AU and at 54°S, was also observed by *IMP-8* at 1 AU in the ecliptic (Gosling et al., 1995). An approximate launch time for the CME was available due to the X-ray observation of an associated solar flare. At the time of the event the separation of the spacecraft was 47° in latitude and 11° in longitude. *Ulysses* was at a high enough latitude to be sampling nearly continuous high-speed wind $\sim750\,\mathrm{km\,s^{-1}}$, and at *Ulysses* overexpansion of the ICME produced a forward-reverse shock pair. Figure 3.29 shows various plasma and magnetic-field parameters for this event plotted for the two spacecraft. The event at *Ulysses* (on the right) shows the typical symmetric profile in most of the properties already described for overexpanding ICMEs. In this case there was no apparent embedded magnetic flux rope. The speed of the ICME at *Ulysses* was $760\,\mathrm{km\,s^{-1}}$, similar to that of the ambient high-latitude solar wind. At *IMP-8* (on the left) the speed of the leading edge of the ICME was $\sim800\,\mathrm{km\,s^{-1}}$, and it was driving a strong shock wave into the slow solar wind ahead. The average speed between launch and detection at 1 AU was estimated as $\sim992\,\mathrm{km\,s^{-1}}$, suggesting that the ICME decelerated as it interacted with the slow wind during its journey out from the Sun. At *IMP-8* the event did contain a magnetic field rotation characteristic of a flux rope. Thus the event was very different at the two spacecraft, the signature at *IMP-8* being that of a typical near-ecliptic fast ICME driving a strong forward shock wave into the slow wind ahead, while at *Ulysses* a pair of relatively weak forward and reverse shocks were driven by the overexpansion of the ICME.

These very different characteristics were attributed to the very different solar wind into which the ICME was propagating at the different latitudes. Clearly this was a fast CME when it left the Sun, so that at high latitudes its speed was fast enough to allow the reverse wave due to overexpansion to be carried out into interplanetary space and be observed at *Ulysses*. At low latitudes, the disturbance was dominated by effects associated with the relative speed between the fast ICME and the slower wind ahead. This may be why overexpanding ICMEs with associated shock pairs have not yet been observed or at least identified at low latitudes.

The different evolution of the ICME-driven disturbances at high and low latitudes raises interesting questions about the latitudinal structure of such ICMEs. This topic has been explored using both two-dimensional (Riley et al., 1997)

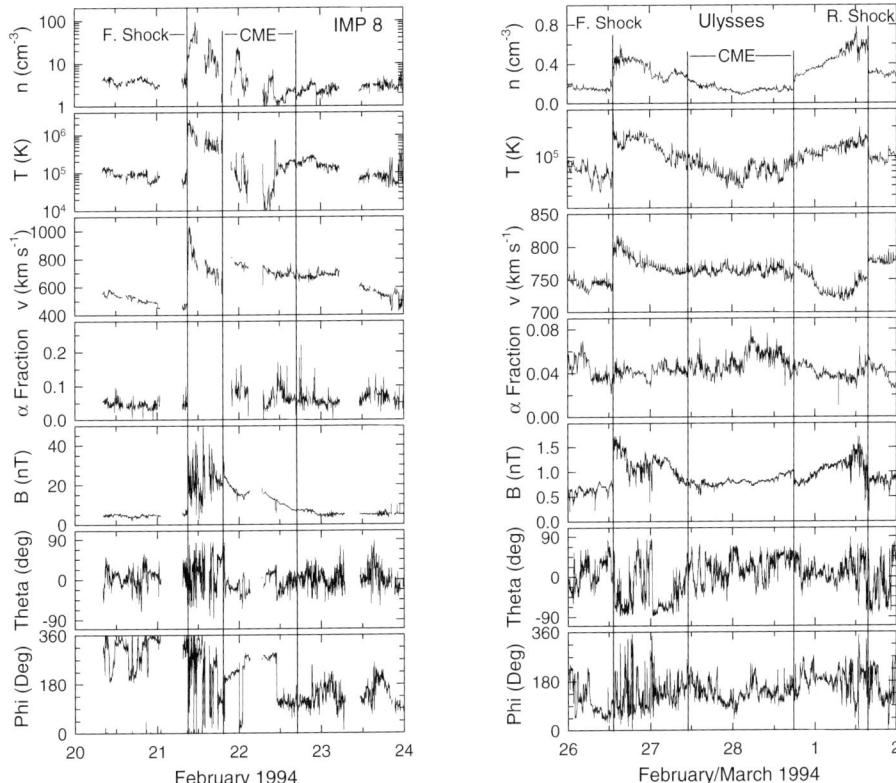

Figure 3.29. Plasma and field parameters obtained from both *Ulysses* and *IMP-8* for an ICME observed in February 1994. From the top, the panels show the proton density, proton temperature, solar wind speed, alpha to proton ratio, magnetic field strength, and magnetic field meridional and azimuthal angles. (From Gosling *et al.*, 1995e).

and three-dimensional (Odstrcil and Pizzo, 1999a,b) hydrodynamic simulations. Figure 3.30 shows a meridional cut through a two dimensional simulation (Riley *et al.*, 1997). Here a background solar-wind flow initially consisted of dense, slow radial flow from 0° to 20° latitude, and fast, tenuous radial flow from 20° to 90° latitude, thus simulating the idealized fast and slow pattern found by *Ulysses* close to solar minimum. The CME disturbance was initiated as a bell shaped, 6-fold increase in pressure at the inner boundary (0.14 AU), of 10 hours duration, extending in latitude from 0° to 45°, and having speed and temperature equal to that of the high-latitude wind. This mimics a high-speed CME with high internal pressure simultaneously propagating into both the slow and fast wind, with a speed considerably faster than that of the slow wind but the same speed as the fast wind. The three panels in the figure show snapshots of the radial velocity, the meridional velocity, and the pressure, seven days after initiation of the disturbance at the inner boundary.

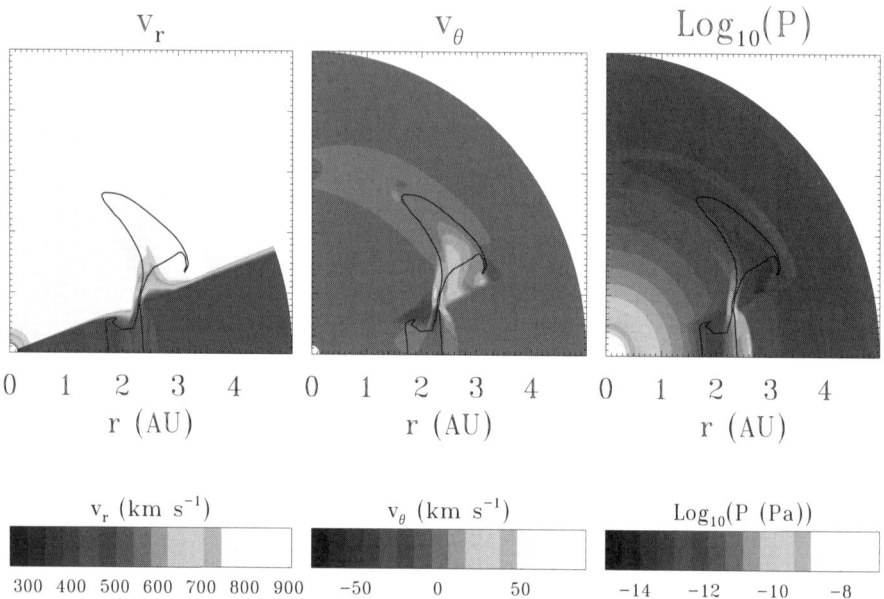

Figure 3.30. Plots in the meridional plane of the radial velocity component, meridional velocity component, and pressure within a two-dimensional simulation of an ICME, from Riley *et al.* (1997). The black line outlines the boundary of the ICME material.

The boundary of the material introduced within the disturbance at the inner boundary is marked on all three panels by a solid curve.

Consistent with the observations discussed above and with one-dimensional simulations, the parts of the disturbance propagating in the slow and fast wind interact with the solar wind quite differently. In the slow wind, a region of high pressure fronted by a shock develops ahead of the ICME, which slows as a result of the interaction with the slow wind. In the fast wind, expansion of the initial high-internal-pressure region inside the ICME produces a relatively weak shock that surrounds the ICME. After 7 days the ICME has effectively separated into two parts because of the velocity shear between the slow and fast ambient flows. The higher latitude portion, propagating at a speed similar to the fast wind, reaches a substantially greater radial distance after 7 days than does the lower latitude portion, which is decelerated. The two portions also separate somewhat in latitude due to meridional flows generated by two rarefaction regions, one created within the high-latitude-portion of the ICME due to the over-expansion, and the other behind the low-latitude portion of the ICME as it outruns the slow ambient solar wind behind.

Three-dimensional hydrodynamic simulations (Odstrcil and Pizzo, 1999a,b) have been used to explore the evolution of ICME-driven disturbances propagating into a corotating fast and slow solar wind structure, similar to that used to study the

Sec. 3.3] Transient flows 157

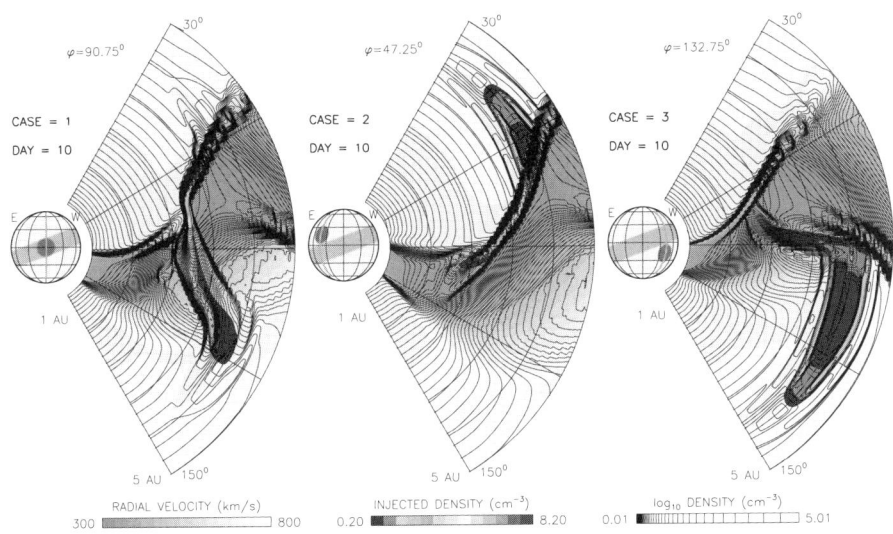

Figure 3.31. A meridional plane cross section of ICMEs propagating in a corotating fast and slow solar wind flow structure, 10 days after CME launch, from the three-dimensional simulation of (Odstrcil and Pizzo, 1999b). The grey-scale shows the radial velocity component; the colour scale shows the density within the injected CME plasma; and the contour scale shows the log of the density throughout the simulation. The three cases shown are (1) a CME injected within the streamer belt, (2) a CME injected to the east of the streamer belt, and (3) a CME injected to the west of the streamer belt, as indicated schematically on the small circular diagrams. The green line on these shows the location of the meridional slice.

radial evolution of CIRs. The disturbances injected at the inner boundary then undergo interaction with the compression and rarefaction regions and flow shears generated by the ambient flow pattern. The disturbances were found to distort in similar, but more complex, ways than those in the two-dimensional simulation above. In certain circumstances ICME driven shocks can merge with CIR driven shocks depending on their relative speeds. By way of example, Figure 3.31, from Odstrcil and Pizzo, 1996b), shows meridional cuts through three cases from these simulations, where the streamer belt tilt was set at 20° to the equator. The three cases correspond to the CME plasma being injected at the inner boundary of the simulation (0.14 AU) in different locations relative to the streamer belt, as indicated by the circular inserts at the centre of each diagram. In all three cases the CME undergoes a significant lateral expansion close to the Sun. Following this, in case 1, where the CME is injected at the centre of the streamer belt, the low-latitude portion becomes caught up within the interacting ambient flows, whereas the high-latitude portions are advected out faster within the high-latitude ambient fast solar-wind flow. In case 2 the CME is injected into the fast wind just to east of the streamer belt. In this case the ICME interaction with the ambient solar wind is initially dominated by the overexpansion due to the high pressure within the ICME. However, the low latitude portion of the ICME is overtaken by the leading edge of the fast wind, and the forward and reverse shocks due to the overexpansion merge with forward and reverse shocks of the resulting CIR. Thus most of the ICME plasma is trapped between the CIR shocks, apart from the highest latitude portion of the ICME, which continues to be advected out in the fast wind. Finally, in case 3 where the CME is injected into the fast wind just to the west of the streamer belt, most of the ICME is advected outwards within the high-latitude fast wind, and only the lowest latitude portion gets caught up in the low latitude interacting flows. In cases 2 and 3 the high-latitude portions of the CMEs which remain within the fast wind exhibit the overexpanding characteristics at 5 AU in agreement with the Ulysses observations.

Neither of the above simulations include forces due to the magnetic field within the ICME and ambient wind. It is thus an open question what additional effect magnetic forces would have on the separation of the high and low latitude portions of an ICME propagating in both the slow and fast solar wind, although it is expected that magnetic forces produce relatively small effects.

3.4 SUMMARY AND CONCLUSION

This chapter has discussed the *Ulysses* observations of corotating and transient flows in the heliosphere, highlighting the new results that have emerged from observations at high latitudes, and placing them in the context of what had been learned previously from the many spacecraft which have explored the heliosphere near the ecliptic.

By surveying CIRs over an extended range of latitudes in both the north and south hemispheres of the heliosphere, the *Ulysses* observations have provided a firm basis for understanding their three-dimensional structure during the declining and minimum phases of the solar cycle. The key observations that reverse shocks persist to higher latitudes than forward shocks and that forward shocks propagate equatorwards and reverse shocks polewards, led to the conclusion that the CIRs have opposite north–south tilts in the northern and southern hemispheres. Modelling has provided clear understanding of how the orientation of the boundary between fast and slow wind near the Sun leads to SIs and CIRs having these opposed north–south tilts.

In Section 3.3 we presented a summary of the *Ulysses* observations of the transient structures that propagate through interplanetary space as a result of mass ejections from the solar corona, building on a brief review of previous knowledge gained from solar and *in-situ* observations near 1 AU. The key results that have emerged from *Ulysses*' journey to high latitudes are (1) that ICMEs attain a minimum speed equal to, or nearly equal to that of the solar wind in which they are embedded, and (2) that the majority of ICMEs observed at high latitudes were overexpanding as a result of high internal pressure. Forward and reverse shocks driven by overexpansion had not previously been detected prior to the *Ulysses* mission. A further result which emerged from both *Ulysses* observations and associated simulations is that the disturbance produced by a fast, overpressured ICME evolves quite differently in the high and low-speed wind. Moreover, the flows shear at the boundary between the fast high-latitude wind and the slow, low-latitude wind can produce significant distortion in ICMEs.

These observations were made during *Ulysses'* first polar orbit at a time when solar-activity was in its declining and minimum phase. At this time the flow pattern in the heliosphere was generated by a relatively narrow band of slow wind symmetrical about the equator of a tilted dipole streamer belt, and a much broader band of fast wind from the polar coronal holes at high latitudes. Thus this period was ideal for the formation of the stable corotating interaction regions observed by *Ulysses* and discussed in Section 3.2. A small number of transient flows due to CMEs were also encountered during *Ulysses'* first polar orbit and were discussed in Section 3.3.

At the time of this writing, *Ulysses* is on its way to high southerly latitudes for the second time and solar activity is rapidly rising towards its maximum. The polar coronal holes have diminished in size and are less stable. Consequently it is unlikely that long-lived sequences of CIRs such as encountered on the first orbit will be found on the second, although interaction regions due to shorter lived high-speed streams from smaller coronal holes are still likely to be observed and will interact with neighbouring flows in the same manner that we have discussed. On the other hand, CMEs should increase in frequency, thus providing many more opportunities to explore the physics of coronal mass ejections at high latitudes.

3.5 ACKNOWLEDGEMENTS

The authors would like to thank C. Foley, R. A. Howard, D. Odstrcil, V. J. Pizzo, and P. Riley for their assistance in the preparation of figures for this chapter. *Ulysses* research at Imperial College is supported by the UK Particle Physics and Astronomy Research Council. Work at Los Alamos was performed under the auspices of the US Department of Energy with support from NASA.

3.6 REFERENCES

Armstrong, T. P., Haggerty, D., Lanzerotti, L. J., Maclennan, C. G., Roelof, E. C., Pick, M., Simnett, G. M., Gold, R. E., Krimigis, S. M., Anderson, K. A., Lin, R. P., Sarris, E. T., Forsyth, R. and Balogh, A. (1994) Observation by Ulysses of hot (~270 keV) coronal particles at 32 south heliolatitude and 4.6 AU. *Geophys. Res. Lett.* **21**, 1747–1750.

Balogh, A., Beek, T. J., Forsyth, R. J., Hedgecock, P. C., Marquedant, R. J., Smith, E. J., Southwood, D. J. and Tsurutani, B. T. (1992) The magnetic field investigation on the Ulysses mission: Instrumentation and preliminary scientific results. *Astron. Astrophys. Suppl. Ser.* **92**, 221–236.

Balogh, A., Forsyth, R. J., Ahuja, A., Southwood, D. J., Smith, E. J. and Tsurutani, B. T. (1993a) The interplanetary magnetic field from 1 to 5 AU: Ulysses observations. *Adv. Space Res.* **13**(6), 15–24.

Balogh, A., Erdös, G. Forsyth, R. J. and Smith, E. J. (1993b) The evolution of the interplanetary sector structure in 1992. *Geophys. Res. Lett.* **20**, 2331–2334.
Balogh, A., Gonzalez-Esparza, J. A., Forsyth, R. J., Burton, M. E., Goldstein, B. E., Smith, E. J. and Bame, S. J. (1995) Interplanetary shock waves: Ulysses observations in and out of the ecliptic plane. *Space Sci. Rev.* **72**, 171–180.
Balogh, A., Gosling, J. T., Jokipii, J. R., Kallenbach, R. and Kunow H. (eds) (1999) Corotating interaction regions. *Space Sciences Series of ISSI, Vol.7*, Kluwer, Dordrecht.
Bame, S. J., McComas, D. J., Barraclough, B. L., Phillips, J. L., Sofaly, K. J., Chavez, J. C., Goldstein, B. E. and Sakurai, R. K. (1992) The Ulysses solar wind plasma experiment. *Astron. Astrophys. Suppl. Ser.* **92**, 237–266.
Bame, S. J., Goldstein, B. E., Gosling, J. T., Harvey, J. W., McComas, D. J., Neugebauer, M. and Phillips, J. L. (1993) Ulysses observations of a recurrent high speed solar wind stream and the heliomagnetic streamer belt. *Geophys. Res. Lett.* **20**, 2323–2326.
Barraclough, B. L., Feldman, W. C. Gosling, J. T. McComas, D. J. Phillips, J. L. and Goldstein, B. E. (1996) He abundance variations in the solar wind: Observations from Ulysses. In Winterhalter, D., Gosling, J. T., Habbal, S. R., Kurth, W. S. and Neugebauer, M. (eds) *Solar Wind Eight*, pp. 277–280, AIP, New York.
Belcher, J. W., and Davis, L. Jr. (1971) Large-amplitude Alfvén waves in the interplanetary medium, 2. *J. Geophys. Res.* **76**, 3534–3563.
Borrini, G., Wilcox, J. M., Gosling, J. T., Bame, S. J. and Feldman, W. C. (1981) Solar wind helium and hydrogen structure near the heliospheric current sheet: A signal of coronal streamers at 1 AU. *J. Geophys. Res.* **86**, 4565–4573.
Borrini, G., Gosling, J. T. Bame, S. J. and Feldman, W. C. (1982) Helium abundance enhancements in the solar wind. *J. Geophys. Res.* **87**, 7370–7378.
Bothmer, V., and Schwenn, R. (1998) The structure and origin of magnetic clouds in the solar wind. *Ann. Geophys.* **16**, 1–24.
Bothmer, V., Desai, M. I., Marsden, R. G., Sanderson, T.R., Trattner, K. J., Wenzel, K.-P., Gosling, J. T., Balogh, A., Forsyth, R. J. and Goldstein, B. E. (1996) Ulysses observations of open and closed magnetic field lines within a coronal mass ejection. *Astron. Astrophys.* **316**, 493–498.
Burlaga, L. F. (1974) Interplanetary stream interfaces. *J. Geophys. Res.* **79**, 3717–3725.
Burlaga, L. F. (1984) MHD processes in the outer heliosphere. *Space Sci. Rev.* **39**, 255–316.
Burlaga, L. F. (1991) Magnetic clouds. In: Schwenn R. and Marsch, E. (eds) *Physics of the inner heliosphere II*, pp. 1–22, Springer-Verlag, Berlin.
Burlaga, L. F. and King, J. H. (1979) Intense interplanetary magnetic fields observed by geocentric spacecraft during 1963–1975. *J. Geophys. Res.* **84**, 6633–6640.
Burlaga, L. F., Sittler, E. Mariani, F. and Schwenn, R. (1981) Magnetic loop behind an interplanetary shock: Voyager, Helios and IMP 8 observations. *J. Geophys. Res.* **86**, 6673–6684.
Burlaga, L. F., McDonald, F. B. and Schwenn, R. (1986) Formation of a compound stream between 0.85 and 6.2 AU and its effects on solar energetic particles and galactic cosmic rays. *J. Geophys. Res.* **91**, 13331–13340.
Burlaga, L. F., McDonald, F. B. and Ness, N. F. (1993) Cosmic ray modulation and the distant heliospheric magnetic field: Voyager 1 and 2 observations from 1986 through 1989. *J. Geophys. Res.* **98**, 1–11.
Burlaga, L. F., Ness, N. F. and Belcher, J. W. (1997) Radial evolution of corotating merged interaction regions and flows between ~14 AU and ~43 AU. *J. Geophys. Res.* **102**, 4661–4671.

References

Burlaga, L., Fitzenreiter, R., Lepping, R., Ogilvie, K., Szabo, A., Lazarus, A., Steinberg, J., Gloeckler, G., Howard, R., Michels, D., Farrugia, C., Lin, R. P. and Larson, D. E. (1998) A magnetic cloud containing prominence material: January 1997. *J. Geophys. Res.* **103**, 277–285.

Burton, M. E., Smith, E. J., Balogh, A., Forsyth, R. J., Bame, S. J., Phillips, J. L. and Goldstein, B. E. (1996) Ulysses out-of-ecliptic observations of interplanetary shocks. *Astron. Astrophys.* **316**, 313–322.

Burton, M. E., Neugebauer, M., Crooker, N. U., von Steiger, R. and Smith, E. J. (1999) Identification of trailing edge solar wind stream interfaces: A comparison of Ulysses plasma and composition measurements. *J. Geophys. Res.* **104**, 9925–9932.

Cargill, P. J., Schmidt, J. Spicer, D. S. and Zalesak, S. T. (2000) The magnetic structure of over-expanding coronal mass ejections: numerical models. *J. Geophys. Res.* **105**, 7509–7519.

Cargill, P.J., Chen, J. Spicer, D. S. and Zalesak, S. T. (1996) MHD simulations of the motion of magnetic flux tubes through a magnetised plasma. *J. Geophys. Res.* **101**, 4855–4870.

Clack, D., Forsyth, R. J. and Dunlop, M. W. (2000) Ulysses observations of the magnetic field structure within CIRs. *Geophys. Res. Lett.* **27**, 625–628.

Crooker, N. U. and Intriligator, D. S. (1996) A magnetic cloud as a distended flux rope occlusion in the heliospheric current sheet. *J. Geophys. Res.* **101**, 24343–24348.

Crooker, N. U., Gosling, J. T. and Kahler, S. W. (1998) Magnetic clouds at sector boundaries. *J. Geophys. Res.* **103**, 301–306.

Crooker, N. U., Gosling, J. T., Horbury, T. S., Wimmer-Schweingruber, R. F., Bothmer, V., Forsyth, R. J., Gazis, P. R., Hewish, A., Intriligator, D. S., Jokipii, J. R., Kóta, J., Lazarus, A. J., Lee, M. A., Lucek, E., Marsch, E., Posner, A., Richardson, I. G., Roelof, E. C., Schmidt, J. M., Siscoe, G. L. and Tsurutani, B. T. (1999) CIR morphology, turbulence, discontinuities, and energetic particles. *Space Sci. Rev.* **89**, 179–220.

Feldman, W. C., Asbridge, J. R., Bame, S. J., Montgomery, M. D. and Gary, S. P. (1975) Solar wind electrons. *J. Geophys. Res.* **80**, 4181–4196.

Forsyth, R. J., Balogh, A., Smith, E. J. and Gosling, J. T. (1997) Ulysses observations of the northward extension of the heliospheric current sheet. *Geophys. Res. Lett.* **24**, 3101–3104.

Funsten, H. O., Gosling, J. T., Riley, P., St. Cyr, O. C., Forsyth, R. J., Howard, R. A. and Schwenn, R. (1999) Combined Ulysses solar wind and SOHO coronal observations of several west limb coronal mass ejections. *J. Geophys. Res.*, **104**, 6679–6689.

Galvin, A. B. (1997) Minor ion composition in CME-related solar wind. In: Crooker, N., Joselyn, J. A. and Feynman, J. (eds) *Coronal Mass Ejections, Geophys. Monogr. Ser.* vol. 99, pp. 253–260, AGU, Washington D. C.

Geiss, J., Gloeckler, G. and von Steiger, R. (1995) Origin of the solar wind from composition data. *Space Sci. Rev.* **72**, 49–60.

Gloeckler, G., Geiss, J., Balsiger, H., Bedini, P., Cain, J. C., Fischer, J., Fisk, L. A., Galvin, A. B., Gliem, F., Hamilton, D. C., Hollweg, J. V., Ipavich, F. M., Joss, R., Livi, S., Lundgren, R., Mall, U., McKenzie, J. F., Ogilvie, K. W., Ottens, F., Rieck, W., Tums, E. O., von Steiger, R., Weiss, W. and Wilken, B. (1992) The solar wind ion composition spectrometer. *Astron. Astrophys. Suppl. Ser.* **92**, 267–290.

Gonzalez-Esparza, J. A., and Smith, E. J. (1997) Three-dimensional nature of interaction regions: Pioneer, Voyager, and Ulysses solar cycle variations from 1 to 5 AU. *J. Geophys. Res.* **102**, 9781–9792.

Gonzalez-Esparza, J. A., Balogh, A., Forsyth, R. J., Neugebauer, M., Smith, E. J. and Phillips, E. J. (1996) Interplanetary shock waves and large-scale structures: Ulysses observations in and out of the ecliptic plane. *J. Geophys. Res.* **101**, 17057–17071.

Gonzalez-Esparza, J. A., Neugebauer, M., Smith, E. J., and Phillips, J. L. (1998) Radial evolution of ejecta characteristics and transient shocks: Ulysses in-ecliptic observations. *J. Geophys. Res.* **103**, 4767–4773.

Gopalswamy, N., Hamaoka, Y. Kosugi, T. Lepping, R. P. Steinberg, J. T., Plunkett, S., Howard, R. A., Thomson, B. J., Gurman, J., Ho, G., Nitta, N. and Hudson, H. S. (1998) On the relationship between coronal mass ejections and magnetic clouds. *Geophys. Res. Lett.* **25**, 2485–2488.

Gosling, J. T. (1990) Coronal mass ejections and magnetic flux ropes in interplanetary space. In: Russell, C. T., Priest, E. R. and Lee, L. C. (eds) *Physics of Magnetic Flux Ropes. Geophys. Monogr. Ser.* vol. 58, pp. 343–364, AGU, Washington, D. C.

Gosling, J. T. (1993) The solar flare myth. *J. Geophys. Res.* **98**, 18937–18949.

Gosling, J. T. (1996) Corotating and Transient solar wind flows in three dimensions. *Ann. Rev. Astron. Astrophys.* **34**, 35–73.

Gosling, J. T. (1997) Coronal mass ejections: An overview. In Crooker, N. Joselyn, J. A. and Feynman, J. (eds) *Coronal Mass Ejections, Geophys. Monogr. Ser.* vol. 99, pp. 9–16. AGU, Washington D. C.

Gosling, J. T. (1999) The role of reconnection in the formation of flux ropes in the solar wind. In: Pevtsov, A. A. Canfield, R. and Brown, M. (eds) *Magnetic Helicity in Space and Laboratory Plasmas. Geophys. Monogr. Ser.*, vol. 111, pp. 205–212, AGU, Washington D. C.

Gosling J. T., and Riley, P. (1996) The acceleration of slow coronal mass ejections in the high-speed solar wind. *Geophys. Res. Lett.* **23**, 2867–2870.

Gosling, J. T. and Pizzo V. J. (1999) Formation and evolution of corotating interaction regions and their three dimensional structure. *Space Sci. Rev.* **89**, 21–52.

Gosling, J. T., Hundhausen, A. J. and Bame, S. J. (1976a) Solar wind stream evolution at large heliocentric distances: Experimental demonstration and the test of a model. *Geophys. Res.* **81**, 2111–2122.

Gosling, J. T., Pizzo, V. and Bame, S. J. (1973) Anomalously low proton temperatures in the solar wind following interplanetary shock waves – Evidence for magnetic bottles?. *J. Geophys. Res.* **78**, 2001–2009.

Gosling, J. T., Asbridge, J. R., Bame, S. J. and Feldman, W. C. (1978) Solar wind stream interfaces. *J. Geophys. Res.* **83**, 1401–1412.

Gosling, J. T., Hildner, E., MacQueen, R. M., Munro, R. H., Poland, A. I. and Ross, C. L. (1974) Mass ejections from the Sun: A view from Skylab. *J. Geophys. Res.* **79**, 4581–4587.

Gosling, J. T., Hundhausen, A. J., Pizzo, V. and Asbridge, J. R. (1972) Compressions and rarefactions in the solar wind: Vela 3. *J. Geophys. Res.* **77**, 5442–5454.

Gosling, J. T., Hildner, E., MacQueen, R. M., Munro, R. H., Poland, A. I. and Ross, C. L. (1976b) The speeds of coronal mass ejection events. *Solar Phys.* **48**, 389–397.

Gosling, J. T., Asbridge, J. R. Bame, S. J. Hundhausen, A. J. and Strong, I. B. (1968) Satellite observations of interplanetary shock waves. *J. Geophys. Res.* **73**, 43–50.

Gosling, J. T., Borrini, G., Asbridge, J. R., Bame, S. J., Feldman, W. C. and Hansen, R. T. (1981) Coronal streamers in the solar wind at 1 AU. *J. Geophys. Res.* **86**, 5438–5448.

Gosling, J. T., Baker, D. N., Bame, S. J., Feldman, W. C., Zwickl, R. D. and Smith, E. J. (1987) Bidirectional solar wind electron heat flux events. *J. Geophys. Res.* **92**, 8519–8535.

Gosling, J. T., McComas, D. J., Phillips, J. L. and Bame, S. J. (1991) Geomagnetic activity associated with Earth passage of interplanetary shock disturbances and coronal mass ejections. *J. Geophys. Res.* **96**, 7831–7839.

Gosling, J. T., Bame, S. J., McComas, D. J., Phillips, J. L., Pizzo, V. J., Goldstein, B. E. and Neugebauer, M. (1993a) Latitudinal variation of solar wind corotating stream interaction regions: Ulysses. *Geophys. Res. Lett.* **20**, 2789–2792.

Gosling, J. T., Bame, S. J., Feldman, W. C., McComas, D. J., Phillips, J. L. and Goldstein, B. E. (1993b) Counterstreaming suprathermal electron events upstream of corotating shocks in the solar wind beyond ~2 AU: Ulysses. *Geophys. Res. Lett.* **20**, 2335–2338.

Gosling, J. T., McComas, D. J., Phillips, J. L., Weiss, L. A., Pizzo, V. J., Goldstein, B. E. and Forsyth, R. J. (1994a) A new class of forward-reverse shock pairs in the solar wind. *Geophys. Res. Lett.* **21**, 2271–2274.

Gosling, J. T., Bame, S. J., McComas, D. J., Phillips, J. L., Scime, E. E., Pizzo, V. J., Goldstein, B. E. and Balogh, A. (1994b) A forward-reverse shock pair in the solar wind driven by over-expansion of a coronal mass ejection: Ulysses observations. *Geophys. Res. Lett.* **21**, 237–240.

Gosling, J. T., Bame, S. J., McComas, D. J., Phillips, J. L., Goldstein, B. E. and Neugebauer, M. (1994c) The speeds of coronal mass ejections in the solar wind at mid heliographic latitudes: Ulysses. *Geophys. Res. Lett.* **21**, 1109–1112.

Gosling, J. T., Bame, S. J., Feldman, W. C., McComas, D. J., Phillips, D. J., Goldstein, B. E., Neugebauer, M., Burkepile, J., Hundhausen A. J. and Acton, L. (1995a) The band of solar wind variability at low heliographic latitudes near solar activity minimum – Plasma results from the Ulysses rapid latitude scan. *Geophys. Res. Lett.* **22**, 3329–3332.

Gosling, J. T., Bame, S. J., McComas, D. J., Phillips, J. L., Pizzo, V. J., Goldstein, B. E. and Neugebauer, M. (1995b) Solar wind corotating stream interaction regions out of the ecliptic plane: Ulysses. *Space Sci. Rev.* **72**, 99–104.

Gosling, J. T., Feldman, W. C., McComas, D. J., Phillips, J. L., Pizzo, V. J. and Forsyth, R. J. (1995c) Ulysses observations of opposed tilts of solar wind corotating interaction regions in the northern and southern solar hemispheres. *Geophys. Res. Lett.* **22**, 3333–3336.

Gosling, J. T., Birn, J. and Hesse, M. (1995d) Three-dimensional reconnection and the magnetic topology of coronal mass ejection events. *Geophys. Res. Lett.* **22**, 869–872.

Gosling, J. T., McComas, D. J. Phillips, J. L. Pizzo, V. J. Goldstein, B. E. Forsyth, R. J. and Lepping, R. P. (1995e) A CME-driven solar wind disturbance observed at both high and low heliographic latitudes. *Geophys. Res. Lett.* **22**, 1753–1756.

Gosling, J. T., Bame, S. J., Feldman, W. C., McComas, D. J., Riley, P., Goldstein, B. E. and Neugebauer, M. (1997) The northern edge of the band of solar wind variability: Ulysses at ~4.5 AU. *Geophys. Res. Lett.* **24**, 309–312.

Gosling, J. T., Riley, P., McComas, D. J. and Pizzo, V. J. (1998) Over-expanding coronal mass ejections at high heliographic latitudes: Observations and simulations. *J. Geophys. Res.* **103**, 1941–1954.

Hammond, C. M., Crawford, G. K., Gosling, J. T., Kojima, H. Phillips, J. L., Matsumoto, H., Balogh, A., Frank, L. A., Kokobun, S. and Yamamoto, T. (1995) Latitudinal structure of a coronal mass ejection inferred from Ulysses and Geotail observations. *Geophys. Res. Lett.* **22**, 1169–1172.

Henke, T., Woch, J., Mall, U., Livi, S., Wilken, B., Schwenn, R., Gloeckler, G., von Steiger, R., Forsyth, R. J. and Balogh, A. (1998) Differences in the O^{+7}/O^{+6} ratio of magnetic cloud and non-cloud coronal mass ejections. *Geophys. Res. Lett.* **25**, 3465–3468.

Hiei, E., Hundhausen, A. J. and Sime, D. G. (1993) Reformation of a coronal helmet streamer by magnetic reconnection after a coronal mass ejection. Geophys. Res. Lett., **20**, 2785–2788.

Hirshberg, J., Bame, S. J. and Robbins, D. E. (1972) Solar flares and solar wind helium enrichments: July 1965–July 1967. *Solar Phys.* **23**, 467–486.

Hoeksema, J. T. (1995) The large-scale structure of the heliospheric current sheet during the Ulysses epoch. *Space Sci. Rev.* **72**, 137–148.

Howard, R. A., Sheeley, N. R., Koomen, M. J. and Michels, D. J. (1985) Coronal mass ejections: 1979–1981. *J. Geophys. Res.* **90**, 8173–8191.

Howard, R. A., Brueckner, G. E., St. Cyr, O. C., Biesecker, D. A., Dere, K. P., Koomen, M. J., Korendyke, C. M., Lamy, P. L., Llebaria, A., Bout, M. V., Michels, D. J., Moses, J. D., Paswaters, S. E., Plunkett, S. E., Schwenn, R., Simnett, G. M., Socker, D. G., Tappin, S. J. and Wang, D. (1997) Observations of CMEs from SOHO/LASCO. In Crooker, N., Joselyn, J. A. and Feynman, J. (eds) *Coronal Mass Ejections*, Geophys. Monogr. Ser. vol. 99, pp. 17–26. AGU, Washington D. C..

Hudson, H. S. and Webb, D. F. (1997) Soft X-ray signatures of coronal ejections. In Crooker, N. Joselyn, J. A. and Feynman, J. (eds) *Coronal Mass Ejections, Geophys. Monogr. Ser.*, vol. 99, pp. 27–38. AGU, Washington D. C.

Hundhausen, A. J. (1973) Nonlinear model of high-speed solar wind streams. *J. Geophys. Res.* **78**, 1528–1542.

Hundhausen, A. J. (1993) Sizes and locations of coronal mass ejections: SMM observations from 1980 and 1984–1989. *J. Geophys. Res.* **98**, 13177–13200.

Hundhausen, A. J. (1997) Coronal mass ejections. In: Jokipii, J. R. Sonett, C. P. and Giampapa, M. S. (eds) *Cosmic Winds and the Heliosphere*, pp. 259–296, University of Arizona Press, Tucson.

Hundhausen, A. J. and Gosling, J. T. (1976) Solar wind structure at large heliocentric distances: An interpretation of Pioneer 10 observations. *J. Geophys. Res.* **81**, 1436–1440.

Hundhausen, A. J. (1988) The origin and propagation of coronal mass ejections. *Proceedings of the Sixth International Solar Wind Conference*, Report NCAR/TN-306, edited by Pizzo, V. J., Holzer, T. E. and Sime, D. G., pp. 181–214. National Center for Atmospheric Research, Boulder, Colorado.

Hundhausen, A. J., Burkepile, J. T. and St. Cyr, O. C. (1994) Speeds of coronal mass ejections: SMM observations from 1980 and 1984–1989. *J. Geophys. Res.* **99**, 6543–6552.

Kahler, S. W. (1977) The morphological and statistical properties of solar x-ray events with long decay times. *Astrophys. J.* **214**, 891.

Kopp, R. A. and Pneuman, G. W. (1976) Magnetic reconnection in the corona and loop prominence phenomenon. *Solar Phys.* **50**, 85–98.

Larson, D. E., Lin, R. P., McTiernan, J. M., McFadden, J. P., Ergun, R. E., McCarthy, M., Reme, H., Sanderson, T. R., Kaiser, M., Lepping, R. P. and Mazur, J. (1997) Tracing the topology of the October 18-20, 1995, magnetic cloud with ~ 0.1–10^2 keV electrons. *Geophys. Res. Lett.* **24**, 1911–1914.

Lee, M. A. (2000) An analytical theory of the morphology, flows, and shock compressions at corotating interaction regions in the solar wind. *J. Geophys. Res.* **105**, 10491–10500.

Low, B. C. (1997) The role of coronal mass ejections in solar activity. In Crooker, N., Joselyn, J. A. and Feynman, J. (eds) *Coronal Mass Ejections, Geophys. Monogr. Ser.* vol. 99, pp. 39–47, AGU, Washington D. C..

MacQueen, R. M., Eddy, J. A. Gosling, J. T. Hildner, E. Munro, R. H. Newkirk, G. A. Poland, A. I. and Ross, C. L. (1974) The outer solar corona as observed from Skylab: Preliminary results. *Astrophys. J. Lett.* **187**, 85–88.

McComas, D. J., Gosling, J. T., Hammond, C. M., Moldwin, M. B., Phillips, J. L. and Forsyth, R. J. (1994) Magnetic reconnection ahead of coronal mass ejection. *Geophys. Res. Lett.*, **21**, 1751–1754.

McComas, D. J., Bame, S. J., Barraclough, B. L., Feldman, W. C., Funsten, H. O., Gosling, J. T., Riley, P., Skoug, R., Balogh, A., Forsyth, R., Goldstein, B. E. and Neugebauer, M., (1998) Ulysses' return to the slow solar wind. *Geophys. Res. Lett.* **25**, 1–4.

Montgomery, M. D., Asbridge, J. R., Bame, S. J. and Feldman, W. C. (1974) Solar wind electron temperature depressions following some interplanetary shock waves: Evidence for magnetic merging? *J. Geophys. Res.* **79**, 3103–3110.

Munro, R. H., Gosling, J. T., Hildner, E., MacQueen, R. M., Poland, A. I. and Ross, C. L. (1979) The association of coronal mass ejection transients with other forms of solar activity. *Solar Phys.* **61**, 201–215.

Neugebauer, M. and Goldstein, R. (1997) Particle and field signatures of coronal mass ejections in the heliosphere. In Crooker, N. Joselyn, J. A. and Feynman, J. (eds) *Coronal Mass Ejections*, Geophys. Monogr. Ser. vol. 99, pp. 245–251, AGU, Washington D. C.

Odstrcil, D. and Pizzo, V. J. (1999a) Three-dimensional propagation of coronal mass ejections (CMEs) in a structured solar wind flow 1. CME launched within the streamer belt. *J. Geophys. Res.* **104**, 483–492.

Odstrcil, D. and Pizzo, V. J. (1999b) Three-dimensional propagation of coronal mass ejections (CMEs) in a structured solar wind flow 2. CME launched adjacent to the streamer belt. *J. Geophys. Res.* **104**, 493–503.

Palmer, I. D. and Gosling, J. T. (1978) Shock-associated energetic proton events at large heliocentric distances. *J. Geophys. Res.* **83**, 2037–2046.

Phillips, J. L. (1997) Coronal mass elections encountered by the Ulysses spacecraft during the in-ecliptic mission phase, Unclassified Report LA-UR-97-1087, Los Alamos National Laboratory, Los Alamos, NM.

Phillips, J. L., Bame, S. J. Gosling, J. T. McComas, D. J. Goldstein, B. E. Smith, E. J. Balogh, A. and Forsyth, R. J. (1992) Ulysses plasma observations of coronal mass ejections near 2.5 AU. *Geophys. Res. Lett.* **19**, 1239–1242.

Phillips, J. L., Balogh, A., Bame, S. J., Goldstein, B. E., Gosling, J. T., Hoeksema, J. T., McComas, D. J., Neugebauer, M., Sheeley, N. R. Jr. and Wang, Y.-M. (1994) Ulysses at 50°S: Constant immersion in the high-speed solar wind. *Geophys. Res. Lett.* **21**, 1105–1108.

Phillips, J. L., Bame, S. J., Feldman, W. C., Goldstein, B. E., Gosling, J. T., Hammond, C. M., McComas, D. J., Neugebauer, M., Scime, E. E. and Suess, S. T. (1995) Ulysses solar wind plasma observations at high southerly latitudes. *Science*, **268**, 1030–1033.

Pizzo, V. J. (1978) A three-dimensional model of corotating streams in the solar wind. 1. Theoretical foundations. *J. Geophys. Res.* **83**, 5563–5572.

Pizzo, V. J. (1982) A three-dimensional model of corotating streams in the solar wind. 3. Magnetohydrodynamic streams. *J. Geophys. Res.* **87**, 4374–4394.

Pizzo, V. J. (1991) The evolution of corotating stream fronts near the ecliptic plane in the inner solar system 2. Three-dimensional tilted-dipole fronts. *J. Geophys. Res.* **96**, 5405–5420.

Pizzo, V. J. (1994) Global, quasi-steady dynamics of the distant solar wind 1. Origins of north-south flows in the outer heliosphere. *J. Geophys. Res.* **99**, 4173–4183.

Pizzo, V. J., and Gosling, J. T. (1994) 3-D simulation of high-latitude interaction regions: Comparison with Ulysses results. *Geophys. Res. Lett.* **21**, 2063–2066.

Riley, P. and Gosling, J. T. (1998) Do coronal mass ejections implode in the solar wind?, *Geophys. Res. Lett.* **25**, 1529–1532.

Riley, P., Gosling, J. T. and Pizzo, V. J. (1997) A two-dimensional simulation of the radial and latitudinal evolution of a solar wind disturbance driven by a fast, high-pressure coronal mass ejection. *J. Geophys. Res.* **102**, 14677–14685.

Riley, P., Gosling, J. T., Weiss, L. A. and Pizzo, V. J. (1996) The tilts of corotating interaction regions at mid-heliographic latitudes. *J. Geophys. Res.* **101**, 24349–24357.

Riley, P., Gosling, J. T., McComas, D. J., Pizzo, V. J., Luhmann, J. G., Biesecker, D., Forsyth, R. J., Hoeksema, J. T., Lecinski, A. and Thompson, B. J. (1999) Relationship between Ulysses plasma observations and solar observations during the Whole Sun Month campaign. *J. Geophys. Res.* **104**, 9871–9879.

Rust, D. M. (1994) Spawning and shedding helical magnetic fields in the solar atmosphere. *Geophys. Res. Lett.* **21**, 241–244.

Schwenn, R. (1990) Large-scale structure of the interplanetary medium. In: Schwenn R. and Marsch E. (eds) *Physics of the Inner Heliosphere 1*, pp. 99–181. Springer-Verlag, Berlin.

Sheeley, N. R., Howard, R. A., Koomen, M. J. and Michels, D. J. (1983) Associations between coronal mass ejection events and soft x-ray events. *Astrophys. J.*, **272**, 349–354.

Sheeley, N. R., Howard, R. A., Koomen, M. J., Michels, D. J., Schwenn, R., Mulhauser, K.-H. and Rosenbauer, H. (1985) Coronal mass ejections and interplanetary shocks. *J. Geophys. Res.* **90**, 163–175.

Siscoe, G. L., Goldstein, B. and Lazarus, A. J. (1969) An east-west asymmetry in the solar wind velocity. *J. Geophys. Res.* **74**, 1759–1762.

Smith, E. J. and Wolfe, J. H. (1976) Observations of interaction regions and corotating shocks between one and five AU: Pioneers 10 and 11, *Geophys. Res. Lett.* **3**, 137–140.

Smith, E. J., Neugebauer, M., Balogh, A., Bame, S. J., Erdös, G., Forsyth, R. J., Goldstein, B. E., Phillips, J. L. and Tsurutani, B.T. (1993) Disappearance of the heliospheric sector structure at Ulysses. *Geophys. Res. Lett.* **20**, 2327–2330.

Smith, E. J., Balogh, A. Burton, M. E. Erdös G. and Forsyth, R. J. (1995) Results of the Ulysses fast latitude scan: Magnetic field observations. *Geophys. Res. Lett.* **22**, 3325–3328.

Suess, S. T., Phillips, J. L., McComas, D. J., Goldstein, B. E., Neugebauer, M. and Nerney, S. (1998) The solar wind – inner heliosphere. *Space Sci. Rev.* **83**, 75–86.

Thomas, B. T. and Smith, E. J. (1981) The structure and dynamics of the heliospheric current sheet. *J. Geophys. Res.* **86**, 11105–11110.

Webb, D. F., and Howard, R. A. (1994) The solar cycle variation of coronal mass ejections and the solar wind mass flux. *J. Geophys. Res.* **99**, 4201–4220.

Webb, D. F., Krieger, A. S. and Rust, D. M. (1976) Coronal x-ray enhancements associated with Hα filament disappearances. *Solar Phys.* **48**, 159–186.

Weiss, L. A., Gosling, J. T., McAllister, A. H., Hundhausen, A. J., Burkepile, J. T., Phillips, J. L., Strong, K. T. and Forsyth, R. J. (1996) A comparison of interplanetary coronal mass ejections at Ulysses with Yohkoh soft x-ray coronal events. *Astron. Astrophys.* **316**, 384–395.

Wimmer-Schweingruber, R. F., von Steiger, R. and Paerli, R. (1997) Solar wind stream interfaces in corotating interaction regions: SWICS/Ulysses results. *J. Geophys. Res.* **102**, 17407–17417.

Wimmer-Schweingruber, R. F., von Steiger, R. and Paerli, R. (1999) Solar wind stream interfaces in corotating interaction regions: New SWICS/Ulysses results. *J. Geophys. Res.* **104**, 9933–9945.

4

Ulysses measurements of waves, turbulence and discontinuities

Tim Horbury and Bruce Tsurutani

4.1 INTRODUCTION

The *Ulysses* dataset has proved to be a unique resource in the study of waves, turbulence, and microstructures in the solar wind. In this chapter we introduce the work to date on turbulence and waves (which we generically term 'fluctuations') and discontinuities using *Ulysses* data and discuss how these results relate to earlier work. Some of the more important results are: the large amplitude Alfvén waves present in the polar heliosphere which enhance cosmic-ray diffusion; the large number of discontinuities in the polar solar wind; the phase steepening of Alfvén waves into rotational discontinuities; the slower turbulent evolution at high latitudes due to the absence of large-scale stream shear; precise measurements of turbulent intermittency; and the presence of magnetic 'holes' in high-speed wind. Finally, we discuss the possibilities for future work using *Ulysses* data, and the relevance of the upcoming solar-maximum polar passes to these subjects.

4.1.1 Why are *Ulysses* data unique?

While it is clear that the orbit of *Ulysses*, sending the spacecraft into previously unsampled regions of the heliosphere, is the principal reason that the resulting dataset is of interest, the relevance of this to magnetic-field and plasma fluctuations is perhaps less clear. In fact, there are several important consequences of the orbit.

The first and most obvious reason to study the population of discontinuities, waves, and turbulence in the polar heliosphere is to characterize their latitude and distance dependence. Magnetic-field fluctuations are important in the propagation of energetic particles and cosmic rays in the heliosphere and the polar heliosphere is particularly important in this regard near solar minimum. Consequently, an understanding of magnetic-field fluctuations, on a wide range of scales, in the polar heliosphere near solar minimum is vital if we are to understand energetic-particle propagation in the heliosphere.

Second, polar measurements of fluctuations are useful for comparison with those at low-latitudes. There is a wealth of low latitude measurements of solar-wind fluctuations, including their radial dependence. This radial variation is actually caused by a temporal development of the fluctuations as they are carried anti-sunward by the solar wind. As such, radial measurements can be used to study the time development of magnetohydrodynamic (MHD) waves, turbulence, and discontinuities. However, the highly disturbed low-latitude environment, with shocks, compressions, and rarefactions, makes it difficult to study the fluctuation development unaffected by these large-scale inhomogeneities. Indeed, it is not clear what effect these have on the development of the turbulence. As a result, *Ulysses* measurements of the radial variation of fluctuations in uniformly high-speed wind allow the study of this development largely unaffected by large-scale inhomogeneities.

Finally, the fact that *Ulysses* stayed within the same high-speed streams for many months makes it possible to study long periods of data that are nearly stationary in a statistical sense. At low latitudes spacecraft stay within individual streams for several days at most, so the *Ulysses* data offer the first opportunity to study datasets of weeks or more. These make it possible to study the characteristics of the fluctuations in much more detail than is possible at low latitudes. It is also worth noting that the fact that *Ulysses* is a 'survey' mission and therefore returns near-continuous data, with remarkably few gaps, makes it possible to construct datasets with very few missing points: this is vital for some analysis techniques.

In summary, then, the polar *Ulysses* dataset is useful because it is taken in a physically important region; it is a 'control' compared with the disturbed low-latitude environment; and it contains long intervals of near-stationary data. All of these properties are exploited in the work described in this chapter.

We approach the study of fluctuations from two perspectives: MHD turbulence, and waves. The Alfvén wave is the dominant MHD mode in the solar wind and we show for the first time that these non-linear waves are related to rotational discontinuities.

4.2 FLUCTUATIONS IN THE POLAR HELIOSPHERE

High-speed polar solar wind near solar minimum is remarkably homogeneous in comparison with low-latitude wind on the large scale. However, it is also very variable at small scales. Figure 4.1 shows bulk field and plasma data from a typical day of polar solar wind. There is variation, on essentially all scales shown, in all parameters. However, it is clear that variations in magnetic-field components are particularly large and indeed typically $\delta \mathbf{B}/|\mathbf{B}|$ is greater than 1. There are also a large number of discontinuities: sharp changes in field direction, sometimes accompanied by changes in field magnitude. We return to the topic of 'microstructure',

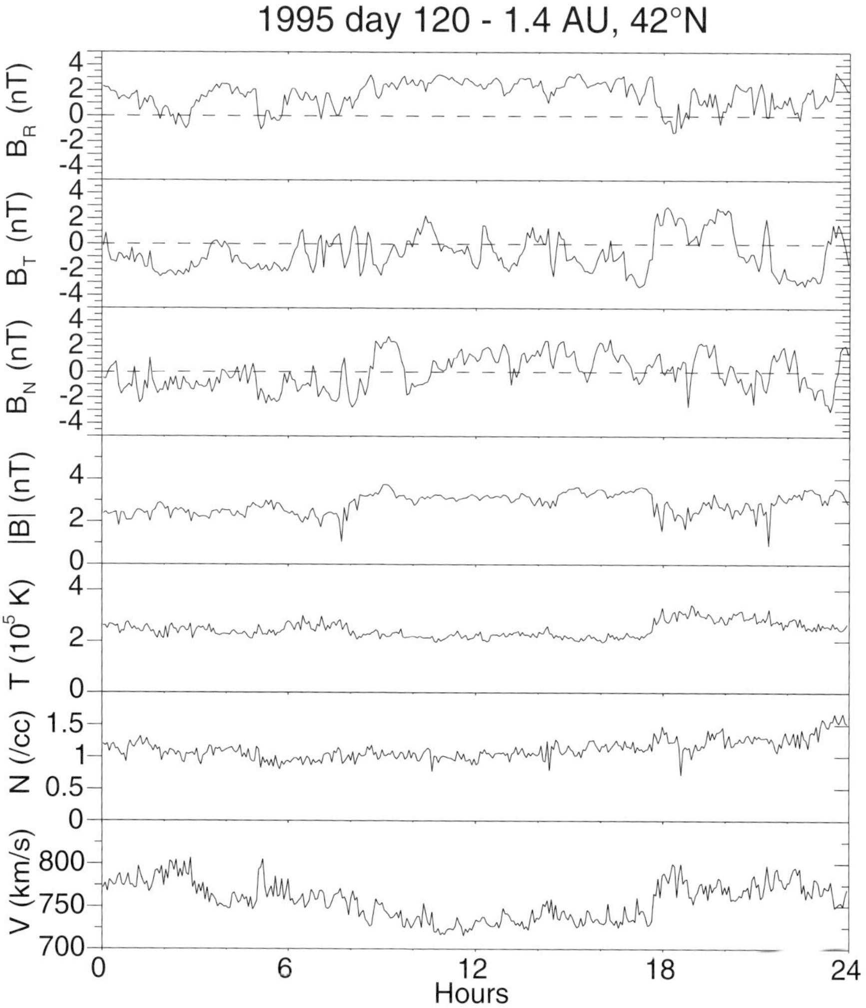

Figure 4.1. A typical day (1995 day 120, 1.4 AU and 42°N) of magnetic-field and plasma data in the polar heliosphere. The panels show variation in the three components and magnitude of the magnetic field, radial velocity, proton temperature, and proton density.

notably discontinuities, later in this chapter, but for now consider large scale, average properties of fluctuations in the polar heliosphere.

In general, fluctuations in high-speed polar wind have proved to be similar to those in high-speed streams at low latitudes (see Tu and Marsch, 1995; Goldstein and Roberts, 1995; and Matthaeus *et al.*, 1995, for recent reviews of observations of MHD waves and turbulence in the heliosphere). This is unsurprising and was predicted by several authors before the *Ulysses* observations (Roberts, 1990; Grappin *et al.*, 1991; Bruno, 1992). Quantitatively, however, it was unclear how polar turbulence, in undisturbed high speed wind streams, would develop relative to those at low latitudes: indeed, high-speed streams travelling at over $700\,\mathrm{km\,s^{-1}}$ are rarely observed at several AU at low latitudes.

In Section 4.2.1 we discuss the general properties of waves and turbulence in the polar solar wind near solar minimum, and compare them with low-latitude observations. In Section 4.2.2 we show how these fluctuations change with distance and latitude, while Section 4.2.3 deals with the processes within MHD turbulence and Section 4.2.4 discusses the relationship between fluctuations and other solar-wind structures.

4.2.1 General character of fluctuations

4.2.1.1 Magnetic-field power spectra

Power spectra of total component power (the trace of the vector power spectral matrix) and field-magnitude power for the same day of data as Figure 4.1 are shown in the left panel of Figure 4.2. As is clear from the time-series data in Figure 4.1, field magnitude variations are small compared with that in the field components. In addition, there are two distinct ranges of scales, which have different spectral indices (gradients of the log–log plots) in total component power in Figure 4.2.

At large scales, on frequencies below around $10^{-3}\,\mathrm{Hz}$, the spectral index is near -1. That is, power scales as $P \propto f^{-1}$. At higher frequencies the spectral index is

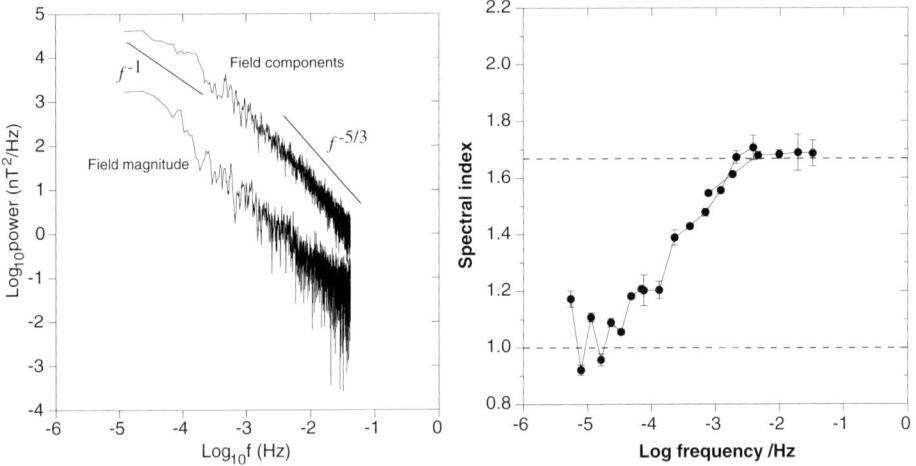

Figure 4.2. (left) Power spectra of total component and field-magnitude power for the same day as Figure 4.1. (right) Spectral index of field-component fluctuations as a function of frequency (from Horbury *et al.*, 1996b).

near $-5/3$. Such variation is difficult to identify by eye, however. It is shown more clearly in the right panel of Figure 4.2, taken from Horbury et al. (1996b), where spectral index values are plotted as a function of frequency for the N (normal) component of the magnetic field – other components are similar.

The presence of three regimes: $1/f$ fluctuations at large scales; $f^{-5/3}$ fluctuations at small scales; and a transition range between them, was identified in *Helios* measurements of high-speed wind at 0.3–1 AU (e.g. see Bavassano et al. (1982b)). Clearly, therefore, polar turbulence is similar to that at low latitudes in high-speed streams.

We note that frequencies above around 10^{-1} Hz are not easily accessible using the *Ulysses* magnetometer, due to influences of spacecraft background effects contaminating small-amplitude fluctuations at such small scales. However, fluctuations on these scales, near proton gyroscales, can be studied with relatively short intervals of data and therefore low-latitude spacecraft with higher cadence magnetometers (e.g. *WIND*) are well suited to such an analysis. Leamon et al. (1998) recently discussed the structure of fluctuations on these scales.

Fluctuations with a spectral index near $-5/3$ are often the result of a turbulent cascade – that is, non-linear transfer of fluctuation energy between scales – in neutral fluids (e.g. Frisch, 1995). The population of fluctuations in high-speed streams with a spectral index $\alpha \sim -5/3$ also appear to be turbulent, with an ongoing turbulent cascade. The strongest evidence for this is the change in shape of the power spectrum with distance. Bavassano et al. (1982b) showed that the range of scales where $\alpha \sim -5/3$ extended to lower frequencies with increasing distance from the Sun in high-speed streams. This is convincing evidence for energy transfer between scales and is interpreted as being due to the transfer of energy from large-scale fluctuations to those at smaller scales, with eventual dissipation as heating of the plasma.

The existence of fluctuations with a spectral index near -1 at large scales in high speed wind shows that on these scales significant energy has yet to be transferred to smaller scales. These large-scale fluctuations appear to be 'waves' which have not interacted since leaving the corona. At smaller scales there is significant non-linear interaction taking place and a cascade develops. We discuss this process in more detail in Sections 4.2.2 and 4.2.3.

4.2.1.2 Elsässer variables

The identification by Belcher and Davis (1971) of a highly correlated variation in the magnetic field and velocity in the solar wind established the existence of Alfvén waves in interplanetary space. Belcher and Davis found that the intervals of high correlation tended to occur after corotating interaction regions and indeed later work, particularly using *Helios* data (see Schwenn and Marsch, (1991) for a comprehensive review of *Helios* results) established that the correlation was highest in high-speed wind (which follows Corotating Interaction Regions (CIRs)) and often much lower in slow wind, although conditions in slow-speed streams are much more variable than those in fast streams, and intervals with high field-velocity correlations can occasionally persist to large distances (e.g. Roberts et al., 1987; Lucek and Balogh, 1998).

As is well known (e.g. Boyd and Sanderson, 1969), variation in the magnetic field $\delta\mathbf{B}$ and velocity $\delta\mathbf{v}$ due to an Alfvén wave are correlated:

$$\delta\mathbf{v} = \pm\delta\mathbf{b} \tag{4.1}$$

where

$$\delta\mathbf{b} = \delta\mathbf{B} \cdot \frac{A}{\sqrt{\mu_0 \rho}} \tag{4.2}$$

where ρ is the mass density and the factor A is related to pressure anisotropy in the plasma:

$$A^2 = 1 - \frac{4\pi(p_\parallel - p_\perp)}{\mathbf{B}^2} \tag{4.3}$$

where $p_\parallel$ and $p_\perp$ are the field-parallel and perpendicular pressures. Streaming pickup ions can also contribute to the pressure – Goldstein et al. (1995a) discussed this issue in some detail for *Ulysses* polar data. Here we assume $A = 1$ although in practice it is somewhat lower.

The field and velocity variation is anti-correlated when the wavevector is parallel to the mean field, and correlated when the wavevector is anti-parallel. While the correlation is never perfect in the solar wind, due to the presence of other wave modes at lower amplitudes, it can be as high as 0.9 at times. The sense of the correlation is typically such that the dominant wave population is propagating away from the Sun in the plasma frame, indicating a solar origin for these waves.

The identification of Alfvén waves in high-speed streams at low latitudes suggested that they would also be present in polar high-speed streams. This proved to be the case: Smith et al. (1995) showed that field and velocity variation was well correlated at high latitudes on spacecraft scales of several hours.

The presence of Alfvén waves is illustrated in Figure 4.3 which shows 24 hours of field and velocity data taken from north- and south-polar flows. During the *Ulysses* polar passes in 1994 and 1995, the southern polar coronal hole, and hence the southern polar wind, had an inward sense of the magnetic field. Outward-propagating Alfvén waves would therefore produce a positive correlation between the field and velocity, which is clear in the left panel of Figure 4.3. In the north, with the opposite sense of the large-scale field, the field and velocity were anti-correlated, as seen in the right panel. Both hemispheres therefore support a population of waves propagating away from the Sun.

The high field-velocity correlation in high-speed streams makes it useful to calculate Elsässer variables:

$$\mathbf{e}^\pm = \delta\mathbf{v} \pm \delta\mathbf{b} \tag{4.4}$$

and their resulting power spectra, usually termed $\mathbf{Z}^\pm$. For a pure Alfvén wave propagating parallel to the magnetic field, $\mathbf{e}^+$ would be zero and $\mathbf{e}^-$ would be twice the variation in $\mathbf{v}$ or $\mathbf{b}$, and the opposite would be true for an anti-parallel propagating wave. Elsässer variables therefore provide a simple and powerful tool

Sec. 4.2] Fluctuations in the polar heliosphere 173

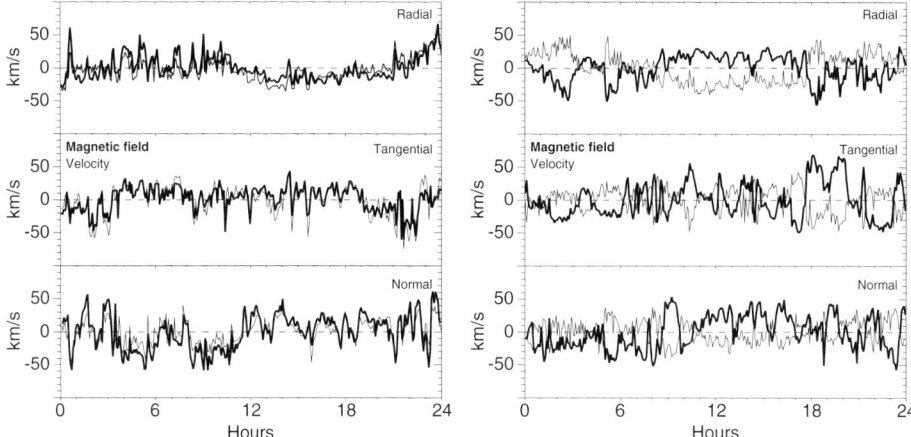

Figure 4.3. Time series of velocity (thin lines) and magnetic field (thick lines) fluctuations in high-speed polar wind. (left) 1995 day 20; (right) 1995 day 120. The highly Alfvénic nature of the fluctuations results in a high correlation or anti-correlation between variation in the magnetic field and velocity.

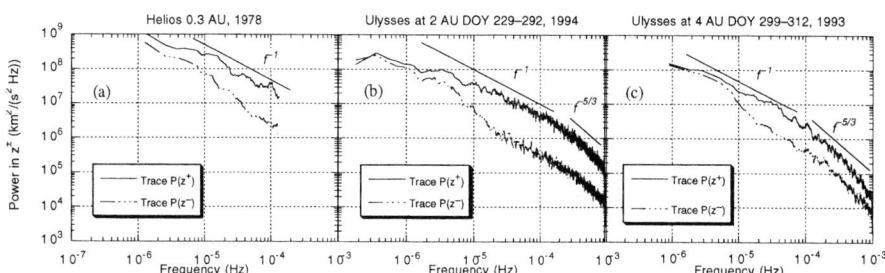

Figure 4.4. Elsässer variable power spectra for three intervals of high-speed solar wind: (a) 0.3 AU (*Helios*); (b) 2 AU (*Ulysses* polar), and (c) 4 AU (*Ulysses* polar) (from Goldstein et al., 1995b).

for selecting fluctuations of an Alfvénic character from time series. However, when power in the two variables is similar, their interpretation is more difficult: this could be due to equal energy in Alfvén waves propagating parallel and anti-parallel to the field, or the presence of non-Alfvénic modes.

It is conventional to 'flip' the sense of the mean magnetic field when necessary so that e^+ and hence Z^+ correspond to outward-propagating waves and e^- and Z^- inward. Then, in high-speed streams we would expect Z^+ to dominate Z^- and this is indeed the case (e.g. Marsch, and Tu, 1990; Roberts et al., 1987).

Figure 4.4 shows $Z^{\pm}$ power spectra in high-speed solar-wind streams at three solar distances: from *Helios* data at 0.3 AU and *Ulysses* polar data at 2 and 4 AU.

This figure, from Goldstein et al. (1995b), illustrates the well-established result that power in $\mathbf{Z}^+$ (the outward-propagating mode) dominates that in $\mathbf{Z}^-$ in high-speed wind, by up to an order of magnitude. When $\mathbf{Z}^+$ is much larger than $\mathbf{Z}^-$ the $\mathbf{Z}^+$ power spectrum approximates that of the magnetic-field power spectrum, and as a result the f^{-1} and $f^{-5/3}$ frequency regions in Figure 4.2 are also visible in Figure 4.4.

Goldstein et al. showed that, as is also observed at low latitudes, power in $\mathbf{Z}^+$ becomes less dominant over that in $\mathbf{Z}^-$ with distance. This is more easily seen using the normalized cross helicity, $\sigma_c = (\mathbf{Z}^+ - \mathbf{Z}^-)/(\mathbf{Z}^+ + \mathbf{Z}^-)$, which measures the relative dominance of the two propagation senses. When σ_c is near $+1$ (-1), $\mathbf{Z}^+$ ($\mathbf{Z}^-$) is the dominant sense. Goldstein et al. showed that σ_c decreased between 2 and 4 AU, again indicating a gradual equalization of the two modes, as a result of the development of the turbulent cascade.

Bavassano et al. (2000) calculated correlations between field and velocity variation, as well as σ_c, on 4-min scales for Ulysses polar data, and showed that the decline in σ_c with distance that was observed by Goldstein et al. (1995) stopped by around 2 AU, and σ_c remained approximately constant, near 0.4, at larger distances. Bavassano et al. (2000) argued that this surprising result meant that σ_c would remain significantly above zero in the distant heliosphere.

Goldstein et al. (1995b) also considered the ratio of kinetic to magnetic energy in fluctuations, the so-called Alfvén ratio, r_A. They showed that this was near $\frac{1}{2}$ for all frequencies above around 10^{-5} Hz, also as seen at low latitudes, showing that magnetic energy dominates. This result is well established in the solar wind, but is not well understood: we would expect $r_A \sim 1$ for a pure Alfvén wave. Goldstein et al. (1995b) have suggested that pressure anisotropy produced by interstellar pickup ions may act to reduce r_A, although not sufficiently to explain the values observed. Simulations (e.g. Schmidt, 1995) can also reproduce the observed reduction in r_A with increasing solar distance by including the effects of both turbulence and convected 'structures', without including pickup-ion effects.

4.2.1.3 Anisotropy

Field-aligned anisotropy is well known in MHD turbulence, with more power perpendicular to the background field than parallel to it. Anisotropy has also been studied in the solar wind (e.g. Coleman, 1968; Bavassano et al., 1982a; Belcher and Davis, 1971; Klein et al., 1991, 1993), where power has also been found to be largely field perpendicular.

As we saw at the beginning of Section 4.2, the amplitude of low-frequency fluctuations in the magnetic field are often comparable with the field magnitude, and so the field direction changes significantly as a result of these fluctuations. It is important, therefore, to consider the *local* background field direction when calculating field-aligned anisotropy, and not a large-scale one, or the Parker spiral direction. This is particularly important when considering small-scale fluctuations. If we calculate the minimum-variance direction (i.e. the direction in which the vector field varies least: see Sonnerup and Cahill (1967) for an introduction to minimum-variance analysis) over successive 5-min intervals, the mean field direction can

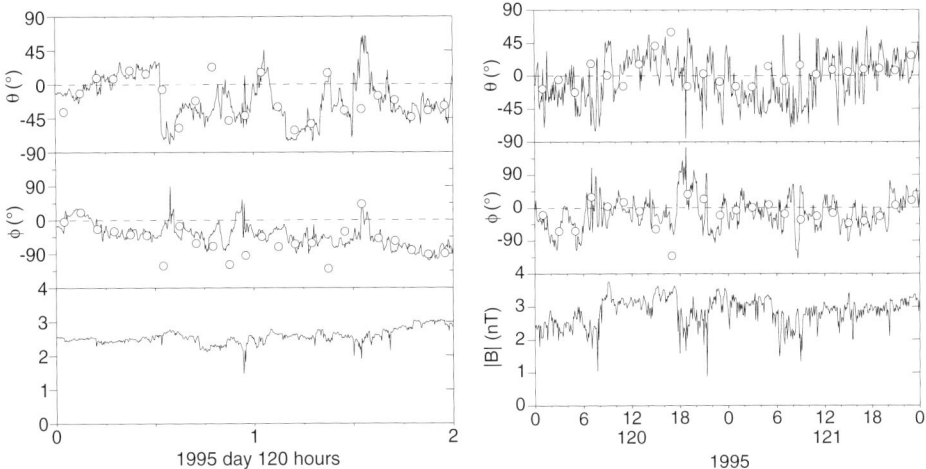

Figure 4.5. Minimum-variance directions (white circles) superimposed on the magnetic-field direction. (left) 2 hours of data from 1995 day 120. Each minimum-variance direction is calculated from 5 min of 12 s averaged data. (right) 2 days of data from 1995. Each minimum-variance direction is calculated from 2 hours of 5-min averaged data. At both scales the minimum-variance direction tends to follow the local field direction, indicating that the fluctuations are predominantly perpendicular to the mean field.

change significantly from one interval to the next, as a result of low-frequency, large-amplitude waves. Horbury *et al.*, (1995a) considered the anisotropy of such small intervals and showed that the minimum-variance direction of fluctuations closely followed the local background-field direction. This is illustrated in Figure 4.5 which shows how the minimum variance direction follows the background field on different scales, even when the field direction is highly variable. The small-scale, turbulent fluctuations 'ride' on the large-scale, lower frequency waves.

It is important, therefore, to consider the scale of fluctuations when calculating anisotropies. At the scale of hours the fluctuations have a radial minimum-variance direction (Forsyth *et al.*, 1996), while at small scales it is field aligned.

4.2.2 Large-scale evolution in the polar heliosphere

There are around 770 days of *Ulysses* data taken entirely within undisturbed high-speed ($\sim$700–800 km s^{-1}) solar wind in the polar heliosphere, covering a solar-distance range of around 1.3–4 AU and 30–80° heliolatitude. Using this extended dataset it is possible to study variation in fluctuations with distance and latitude. Knowledge of such variation is useful both for understanding turbulent energy transfer in magnetofluids and for estimating energetic-particle propagation throughout the polar heliosphere. In addition, such variation can yield information about conditions within the corona, as we will see.

4.2.2.1 Radial spectral-index variation

As discussed in Section 2.1.1 changes in the scale-dependent spectral index of magnetic-field fluctuations have been observed in high-speed streams at low latitudes, primarily in *Helios* data. These are interpreted as indicative of a turbulent cascade, as shown in Figure 4.6. The high-speed solar wind is seeded with a population of fluctuations in the corona. Above the Alfvén critical point this population is composed of outward-propagating Alfvén waves as well as compressive fast- and slow-mode waves. The compressive modes rapidly damp, leaving just outward-propagating Alfvén waves. As these waves are carried away from the Sun, those at the highest frequencies interact non-linearly with each other more quickly than those at lower frequencies. These non-linear interactions transfer energy to even smaller scales, and a cascade is produced. The small-scale limit of the cascade is the end of the MHD regime, where fluctuation energy is dissipated in heating the plasma.

Over time, progressively lower frequency waves interact, transferring energy to smaller scales, hence reducing power at their scale and steepening the power spectrum. The range of scales where the spectral index α is near $-5/3$ is, therefore, the turbulent inertial range; the energy transfer is also occurring in the transition scale, although more slowly; and there is effectively no energy transfer

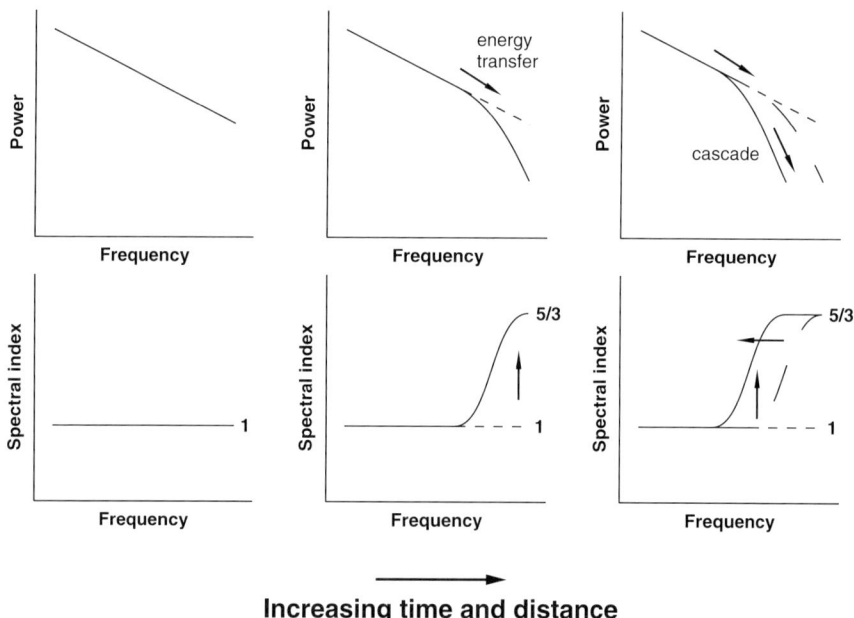

Figure 4.6. Turbulent energy and changes in the shape of the fluctuation power spectrum in high-speed wind. Power levels (top) and spectral index variation (bottom) are shown for three solar distances, with the left being closest to the Sun and the right furthest away. See text for discussion (from Horbury, 1996c).

within the population of waves at low frequencies where α is near -1. Indeed, the small-scale edge of the f^{-1} regime is effectively the largest scale at which no significant energy transfer has yet occurred. The fluctuations are the energy source for the ongoing turbulent cascade. We note in passing that the physical mechanism which produces the original $1/f$ Alfvén waves within the corona is not known, although various mechanisms have been suggested (e.g. see Matthaeus and Goldstein, 1986).

While it is well established that there is an active turbulent cascade at low latitudes, the role of large-scale dynamics in driving this process is not well understood. Roberts (1990), Grappin et al. (1991) and Bruno (1992) suggested that the increase in extent of the inertial range with distance observed at low latitudes might happen more slowly in the polar heliosphere, where the absence of large-scale stream structure meant that the fluctuations were less disturbed.

Horbury et al. (1996c) showed that, based on an analysis of data in 1994 while Ulysses was travelling towards 80°S, there was a slight change in spectral indices with time. It was not clear whether this change was a latitudinal or radial one, however. Horbury et al. (1995b) extended this analysis past 80°S, and showed that the trend continued past the highest latitude and that it was therefore radial, not latitudinal. They concluded that the turbulent cascade in the polar heliosphere was an active, ongoing one, and therefore that MHD turbulence would develop from Alfvén waves in the absence of the large-scale stream structure that is present at low latitudes.

To measure these changes more precisely and quantitatively compare Ulysses observations with earlier results, Horbury et al. (1996b) used structure function measurements of the magnetic-field from both the northern and southern hemisphere to study the variation with solar distance of the magnetic-field spectral index. For many intervals of polar data, they calculated the scale dependence of the spectral index, and estimated the 'breakpoint', the transition scale between the inertial range and the $1/f$ waves, defining this scale as that where the spectral index was -1.4, between the turbulence ($\alpha \sim -5/3$) and 'wave' ($\alpha \sim -1$) values. In this way they measured the distance dependence of this scale and hence of the inertial range. They found no significant difference between fluctuations in the northern and southern hemispheres, as expected.

Horbury et al. found that the breakpoint scale varied linearly with solar distance; that is, the timescale of the transition τ varies with solar distance r as $\tau \propto r^x$, where the exponent $x = 1.1 \pm 0.1$. This result is consistent with Alfvén waves decaying – that is, transferring significant energy to smaller scales – in a certain number of wave periods. If this number is n then the time taken for a wave of period t to decay is nt, which corresponds to a solar distance $d = nt/v_{sw}$. Therefore, the edge of the $1/f$ region – or equivalently the breakpoint scale, which is proportional to t – would be a linear function of distance, as observed.

4.2.2.2 Radial and latitudinal variation in power and spectral index

The nature of the Ulysses polar orbit, with only two distance samples at each latitude, means that it is not possible to unambiguously determine the latitude and distance dependencies of fluctuations without some assumption as to their form.

Jokipii *et al.* (1995) assumed that variation in magnetic-field variances along *Ulysses*' orbit within the high-speed polar solar wind were entirely due to changes in the solar distance of the spacecraft that is, that there was no latitude or temporal variation. Jokipii *et al.* considered fluctuations on two spacecraft scales: 1 hour and 20 days. The smaller scale is short enough such that fluctuations on this scale can be considered to be waves within the plasma, on scales of around 3×10^6 km in the plasma frame. We have seen that waves on these scales are not turbulent in the polar solar wind, but have a power spectral index $\alpha \sim -1$. These are rather pure Alfvén waves propagating anti-sunward in the solar wind frame. Such non-interacting waves, propagating through a slowly varying medium, can be approximated by a WKB (Wentzel, Kramers, Brillouin) model (e.g. Bazer and Hurley, 1963) which predicts a power decrease with distance of $P(r) \propto r^{-3}$. Jokipii *et al.* showed that magnetic-field power levels on hourly scales did indeed decrease with distance as r^{-3}, consistent with the waves being non-interacting (i.e. with no energy transfer between scales) on the range of distances covered by *Ulysses*.

On 20-day scales, a significant fraction of a solar rotation, Jokipii *et al.* (1995) argued that fluctuations could not be considered as waves in the plasma. Rather, they argued, fluctuations are caused by 'field line random walk' on the Sun (e.g. produced by supergranular motion dragging field lines in the photosphere). In this case, as shown by Jokipii and Parker (1968), power levels should decrease as r^{-2}, more slowly than the WKB case. Jokipii *et al.* (1995) showed that *Ulysses* measurements of power levels on the 20-day scale were consistent with a decrease as r^{-2}. As Jokipii *et al.* pointed out, this slow decrease in power with distance at large scales results in larger power levels far from the Sun than would otherwise be the case, resulting in enhanced effective diffusion coefficients for energetic cosmic rays.

Using a similar method to Jokipii *et al.* (1995), Forsyth *et al.* (1996) considered the radial variation in magnetic-field variance on hourly scales in the undisturbed polar solar wind. Forsyth *et al.* stressed the importance of removing all large-scale compressive structures from the analysed data, which otherwise produce spurious power and can alter the apparent radial scaling. They showed that the hourly variance of magnetic field components varied with distance as $r^{-\alpha}$ with $\alpha \sim 3.4$, a faster radial decrease than WKB. Forsyth *et al.* interpreted this as evidence of energy transfer in a turbulent cascade and concluded that measurements of radial power scaling at low latitudes, which often found r^{-3} variation (e.g. Bavassano and Smith, 1986), could, be influenced by compression effects. As we will see later in this section, however, fluctuations on different scales vary differently with distance, producing a more complex scenario than that discussed by Forsyth *et al.*

Forsyth *et al.* (1996) considered variance of all three magnetic-field components, and the field magnitude, separately on hourly scales, in contrast to Jokipii *et al.* (1995) who only considered total component variance. Forsyth *et al.* showed that there was considerably less power in the radial field component than either of the two perpendicular components. This radial minimum-variance direction is consistent with expectations based on WKB theory, but not usually observed at low latitudes (e.g. Klein et al., 1991). Forsyth et al. also showed that, while power in the field magnitude was much lower than in the components, it decreased much less rapidly with distance, as $r^{-2.5}$.

More recently, Horbury and Balogh (2000) have analysed the entire north- and south-polar magnetic-field datasets with the aim of calculating the latitude and distance dependence of power levels and spectral indices on a number of different scales. Taking successive 5-day intervals of field data they estimated power levels and spectral indices in several ranges of wavenumber (a wavenumber k is related to spacecraft frequency f and solar-wind speed V_{SW} as $k = 2\pi f/V_{SW}$). They then fitted the variation in power P and spectral index α, along the *Ulysses* polar orbit for each wavenumber band, to a distance (r) and latitude (θ) dependence of the form:

$$\log_{10} P(k) \propto A(k) \log_{10} r + B(k) \cos\theta \tag{4.5}$$

$$\alpha(k) \propto C(k) \log_{10} r + D(k) \cos\theta \tag{4.6}$$

The results of this procedure are shown in Figure 4.7 for the field components – the panels, from top to bottom, show the parameters A, B, C and D from equations (4.5) and (4.6), and finally the spectral index $\alpha(k)$ at 2.5 AU for reference.

There is not space in this text to discuss all the implications of the data in Figure 4.7, so we only point out some of the more obvious features. The bottom panel shows that the transition from f^{-1} Alfvén waves to $f^{-5/3}$ turbulence occurs at around $k = 10^{-6}\,\mathrm{km}^{-1}$ – although all scales have rather poorly defined boundaries, and it is not possible to identify a boundary to better than around a factor of 2 in wavenumber. For wavenumbers between 4×10^{-7} and $10^{-6}\,\mathrm{km}^{-1}$, the scales on which the Alfvén waves are present, the top panel shows that field component power scales with distance as $P \propto r^{-3}$, as expected from WKB theory (e.g. Bazer and Hurley, 1963) for non-interacting waves in the solar wind. In contrast, at smaller scales (larger wavenumbers) power decreases more rapidly with distance, as a result of the turbulent cascade which transfers power to smaller scales. Power at wavenumbers below around $10^{-7}\,\mathrm{km}^{-1}$ actually decreases more slowly than the WKB result, consistent with the results of Jokipii et al. (1995) and their interpretation of these as 'structures' passing the spacecraft.

It is rather surprising that the range of scales where WKB scaling holds is so small – around a factor of 4 or so. However, there is likely to be some 'leakage' of power from one scale to another as shown in Figure 4.7, smearing out the edges between different scales. As a result, each range probably appears smaller in Figure 4.7 than it really is, and transitions between them occur over a wider range of scales. Nevertheless, the WKB scale range still covers less than an order of magnitude. This may be simply due to the fact that the highest frequency of these fluctuations is gradually decaying, moving the small-scale (high wavenumber) end of the WKB range to lower wavenumbers with increasing distance, and that by several AU most of these waves have decayed. However, the spacecraft scale of the large-scale end of the WKB regime, which is the scale on which 'structures' become important, is also likely to be a function of distance (e.g. Jokipii et al., 1995), so it may be that

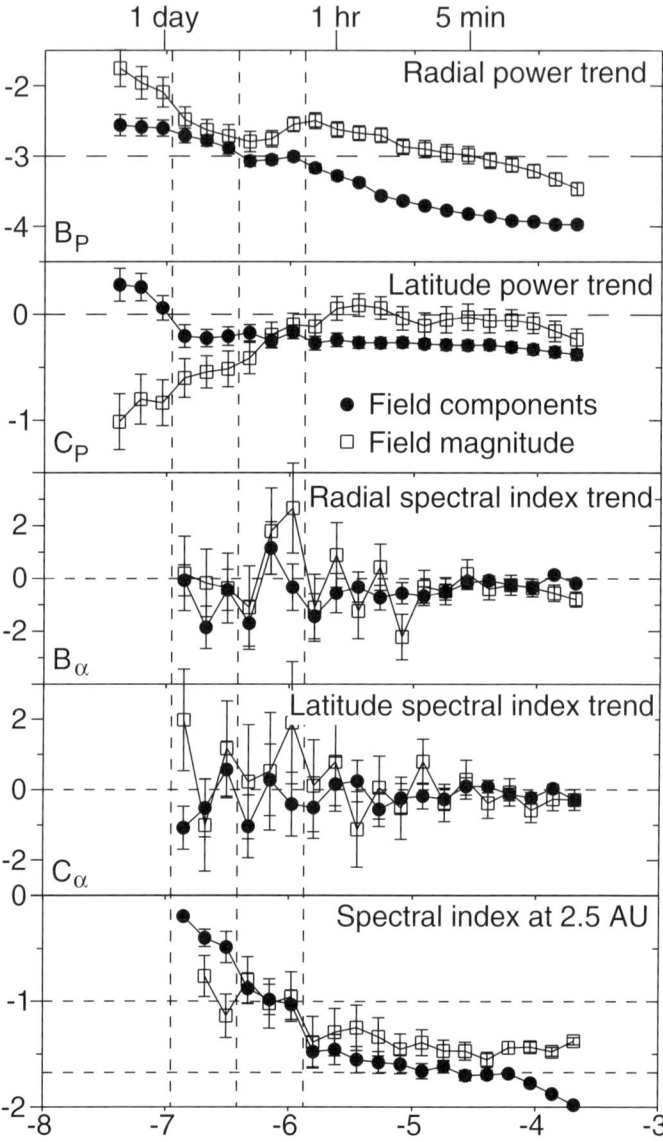

Figure 4.7. Distance and latitude dependence of power and spectral index for magnetic-field components as a function of wavenumber in the polar solar wind. Panels show, from top to bottom, exponent of radial-power variation; coefficient of latitudinal-power variation; coefficients of radial and latitudinal spectral-index variation; and spectral index at 2.5 AU for reference (from Horbury and Balogh, 2000).

both the large- and small-scale ends of the WKB-like regime move to progressively lower wavenumbers with solar distance, in the spacecraft frame.

The second panel of Figure 4.7 shows that, for wavenumbers above around 4×10^{-7} km^{-1} – (i.e. all scales where fluctuations sampled are waves or turbulence, rather than 'structures') there is a significant, scale-independent variation in power with latitude, with lower power at higher latitudes. Horbury and Balogh (2000) interpreted this latitude variation as being due to latitude-dependent overexpansion of the polar coronal hole from which the solar wind and fluctuations originate: the approximately dipolar polar magnetic field, with a higher magnitude at higher latitudes, combined with overexpansion which eventually equalizes field magnitude (Smith and Balogh, 1995), produces a larger expansion factor at higher latitudes and hence lower amplitude fluctuations. The absence of this effect – and indeed its reversal – for wavenumbers below $\sim 4 \times 10^{-7}$ km^{-1} is, again, consistent with the presence of structures at these large scales.

Estimates of radial variation in spectral index (shown in the third panel), while subject to large uncertainties, show a significant negative trend for wavenumbers above 10^{-6} km^{-1}, as expected, in the turbulent cascade: this is evidence of the steepening of the power spectrum with distance, as depicted in Figure 4.6. This radial steepening is not present at the smallest scales depicted in Figure 4.7, indicating that the spectral index has reached the final inertial range value, near $-5/3$. Finally, the fourth panel of Figure 4.7 shows that there is no significant latitude variation in spectral index in the polar heliosphere: the dynamical 'age' of the turbulence (i.e. how developed it is) does not vary with latitude.

The results in Figure 4.7 are consistent with those of earlier work, from both low-latitude and *Ulysses* data. The gradual 'aging' of the turbulence (i.e. the decay of Alfvén waves generated in the lower corona as they are carried outwards in the solar wind and the subsequent turbulent cascade) is reflected by changes in the radial power scaling and the spectral index. While *Ulysses* data allows the measurement of this process to around 4 AU, it is likely to continue into the distant heliosphere.

The measurements in Figure 4.7 are important for calculations of energetic-particle propagation in the polar heliosphere. Using such data it is possible to estimate power levels on many scales over a wide range of distance and latitude near solar minimum. Indeed, combining these data with the results of Horbury *et al.* (1995) regarding the rate of change of the 'breakpoint' scale with distance, we can estimate, at least crudely, the power spectrum over a wide range of scales at essentially all solar distances in the heliosphere. Such estimates are essential for calculating energetic-particle diffusion rates in the heliosphere (e.g. see Fisk *et al.* 1998).

Bavassano *et al.* (1998) considered magnetic-field and plasma data from *Ulysses* on 1-, 4- and 12-hour scales, at low, mid- and high latitudes. They showed that σ_c, the normalized cross-helicity, was lowest at mid-, not low or high, latitudes. This initially surprising result was interpreted by Bavassano *et al.* as being due to the very strong velocity shear present at mid-latitudes. Velocity shear is so strong at mid-latitudes during the declining phase of the solar cycle, when these observations were made, because of the presence of fast, polar solar wind ($V_{SW} \sim 750$ km s^{-1}) and slow, low-latitude wind ($V_{SW} \sim 350$ km s^{-1}): both are present only at mid-latitudes.

Bavassano et al. argued that this shear drove the dynamical evolution of the fluctuations.

Bavassano et al. (1998) also showed that magnetic-field energy dominated kinetic energy in the fluctuations at all latitudes including the high-speed polar solar wind. In other words, the Alfvén ratio was less than 1 (see Section 4.2.1.2), as previously reported by Goldstein et al. (1995b). However, Bavassano et al. showed that this dominance was greatest in the polar solar wind, which was dynamically less evolved. Therefore, they argued that the view that dynamical evolution of the fluctuations drives the reduction of the Alfvén ratio, from near 1 close to the Sun to near 0.5 at several AU, was incorrect, because the least evolved plasma had the lowest value of the Alfvén ratio. Rather, Bavassano et al. argued that, as suggested by Goldstein et al. (1995a), pickup ions could alter the plasma pressure anisotropy and produce an apparent increase in the dominance of magnetic energy. This is likely to be negligible at low latitudes, particularly between 0.3 and 1 AU, but is significant at high latitudes between 2 and 4 AU.

4.2.2.3 Comparison with low latitudes

Ulysses measurements of fluctuations at high latitudes complement those recorded by earlier low-latitude spacecraft. As we discussed in Section 4.1.1, the presence of both fast and slow solar-wind streams at low latitudes means that streams interact as they travel away from the Sun. In this way, high-speed streams are slowed – and the fluctuations within them are inevitably altered by this process. By 1 AU most high-speed streams have been slowed significantly by stream–stream interactions: it is rare to observe solar-wind speeds over $700\,\mathrm{km\,s}^{-1}$ at 1 AU, while *Ulysses* measurements show that the high-speed polar wind travels at around $750\,\mathrm{km\,s}^{-1}$. We must therefore travel to less than 1 AU to sample undisturbed solar wind at low latitudes. To study fluctuations emanating from polar and equatorial coronal holes, it is of interest to compare undisturbed fluctuations at low and high latitudes and therefore low-latitude high-speed streams at less than 1 AU.

The two *Helios* spacecraft travelled repeatedly between 1 and 0.3 AU – the closest solar approach of any spacecraft – for several years. Results from these missions showed that at 0.3 AU high-speed streams ($V_{SW} > 700\,\mathrm{km\,s}^{-1}$) existed for several days during each solar rotation. Stream–stream interactions are not, therefore, significant at 0.3 AU, but develop between this distance and 1 AU.

The *Helios* data have been studied in considerable detail: see Schwenn and Marsch (1991) for a comprehensive review. In particular, the development of turbulence within high- and low-speed streams has been a topic of active interest. As discussed in Section 4.2.2.1 for the polar solar wind, fluctuations in high-speed streams at low latitudes 'evolve' as they travel away from the Sun, with the break-point scale between $1/f$ waves and $f^{-5/3}$ turbulence moving to larger scales with distance.

Marsch and Tu (1996) compared measurements of the scale dependence of the spectral index measured by Horbury et al. (1995c) with *Helios* observations at 0.3 AU and showed that turbulence observed by *Ulysses* further from the Sun was

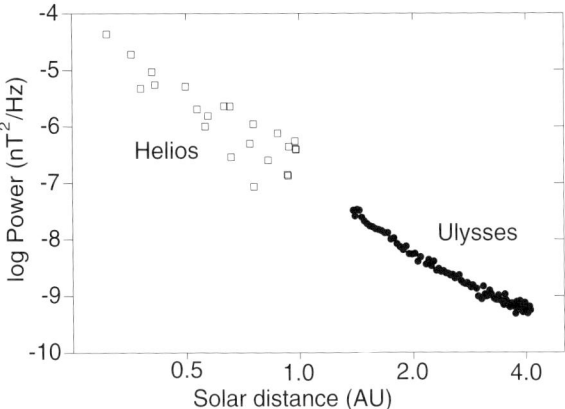

Figure 4.8. Comparison of power levels measured in high-speed streams between 0.3 and 1 AU by *Helios 1* and *Ulysses* polar measurements.

indeed more evolved than that seen by *Helios*, as expected. However, it was not clear from this work whether the turbulence was evolving more rapidly (or at the same rate) at high or low latitudes. The presence of stream–stream interactions such as shocks, compressions and rarefactions at low latitudes could 'drive' the fluctuations, producing a faster development and hence a more rapid change in the power spectrum than that seen at high latitudes where these interactions are not present. However, the presence of microstreams at high latitudes (Neugebauer *et al.*, 1995) may provide sufficient velocity shear and compression to force the development of the fluctuations. A comparison of power levels and development rates at high and low latitudes would help to address these issues.

Figure 4.8 shows power levels in magnetic-field fluctuations for low-latitude *Helios* measurements of high-speed streams between 0.3 and 1 AU and *Ulysses* polar measurements out to 4 AU. The decrease in power levels with distance, largely due to the expansion of the solar wind, is clear, as is the close agreement between the two datasets. This is despite the data being taken around 16 years apart, from solar wind emanating from two different coronal holes, and illustrates the remarkable consistency of fluctuations in coronal holes.

The large scatter in *Helios* measurements, which is larger at larger solar distances, is due to stream–stream variation, probably caused by compressions or rarefactions. The reduced scatter near 0.3 AU is due to the lack of such interactions at these distances. *Ulysses* data have a small scatter by comparison, as a result of the homogeneity of power levels in the polar heliosphere.

Clearly, power levels are similar at high and low latitudes in high-speed streams within the accuracy of the measurements. We turn our attention now to variation in the dynamical 'age' of the fluctuations (i.e. how developed they are) and

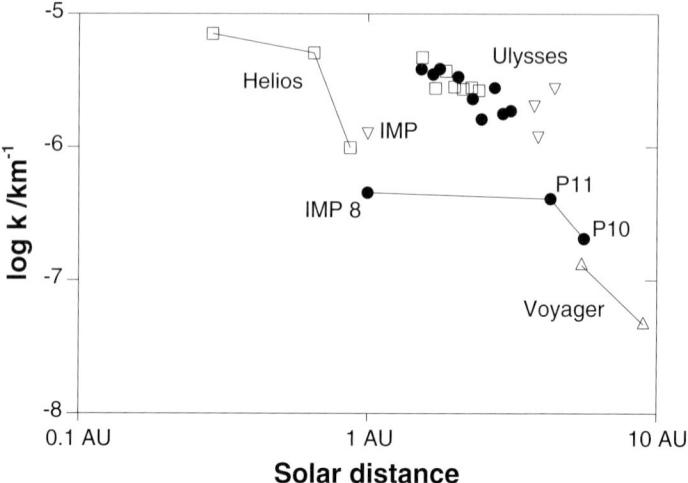

Figure 4.9. Variation of spectral breakpoint with solar distance in and out of the ecliptic. *Ulysses* measurements in high-speed flows are marked by circles (south-polar pass), squares (north-polar pass) and triangles (three high-speed velocity declines at mid-latitudes). Published data are also shown from *Helios* high-speed streams (squares); *IMP* (triangle); *Voyager* (joined triangles) and a three spacecraft study of the same plasma region at different solar distances (joined circles) (from Horbury et al., 1996b).

whether low-latitude fluctuations are driven or forced by their inhomogeneous environment.

Horbury et al. (1996b) compared their measurements of the 'breakpoint' in power spectra, discussed in Section 4.2.2.1, with previously published measurements of the breakpoint at different distances at low latitudes. These results are summarized in Figure 4.9, which shows the breakpoint wavenumber (which represents the scale of the smallest Alfvén wave that has yet to transfer significant energy into the turbulent cascade) as a function of solar distance – evolution corresponds to a decrease in this value with distance. It is clear from Figure 4.9 that *Ulysses* polar measurements show the presence of a significantly less evolved population than is present at low latitudes at similar distances. However, care is needed when interpreting Figure 4.9: different estimates of the breakpoint scale were made in different ways, so a precise comparison is difficult. The simplest comparison is with the three *Helios* data points from high-speed streams – even with just these three points, the slower polar evolution is clear.

The faster low-latitude development seen in Figure 4.9 was interpreted by Horbury et al. (1996b) as evidence for the driven evolution of low-latitude fluctuations. With this interpretation, high-latitude fluctuations represent the behaviour of 'unforced' magnetofluid turbulence, again emphasizing the importance of *Ulysses* data in understanding MHD turbulence throughout the Universe. Nevertheless, the polar solar wind is not entirely without velocity shear, compression, and rarefaction.

The most obvious structures are microstreams (e.g. Neugebauer et al., 1995) with velocity shears of several tens of km s^{-1} over scales of a few days. It is not clear how fluctuations interact with this relatively weak variation, or indeed with smaller scale structures such as discontinuities. This is a subject of considerable importance, since understanding how MHD turbulence develops is vital for predicting its amplitude in many poorly sampled heliospheric and astrophysical environments where it is dynamically important. In addition, it is clear that the different turbulent evolution at low and high latitudes has implications for energetic-particle transport in the heliosphere: models should not assume the same fluctuations at all latitudes at the same distance.

Even though low-latitude fluctuations are forced by stream–stream interactions, close to the Sun, before these interactions occur, fluctuations in low- and high-latitude high-speed streams should be the same. Horbury et al. (1996b) tested this hypothesis by comparing the shape of power spectra calculated by Bavassano et al. (1982b) from *Helios* data at 0.3 AU, where stream–stream interactions are undeveloped, with *Ulysses* polar data at 2.4 AU. As expected, the fluctuations at 0.3 AU were significantly less evolved than those at 2.4 AU, with a much smaller inertial range and a breakpoint at a higher frequency. Horbury et al. then 'shifted' the *Helios* spectrum to 2.4 AU using the measured polar evolution rate; that is, given their measurements of how the power spectrum changed with distance in the polar heliosphere, they altered the unevolved 0.3 AU spectrum in the same way to 'evolve' it to 2.4 AU. After this simple procedure, the *Ulysses* spectrum and the shifted *Helios* spectrum were consistent, confirming the hypothesis that fluctuations in high-speed solar wind from coronal holes are similar at 0.3 AU, and therefore are likely to be similar within the coronal holes themselves.

4.2.3 Small-scale turbulent processes

Using the long time periods of nearly stationary data taken within high-speed polar-wind streams it is possible to study the dynamics of magnetofluid turbulence with considerably greater precision than is possible with low-latitude spacecraft data.

4.2.3.1 Intermittency

Intermittency is a well-established property of neutral fluid turbulence – it is simply a measure of the spatial inhomogeneity of energy transfer in a turbulent medium (see Frisch (1995) for a general introduction to intermittency in neutral fluids), with some parts of the fluid dynamically active and other parts relatively quiet. Energy is continually transferred between scales (usually from large to small) in a turbulent fluid, but this process does not occur at an equal rate throughout the fluid. In neutral fluids this has a relatively straightforward interpretation because energy transfer is produced by velocity shear: shear is greatest at the edges of 'eddies' or vortices in the fluid, but relatively low in their centres.

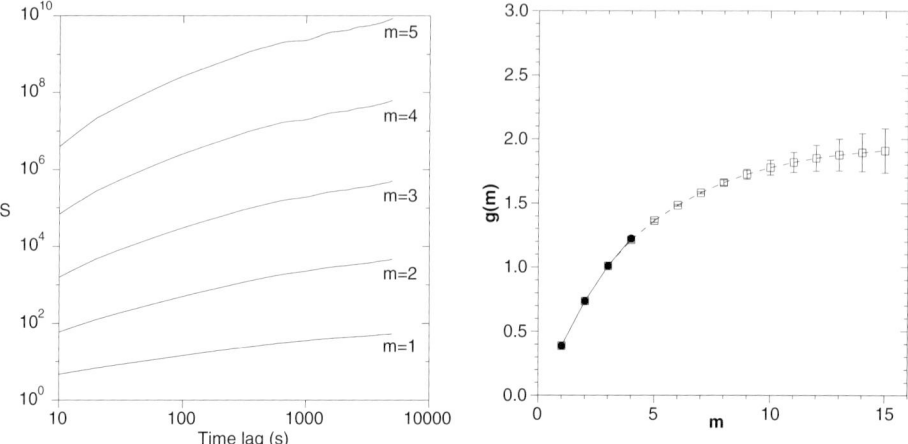

Figure 4.10. (left) Structure-function values for a range of scales and moments, calculated from 5 days of polar magnetic-field data. (right) Gradients of structure functions on spacecraft scales of 20–60 s, corresponding to solar-wind scales of about 1.5×10^4 to 4.5×10^4 km. Error analysis rejects all but the first four (filled) moments (from Horbury and Balogh, 1997).

Intermittency can be measured using structure functions. For a time series of velocity measurements $u(t)$ structure functions are defined as:

$$s(\tau, m) = \langle |u_i(t+\tau) - u_i(t)|^m \rangle \tag{4.7}$$

where i is a component of the velocity and $\langle \; \rangle$ denotes an average over the dataset. Structure functions are functions of a time scale τ and a moment m. They measure the dependence of fluctuations on scale, like a power spectrum, but while a power spectrum is effectively a second-order analysis, structure functions analyse a range of orders m. Structure functions calculated from a typical interval of magnetic-field data within high-speed solar wind are shown in Figure 4.10.

In general, we are interested in the scaling properties of structure functions of different moments. In the inertial range of a turbulent fluid (i.e. far from energy input and output scales and when the average energy transfer between scales is steady in time, so the transfer rate at every scale is equal) structure functions are power-law functions of scale:

$$s(\tau, m) \propto \tau^{g(m)} \tag{4.8}$$

It is the dependence of the $g(m)$ functions on the moment m that is of interest.

In neutral fluids, the energy transfer rate at a scale l, $\langle \varepsilon(l) \rangle \propto l \cdot \langle \mu(l)^3 \rangle$ (Kolmogorov, 1941, hereafter K41 – see also Frisch, 1995). In the inertial range, where ε is independent of scale, we would therefore expect u^3 to scale linearly with l – or, from equations (4.7) and (4.8), $g(3) = 1$. This is indeed the case (e.g. Frisch, 1995). If $\varepsilon(l)$ was spatially invariant, then structure functions of moments other than 3 would be simple powers of $\langle u(l)^3 \rangle$, and so their gradients would be linearly related to $g(3)$, and in general $g(m) = m/3$. Experimentally, significant deviations of $g(m)$ from these values are found for inertial range neutral fluid turbulence for $m \geqslant 3$. This variation is interpreted as the result of spatial variation in ε (i.e. intermittency). Several models of intermittency exist, although most are almost entirely phenomenological: they assume particular forms for the distribution of energy transfer within the fluid (e.g. Borgas, 1992; Frisch, 1995).

A number of authors have calculated structure functions of solar-wind magnetic field and/or velocity data (e.g. Burlaga, 1991; Marsch and Liu, 1993; Marsch and Tu, 1994; Carbone, 1994): see Marsch and Tu (1997) for an excellent recent review. Ruzmaikin et al. (1995) and Horbury et al. (1996c) calculated structure functions from Ulysses polar magnetic-field data and compared their results with several models of intermittent turbulence. As discussed in Section 4.1.1, long intervals of polar data are available, making more precise calculations possible than is the case with low-latitude spacecraft. In addition, the lack of large-scale stream structure means that we can be sure that the fluctuations are developing undisturbed.

Horbury and Balogh (1997) showed that structure function estimates for larger moments (typically above around 4 or 5) are strongly influenced by a small number of data points and are therefore statistically unreliable: a typical example is shown in Figure 4.10. The number of reliable moments is not a strong function of the length of the dataset. Previously published structure functions were affected by this problem, making some conclusions about the intermittency of solar-wind turbulence, and particularly the accuracy of models of the intermittency, unreliable.

Horbury and Balogh showed there was still useful information in the remaining structure functions after allowing for these limitations. After making a careful comparison of several models with their data, they concluded that the p model of Meneveau and Sreenivasan (1987) was a good description of the data. This model has one free parameter p which describes the inequality of energy transferred to daughter eddies by a parent: $p = \frac{1}{2}$ corresponds to no intermittency, when $g(m) = m/3$, while $p = 1$ is the maximally intermittent case. Values of the structure function scaling parameters for the p model are given by:

$$g(m) = 1 - \log_2(p^{m/3} + (1-p)^{m/3}) \qquad (4.9)$$

Typically, $p \sim 0.7$ in terrestrial fluids (Meneveau and Sreenivasan, 1887) and Horbury and Balogh, (1997) found $p \sim 0.7$ was a good description of polar turbulence – an example is shown in the left panel of Figure 4.11 – although there was large variation between data intervals. We discuss this variation in the next section.

In neutral fluid turbulence, intermittency is relatively simple to interpret as the spatial inhomogeneity of energy transfer, as we discussed at the beginning of this section. As we have seen, the signatures of intermittency seen in neutral fluids are also observed in MHD turbulence. In this case, however, the interpretation is considerably more difficult. There is no equivalent of the K41 relation for MHD fluids, and it is therefore difficult to link velocity or magnetic-field fluctuations with energy transfer.

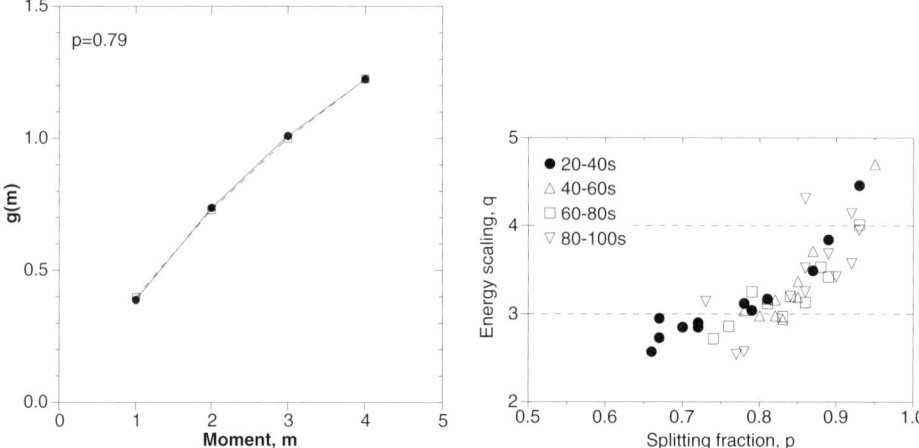

Figure 4.11. (left) Observed $g(m)$ values (black circles; errors are smaller than the symbol height) and a least-squares fit of the generalised p model (open squares). (right) Estimates of the intermittency measure p and energy scaling q from several intervals of data and several time scales in polar data. The left hand termination of the $p - q$ curve, at $p \sim 0.7$, $q \sim 3$, is likely to reflect the underlying inertial range values (from Horbury and Balogh, 1997).

Intermittency is related to temporal or spatial changes in the level of fluctuations, and the presence of large 'jumps'. In the solar wind the presence of large numbers of discontinuities, as discussed in detail in Section 4.3, naturally introduces 'jumps' in the magnetic field and could contribute to apparent intermittency. In addition, the known field-aligned anisotropy of MHD turbulence, with higher power perpendicular to the field, could produce apparent changes in fluctuation levels in one direction over time as the background field direction changed. However, these two effects do not seem to contribute significantly to observed intermittency levels. While the role of discontinuities in the energy transfer process is not clear, they undoubtedly interact with and alter the turbulence which propagates around and through them.

What, then, is the physical cause of intermittency in MHD turbulence and what is its significance? At this time, there is no satisfactory answer to this problem. It is apparent that intermittency is generated by the turbulent cascade. The remarkable similarity of the observed intermittency in MHD and neutral fluids suggests that this is a universal property of turbulence wherever it occurs, but until a more satisfactory theory of MHD turbulence, including intermittency, can be developed the role of intermittency will remain unclear.

4.2.3.2 Energy transfer

As discussed in the previous section, in neutral fluids there exists an exact relationship between the energy transfer rate and velocity shear, with the consequence that the third-order structure function scaling parameter $g(3) = 1$ in the inertial range. This result is independent of any spatial inhomogeneity in the energy transfer, and therefore in intermittency, and has been confirmed by experiment (e.g. see Frisch, 1995 for a review of the theory and observations). No equivalent relationship is known for magnetofluids, let alone a medium as complex as the collisionless, non-thermal solar wind which is at best only approximately fluid-like. However, Kraichnan (1965), hereafter K65, showed that a similar relationship could exist in MHD under restricted circumstances. The K65 theory assumed equal populations of Alfvén waves travelling parallel and anti-parallel to the background magnetic field. In this case, decay of the fluctuations and energy transfer to smaller scales occurs due to the interaction of oppositely propagating waves. However, any two such waves rapidly decorrelate because they propagate away from each other, leading to a limited interaction time. This 'Alfvén decorrelation effect' results in a slower energy cascade than in the neutral fluid case, and therefore a shallower power spectrum with a spectral index $\alpha = -3/2$ – and, of course, shallower structure functions. In fact, a consequence of K65 is that it is the fourth-order structure function scaling that is unity: $g(4) = 1$ independent of intermittency corrections.

We stress that several assumptions of the K65 model are not satisfied in the solar wind: the amplitudes of field parallel and anti-parallel propagating Alfvén waves are not equal in general, and compressive modes are also present, for example. It would not therefore be surprising to find that observations were not consistent with the model. However, the difficulty of accurately measuring the power spectral index α sufficiently accurately in the solar wind to distinguish between the $\alpha = -5/3$ (K41) and $\alpha = -3/2$ (K65) cases has made it difficult to determine whether Kraichnan's model is a good description of solar-wind turbulence, although careful measurements (see e.g. Goldstein and Roberts, 1995 for a review) tend to result in values nearer $\alpha \sim -5/3$ than $\alpha \sim -3/2$. A key assumption of K65 is that the Alfvén decorrelation time is much larger than the energy transfer time when the decorrelation effect is not present, although this is not necessarily the case in the solar wind, so in reality K41 and K65 represent two extremes of a continuum of possibilities for the energy transfer mechanism.

Carbone (1994) suggested that measurements of $g(3)$ and $g(4)$ in the inertial range in the solar wind could be used as an additional method to distinguish between K41 and K65, based on which of the two scalings was closer to 1. As we have seen, *Ulysses* polar solar-wind measurements provide the long, stationary time series necessary to make the precise measurements needed. Horbury and Balogh (1997) performed this test using *Ulysses* data and showed that in the inertial range $g(3)$ was much closer to 1 than $g(4)$ and on this basis they concluded that, at least in the polar solar wind, the Alfvén decorrelation effect was not significant and that therefore the cascade appeared to be 'Kolmogorov'-like in character.

The left panel of Figure 4.11 shows observed $g(m)$ values for the inertial range of an interval of polar data compared with the *p* model with $p = 0.79$. Agreement with the model is striking, and it is noteworthy that the observed $g(3)$ is very close to unity, as expected for a K41-like cascade.

Horbury and Balogh (1997) discussed a 'generalized' p model which allowed for a K41-like or K65-like cascade. By introducing a new parameter q, they could vary the energy scaling of the cascade, producing $g(m)$ values given by:

$$g(m) = 1 - \log_2(p^{m/q} + (1-p)^{m/q}) \tag{4.10}$$

where $q = 3$ corresponds to the K41 cascade, as in equation (4.9) for the original Meneveau and Sreenivasan (1987) p model with a K41-like cascade, while $q = 4$ corresponds to the K65-like case. By finding values of p and q which best fit the observed $g(m)$ values, Horbury and Balogh (1997) could estimate both the intermittency and the energy scaling of the turbulence.

The right panel of Figure 4.11 shows the result of such fits to several intervals of polar data. There is a clear correlation between values of p and q. Horbury and Balogh argued that this is due to some sampling of non-inertial range fluctuations, and that the left-hand termination of the p–q curve, at $p \sim 0.7$, $q \sim 3$, reflected inertial range values in the solar wind. On the basis of these values, the intermittency and energy transfer scaling of solar-wind fluctuations appear to be remarkably similar to those in terrestrial neutral fluid turbulence.

4.2.3.3 The transition scale

In the discussion above, we considered only the intermittency and energy transfer within the inertial range. Indeed, the models described implicitly assume inertial range conditions. Horbury et al. (1997) showed that we could identify the large-scale edge of the inertial range using structure function gradients, and that fitting the generalized p model to structure function values outside this range produced very high estimates of q and p. They suggested that these high values were artefacts of the analysis fitting procedure, in that the p model would produce these results when fitted to non-inertial range turbulence.

Horbury et al. (1997) tested this conjecture using a recently published model by Tu et al. (1996). This model extended the p model (Meneveau and Sreenivasan, 1987) to non-inertial scales, introducing the concept of a non-intermittent spectral index which could vary with scale. It could be used with both K41 and K65-like energy transfer scalings. The Tu et al. model predicted a slightly different shape of the $g(m)$ curve than the generalized p model used by Horbury and Balogh (1997). Horbury et al. (1997) fitted these three models (Tu et al., 1996 with both K41 and K65 scaling and the generalized p model) to structure function measurements on a range of scales from just within to far outside the inertial range. Their results are shown in Figure 4.12 for magnetic-field data taken at 1.9 AU and 71°S in a high-speed stream. They found that, while fitting the generalized p model produced large values of p and q outside the inertial range, fits to the Tu et al. models produced estimates of p which were similar to those in the inertial range, and steadily decreasing spectral index values with increasing scale. This is exactly what we would expect, since the Tu et al. model is applicable to scales outside the inertial range (the 'transition' scale) but the p model is not.

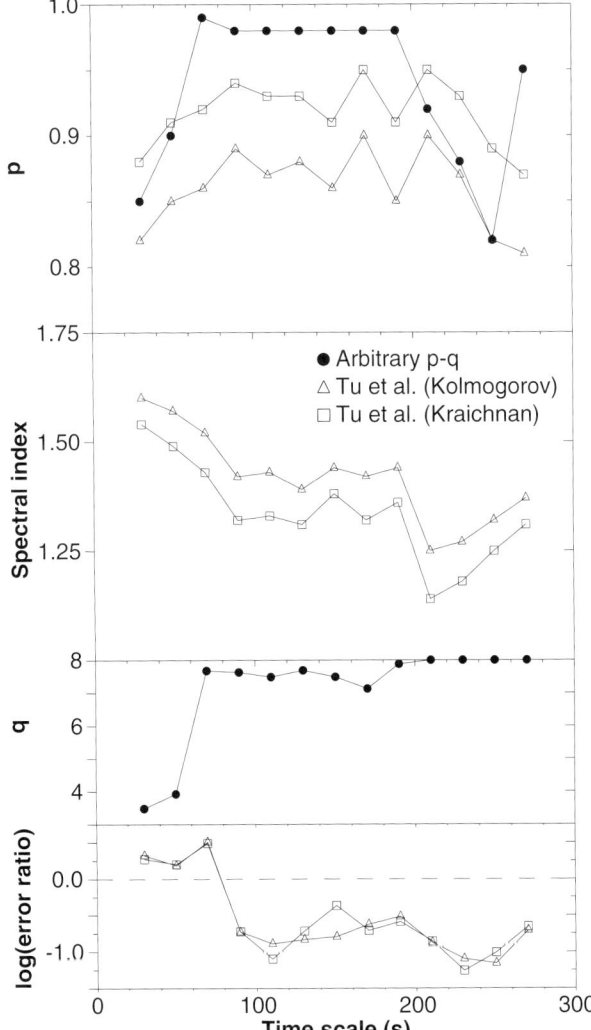

Figure 4.12. Variation in intermittency outside the inertial range. The large-scale end of the inertial range lies at around 80 s for the data interval used here, so the leftmost three data points lie within it, but the rest are within the transition scale. The top panel shows the p value (intermittency level) estimated using the general p model, and the Tu *et al.* (1996) non-inertial p model: values 'saturate' near 1 for the general p model outside the inertial range, but very much less for the Tu *et al.* models. The second panel shows spectral index values, derived from the Tu *et al.* models, which gradually decline away from the inertial range as expected. The third panel shows q values, calculated from the generalized p model, which also saturate outside the inertial range. Finally, the bottom panel shows the relative goodness of fit of the Tu *et al.* models to the generalized p model, showing that the general model is better – log(error ratio) is positive – in the inertial range, but the Tu *et al.* models are a better description of the data – log (error ratio) is negative – outside it (from Horbury *et al.*, 1997).

The Horbury et al. (1997) result shows that intermittency levels are relatively constant with scale for at least scales up to around three times larger than the edge of the inertial range. It is not possible to measure intermittency using structure functions when the spectral index is near -1, so the intermittency of large-scale Alfvén waves is not known. However, while the interpretation of intermittency in MHD turbulence is not clear, so the meaning of intermittency in non-interacting Alfvén waves is even less so. At this time, it is not known whether intermittency develops with the energy cascade or whether it is already present in the large-scale Alfvén waves.

4.2.4 Fluctuations and solar-wind structure

4.2.4.1 Mid-latitude trailing edges

Previous sections of this chapter have dealt with fluctuations within high-speed polar solar-wind streams. However, Ulysses has sampled other regions of the heliosphere that are of interest in the study of waves and turbulence. The most important of these is the mid-latitude streamer belt, where fast polar wind interacts with slower low-latitude wind. From late 1992 until early 1994, Ulysses travelled southwards at around 5 AU, sampling both high speed (>700 km s^{-1}) and slow speed (<500 km s^{-1}) wind every solar rotation. This period was during the declining phase of the solar cycle, when the current sheet was still inclined relative to the equatorial plane and conditions were remarkably steady from one rotation to the next. As a result, interactions between fast and slow wind covered a wide range of latitudes (from around 10 to 30°) and indeed this is likely to be the case during the declining phase of every solar cycle. At lower latitudes, wind at speeds of 700 km s^{-1} or more is rarely observed at several AU, because it has all collided with slower wind and hence been slowed. The Ulysses observations of CIRs near 5 AU between 1992 and 1994 are therefore a unique record of this important region of the heliosphere.

As discussed in Section 4.2.2.2, Bavassano et al. (1998) showed that the normalized cross-helicity (σ_c) dropped more rapidly at mid-latitudes in Ulysses data than at lower or higher latitudes. They interpreted this result as highlighting the importance of stream shear, which is high at mid-latitudes where high-speed polar wind streams interact with slower low-latitude streams, in driving the turbulence and hence reducing σ_c.

Lucek and Balogh (1998) used a wavelet analysis of hourly sampled Elsässer variables (and hence σ_c) to study fluctuations during several months when Ulysses was at mid-latitudes. Using this novel technique they could identify periods of high positive σ_c immediately after CIRs which, as expected, were effectively high-speed wind which had yet to interact with the CIR. There was often a sharp boundary in σ_c in the trailing edge (velocity decline) a few days after the CIR, which could be interpreted as the transition from wind which was originally fast to that which was originally slow.

Fluctuations in slow wind, on the other hand, were much more variable and Lucek and Balogh (1998) identified periods of Alfvénic fluctuations with a predominantly sunward-propagation sense, and others with no dominant Alfvénic component. Power levels were also more variable. Although these short-term changes in fluctuation properties were observed in slow wind closer to the Sun is it interesting to note that while the large-scale structure of the corotating structures observed by *Ulysses* at mid-latitudes changed little from one rotation to the next, Lucek and Balogh showed that fluctuation properties, particularly in slow wind, changed considerably.

We note that although Lucek and Balogh (1998) considered scales of an hour and above in the spacecraft frame, which is well outside the inertial range, nevertheless the generally 'broadband' nature of solar-wind fluctuations, where properties are often similar over a wide range of scales – a natural consequence of the turbulent cascade – suggests that similar variation in fluctuation properties probably occur at smaller scales and within the inertial range.

In 1996 and 1997 the sequence of CIRs observed by *Ulysses* as it travelled to lower latitudes in the northern hemisphere were less regular than those in 1992–1994 and began at a lower latitude. This was due to the reduced inclination of the current sheet in 1996, and has the unfortunate consequence of making the later set of mid-latitude CIR encounters considerably less useful for fluctuation studies.

4.2.4.2 Small-scale fluctuation variation

As we saw in Figure 4.1, there is considerable 'microstructure' in the polar solar wind, with variation in essentially all bulk parameters on a wide range of scales. One established class of polar wind structure is 'microstreams', velocity decreases or increases, typically a few tens of $km\,s^{-1}$ in amplitude and lasting a day or more. Microstreams have also been seen in *Helios* measurements of low-latitude high-speed streams. Neugebauer *et al.* (1995), in an effort to characterize conditions within microstreams, examined the normalized cross-helicity σ_c around them using a superposed epoch analysis. Normalized cross-helicity is a measure of the 'Alfvénicity' of the fluctuations (see Section 4.2.1.2). Neugebauer *et al.* showed that values of σ_c were lower in the leading edges of microstreams that were faster than the ambient wind, and in the trailing edges of those that were slower; that is, σ_c values were lower in compression regions, suggesting that fluctuations may be 'driven' or 'aged' by these slight compressions.

Recent work to identify individual cases of such Alfvénicity reductions failed to find events with dramatic reductions in σ_c, emphasizing the small effects that these compressions can have. Further work is planned in this area, as this is a potentially important result in showing how stream structure can alter fluctuations.

Recently, Horbury *et al.* (1998) and Lucek *et al.* (1999) reported the existence of a different class of structure with a related turbulence signature. These events, of only a few hours in duration, tend to have radially directed magnetic fields and dramatically reduced fluctuations levels. Figure 4.13 shows two examples of such events, where power levels drop by around an order of magnitude while the field points nearly radially. Lucek *et al.* (1999) showed that these events also have a signature in plasma parameters, with slightly reduced proton temperatures and velocities on

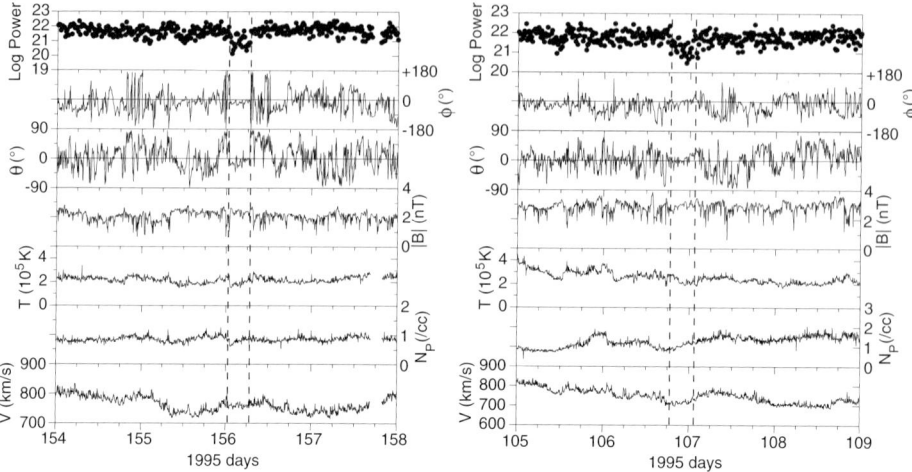

Figure 4.13. Two examples of events when near-radial magnetic fields are accompanied by dramatically reduced magnetic-field power levels. The events are delineated by dashed vertical lines (from Horbury *et al.*, 1998).

average. The origin of these structures is unknown, but they may be the result of inhomogeneities in the corona.

4.2.4.3 Turbulence as a diagnostic

It is possible to use turbulence as a tool for examining other phenomena in the solar wind. Balogh *et al.* (1999) recently used σ_c values to explain the structure of magnetic-field polarity reversals in the polar solar wind. These events, when the magnetic field is directed more than 90° away from the Parker spiral direction for several hours or more, could have two origins: they could be 'mini-sectors' caused by a change of polarity in the magnetic field in the lower corona that is simply propagated into the solar wind; or they could be the result of large-amplitude waves causing the magnetic field to loop back on itself.

Balogh *et al.* (1999) showed that the sign of the cross-helicity reversed during these polarity reversals. Regions where σ_c is greater than zero are dominated by Alfvénic fluctuations propagating anti-sunward, and dominate the polar solar wind, as expected: above the Alfvén critical point in the upper corona σ_c should be positive and this value is propagated into the solar wind. The fact that within polarity reversals σ_c was negative showed that Alfvénic fluctuations were propagating sunward (in the solar-wind frame) in these regions. If polarity reversals were caused by small regions of the lower corona of opposite polarity to the bulk of the coronal hole from which they emanated, we would expect no change in the sign of σ_c. The observed change in σ_c is strong evidence, therefore, for these events being regions of 'folded over' field, as these folds make the small-scale Alfvénic fluctuations travel back sunward in these regions.

4.3 DISCONTINUITIES AND ALFVÉN WAVES

Among fundamental microstructures present in the solar-wind plasma are Directional Discontinuities (DDs): sharp angular changes in the interplanetary magnetic-field directionality (Colburn and Sonett, 1966; Burlaga, 1971). Theoretically, there can be several different types: rotational (RDs), tangential (TDs), shocks (fast, intermediate and slow), and Contact Discontinuities (CDs). In this review RDs, TDs and slow shocks will be covered. Fast shocks is a very broad topic and is best left for a separate review article.

Table 4.1 gives the properties of the four types of discontinuities, as discussed in Landau and Lifschitz (1960). This table illustrates the properties that can be used to distinguish the various types of discontinuities from one another. Tangential discontinuities have no mass flow across their surfaces. TDs can be thought of as surfaces separating two different types of plasmas and for the ideal case are infinite in extent. The field direction and magnitude can be different on the two sides as well as plasma densities, temperatures and even composition (H^+, He^{++}, O^{+6}/O^{+7}). TDs can be identified by the lack of a magnetic-field normal ($B_N = 0$) across their surface. RDs have mass flow across their surfaces, non-zero magnetic-field normal components, and for isotropic plasmas (see Hudson, 1970 for the more complex anisotropic case) the transverse component of B is constant ($[B_T] = 0$). RDs ideally have large-field normal components and no changes in magnetic-field magnitude. Shocks have mass flow across their surfaces, and neither the magnetic field normal component nor the transverse component need be conserved.

Why are discontinuities and shocks of interest to plasma physicists? RDs are essentially sharply kinked fields or short-wavelength Alfvén waves. There may be 'dissipation' associated with such waves (e.g. conversion of wave energy into plasma energy). So, if RDs are dissipative, this energy could go into heating of the ambient plasma. TDs, as mentioned above, separate plasmas and fields of different types. They are assumed to be convected from the solar corona all the way out to *Ulysses* distances and beyond. What are they and what do they represent near the Sun? This is one of the fundamental questions. Slow shocks play a fundamental role in the classic picture of magnetic reconnection as theorized by Petschek (1964). Do slow shocks occur in interplanetary space and if so how often and what is their role?

Table 4.1. Criteria for mass flux and magnetic-field changes across idealized discontinuities. Here [ ] denotes a change in the magnitude of a component of the magnetic field vector $\vec{B}$ across the discontinuity. Subscripts T and N refer to directions tangential (parallel) and normal (perpendicular) to the discontinuity plane.

Type of discontinuity	Mass flux. ρv_n	Change in magnetic field	$[\vec{B}]$
Contact discontinuity	0	$[\vec{B}_T] = 0$	$B_N \neq 0$
Tangential discontinuity	0	$[\vec{B}_T] \neq 0$	$B_N = 0$
Rotational discontinuity	$\neq 0$	$[\vec{B}_T] = 0$	$B_N \neq 0$
Shock	$\neq 0$	$[\vec{B}_T] \neq 0$; $[B_T] \neq 0$	$B_N \neq 0$

4.3.1 DD radial and latitudinal gradients

The topics of radial and latitudinal gradients of DDs are important towards understanding their generation mechanisms and their evolution. If TDs are dissipated/broadened through magnetic reconnection, or if RDs are dissipated through coupling to other electromagnetic or electrostatic plasma-wave modes (which are then in turn damped), these processes could have profound effects on the evolution of the solar wind plasma, and hence the heliosphere. TDs are thought to separate dissimilar plasmas and can be interplanetary markers of structures convected from the Sun or its corona. RDs and Alfvén waves contribute to the scattering of solar energetic particles and cosmic rays.

There has been a controversy about the gradients in occurrence rate of DDs (Burlaga, 1971; Mariani *et al.*, 1973; Behannon, 1978; Tsurutani and Smith, 1979; Lepping and Behannon, 1986). Much of the problem (and dispute) comes from the difficulty in separating latitudinal effects from radial distance effects and the difference in definition used. The *Ulysses* spacecraft, which first went from 1 to 5 AU, and with a Jupiter gravity assist, to 80°S latitude and finally to 80°N latitude, is ideal to address this problem. *Ulysses* provided a full coverage of heliographic latitudes, and, as we shall see, will allow a resolution of this problem.

There are two computerized methods of detecting DDs, one by Tsurutani and Smith (1979) hereafter called TS and another by Lepping and Behannon (1986) hereafter called LB. For large statistical studies it is necessary to identify DDs by computer analyses, as tens of thousands of discontinuities are often used (however, for a determination of the discontinuity type, no one has been able to automate this yet). The TS criteria require a field directional change of $\Delta \mathbf{B}/|\mathbf{B}| \geq 0.5$ and $\Delta|\mathbf{B}|$ greater than 2σ when σ^2 is the variance on either side of the discontinuity. These criteria are applied to 1-min vectors when the two 1-min vectors being intercompared are separated by 3 min. This vector separation allows for discontinuities with 'thicknesses' as large as 60 s to be detected without bias. The LB criteria (slightly modified from the original method) require that field directional change between vectors be at least 30°, or $\theta = \cos^{-1}(\mathbf{B}_1 \times \mathbf{B}_2)/|\mathbf{B}_1||\mathbf{B}_2| \geq 30°$. B_1 and B_2 are the upstream and downstream field vectors by standard convention. These criteria were applied to 1-min averages with B_1 and B_2 separated by 2 min. For more details concerning the methods of selection, we refer the reader to the original articles.

TDs and RDs are schematically shown in Figure 4.14. As mentioned before, TDs can be thought of as infinite sheets separating dissimilar plasmas and fields. RDs can be thought of as sharply kinked Alfvén waves, a topic which we will come back to later.

Figure 4.15, from Tsurutani *et al.* (1996a), shows the radial gradient of DD occurrence rate from 1 to 5 AU. *Ulysses* was in the ecliptic plane. The vertical scale is the normalized number of discontinuities per day, using 0.2 AU interval bins (the normalization takes data gaps into account). The TS and LB Rate of Occurences of Interplanetary Discontinuities (ROID) values are different (because of the slightly different criteria), but show the same general trend. Both plots show a gradual decrease in rate of occurrence with increasing radial distance. Both plots

Sec. 4.3] **Discontinuities and Alfvén waves** 197

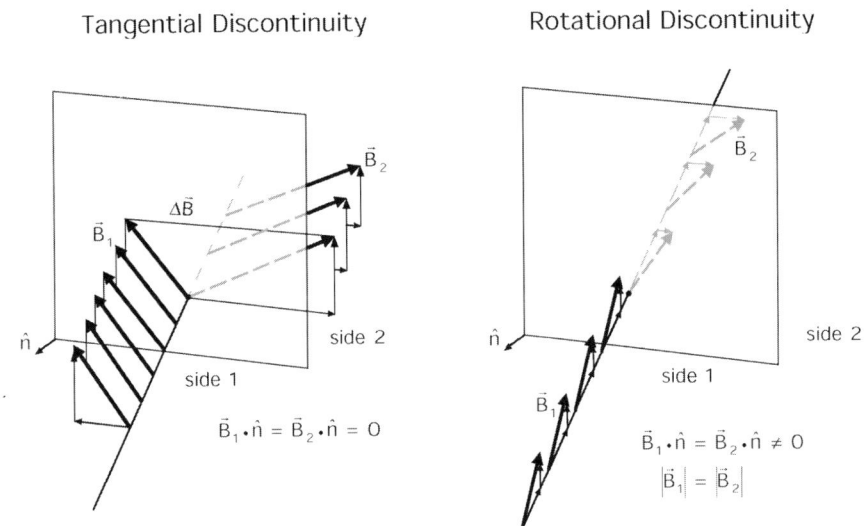

Figure 4.14. Schematics of idealized rotational and tangential discontinuities (from Tsurutani et al., 1997a).

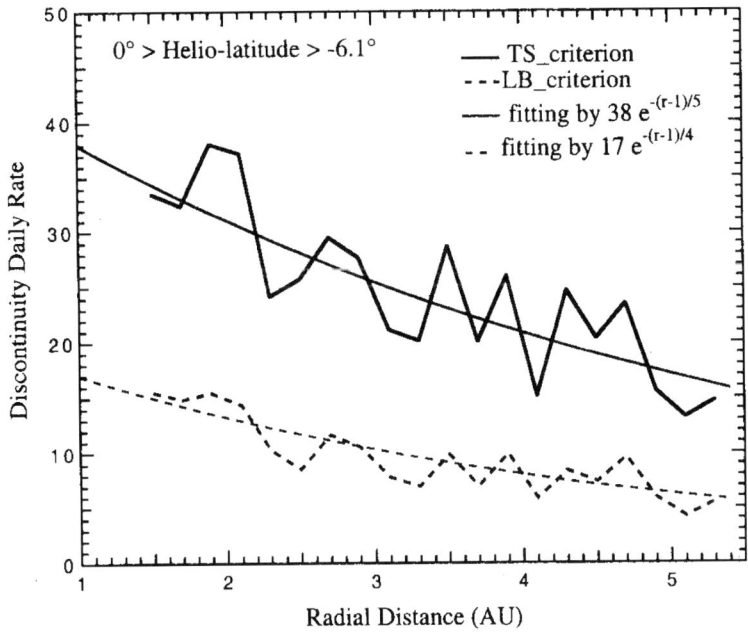

Figure 4.15. The in-ecliptic rate of occurrences of interplanetary discontinuities (ROID) values from 1 to 5 AU. The solid curve is the result of the TS criteria and the dashed curve that of the LB criteria (from Tsurutani et al., 1996a).

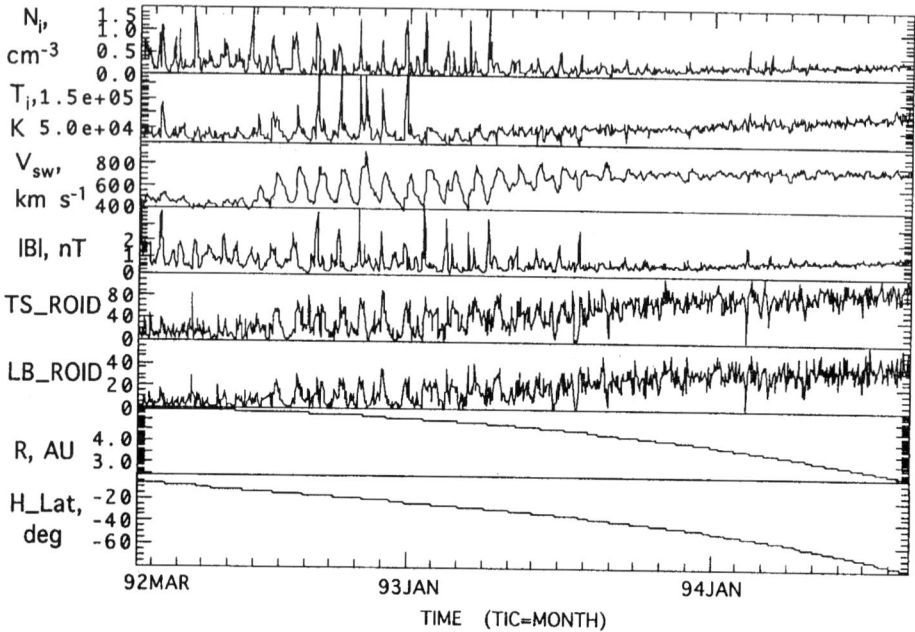

Figure 4.16. The solar-wind plasma and magnetic field and ROID values for the TS and LB criteria, from Tsurutani et al. (1996b).

have been fitted by exponential curves. TS is best fit with $e^{-(r-1)/5}$ dependence where r is measured in AU. LB is best fit by a $e^{-(r-1)/4}$ dependence. This result is a confirmation of previous radial gradient results of TS and LB, and has been explained partially as effects of increasing discontinuity thickness with decreasing magnetic-field strength (with increasing radial distance from the Sun). Some of the 'thickened' discontinuities therefore fall outside the computerized selection criteria. Some part of this ROID decrease might also be due to the dissipation of discontinuities.

The TS and LB ROID values from the ecliptic plane to 80°S latitude are given in Figure 4.16. The proton densities, temperatures and solar-wind velocities are shown in the top three panels. Next are the interplanetary magnetic field (IMF) $|B|$, TS and LB ROID values, and at the bottom the Ulysses position in radial distance and heliolatitude. The spacecraft moves closer to the Sun as it goes toward the pole (we can observe some radial dependence in the TS and LB ROID plots).

An important point to note in this figure is that the ROID value depends strongly on the type of stream structure that Ulysses is engulfed in. From July 1992 through April 1993 there are recurring 'high-speed' streams (Phillips et al., 1994) associated with a coronal hole. Here the term 'high-speed' is used by interplanetary scientists to mean solar wind with more or less continuous speeds

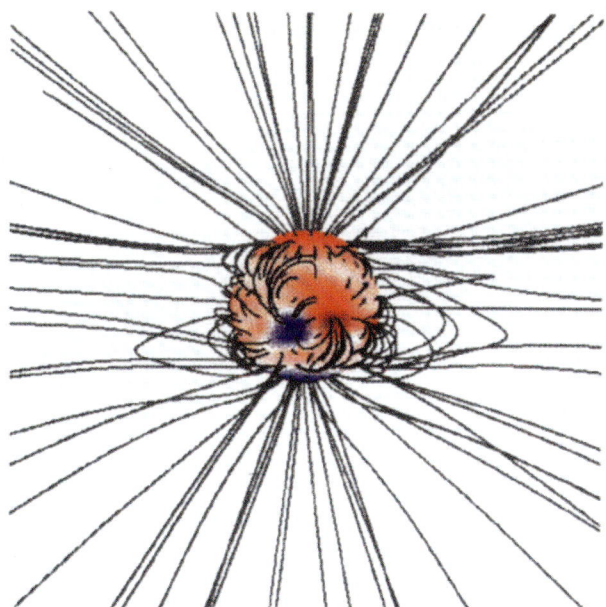

Plate 1. A model of the external field lines of the solar magnetic field at the time when *Ulysses* crossed the equator from south to north in early 1995. The figure was generated by P. C. Liewer from a magnetohydrodynamic model of the solar -ind flow computed by Z. Mikic and J. Linker (from Neugebauer, 1999).

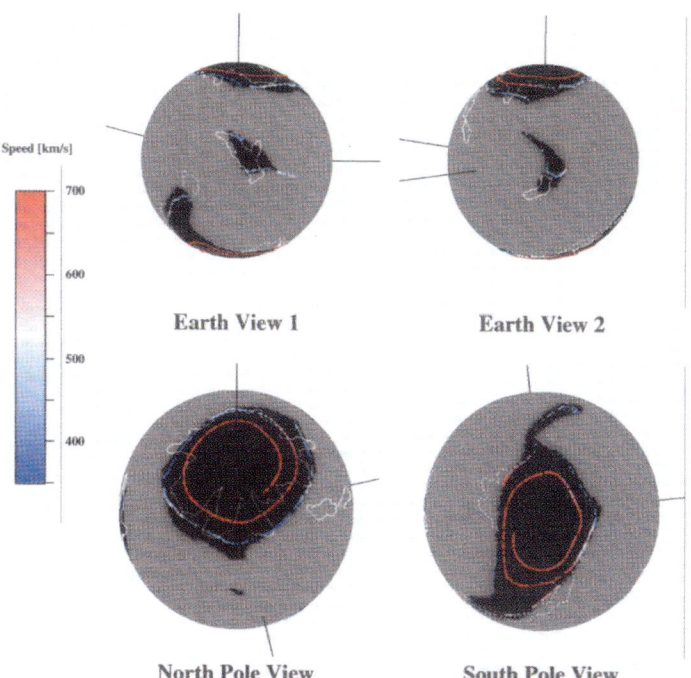

Plate 4. Four views of the Sun during the *Ulysses* fast latitude scan. Dashed white lines are the boundaries of coronal holes observed from the ground in He 10830 Å; grey shading indicates open field lines; and blue to red colour coding indicates solar-wind speed. The open field lines and the footpoints of field lines threading *Ulysses* were calculated by Z. Mikic and J. Linker using an MHD model.

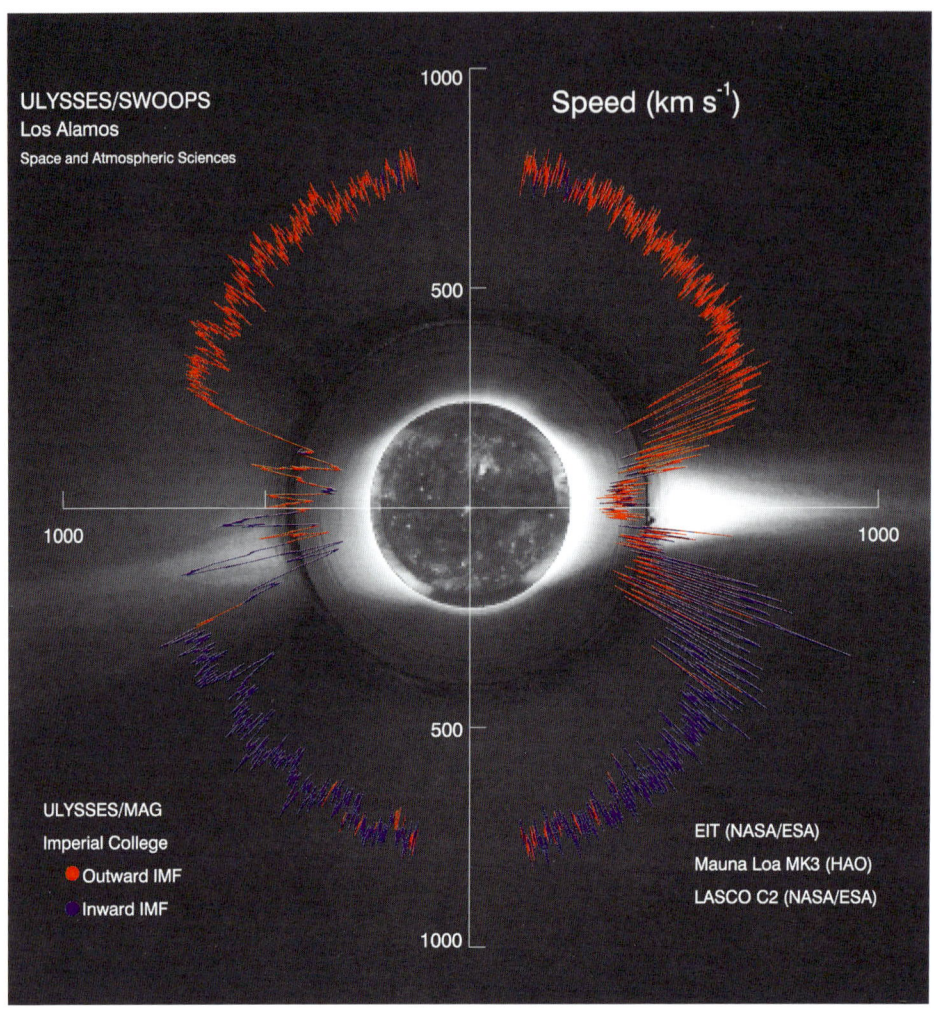

Plate 2. A polar plot of solar-wind speed versus heliographic latitude observed by *Ulysses*. The data acquired outside ∼2.2 AU are plotted on the right, while the data acquired closer to the Sun are on the left. The difference in appearance between the two equatorial crossings results from the different distance from the Sun, the different angular speed of the spacecraft, and the different tilt of the heliospheric current sheet. Solar-wind speed is colour-coded according to the polarity of the interplanetary magnetic field (IMF), with fields pointing away from the Sun indicated by red and inward fields indicated by blue. Also included are images of the Sun as viewed near solar minimum by the Extreme-Ultraviolet Imaging Telescope (EIT) on *SOHO* (disk image), the Mauna Loa white-light coronameter (low corona) operated by the High Altitude Observatory, and the C2 white-light coronagraph of the LASCO instrument on *SOHO* (outer corona) (from McComas et al., 1998).

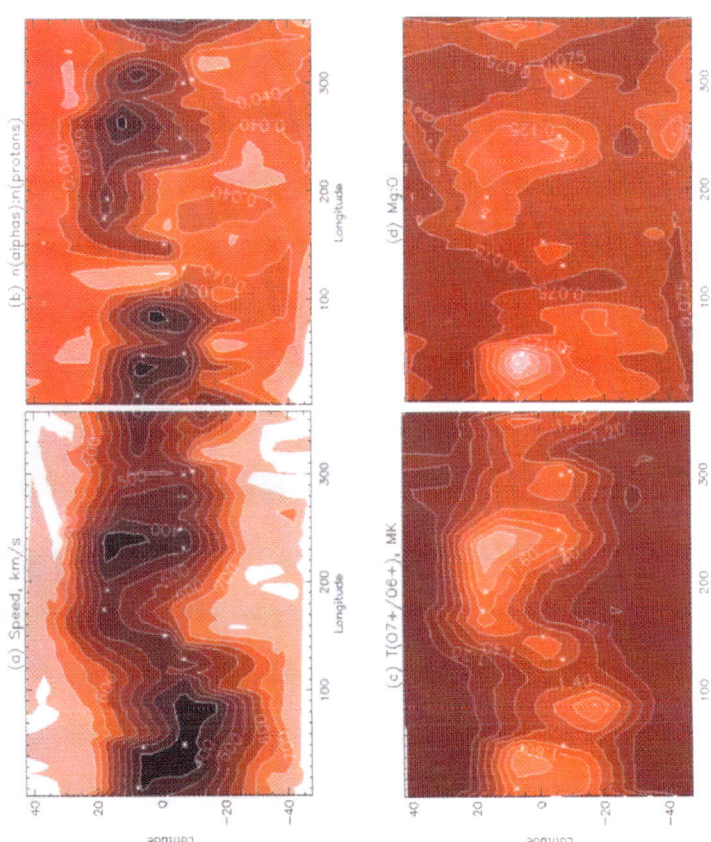

Plate 3. (a) Contours of solar-wind speed as a function of latitude and Carrington longitude. The white areas denote speeds $>750\,\mathrm{km\,s^{-1}}$. The white asterisks indicate crossings of the heliospheric current sheet. The contours are calculated from averages over $5 \times 5°$ bins in latitude and longitude from five rotations (Carrington rotations 1891–1895) of *Ulysses* data and from *WIND* data for Carrington rotation 1893. (b) Contours of the alpha-particle abundance N_α/N_p computed from running 5-bin averages along both the *WIND* and *Ulysses* trajectories. (c) Contours of the ionization temperature calculated from the density ratio of O^{7+} to O^{6+} ions, in units of 10^6 K, computed from running 5-bin averages along the *Ulysses* trajectory (d) Contours of density ratios of Mg to O computed from running 5-bin averages along the *Ulysses* trajectory (from Neugebauer *et al.*, 1998).

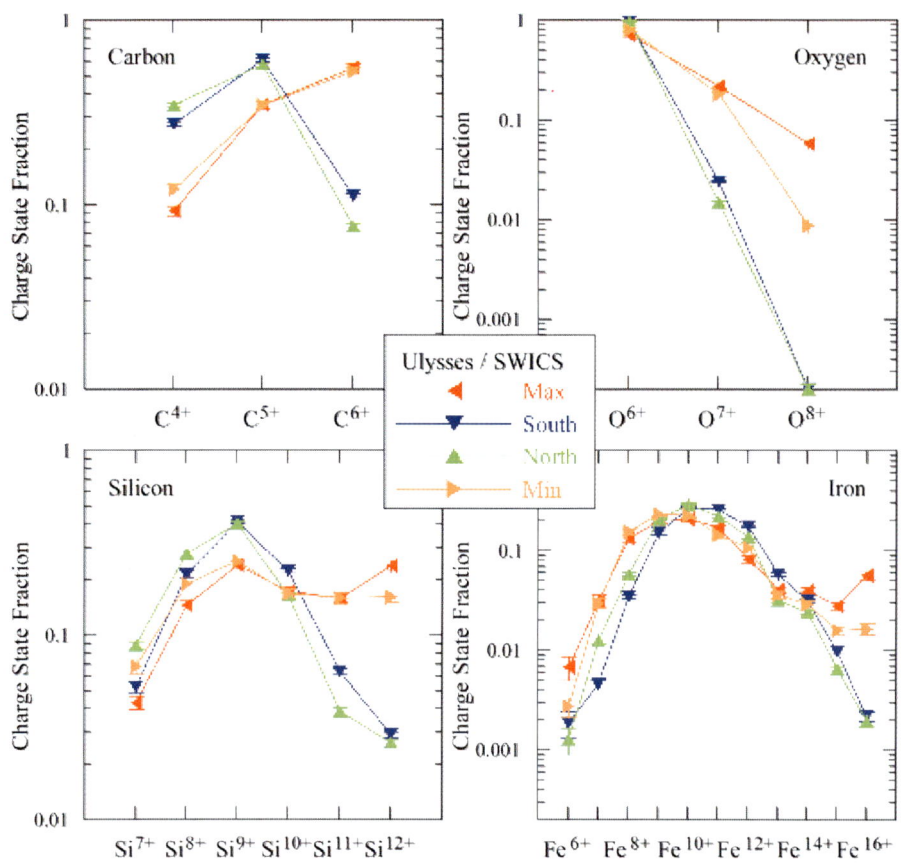

Plate 5. Charge-state distributions for solar-wind carbon, oxygen, silicon, and iron ions accumulated during four 300-day intervals (see text for start and stop times). The 'South' and 'North' intervals were in the fast, high-latitude wind and the 'Max' and 'Min' intervals were in the slow, low-latitude wind. The data points are averages of daily values and the bars indicate the probable errors of the mean values (from von Steiger *et al.*, 2000).

>700 km s^{-1}). The TS and LB ROID values increase and decrease in coincidence with the speed of the solar wind. A typical ratio of the ROID value in a high-speed stream to that outside the stream is ~5:1 to 10:1.

After August 1993 *Ulysses* was permanently embedded in a high-speed stream with V_{SW} = 700–800 km s^{-1}. The TS ROID value remained between 80 and 100 discontinuities day^{-1}. Thus there is a strong stream dependence on the ROID values which masks any subtle latitudinal gradient that may be present. In this review we will focus on this first-order dependence.

4.3.2 Discontinuity 'thicknesses'

The greater number of discontinuities in high-speed flows can be related to either some intrinsic relationship between Alfvén waves and discontinuities or to the fast convective flow of discontinuities past the spacecraft. For the latter case, a faster flow means that thicker discontinuities will be picked up by the computer selection criteria, and more of the discontinuity thickness 'distribution' will be detected. To explore this latter effect first, we examine discontinuity thicknesses.

The top part of Figure 4.17 gives the discontinuity 'temporal' thicknesses for discontinuities at 67°S latitude at 3.0 AU. This is for the interval 1994 days 154–155. We use 'temporal' thicknesses rather than spatial values because it is the 'temporal' thickness distributions that is the issue here for the computer selection criteria. Each discontinuity thickness was determined by hand analysis of high-resolution magnetic-field data. The $2/e$ (65 per cent) value of the total angular change was determined to define the temporal thickness. The temporal thickness distribution for all the TS discontinuities detected during the 2-day interval is given in the top panel of Figure 4.17.

The discontinuity spatial thicknesses can be determined by performing minimum-variance analyses to determine the minimum-variance direction (or normal direction) and calculating the thickness by the expression $V_{SW} \Delta T \cos \theta_{nV}$ where V_{SW} is the solar-wind speed, ΔT the temporal thickness, and θ_{nV} the angle between the discontinuity normal and the solar-wind velocity vector. To simplify the calculation, we assume that the solar-wind velocity vector lies along the $-R$ direction (in reality the velocity may deviate a few degrees from this direction in polar coronal-hole high-speed flows). Making this simplifying assumption only leads to minor errors. The spatial thickness distribution for days 154 to 155 1994 are shown in the bottom panel of Figure 4.17.

In the above calculations it is assumed that the discontinuities are simply convected by the solar-wind. This is true if the discontinuities are TDs. On the other hand, if most of the events are RDs (we will argue this case later), then the discontinuity structures will propagate relative to the solar-wind plasma. The Alfvén speed varies from 40 to 70 km s^{-1}, whereas the solar-wind speed varies from 400 to 800 km s^{-1}. Thus the RD propagation is only a ~10 per cent correction factor, and

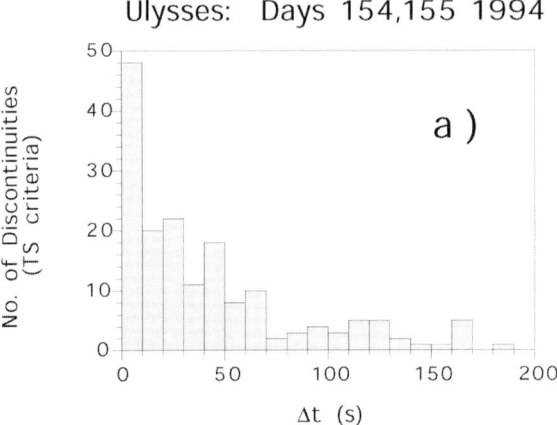

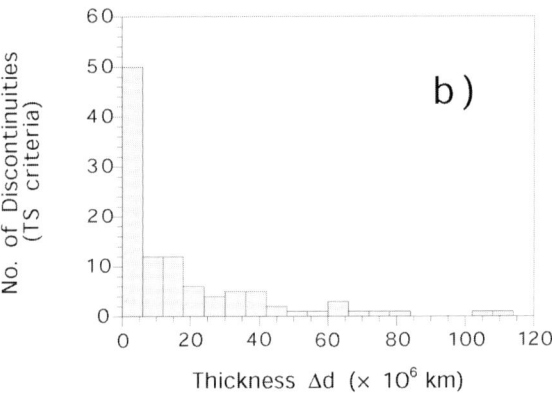

Figure 4.17. The temporal (top panel) and spatial (bottom panel) 'thicknesses' of interplanetary discontinuities at 67°S latitude and 3.0 AU. The spatial thickness for each individual discontinuity was determined by dividing the temporal 'thickness' by the solar-wind velocity. The discontinuity normal direction was also taken into account (from Tsurutani et al., 1996a).

to first order can be neglected. For our purposes we can assume that both RDs and TDs are simply convected by the solar wind.

The distributions in Figure 4.17 indicate that most discontinuities are small-scale features. There is a 'tail' to the distribution but the number of discontinuities falls off with increasing thickness.

A simple calculation, using the solar-wind speeds and the values in the top panels of Figure 4.17, can be made to demonstrate that increased solar-wind speeds cannot be the total cause of the ROID value increase in high-speed streams. The solar-wind speed during days 154 and 155 was a near-constant $\sim$750 km s^{-1}. If the solar-wind speed were lowered to slow-speed values of $\sim$400 km s^{-1}, then some of the thicker discontinuities would be missed. An upper bound on this number can be easily calculated in the following manner. The TS selection method detects all discontinuities with thicknesses less than 60 s. Discontinuities with thicknesses between 60 s and 120 s can be detected without bias if their location is well placed (start and stop times) within the time interval. Discontinuities thicker than 120s can also be detected if their vector changes ($\Delta \mathbf{B}/B$) are sufficiently large (>0.5). Thus the cutoff on the 60-s thickness discontinuities would be 400 km s^{-1}/750 km s$^{-1} \times$ 60 s, or 32 s. All discontinuities with thicknesses less than 32 s in Figure 4.17 would be detected in a 400 km s^{-1} solar-wind stream. From Figure 4.17a that would correspond to 89 of the 171 discontinuities, or a ratio of $\sim$2.0. Some of the thicker discontinuities would also be detected in a slower speed stream, so as previously mentioned this is only an upper limit on the ROID value variability. Similar analyses have been done with LB discontinuities, and similar results were found.

Solar speed variation cannot account for ROID value variation of $\sim$5 to $\sim$10. An upper limit to the ROID value variation due to thicker discontinuities being detected would be a factor of 2. Thus, some other explanations must be found.

4.3.3 The relationship between discontinuities and Alfvén waves

Figure 4.18, taken from Tsurutani *et al.* (1997a), illustrates one 10-hour period of the interplanetary magnetic field at low heliographic latitudes. The interval is day 17 1992. *Ulysses* was near the ecliptic plane (6.0°S latitude), 5.2 AU from the Sun. The field is displayed in RTN coordinates. The TS selected discontinuities are identified by vertical lines. By visual inspection, we can note a relationship between slowly rotating fields (Alfvén waves) and the sharp directional changes (DDs). Most of the DDs exhibit little or no magnetic-field magnitude changes, indicating that they are most likely rotational in nature.

Discontinuities often occur at the edge of slowly rotating fields. Figure 4.19 shows such an example. This interval occurs on day 210 1995 when *Ulysses* was at +80.2° heliographic latitude. Other examples have been noted in the ecliptic plane and at mid-latitudes. The results were essentially the same.

Riley *et al.* (1996) have shown that in the ecliptic plane the arc-polarized waves account for 5–10 per cent of the *Ulysses* dataset. The waves are propagating outward in the rest frame of the solar-wind plasma.

The magnetic field is plotted in minimum-variance coordinates where B_1, B_2 and B_3 correspond to the field components along the maximum, intermediate, and minimum-variance directions. Point 1 at 2325:31 UT designates the start of a slow field rotation detected primarily in the B_1 component. Point 2 at $\sim$2340:47 UT is the end of the slow rotation and the start of a faster field-rotation interval. Point 3 is the

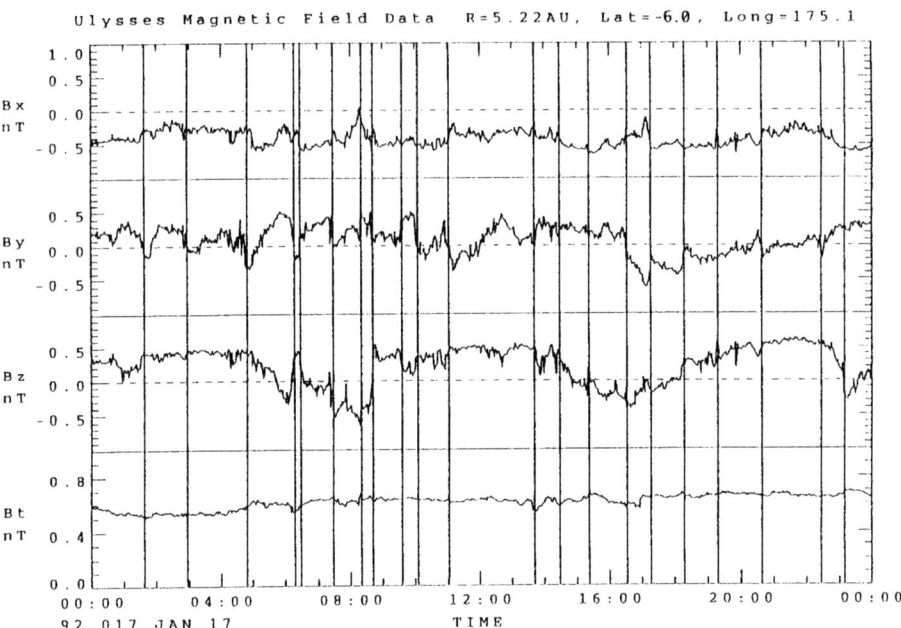

Figure 4.18. The magnetic field at 5.2 AU at 6.0°S latitude. Many of the TS selected discontinuities (vertical lines) occur at the edges of the more slowly rotating Alfvén waves (from Tsurutani et al., 1997a).

end of the fast field rotation and the end of the interval of analysis. The top panel displays the field in minimum-variance coordinates. The bottom panel is a hodograph of the B_1–B_2 components. In the latter panel the field rotates from left to right in an arc. From point 2 to 3 the field rotates back completing the 'arc' rotation. The field rotation is, to first order, non-compressive. The rotation occurs in a plane (not shown). Thus the discontinuity is believed to be part of the Alfvén wave, and thus we say that the Alfvén wave is 'phase steepened' (i.e. there is more phase rotation at one edge of the wave – see also Tsurutani et al., 1994).

A schematic illustrating the difference between planar waves and spherical waves (when the field magnitude is preserved) is shown in Figure 4.20. For planar waves (left) there are two fundamental polarizations, circular and elliptical, with left-handed and right-handed rotations. Linear polarization is a mix of equal amplitude left-and right-circular polarized waves. For large amplitude waves where the magnetic field is constant (right) there are circular and elliptic polarizations possible, and also an arc polarization. The circular- and arc-polarization perturbation vectors occur in a plane, while elliptically polarized waves are three dimensional (not planar). Note that arc polarization is the large-amplitude equivalent of the small-amplitude linear polarized case. Tsurutani et al. (1997b) have suggested that interplanetary waves are spherical arc polarized structures.

Sec. 4.3] Discontinuities and Alfvén waves 203

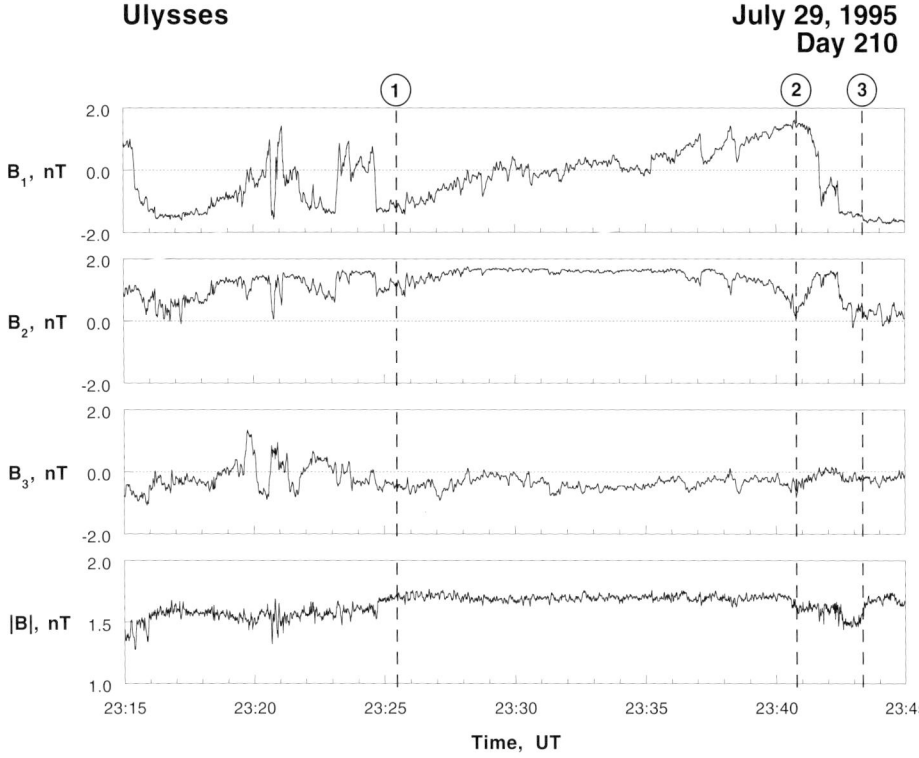

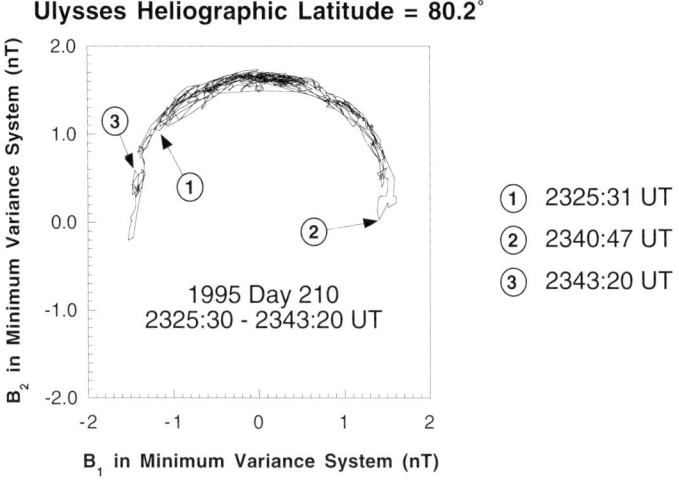

Figure 4.19. The relationship between a rotational discontinuity (between points 2 and 3) and slowly rotating Alfvén waves (between points 1 and 2). The rotational discontinuity is the phase-steepened edge of the Alfvén wave. The wave is arc-polarized and to first order, non-compressive (from Tsurutani *et al.*, 1997b).

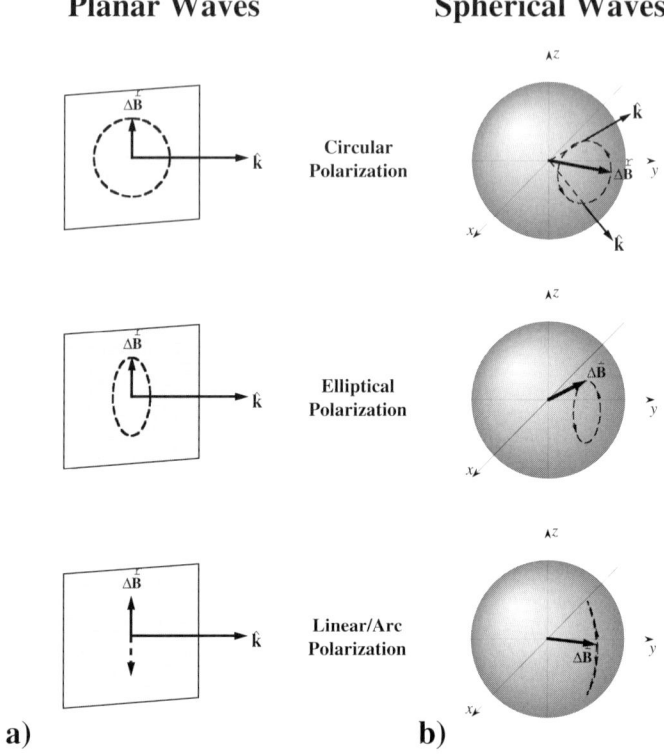

Figure 4.20. The different possible polarizations for plane waves (left) and spherical waves (right) (from Tsurutani et al., 1997b).

The presence of such large numbers of rotational discontinuities in high-speed wind clearly adds significant power to magnetic-field spectra. While these are still 'Alfvénic' with an anti-sunward sense of propagation, and so we can think of them as steepened Alfvén waves, nevertheless they are clearly different from the general population of waves and turbulence in the solar wind. In particular, a population of sharp discontinuities can contribute power to a spectrum over a wide range of scales, thereby altering spectral-index estimates, for example. Similarly, they are large steps in the magnetic field on small scales, and can therefore appear as 'intermittent' (see Section 4.2.3.1). It is not clear at this time what role discontinuities play in the turbulent cascade (see Section 4.2) and whether they are active in the energy-transfer process. If they interact with other fluctuations, then they should be considered as a particular constituent of the turbulence. However, if fluctuations are not significantly altered by discontinuities, their effect must be removed from diagnostics such as power spectra to determine the behaviour of the active component of the turbulence cascade. This is a topic of active study at this time.

4.3.4 Alfvénic shocks?

Because non-compressional Alfvén waves have speeds intermediate between fast (magnetosonic) and slow (sound) waves in the solar wind, in MHD they are considered 'intermediate' waves. The strong phase steepening indicates that dispersion and perhaps dissipation is occurring. To investigate the possibility that these discontinuities are Alfvénic (or intermediate) shocks, we examine the jump (upstream versus downstream) conditions across the discontinuities.

Since the Alfvén waves in the polar regions above the Sun are almost propagating purely outwards, this is an easy test to make. We simply assume that all of the waves are propagated radially outward. Thus, in standard notation the anti-sunward region is region 1 or the 'upstream' region, and the sunward region is region 2 or the 'downstream' region.

Figure 4.21 shows the relative jumps in temperature, density, and magnetic-field strength. In each case we use a standard notation: ΔX is equal to $X_2 - X_1$, where X_1 and X_2 are the upstream and downstream values, respectively. Basically, we find the solar-wind parameters typically jump by ~10 per cent or less. The changes do not occur in any consistent fashion (i.e. the variation appears to be random. Thus we find no obvious dissipation effects from the statistical analyses.

The lack of a pattern to the jump conditions does not mean that these phase-steepened Alfvén waves are not intermediate shocks. It simply implies that if there is dissipation, the process is much slower than the time separation of the measurements.

4.3.5 North–south asymmetries?

One fundamental question concerning the heliosphere is: Are there north–south asymmetries in discontinuities? This question is related to cosmic-ray-particle transport, size of the polar coronal holes and magnetic fields in the two regions. To answer the question of possible asymmetries, all discontinuities within 1 month near the north-polar pass (days 201–231 1995, heliographic latitude 80°S) were compared with 1 month near the south pole (days 226–256 1994, heliograph latitude 80°S). Figure 4.22 (Tsurutani et al., 1996b) shows the TS ROID value plus other interplanetary parameters. It is noted that the plasma densities, temperatures, magnetic field-magnitudes and ROID values are slightly higher at the north pole compared with those at the south pole. However, the two polar passes occurred at slightly different radial distances, a feature that must be taken into account. At the north pole at 2.0 AU the ROID value is 115 DDs/day. At the south pole at 2.3 AU the ROID value was 105 DDs/day. If we normalize the south-polar data by the previously determined radial gradient of $e^{-(r-1)/5}$ AU, the south-pole value would be 113 DDs/day at 2.0 AU. Thus there is no obvious north–south asymmetry in ROID values. This is also remarkable given that the two polar passes occurred nearly 1 year apart.

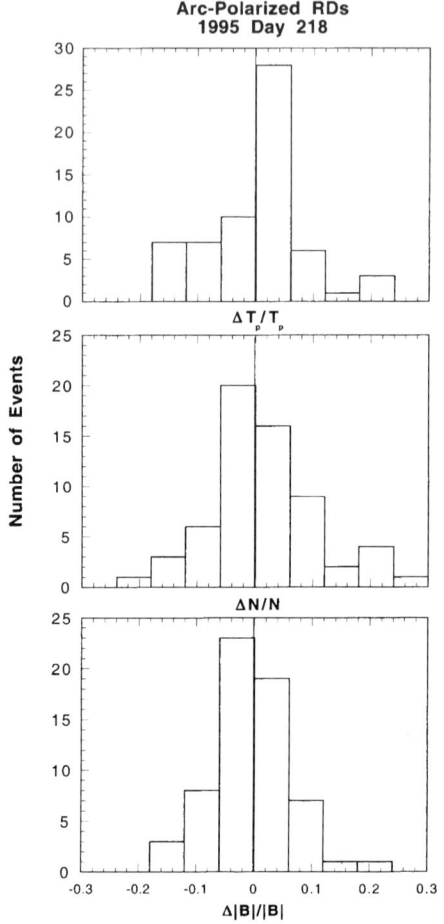

Figure 4.21. The normalized jumps in field magnitude, density, and temperature across rotational discontinuities. In each case the difference value is normalized by dividing by the sunward 'downstream' value (from Tsurutani and Ho, 1999).

4.3.6 Evolution of non-linear Alfvén waves and rotational discontinuities

Medvedev *et al.* (1998) studied the non-linear dynamics of Alfvén wave trains in a $\beta \sim 1$ ($\beta = 8\pi \sum N_i k T_i / |B|^2$) isothermal plasma, using the Kinetic Non-Linear Schrödinger (KNLS) model. They found that the combined effects of wave non-linearity and Landau damping result in an 'S'-shaped B_1–B_2 hodograph and arc-polarized rotational discontinuities. Buti *et al.* (1999), analysing Alfvén waves propagating in a medium with non-uniform densities and inhomogeneous magnetic fields, find that waves steepen and can be accelerated. The numerical solution, a form of a Modified Derivative Non-linear Schrödinger equation (MDLS) evolves into

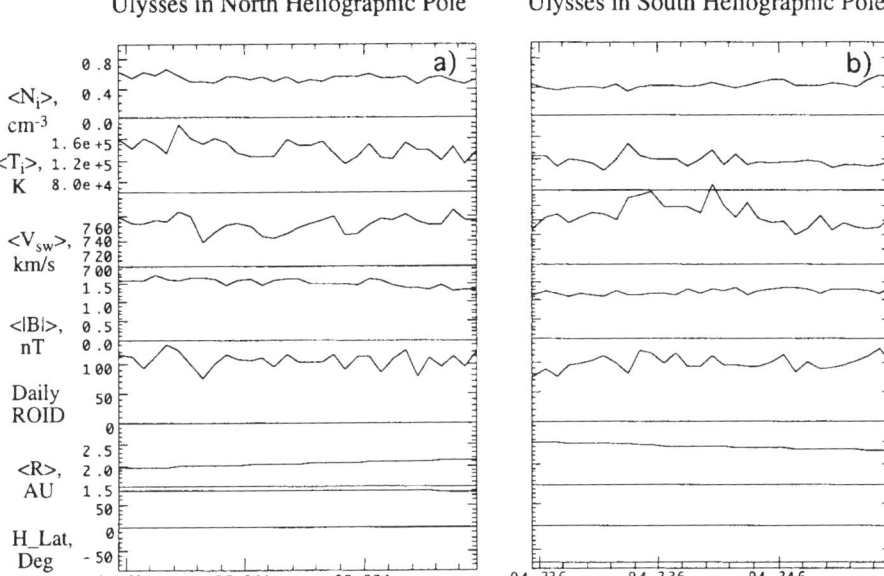

Figure 4.22. A comparison of discontinuity-occurrence rates over the north pole and over the south pole. From top to bottom are the average solar-wind densities, temperatures, velocities, magnetic-field magnitudes, TS ROID values, and *Ulysses* heliocentric radial distances and latitudes. All ROID values are relatively constant. If the different radial distances of the two *Ulysses* passes are taken into account, there is no obvious asymmetry present (from Tsurutani *et al*, 1996b).

turbulence due to inhomogeneities. Thus, both works predict dissipation of non-linear Alfvén waves.

Vasquez and Hollweg (1998a, b) also attempted to explain the formation of arc-polarized rotational discontinuities associated with Alfvén waves. From $2\frac{1}{2}$D hybrid numerical simulations of oblique Alfvén waves with linear polarizations, it is found that Alfvén waves evolve into spherical waves with arc polarization. Vasquez and Hollweg (1998a) suggest that the RDs are steady with their normals oblique to $\mathbf{B}_0$. Vasquez and Hollweg (1998b) attempted to produce RDs with small normal angles to $\mathbf{B}_0$. To do this they started with linearly polarized waves with propagation angles $\leqslant 10°$ to $\mathbf{B}_0$. The waves, even for small amplitudes, tend to steepen and form RDs with normals nearly along $\mathbf{B}_0$. However, they found that the RDs constantly widen due to dispersion effects and concluded that Alfvén waves with embedded RDs 'must evolve through a succession of arc-polarized or spherically polarized waveforms as they travel outward from the sun'.

4.3.7 TDs versus RDs

From the previous discussion, because there is such a large increase in DD rates within high-speed streams, we might expect all increases to be associated with increases in the occurrence of RDs. Also since there are no major streams present (but there are 'microstreams': Neugebauer *et al.*, 1995), there should not be many TDs found associated with stream–stream interactions. During 1994–1995, a period near solar minimum, the Heliospheric Current Sheet (HCS) was in the ecliptic plane (Smith *et al.*, 1993). Thus there should not be HCS-associated TDs at high latitude as well. We might therefore expect an absence of TDs in the polar regions.

To identify the discontinuity type we follow the Smith (1973a, b) method of using discontinuity 'phase space' regions. The minimum-variance method is again used to rotate the IMF into this system to determine B_3, the normal to the discontinuity. We divide by the larger field magnitude on either side of the discontinuity (B_L) and also measure the field-magnitude jump across the structure. The particular 'phase space' we use is B_3/B_L and $\Delta|B|/B_L$ where $\Delta|B|$ is the magnitude of the field vector change across the discontinuity.

For 4 days where *Ulysses* was over the north heliographic pole (days 218–221 1995), all 416 discontinuities that were selected by the TS criteria were analysed. These are displayed in discontinuity phase space in Figure 4.23. Events with large $\Delta|B|/B_L$ and small B_3/B_L are identified as TDs. The cut-off for this type of discontinuity is arbitrary, but we use $\Delta|B|/B_L > 0.2$ and $B_3/B_L < 0.2$. The number of clear TDs is 6.1 per cent of the total number of events.

Ho *et al.* (1995) examined 1,486 discontinuities identified during 15 days near the south pole. They identify 78 clear TDs or an occurrence rate of 5.2 per cent relative

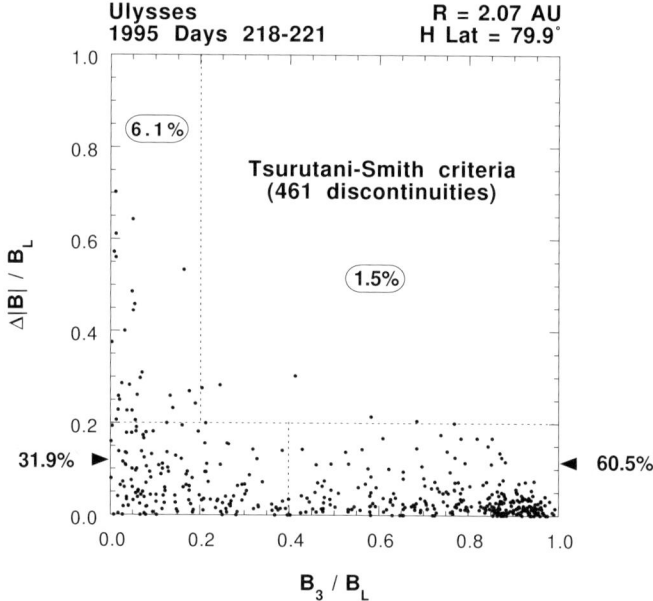

Figure 4.23. Discontinuity phase spaces (B_3/B_L, $\Delta|B|/B_L$) distributions for days 218–221 1993 at the north heliographic pole (from Tsurutani *et al.*, 1996b).

to all DDs. For in-ecliptic studies a broad range of values from 2.5 to 12.7 per cent were found in the reports of Smith (1973a), Neugebauer *et al.* (1984) and Lepping and Behannon (1986). This variability in the ecliptic plane may reflect true variability, and not statistical variation. The Ho *et al.* (1995) value lies in the centre of the spread of the near-ecliptic values. However, the surprising thing is that there are any TDs in the polar region at all.

4.3.8 Nature of tangential discontinuities at high latitudes

At high latitudes TDs have been detected at the edges of decreases in the magnetic-field magnitude. For simplicity these are called magnetic decreases (MDs; Tsurutani *et al.*, 1999; Tsurutani and Lakhina, 2000). Several types of MDs are shown in Figure 4.24, taken near the south pole. The co-ordinate system is Solar-Heliospheric. Panel (a) shows an MD with a field-magnitude decrease from $\sim$1.5 nT to as low as 0.2 nT from $\sim$0942:40 to 0944:10 UT. The field decrease is bounded by two sharp discontinuities. This is often the case. The normal of the first discontinuity is oriented at 80° relative to the ambient magnetic field and 90° for the second discontinuity. The $\Delta|B|/B_L$ values are 0.35 and 0.8, respectively. The discontinuities are thus TDs. Note in this case the field orientations prior to the MD and after the MD are almost the same.

Another type of MD is shown in panel b). The field direction rotates smoothly throughout the MD, and the field orientation after the MD is considerably different from that prior to the structure. The field magnitude is 1.4 nT prior to the structure and 1.0 afterward. The small decreases at 2151:40 UT has a normal oriented $\sim$49° relative to $\mathbf{B}_o$. This appears to be an RD with a significant magnitude change $\Delta|B|/B_0 \sim 0.25$. The second discontinuity at $\sim$2153 UT has a normal $\theta_{B_3} = 77°$ and a $\Delta|B|/B_0 = 0.55$. This is a TD.

The third example in panel c) again has discontinuities at both boundaries. θ_{B_3} is equal to 89° for the first event and $\theta_{B_3} = 88°$ for the second. Both are clearly TDs.

Figure 4.25 gives the properties of 129 discontinuities which bound MDs. These discontinuities were detected at the south pole and were chosen such that $\Delta|B|/B_L > 0.2$. Most (61 per cent) of DDs are tangential in nature. However, there are clearly a significant number of RDs present as well. The distribution of discontinuity normals (B_3/B_L) is a continuum.

The field decrease within MDs (bounded by the same 129 discontinuities) has been examined. The distribution is displayed in Figure 4.26. The number of events becomes smaller with increasing field-magnitude decreases.

To determine if there is a 'thickness' dependence on the relative decrease of the magnetic fields, normalized distribution (percentages) were calculated for MDs with $\Delta|B|/B_L$ values 0.2–0.3 (top panel), 0.3–0.4, 0.4–0.6 and 0.6–1.0. This is shown in Figure 4.27. From an intercomparison of the four panels, it appears that the temporal thickness is independent of the relative sizes of the field decreases, to first order.

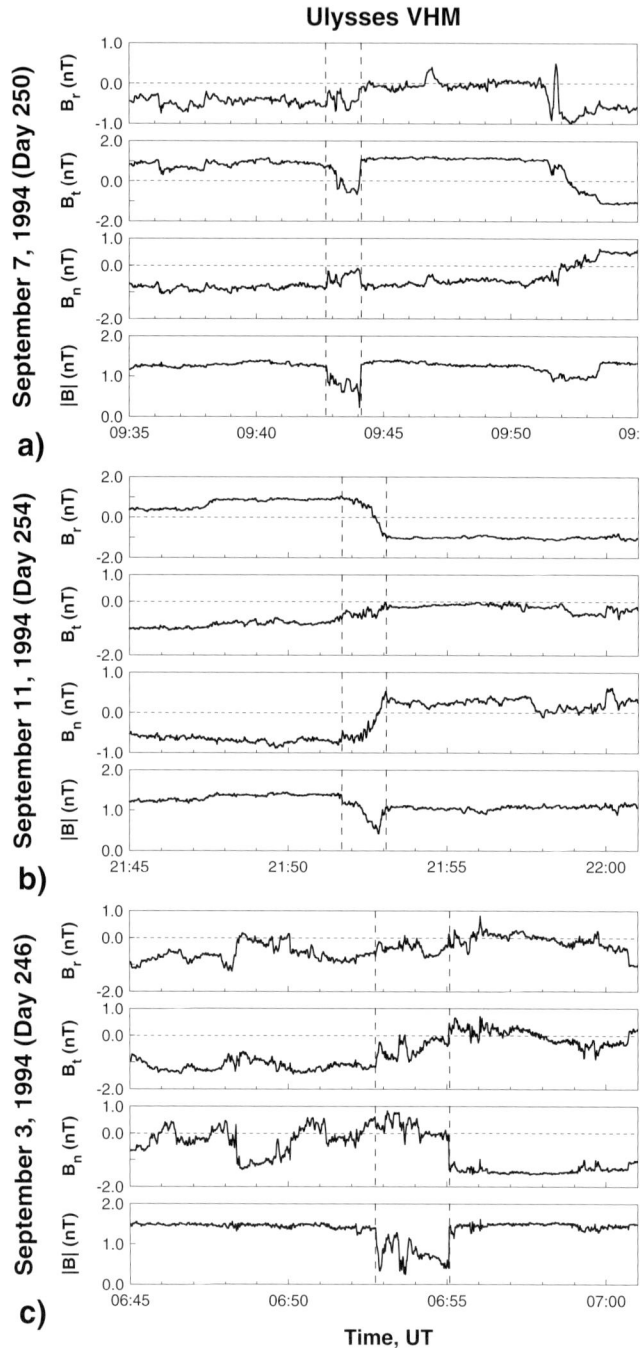

Figure 4.24. Several different types of MDs. These examples were taken at 80°S latitude, 2.3 AU from the Sun (from Tsurutani and Ho, 1999).

Sec. 4.3] **Discontinuities and Alfvén waves** 211

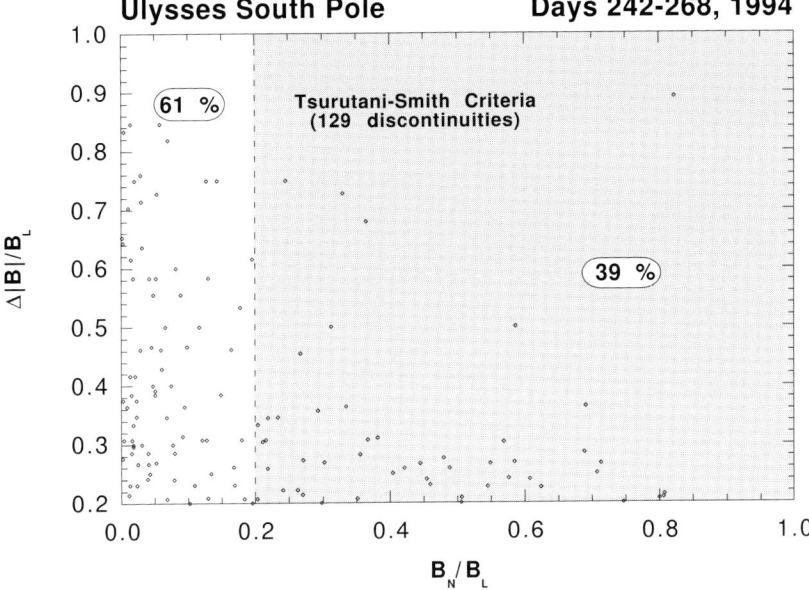

Figure 4.25. Minimum variance results for 129 discontinuities bounding MDs (from Tsurutani and Ho, 1999).

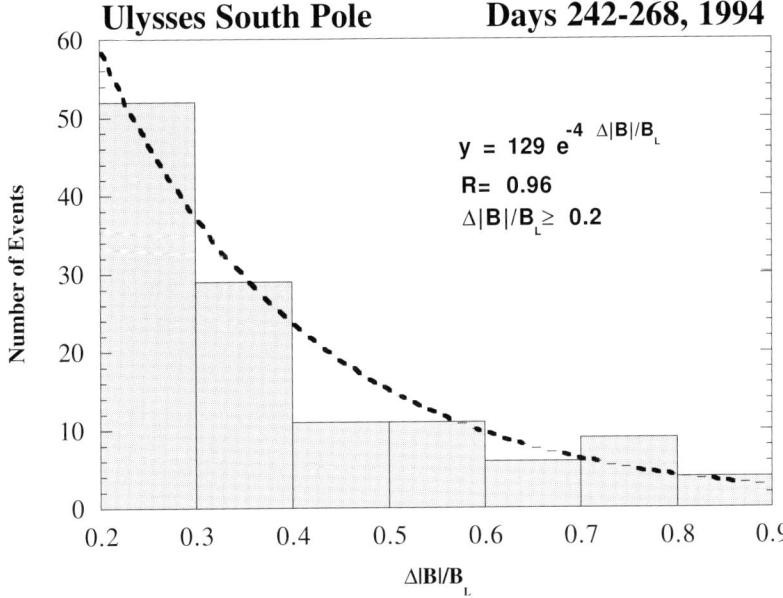

Figure 4.26. A distribution of field-magnitude changes within MDs (from Tsurutani and Ho, 1999).

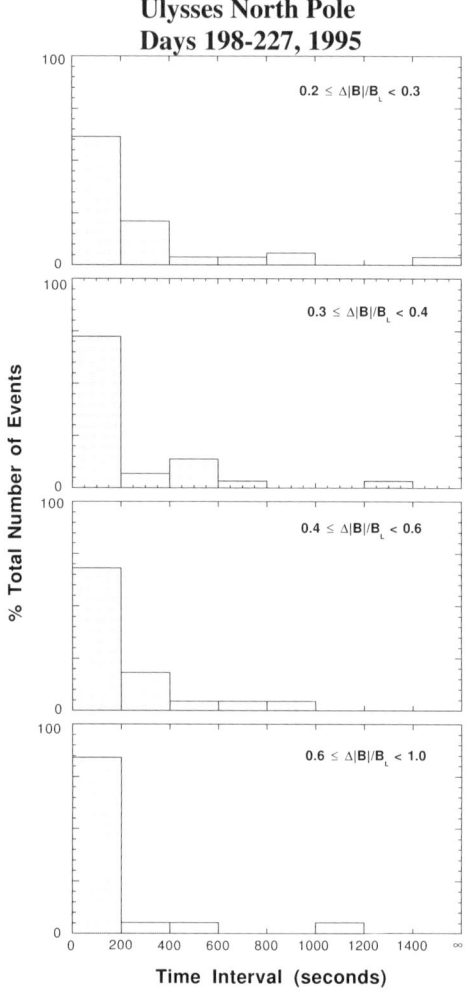

Figure 4.27. The MD thickness distribution for different values of $\Delta|B|/B_L$. The typical widths are $\sim$2–4 proton gyroradii (from Tsurutani and Ho, 1999).

4.3.9 MDs and magnetic holes

MDs have properties somewhat similar to magnetic holes (see Turner et al., 1977 for a discussion of holes). Turner et al. defined 'holes' as regions of low magnetic-field intensities (<1 nT) at 1 AU. Since B_0 is typically $\sim$5 nT at 1 AU, $\Delta|B|/B_L \geqslant 80$ per cent. There are clearly MD events with similarly large field decreases (Figure 4.26). Turner et al. (1977) found that holes have a characteristic dimension of $\sim$20,000 km. For the quiet solar wind with $V_{SW} \sim 400$ km s^{-1}, this corresponds to a timescale of $\sim$50 s. Figure 4.28 illustrates MD spatial scales. These were calculated by $d = \Delta t V_{SW} \cos\theta_{3V}$ where Δt is the thickness in s, V_{SW} is the solar-wind velocity

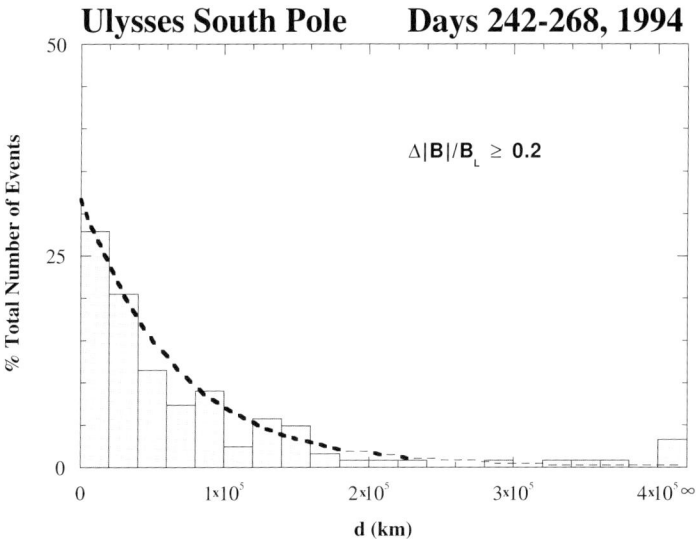

Figure 4.28. A histogram of MD thicknesses (from Tsurutani and Ho, 1999).

and θ_{3V} is angle between the discontinuity normal and the solar-wind velocity. The scale size of MDs is ~20,000–40,000 km. This number is similar to that found by Turner et al. for holes. Turner et al. also noted that there were some magnetic holes where the field directionality was nearly constant within the field decrease (linear holes) and others where the field change was substantial. MDs have similar properties.

Magnetic holes as originally described by Turner et al. (1977) have one property that is significantly different from that of high-latitude MDs. Magnetic holes in the ecliptic plane were not bounded by discontinuities. An example from Turner et al. is given in Figure 4.29.

Winterhalter et al. (1994, 1995) used *Ulysses* near-ecliptic field and plasma data to study linear holes (defined as events where **B** changed by less than 5°). They found that linear holes are preferentially detected in interaction regions at the leading edges of high-speed streams. The plasma surrounding the holes were marginally stable against the mirror mode. Winterhalter et al. (1994) argue that these holes are remnants of mirror mode structures that were created upstream of their observations.

The *Ulysses* high-latitude data have not yet been compared directly with the Winterhalter et al. (1994, 1995) near-ecliptic results. However, Ho et al. (1995) have examined high-latitude TDs and their neighbouring high beta plasma. One such example is given in Figure 4.30.

Typically there is pressure balance across the magnetic decreases. It is generally found that the plasma in adjacent regions to the MDs were stable against the mirror instability. The instability criterion is: $R = (\beta_\perp/\beta_\parallel)/(1 + 1/\beta_\perp) > 1$ where the plasma 'beta' $\beta = 8\pi \Sigma N_i k T_i / |B|^2$ and the sum in the plasma pressure is over all

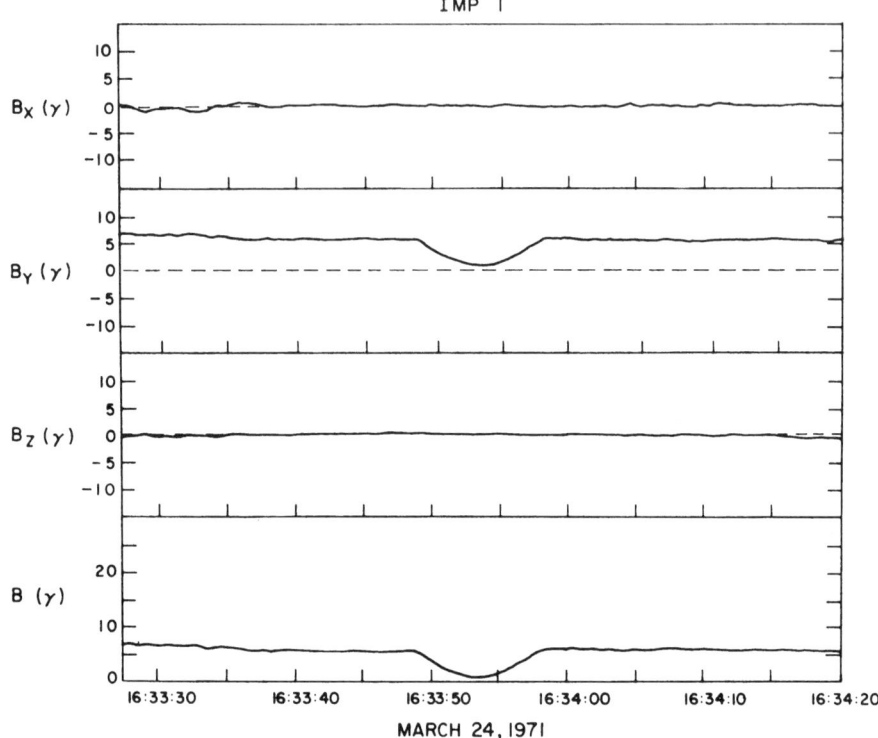

Figure 4.29. An example of a linear magnetic hole (from Turner et al., 1977).

ion species and electrons. '⊥' and '∥' indicate components perpendicular and parallel to the ambient magnetic-field direction.

In the Turner et al. (1977) work their holes were reported as isolated structures. However, we often find more complex structures at high latitudes such as that shown in Figure 4.29. When the MD structures are complex there are typically only discontinuities at the two edges of the major structure. There are generally few, if any, TDs (or RDs) in the centre of the extended MDs.

If these holes are indeed created by the mirror instability closer to the Sun, the presence of discontinuities at the boundaries is not understood. Mirror-mode structures found in planetary magnetosheaths (Tsurutani et al., 1982) do not have such boundary structures. Boundary-layer dynamics must be playing an important role. Another possibility which cannot be ruled out at this time is that these structures at high latitude are generated by a mechanism other than the mirror instability. These 'linear MDs' were not related to any particular stream structure, a result different from the Winterhalter et al. (1994) results.

Ho et al. (1995) also detected current-sheet-related TDs. This is shown in Figure 4.31. In this event the current sheet was located on the gradient of a microstream (however, not all current-sheet-related TDs were found to have such a relationship).

Sec. 4.3] Discontinuities and Alfvén waves 215

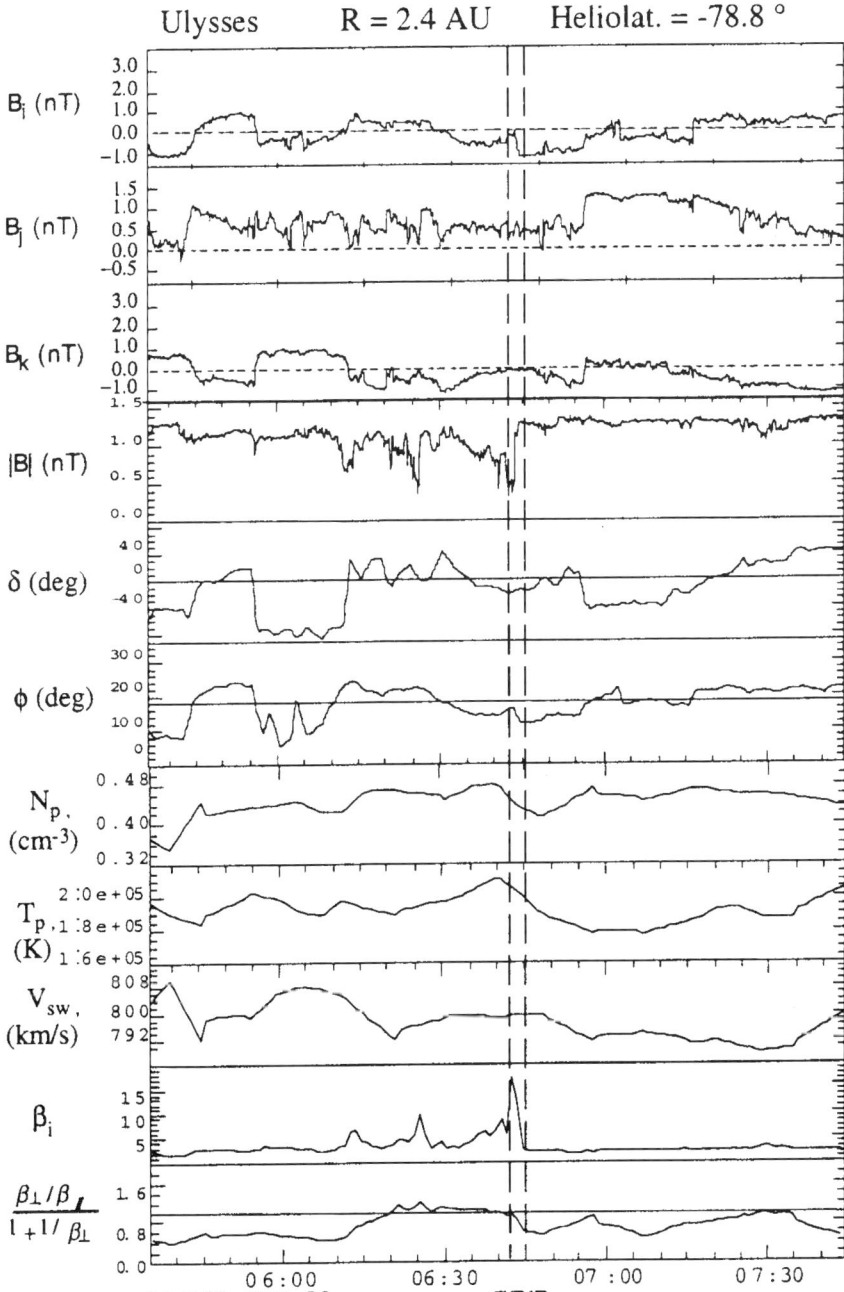

Figure 4.30. An example of a TD at the edge of MD-related field decreases. Note that the field decreases are complex and are significantly different than ordered as mirror mode structures detected in planetary magnetosheaths (from Ho et al., 1995).

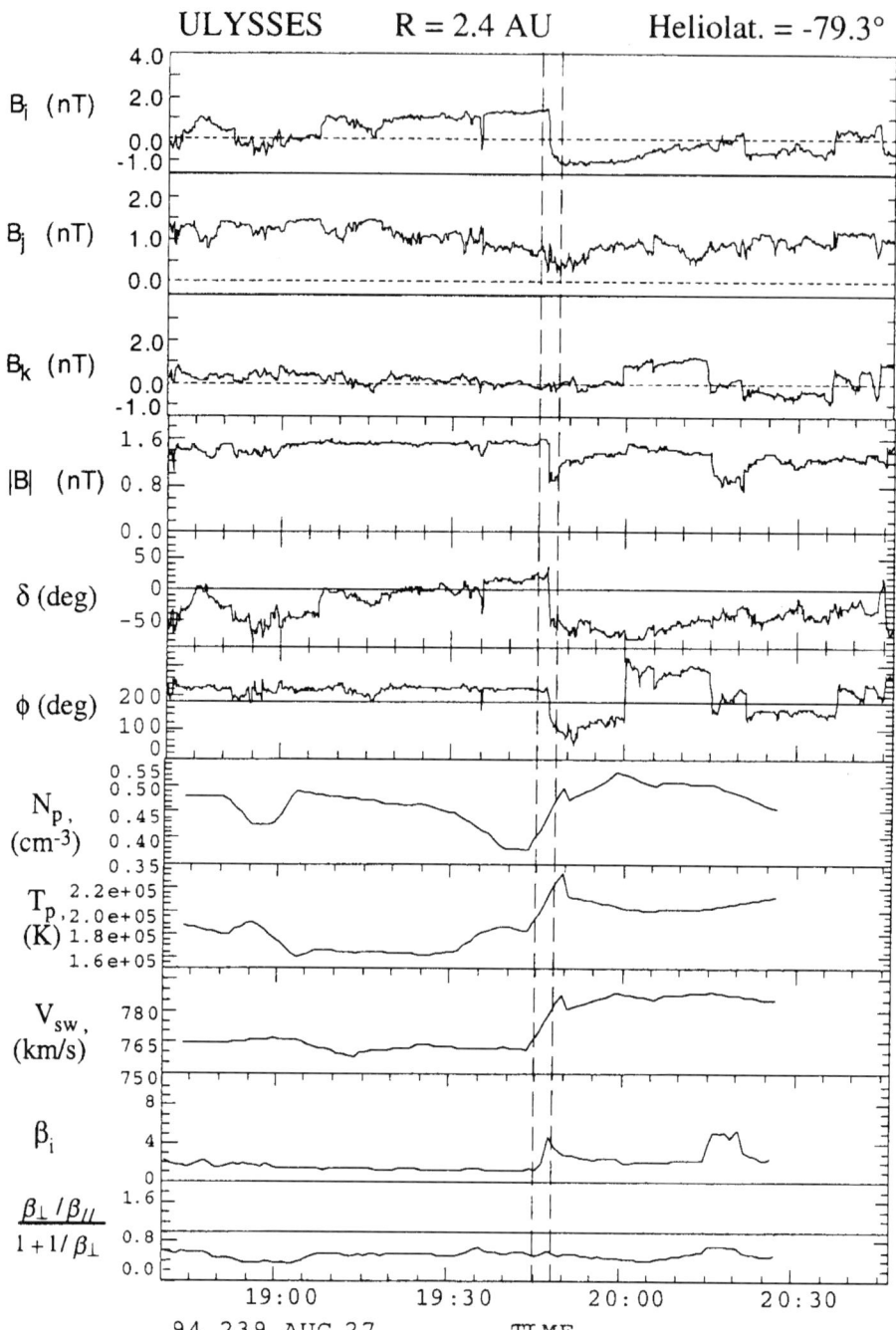

Figure 4.31. A current-sheet-related TD on day 239 1994. *Ulysses* was at 79.3°S heliographic latitude. This event occurred at the gradient of a microstream (from Ho *et al.*, 1995).

Sec. 4.3] Discontinuities and Alfvén waves 217

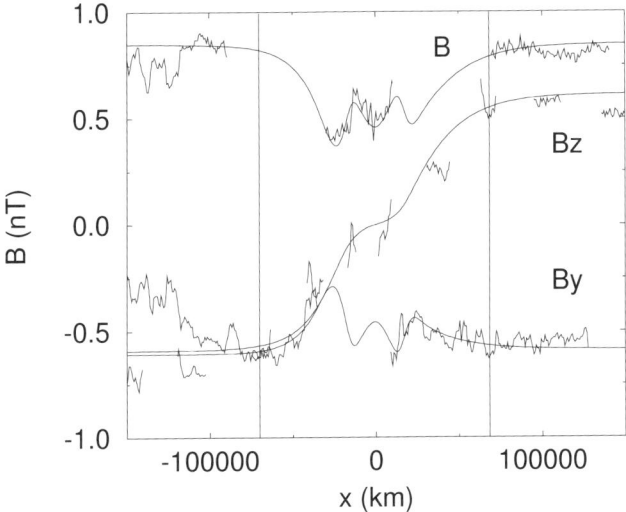

Figure 4.32. Observed (dashed lines) and simulated (solid lines) of two components of the field in minimum-variance co-ordinates (rotated for convenience by $-15°$ along the X-axis). The normal field (B_X) is less than 15 per cent of B magnitude. Two dashed vertical lines delimit the TD crossing centred at 0529 UT 3 July 1993. In the RTN co-ordinate system, the normal is 0.96, -0.05, -0.30, indicating that the TD was essentially orthogonal to the solar-wind velocity vector (from DeKeyser et al. 1996a).

The multilayer structure of a TD has been reproduced by a kinetic model (DeKeyser et al., 1996). A generalized Vlasov model is used. A best choice is selected for the velocity distribution function and therefore 'predicts' the plasma parameters in the absence of high-time-resolution plasma measurements.

Figure 4.32 shows a simulation of a TD. The dashed lines are the magnetic-field data and the solid lines are the results of the simulation. There are several current layers with current density of the order of 3×10^{-11} A m^{-2}. There is an electrostatic potential of less than 50 mV, indicating an extremely weak field along its normal. DeKeyser et al. comment that such models do not address issues of particle accessibility. They also note that the current layers with large shears and/or gradients may be unstable for growth of electromagnetic plasma waves, and thus may be in a state of turbulence rather than equilibrium. The simulation fits the data quite well.

Another work by DeKeyser et al. (1996b) has implied the presence of non-Maxwellian electron and ion velocity distribution functions within the transition layer. The non-constant solar wind bulk velocity is a major reason why the plasma develops current-carrying boundary layers.

4.3.10 Slow shocks

One of the rarest of discontinuities in interplanetary space are shocks. Of the various types of shocks (fast, intermediate, and slow) well-defined intermediate shocks

(Chao et al., 1993) and slow mode shocks are the rarest. Over the last few decades of interplanetary space missions only a few slow shocks have been reported in the literature (Chao and Olbert, 1970; Richter, 1991; Whang et al., 1998a, b). The reason for the rarity is most probably that slow-mode shocks develop in regions where collisionless damping of slow-mode waves is small and the damping time exceeds the wave-steepening time. This requires that plasma $\beta < 1$ and $T_p/T_e < 1$, conditions not typically found in interplanetary space.

However, in the *Ulysses* data some slow shocks have been detected and, in fact, a slow shock pair (forward and reverse) has been found for the first time. This pair of slow-mode shocks is found embedded within a CIR where the above plasma conditions were met (at 5.3 AU and 9°S heliolatitude). It is possible that at even larger radial distances slow-mode shocks may be even more common.

Figure 4.33 shows the plasma and field data for the slow shock pair (from Ho et al., 1998). The two shock discontinuities are denoted by the vertical lines. The magnetic coplanarity theorem is used to calculate the shock normal. Next the upstream plasma-normal velocities were calculated using the Rankine–Hugonot conservation relations. Finally, the normal velocities were checked to determine if the value is intermediate between the Alfvén velocity in the shock-normal direction and the slow-mode velocity. More details can be found in Ho et al. (1996). The forward slow shock occurs at 0710 UT and the reverse slow shock at 1105 UT. At the first event V_{SW} increases from 425 to 463 km s^{-1}, N_p from 0.6 to 1.2 cm^{-3}, and T_p from 0.4 to 0.5 $\times$ 10^5 K. $|B|$ decreases from 1.4 to 1.1 nT. At the second discontinuity V_{SW} increases from 447 to 500 km s^{-1}, N_p decreases from 0.2 to 0.07 cm^{-3}, and T_p decreases from 1.5 to 0.6 $\times$ 10^5 K. $|B|$ increases from 1.2 to 1.7 nT. The shock speeds have been calculated to be 60 km s^{-1} and 115 km s^{-1}, respectively. These speeds are not unlike those of *Ulysses* fast-mode shocks at these distances.

These slow-mode shocks may be associated with magnetic-field reconnection. Energetic 30–90-keV electron and ion acoustic-like plasma waves were associated with this event. This is most probably one form of magnetic energy dissipation. The frequency of occurrence of slow shock events are necessary to determine the overall effectiveness of the dissipation. Perhaps a mission like *Cassini* which goes to 10 AU can be used to assess these possible effects.

4.4 CONCLUSIONS

Many new discoveries regarding waves, discontinuities and turbulence have been made by the *Ulysses* mission. The discovery that RDs are the phase-steepened edges of non-linear Alfvén waves was one of the major finds of the mission. The consequences of such features in the interplanetary medium is currently being explored. In fact, non-linear waves and wave phase steepening are part of a new emerging field in plasma physics. A merger of space-plasma physicists, fusion-plasma physicists, and mathematicians are hard at work grappling with both theory and observations to try to understand these features (Hada and Matsumoto, 1997; Marsch et al., 1999).

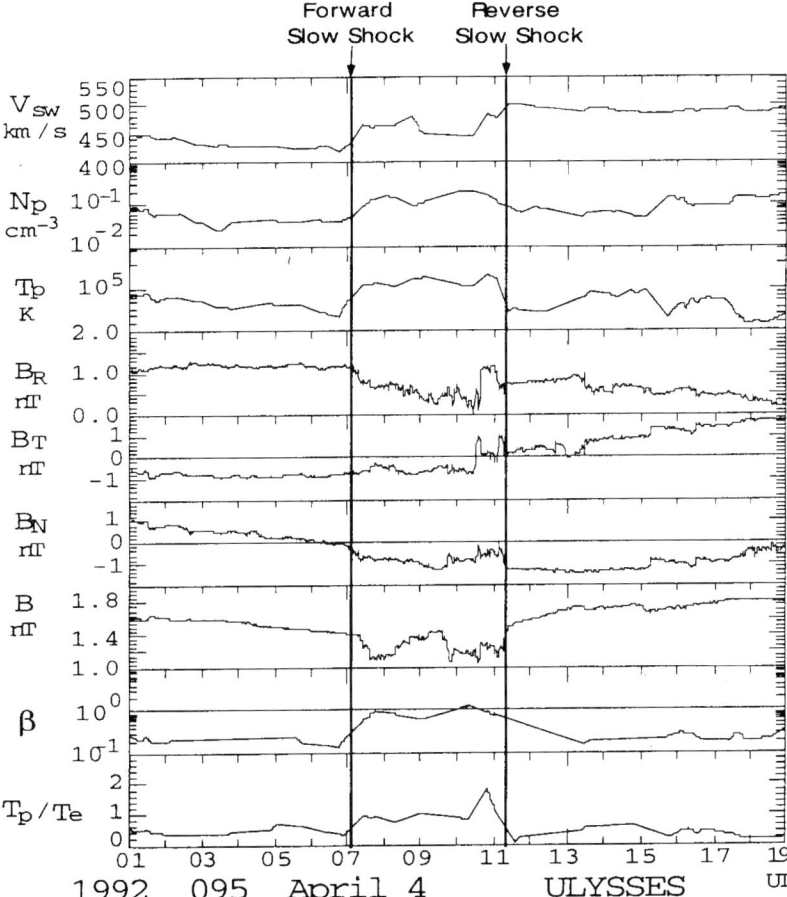

Figure 4.33. A slow shock pair. From top to bottom are: the solar-wind speed, proton density, proton temperature, magnetic-field components in solar–heliospheric co-ordinates, field magnitude, plasma beta, and proton-to-electron temperature ratios (from Ho et al., 1998).

Besides the problems listed above there is still much more to do and learn. It is not understood why perfect, ideal RDs ($\Delta|B|/B_L = 0$ and $B_3/B_L = 1$) and perfect TDs ($\Delta|B|/B_L \neq 0$ and $B_3/B_L = 0$) exist only rarely in nature. Such pure TDs and RDs are generally not found in interplanetary space.

So far only two regions of discontinuity phase space have been dealt with in detail, that of $\Delta|B|/B_L > 0.2$ and $B_3/B_L < 0.2$ (TDs) and $\Delta|B|/B_L < 0.2$ and $B_3/B_L > 0.4$ (RDs). In Figure 4.23 the region $\Delta|B|/B_L > 0.2$ and $B_3/B_L > 0.4$ has not been well studied. This region of phase space is ordinarily thought to be the region where interplanetary shocks are found. However, clearly fast forward shocks are not present at these high latitudes. These discontinuities must be due to some other phenomenon.

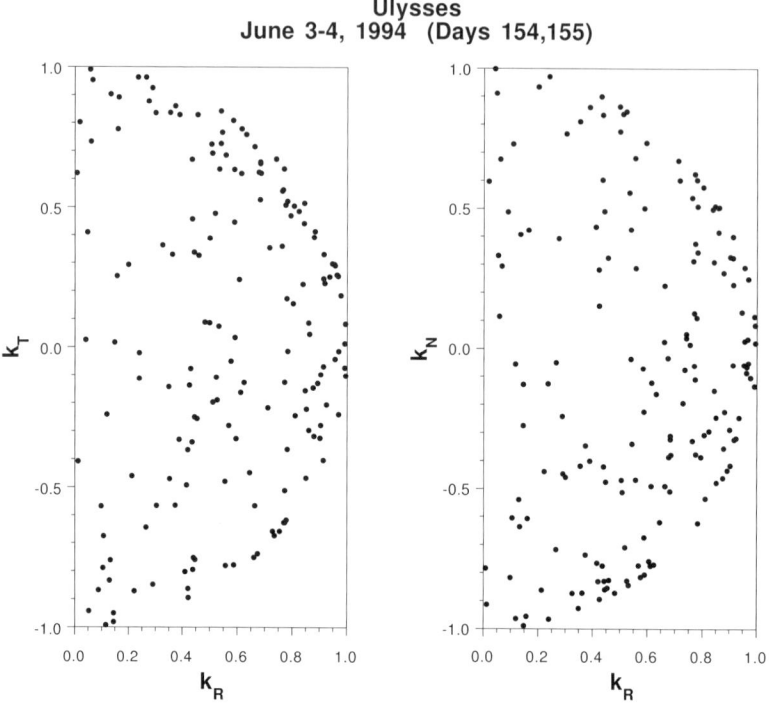

Figure 4.34. The discontinuity normals for days 154–155 1994. The normals are displayed in the RTN co-ordinate system (from Tsurutani *et al.* 1995).

The MDs detected in the polar regions of the heliosphere are not well understood at this time. Why these field decreases are bounded by discontinuities is a mystery. Minimum-variance analyses indicate that there are both TDs and RDs bounding these structures. The field jumps across the TDs are very large. Events with large normals ($B_N/B_L > 0.4$) and large values of $\Delta|B|/B_L$ were noted. This last set of events clearly belong to the little-studied $\Delta|B|/B_L > 0.2$ and $B_N/B_L > 0.4$ region of phase space. A study of these new types of discontinuities is currently being performed.

The region of $\Delta|B|/B_L < 0.2$ and $B_3/B_L < 0.4$ also needs to be studied. Neugebauer *et al.* (1984), using field and plasma data, found that many of the discontinuities with 'properties of both TDs and RDs' were in actuality mainly RDs. However, this conclusion is not agreed upon by all workers in the field.

One clear outstanding problem is the determination of the directions of propagation of RDs. Figure 4.34 shows the minimum variance directions of the discontinuities from days 154–155 1995 in the RTN co-ordinate system. This pattern looks quite isotropic, a somewhat surprising outcome. However, it should be noted that for arc-polarized structures almost all of the variance will be in the B_1 direction with very little in B_2. Thus the ratio of λ_2/λ_3 will be generally close to 1.0 making the uncertainty in determination quite large. There is also a physical problem with this issue as well. If, as everyone has agreed, the Alfvén waves are propagating radially outward, then the leading or trailing edges should be directed in the anti-sunward or sunward directions, respectively. However, the minimum-variance direction is more or less isotropic as has been noted and individual examples have indicated directions orthogonal to the radial.

Tsurutani *et al.* (1997) in examining this issue believe that for spherical waves the minimum-variance direction may not be the direction of propagation, but the intermediate-variance direction instead. Arguments have been made why an orthogonal directionality would imply a left-hand and a right-hand wave coupling, two modes which would soon dispersively separate. Thus the RDs would quickly broaden and parts would quickly move away from the slowly rotating part of the Alfvén wave. Currently, studies using multiple spacecraft are focused on this topic. However, the ultimate answer might come when the Cluster II mission gets launched and becomes operational.

With regard to MHD turbulence, while *Ulysses* data has allowed detailed measurements of evolution rates and small-scale properties such as intermittency, it is still not clear how the energy cascade develops: How is energy transferred, and what role do structures and discontinuities play? Do microstreams play a role? Are phase-steepened waves evidence of this process occurring? Do discontinuities interact with the turbulence, or are they a fundamental constituent of it? Are discontinuities responsible for intermittency in MHD turbulence? If much of the fluctuation power has wavevectors perpendicular to the mean field direction, as appears to be the case (Bieber *et al.*, 1996), is this the result of tangential discontinuities, or some other component of the turbulence, and what effect does this have on the cascade? Some of these questions can be answered using *Ulysses* data, and more detailed analysis of the existing dataset will undoubtedly continue for many years. Indeed, the data from the 1994–1996 polar passes, within high-speed solar wind, are a unique set and, from the perspective of MHD processes, will probably remain the most important output of the *Ulysses* mission.

In the next few years *Ulysses* will again pass over the Sun's poles, this time while the Sun is near solar maximum. Conditions at high latitudes will be very different: it is likely that there will be no long-lasting coronal holes covering the polar regions, and hence no long-lasting high-speed wind streams. Rather, there will be a mixture of wind streams of different velocities, along with many transient events. The resulting velocity shears will drive compressions, rarefactions and shocks, and lead to a much more disturbed environment that that near solar minimum. As a result, small-scale fluctuations will also be very different, both due to variation in source conditions and the environment in which they develop within the solar wind. In general, conditions should be similar to those at low latitudes but with some important differences. At high latitudes the Parker spiral direction is closer to radial than near the ecliptic and stream shear can be less developed. Indeed, there may be strong shears persisting out to several AU, with fast and slow streams flowing side by side rather than colliding. In this case, the shear is expected to decrease the Alfvénicity of the fluctuations (e.g. Goldstein *et al.*, 1999) and *Ulysses* data may provide particularly clear examples of this process, the results of which can be applied to more complex environments at low latitudes.

4.5 ACKNOWLEDGEMENTS

The authors are grateful to members of the *Ulysses* team, particularly A. Balogh, D. J. McComas, E. J. Smith, and R. Marsden. This work was supported in part by PPARC (UK) Grant GR/L29903; T. Horbury is supported by a PPARC Fellowship. Portions of this work were performed at the Jet Propulsion Laboratory, California Institute of Technology, Pasadena, under contract with the National Aeronautics and Space Administration.

4.6 REFERENCES

Balogh, A., Forsyth, R. J., Lucek, E. A., Horbury, T. S. and Smith, E. J. (1999) Heliospheric magnetic field polarity inversions at high heliographic latitudes. *Geophys. Res. Lett.* **26**, 631–634.

Bavassano, B. and Smith, E. J. (1986) Radial variations of interplanetary Alfvénic fluctuations: Pioneer 10 and 11 observations between 1 and 5 AU. *J. Geophys. Res.* **91**, 1706–1710.

Bavassano, B., Dobrowolny, M., Fanfoni, G., Mariani, F. and Ness, N. F. (1982a) Statistical properties of MHD fluctuations associated with high-speed streams from Helios-2 observations. *Sol. Phys.* **78**, 373–384.

Bavassano, B., Dobrowolny, M., Mariani, F. and Ness, N. F. (1982b) Radial evolution of power spectra of interplanetary Alfvénic turbulence. *J. Geophys. Res.* **87**, 3617–3622.

Bavassano, B., E. Pietropaolo and Bruno, R. (1998) Cross-helicity and residual energy in solar wind turbulence: Radial evolution and latitudinal dependence in the region from 1 to 5 AU, *J. Geophys. Res.* **103**, 6521–6529.

Bavassano, B., Pietropaolo, E. and Bruno, R. (2000) Alfvénic turbulence in the polar wind: A statistical study on cross helicity and residual energy variations. *J. Geophys. Res.* **105**, 12697–12704.

Bazer, J. and Hurley, J. (1963) Geometrical hydromagnetics. *J. Geophys. Res.* **68**, 147–174.

Behannon, K. W. (1978) Heliocentric distance dependence of the interplanetary magnetic field. *Rev. Geophys.* **16**, 125.

Belcher, J. W. and Davis, L. Jr. (1971) Large-amplitude Alfvén waves in the interplanetary medium, 2. *J. Geophys. Res.* **76**, 3534–3563.

Bieber, J. W., Wanner, W. and Matthaues, W. H. (1996) Dominant two-dimensional solar wind turbulence with implications for cosmic ray transport. *J. Geophys. Res.* **101**, 2511–2522.

Borgas, M. S. (1992) A comparison of intermittency models in turbulence. *Phys. Fluids A*, **4**, 2055–2061.

Boyd, T. J. M. and Sanderson, J. J. (1969) Plasma dynamics, *Thomas Nelson*, London.

Bruno, R. (1992) Inner heliosphere observations of MHD turbulence in the solar wind. Challenges to theory. In: Marsch, E. and Schwenn, R. (ed.) *Solar Wind 7*, Pergamon Press, Oxford.

Burlaga, L. F. (1971) Nature and origin of directional discontinuities in the solar wind. *J. Geophys. Res.* **76**, 4360.

Burlaga, L. F. (1991) Intermittent turbulence in the solar wind. *J. Geophys. Res.* **96**, 5847–5851.

Buti, B., Galinski, V. L., Shevchenko, V. I., Lakhina, G. S., Tsurutani, B. T., Diamond, P. and Medvedev, M. V. (1999) Evolution of nonlinear Alfvén waves in streaming inhomogeneous plasmas. *Astrophys. J.* **523**, 849.

Carbone, V. (1994) Scaling exponents of the velocity structure functions in the interplanetary medium. *Ann. Geophys.* **12**, 585–590.

Chao, J. K. and Olbert, S. (1970) Observation of slow shocks in interplanetary space. *J. Geophys. Res.* **75**, 6394.

Chao, J. K., Lyu, L. H., Wu, B. H., Lazarus, A. J., Chang, T. S. and Lepping, R. P. (1993) Observations of an intermediate shock. *J. Geophys. Res.* **98**, 17,443.

Colburn, D. S. and Sonett, C. P. (1966) Discontinuities in the solar wind. *Space Sci. Rev.* **5**, 439.

Coleman, P. L. Jr. (1968) Turbulence, viscosity, and dissipation in the solar-wind plasma, *Astrophys. J.* **153**, 371–388.

DeKeyser, J., Roth, M., Lemaire, J., Tsurutani, B. T., Ho, C. M. and Hammond, C. M. (1996a) Theoretical plasma distributions consistent with Ulysses magnetic field observations in a solar wind tangential discontinuity. *Solar Phys.* **66**, 415.

DeKeyser, J., Roth, M., Tsurutani, B. T., Ho, C. M. and Phillips, J. L. (1996b) Solar wind velocity jumps across tangential discontinuities: Ulysses observations and kinetic interpretation. *Astron. Astrophys.* **166**, 415.

Fisk, L. A., Jokipii, J. R., Simnett, G. M., von Steiger, R. and Wenzel, K.-P. (eds.) (1998) *Cosmic rays in the heliosphere*, Kluwer Academic, Dordrecht.

Forsyth, R. J., Horbury, T. S., Balogh, A. and Smith, E. J. (1996) Hourly variances of fluctuations in the heliospheric magnetic field out of the ecliptic plane. *Geophys. Res. Lett.* **23**, 595–598.

Frisch, U. (1995) Turbulence, *Cambridge University Press*, Cambridge.

Goldstein, B. E., Neugebauer, M. and Smith, E. J. (1995a) Alfvén waves, alpha-particles, and pickup ions in the solar-wind, *Geophys. Res. Lett.* **22**, 3389–3392.

Goldstein, B. E., Smith, E. J., Balogh, A., Horbury, T. S., Goldstein, M. L. and Roberts, D. A. (1995b) Properties of magnetohydrodynamic turbulence in the solar wind as observed by Ulysses at high heliographic latitudes. *Geophys. Res. Lett.* **22**, 3393–3396.

Goldstein, M. L. and Roberts, D. A. (1995) Magnetohydrodynamic turbulence in the solar wind. *Ann. Rev. Astron. Astrophys.*, **33**, 283–325.

Goldstein, M. L., Roberts, D. A., Deane, A. E., Ghosh, S. and Wong, H. K. (1999) Numerical simulation of Alfvénic turbulence in the solar wind. *J. Geophys. Res.* **104**, 14437–13351.

Grappin, R., Velli, M. and Mangeney, A. (1991) Alfvénic versus standard turbulence in the solar wind. *Ann. Geophys.* **9**, 416–426.

Hada, T. and Matsumoto, H. (1997) Nonlinear Waves and Chaos in Space Plasmas. Terra Scientific, Tokyo.

Ho, C. M., Tsurutani, B. T., Goldstein, B. E., Phillips, J. L. and Balogh, A. (1995) Tangential discontinuities at high heliographic latitudes ($\sim -80°$), *Geophys. Res. Lett.* **22**, 3409.

Ho, C. M., Tsurutani, B. T., Lin, N., Lanzerotti, L. J., Smith, E. J., Goldstein, B. E., Buti, B., Lakhina, G. S. and Zhou, X. Y. (1998) A pair of forward and reverse slow-mode shocks detected by Ulysses at $\sim$5 AU. *Geophys. Res. Lett.*

Ho, C. M., Tsurutani, B. T., Smith, E. J. and Feldman, W. C. (1996) Properties of slow-mode shocks in the distant (>200 RE) geomagnetic tail. *J. Geophys. Res.* **121**, 15277.

Horbury, T. S. (1996) Ulysses observations of magnetic field fluctuations in the heliosphere. *PhD dissertation*, University of London.

Horbury, T. S. and Balogh, A. (1997) Structure function measurements of the intermittent turbulent cascade. *Nonlinear Processes in Geophysics*, **4**, 185–199.

Horbury, T. S. and Balogh, A. (2001) Evolution of magnetic field fluctuations in high speed solar wind streams: Ulysses and Helios observations. *J. Geophys. Res.* in press.

Horbury, T. S., Balogh, A., Forsyth, R. J. and Smith, E. J. (1995a) Anisotropy of inertial range turbulence in the polar heliosphere. *Geophys. Res. Lett.* **22**, 3405–3408.

Horbury, T. S., Balogh, A., Forsyth, R. J. and Smith, E. J. (1995b) Observations of evolving turbulence in the polar solar wind. *Geophys. Res. Lett.* **22**, 3401–3404.

Horbury, T. S., Balogh, A., Forsyth, R. J. and Smith, E. J. (1995c) Ulysses magnetic field observations of fluctuations within polar coronal flows. *Ann. Geophys.* **13**, 105–107.

Horbury, T. S., Balogh, A., Forsyth, R. J. and Smith, E. J. (1996a) The rate of turbulent evolution over the Sun's poles. *Astron. Astrophys.* **316**, 333–341.

Horbury, T. S., Balogh, A., Forsyth, R. J. and Smith, E. J. (1996b) Magnetic field signatures of unevolved turbulence in solar polar flows. *J. Geophys. Res.* **101**, 405–413.

Horbury, T. S., Balogh, A., Forsyth, R. J. and Smith, E. J. (1997) Ulysses observations of intermittent heliospheric turbulence. *Adv. Space Res.* **19**, 847–850.

Horbury T. S., Lucek, E. A, Balogh, A. and McComas, D. J. (1998) Wave power dropouts associated with radial field intervals in high speed solar wind. *Geophys. Res. Lett.* **25**, 4297–4300.

Hudson, P.D. (1970) Discontinuities in an anisotropic plasma and their identification in the solar wind. *Planet. Space Sci.* **18**, 1611.

Jokipii, J. R. and Parker, E. N. (1968) Random walk of magnetic field lines of force in astrophysics. *Phys. Rev. Lett.* **21**, 44–77.

Jokipii, J. R., Kóta, J., Giacalone, J., Horbury, T. S. and Smith, E. J. (1995) Interpretation and consequences of large scale magnetic variances observed at high heliographic latitude. *Geophys. Res. Lett.* **22**, 3385–3388.

Klein, L. W., Roberts, D. A. and Goldstein, M. L. (1991) Anisotropy and Minimum Variance Directions of Solar Wind Fluctuations in the Outer Heliosphere. *J. Geophys. Res.* **96**, 3779–3788.

Klein, L., Bruno, R., Bavassano, B. and Rosenbauer, H. (1993) Anisotropy and Minimum Variance of Magnetohydrodynamic Fluctuations in the Inner Heliosphere. *J. Geophys. Res.* **98**, 17461–17466.

Kolmogorov, A. (1941) The local structure of turbulence in incompressible viscous fluid for very large Reynolds' numbers. *Comptes rendus (Doklady) de l'Academie des sciences de l'URSS*, **30**, 301–305.

Kraichnan, R. H. (1965) Inertial-range spectrum of hydromagnetic turbulence. *Phys. Fluids*, **8**, 1385–1387.

Landau, L. D. and Lifschitz, E. M. (1960) Electrodynamics of Continuous Media, 255 pp., Pergamon, New York.

Leamon, R. J., Smith, C. W., Ness, N. F., Matthaeus, W. H. and Wong, H. K. (1998) Observational constraints on the dynamics of the interplanetary magnetic field dissipation range. *J. Geophys. Res.* **103**, 4775–4787.

Lepping, R. P. and Behannon, K. W. (1986) Magnetic field directional discontinuities: Characteristics between 0.46 and 1.0 AU. *J. Geophys. Res.* **91**, 8725.

Lucek E. A. and Balogh, A. (1998) The identification and characterization of Alfvénic fluctuations in Ulysses data at midlatitudes. *Astrophys. J.* **507**, 984–990.

Lucek, E. A., Horbury, T. S., Balogh, A. and McComas, D. J. (1999) Plasma signatures of radial field power dropouts, *Proceedings of Solar Wind 9* (Habbal, S. R., Esser, R., Hollweg, J. V. and Isenberg, P. A. eds) *American Institute of Physics*, 475–478.

Mariani, F., Bavassano, B., Villante, U. and Ness, N. F. (1973) Variation of the occurrence rates of discontinuities in the interplanetary magnetic field. *J. Geophys. Res.* **78**, 8011.

Marsch, E. and Liu, S. (1993) Structure functions and intermittency of velocity fluctuations in the solar wind. *Ann. Geophys.* **11**, 227–238.

Marsch, E. and Tu, C.-Y. (1990) On the radial evolution of MHD turbulence in the inner heliosphere. *J. Geophys. Res.* **95**, 8211–8229.

Marsch, E. and Tu, C.-Y. (1994) Non-Gaussian probability distributions of solar wind fluctuations. *Ann. Geophys.* **12**, 1127–1138.

Marsch, E. and Tu, C.-Y. (1996) Spatial evolution of the magnetic field spectral exponent in the solar wind: Helios and Ulysses comparison. *Ann. Geophys.* **101**, 11149–11152.

Marsch, E. and Tu, C.-Y. (1997) Intermittency, non-Gaussian statistics and fractal scaling of MHD fluctuations in the solar wind. *Nonlinear Processes in Geophysics*, **4**, 101–124.

Marsch, E., Tsurutani, B. T. and Diamond, P. H. (eds.) (1999) Nonlinear waves and chaos. *Nonlinear Proc. Geophys.* **6**.

Matthaeus, W. H. and Goldstein, M. L. (1986) Low-frequency $1/f$ noise in the interplanetary magnetic field. *Phys. Rev. Lett.* **57**, 495–502.

Matthaeus, W. H., Bieber, J. W. and Zank, G. P. (1995) Unquiet on any front – anisotropic turbulence in the solar wind. *Rev. Geophys.* **33**, 609–614.

Medvedev, M. V., Diamond, P. H., Shevchenko, V. I. and Galinsky, V. L. (1998) Dissipative dynamics of collisionless nonlinear Alfvén wave trains. *Phys. Rev. Lett.* **78**, 4934.

Meneveau, C. and Sreenivasan, S. R. (1987) Simple multifractal cascade model for fully developed turbulence. *Phys. Rev. Lett.* **59**.

Neugebauer, M., Clay, D. R., Goldstein, B. E., Tsurutani, B. T. and Zwickl, R. D. (1984) A re-examination of rotational and tangential discontinuities in the solar wind. *J. Geophys. Res.* **89**, 5395.

Neugebauer, M., Goldstein, B. E., McComas, D. J., Suess, S. T. and Balogh, A. (1995) Ulysses observations of microstreams in the solar wind from coronal holes. *J. Geophys. Res.* **100**, 23389.

Petschek, H. E. (1964) Magnetic field annihilation, AAS-NASA Symposium on the physics of solar fluids ed. by Ness, W.N. NASA SP-50, 425.

Phillips, J. L., Balogh, A., Bame, S. J., Goldstein, B. E., Gosling, J. T., Hoeksema, J. T., McComas, D. J., Neugebauer, M., Sheeley, N. R. Jr. and Wang, Y. M. (1994) Ulysses at 50 south: Constant immersion in the high speed solar wind. *Geophys. Res. Lett.* **21**, 1105.

Richter, A. K. (1991) Interplanetary slow shocks, Space and Solar Physics, 21, 23, Physics of the Inner Heliosphere II (Schwenn, R. and Marsch, E. eds) Springer-Verlag, Heidelberg.

Riley, P., Sonett, C. P., Tsurutani, B. T., Balogh, A., Forsyth, R. J. and Hoogeveen, G. W. (1996) Properties of arc-polarized Alfvén waves in the ecliptic plane: Ulysses observations. *J. Geophys. Res.* **101**, 19987.

Roberts, D. A. (1990) Turbulent polar heliospheric fields. *Geophys. Res. Lett.* **17**, 567–570.

Roberts, D. A., Klein, L. W., Goldstein, M. L. and Matthaueus, W. H. (1987) The nature and evolution of magnetohydrodynamic fluctuations in the solar wind: Voyager observations. *J. Geophys. Res.* **92**, 11021–11040.

Ruzmaikin, A., Feynman, J., Goldstein, B. E., Smith, E. J. and Balogh, A. (1995) Intermittent turbulence in solar wind from the south polar hole. *J. Geophys. Res.* **100**, 3395.

Schmidt, J. M. (1995) Spatial transport and spectral transfer of solar wind turbulence composed of Alfvén waves and convective structures II: numerical results. *Ann. Geophys.* **13**, 475–493.

Schwenn, R. and Marsch, E. (eds.) (1991) *Phyiscs of the inner heliosphere II – particles, waves and turbulence*, Springer, Berlin.

Smith, E. J. (1973a) Identification of interplanetary tangential and rotational discontinuities. *J. Geophys. Res.* **78**, 2054.

Smith, E. J. (1973b) Observed properties of interplanetary rotational discontinuities. *J. Geophys. Res.* **78**, 2088.

Smith, E. J. and Balogh, A. (1995) Ulysses observations of the radial magnetic field. *Geophys. Res. Lett.* **22**, 3317–3320.

Smith, E. J., Balogh, A., Neugebauer, M. and McComas, D. (1995) Ulysses observations of Alfvén waves in the southern and northern solar hemispheres. *Geophys. Res. Lett.* **22**, 3381–3384.

Smith, E. J., Neugebauer, M., Balogh, A., Bame, S. J., Erdös, G., Forsyth, R. J., Goldstein, B. E., Phillips, J. L. and Tsurutani, B. T. (1993) Disappearance of the heliospheric sector structure at Ulysses. *Geophys. Res. Lett.* **20**, 2327.

Sonnerup, B. U. Ö. and Cahill, L. J. Jr. (1967) Magnetopause Structure and Attitude from Explorer 12 Observations. *J. Geophys. Res.* **72**, 171–183.

Tsurutani, B. T. and Lakhina, G. S. (2000) Plasma microstructure in the solar wind. *Enrico Fermi School of Physics*, in press.

Tsurutani, B. T. and Smith, E. J. (1979) Interplanetary discontinuities: Temporal variations and the radial gradient from 1 to 8.5 AU. *J. Geophys. Res.* **84**, 2773.

Tsurutani, B. T., Glassmeier, K.-H. and Neubauer, F. M. (1997) A review of nonlinear low frequency (LF) wave observations in space plasmas: On the development of plasma turbulence. In: Hada T. and Matsumoto, H. (eds) Nonlinear Waves and Chaos in Space Plasmas, *Terra Scientific*, Tokyo, 1.

Tsurutani, B. T. and Ho, C. M. (1999) A review of discontinuities and Alfvén waves in interplanetary space. *Rev. Geophys.* **37**, 517–541.

Tsurutani, B. T., Ho, C. M., Arballo, J. K., Goldstein, B. E. and Balogh, A. (1995) Large amplitude IMF fluctuations in corotating interaction regions: Ulysses at midlatitudes. *Geophys. Res. Lett.* **22**, 3397.

Tsurutani, B. T., Ho, C. M., Arballo, J. K., Lakhina, G. S., Glassmeier, K.-H. and Neubauer, F. M. (1997b) Nonlinear electromagnetic waves and spherical arc-polarized waves in space plasmas. *Plasma Phys. Controlled Fusion*, **39**, A237.

Tsurutani, B. T., Ho, C. M., Arballo, J. K., Smith, E. J., Goldstein, B. E., Neugebauer, M., Balogh, A. and Feldman, W. C. (1996a) Interplanetary discontinuities and Alfvén waves at high heliographic latitudes: Ulysses. *J. Geophys. Res.* **101**, 11027.

Tsurutani, B. T., Ho, C. M., Sakurai, R., Goldstein, B. E., Balogh, A. and Phillips, J. L. (1996b) Symmetry in discontinuity properties at the north and south heliospheric poles: Ulysses, *Astron. and Astrophys.* **316**, 342.

Tsurutani, B. T., Ho, C. M., Smith, E. J., Neugebauer, M., Goldstein, B. E., Mok, J. S., Arballo, J. K., Balogh, A., Southwood, D. J. and Feldman, W. C. (1994) The relationship between interplanetary discontinuities and Alfvén waves: Ulysses observations. *Geophys. Res. Lett.* **21**, 2267.

Tsurutani, B. T., Lakhina, G. S., Winterhalter, D., Arballo, J. K., Galvan, C. and Sakurai, R. (1999) Energetic particle cross-field diffusion: Interaction with Magnetic Decreases (MDUs). *Nonlinear Proc. Geophys.* **6**, 235–242.

Tsurutani, B.T., Smith, E. J., Anderson, R. R., Ogilvie, K. W., Scudder, J. D., Baker, D. N. and Bame, S. J. (1982) Lion roars and nonoscillatory drift mirror waves in the magnetosheath. *J. Geophys. Res.* **87**, 6060.

Tu, C.-Y and Marsch, E. (1995) MHD structures, waves and turbulence in the solar wind – observations and theories. *Space Sci. Rev.* **73**, 1–210.

Tu, C.-Y., Marsch, E. and Rosenbauer, H. (1996) An extended structure-function model and its application to the analysis of solar wind intermittency properties. *Ann. Geophys.* **14**, 270–285.

Turner, J. M., Burlaga, L. F., Ness, N. F. and Lemaire, J. F. (1977) Magnetic holes in the solar wind. *J. Geophys. Res.* **82**, 1921.

Vasquez, B. J. and Hollweg, J. V. (1998a) Formation of spherically polarized Alfvén waves and embedded rotational discontinuities from a small number of entirely oblique waves. *J. Geophys. Res.* **103**, 335.

Vasquez, B. J. and Hollweg, J. V. (1998b) Formation of embedded rotational discontinuities with nearly field-aligned normals. *J. Geophys. Res.* **103**, 349.

Whang, Y. C., Larson, D., Lin, R. P., Lepping, R. P. and Szabo, N. (1998a) Plasma and magnetic field structure of a slow shock: Wind observations in interplanetary space. *Geophys. Res. Lett.* **25**, 2625.

Whang, Y. C., Zhou, J., Lepping, R. P., Szabo, A., Fairfield, D., Kokubun, S., Ogilvie, K. O. and Fitzenreiter, R. (1998b) Double discontinuity: A compound structure of slow shock and rotational discontinuity. *J. Geophys. Res.* **103**, 6513.

Winterhalter, D., Neugebauer, M., Goldstein, B. E., Smith, E. J., Tsurutani, B. T., Bame, S. J. and Balogh, A. (1995) Magnetic holes in the solar wind and their relation to mirror-mode structures. *Space Sci. Rev.* **72**, 201. Kluwer, Netherlands.

Winterhalter, D., Neugebauer, M., Goldstein, B. E., Smith, E. J., Bame, S. J. and Balogh, A. (1994) Ulysses field and plasma observations of magnetic holes in the solar wind and their relationship to mirror-mode structures. *J. Geophys Res.* **99**, 23371.

5

Waves and instabilities in the three-dimensional heliosphere

Robert J. MacDowall and Paul J. Kellogg

5.1 INTRODUCTION

Observations of plasma waves in the solar wind provide insights into a broad range of kinetic processes. Examples include the waves produced by electrons accelerated by solar flares or by Interplanetary (IP) shocks, a variety of wave modes produced at tangential and rotational discontinuities of the magnetic field, and waves associated with variation in the solar-wind heat flux. In each case, unstable particle distributions cause oscillations of a subset of the particles in the plasma that are detectable by electric or magnetic field antennas. Such data have been obtained from a number of heliospheric missions, such as *Helios 1* and *2*, *Voyager-1* and *-2*, *ISEE-3*, and *WIND*. *Ulysses*, the first mission to make *in-situ* observations at the highest heliolatitudes, has provided a new perspective on plasma waves in the Interplanetary Medium (IPM). In this chapter we concentrate on results from the first orbit of *Ulysses* (1992–1998), a large fraction of which occurs during the minimum of the solar cycle when wave activity was typically at a significantly lower level than is observed at solar maximum.

5.1.1 Wave modes in the solar wind

Table 5.1 lists many of the wave modes likely to occur in the solar wind with their typical frequency ranges and likely source mechanisms. Detection of these waves provides diagnostic information about the unstable particle distributions that interact with the waves. In a telemetry-limited mission like *Ulysses*, it is usually the case that particle measurements cannot be transmitted with sufficient time resolution to permit routine detection of short-lived unstable distributions. Consequently, the wave observations provide the primary evidence for physical processes that grow and saturate on timescales of seconds or less.

Table 5.1. Wave modes in the solar wind.

Mode	Frequency range*	Likely sources
Free space electromagnetic	$f > f_{pe}$	Conversion from Langmuir waves
Langmuir	$f > f_{pe}$	Electron beams
Ion acoustic	$f_{pi} < f < f_{pe}$ (observed)	Electron heat flux, ion-beam instability
Electron acoustic	$f_{pi} < f < f_{pe}$ (observed)	Electron-beam instability
Whistler	$f > f_{ce}$	Doppler-shifted cyclotron resonance, electron heat flux, currents
Lower hybrid waves	$f \leq f_{LHR}$	Electron and ion beams, currents

*f is the observed frequency, f_{pe} is the electron-plasma frequency, f_{pi} is the ion-plasma frequency, f_{ce} is the electron-cyclotron frequency, and f_{LHR} is the lower hybrid resonance.

The dispersion relation $\omega = \omega(\mathbf{k})$ of a given wave mode with wave vector $\mathbf{k}$ provides the phase velocity $v_{ph} = \omega/k$ and the group velocity $v_{gr} = d\omega/dk$ of the wave at a given frequency ω. Waves whose phase velocities are of the order of the solar-wind velocity $\mathbf{V}_{sw}$ or less will be observed with a significant Doppler shift in the spacecraft frame of reference. Their frequencies observed in the spacecraft frame are $\omega_{obs} = \omega + \mathbf{V}_{sw} \cdot \mathbf{k}$, where ω is the frequency in the solar-wind frame. This would be the case, for example, for much of the short-wavelength ion acoustic wave activity observed in the solar wind. Another important parameter for the understanding of plasma waves is the Debye length $\lambda_{De} = v_{Te}/\omega_{pe}$, where v_{Te} is the electron thermal velocity and ω_{pe} is the electron plasma frequency. The Debye length corresponds to the diameter of a volume that can be non-neutral, and so the electrons are not 'locked' to the ions. Consequently, it is the length that divides coherent, fluid-like behaviour at large scales from incoherent, independent particle-like behaviour at small scales. For example, the minimum wavelength of ion acoustic waves in the solar wind is thought to be of order λ_{De}, because shorter wavelengths would be more strongly damped.

In this chapter we focus on waves with frequencies of a few tenths of Hz and above; lower frequency waves (Alfvén, etc.) are discussed elsewhere in this volume. The chapter is organized according to the various sources that excite such waves, such as fast electron beams from solar flares, IP shocks, or other solar-wind discontinuities. This organization permits us to address the various waves in the context of the physical structures and processes that produce them. Although these waves have been studied by heliospheric missions for more than 20 years, the understanding of the emission processes is incomplete and there are still discoveries to be made. For example, the wave activity occurring at the electron plasma frequency f_{pe} is presumed to be Langmuir mode; however, the source location of many of these waves has only recently been identified to be magnetic holes (Lin et al., 1995). The electrostatic waves detected in the frequency range between f_{pe} and the ion-plasma frequency f_{pi} are usually identified as Doppler-shifted ion acoustic waves (Gurnett,

1991 and references therein) however, the temperature ratios observed at these times are typically $T_e/T_p < 3$–5, which would cause ion acoustic waves to be strongly damped. The source of electromagnetic waves at frequencies below the electron-cyclotron frequency f_{ce}, which may play a role in heat flux regulation, is also not satisfactorily understood (Gary *et al.*, 1994; Thejappa *et al.*, 1995). There remain significant questions about the sources and significance of most of the plasma modes found in the solar wind. Only when these questions are answered can the plasma waves be fully utilized as probes of kinetic phenomena occurring in IP space.

5.1.2 Instrumentation for wave observations: *Ulysses* overview

Technology, telemetry, and other restrictions typically prevent measuring and transmitting the electric and magnetic-field vectors with the desired frequency and time resolution. Consequently, a spacecraft wave instrument package is usually designed with several receivers, each optimized for measurements in a certain frequency range. The data are then sampled or averaged in such a way as to provide useful, albeit incomplete, information about the observed waves. It is critical that these limitations be kept in mind when interpreting the wave data.

On *Ulysses*, the Unified Radio and Plasma Wave Investigation (URAP) consists of five sensors (three electric field antennas and two search coils) and a number of receivers designed to work with these sensors. The majority of the URAP wave measurements are obtained by the Radio Astronomy Receiver (RAR), Plasma Frequency Receiver (PFR), Waveform Analyzer (WFA), and Fast Envelope Sampler (FES), which comprise four of the six components making up the URAP investigation. The RAR consists of four superheterodyne radio receivers that cover two bands (1.25–48.5 kHz and 52–940 kHz). Two of the receivers are connected to the spin-axis monopole and two receivers are connected to either the spin-plane dipole antenna or to the summation of signals from the spin-plane and spin-axis antennas. The summation configuration is used to facilitate determining the direction of a remote radio source (Manning and Fainberg, 1980). The spin-plane dipole consists of a pair of wire booms, each 35 m long, which form a dipole of 72 m tip-to-tip length (including the diameter of the spacecraft). The spin-axis monopole is a 7.5-m boom, whose effective electrical length is less than 4 m. The PFR covers the range from 0.57 to 35 kHz with 15 per cent frequency resolution, large dynamic range, and good time resolution. One of the two PFR receivers is connected to the spin-plane dipole; the other is connected to the spin-axis monopole. The WFA provides spectral analysis in the range from ~ 0.1 to 448 Hz for both electric and magnetic fields. The FES captures short duration electric or magnetic-field events with a time resolution up to 1 ms and selects the largest events in a time interval on the order of an hour for inclusion in the telemetry. Table 5.2 lists the frequency ranges, bandwidths, numbers of channels, typical time resolutions, and threshold levels of these receivers. A more complete description of each of the instruments can be found in Stone *et al.* (1992).

Ulysses presents an excellent opportunity to measure electric fields of waves at lower frequencies than previously possible. Measurements of such electric fields are

Table 5.2. *Ulysses* URAP instrument parameters.

Name	Frequency range	Bandwidth	Maximum channels	Typical time resolution	Threshold (@frequency)
Radio Astronomy Receiver – hi band	52–940 kHz	3 kHz	12	144 s	2×10^{-8} V Hz$^{-1/2}$ @ 940 kHz
Radio Astronomy Receiver – low band	1.25–48.5 kHz	0.75 kHz	64	128 s	10^{-8} V Hz$^{-1/2}$ @ 42 kHz
Plasma Frequency Receiver	0.57–35 kHz	14% of centre frequency	32	16 s	10^{-6} V Hz$^{-1/2}$ @ 3 kHz
Waveform Analyzer – E-field	0.08–448 Hz	25% of centre frequency	24	64 s	10^{-5} V Hz$^{-1/2}$ @ 19 Hz
Waveform Analyzer – B-field	0.22–448 Hz	25% of centre frequency	22	64 s	2 pT Hz$^{-1/2}$ @ 3.5 Hz
Fast Envelope sampler	E: 6–60, 2–20, or 0.6–6 kHz B: 1–500 Hz	Full frequency range	2	1 ms	For 1 ms data E: 5×10^{-6} V m^{-1} @ 6–60 kHz B: 3×10^{-3} nT

usually impossible on a spinning spacecraft because the varying photoemission, due to the varying area presented to the Sun, causes the antenna potential to rise and fall by the order of 10 volts, whereas we are trying to measure fields of millivolts or less. Since *Ulysses* spins around an axis pointing near the Sun, the photoemission is much less variable.

To this end, the WFA uses a digital transform (a 16-point wavelet transform) to measure the spectrum of electric and magnetic fields down to the spin frequency, 0.08 Hz. This range is divided into two subranges, 0.08 Hz–5 Hz and 9.3 Hz–448 Hz, each range being defined by hardware filters. In the 9.3–448-Hz range the hardware filters work very well to eliminate photoemission effects, though some contamination of the signal is seen at large solar aspect angles (e.g. see Fig. 2 of (Lin *et al.*, 1998). Waves in this range, principally whistlers and ion acoustic waves, are measured over the entire range of latitudes and solar distances of the *Ulysses* orbit. The identification of these magnetic signals as whistler waves is based on their frequency range, as in most previous experiments (e.g. Beinroth *et al.*, 1981), because the B_z search coil suffered from interference from another experiment and the polarization could usually not be determined. These results have been reviewed by Lin *et al.* (1998; 1997). A few typical electric field spectra are also shown by Kellogg (2000).

In the range from 0.08 Hz to 5 Hz, there is a strong interference signal at the spin period (in addition to the photoelectron variation that has a period of half the spin period). This signal disappears when the spin axis points directly at the Sun and has an average amplitude of about 1.2 sin θ volts (peak), where θ is the angle between the spin axis and the Sun. This effect must be due to a photoelectric effect between the sunlit spacecraft surface and the antennas, and has also been seen on the *ISEE-1* spacecraft (Pedersen *et al.*, 1984). The effect also depends on the density of the ambient plasma because the amplitude is smaller when the plasma density is large. The flux of ambient electrons opposes the effect of the photoelectrons, and is more effective in holding a constant potential when the electron density is larger. The 16-point wavelet transform of the WFA did not have very good discrimination against leakage, thus the large signal at one cycle per spin contaminated all the channels below 5 Hz to some extent, by an amount which depended on solar aspect angle. This can be seen, for example, in fig. 2 of Lin *et al.*, (1998). Nevertheless, some measurements of electric fields in this range were possible. To eliminate the dependence of the background signal on both density and solar aspect angle, Lin *et al.*, (1998, 1997) determined the lowest signal in a certain period and subtracted this from the observations to find signals above the background. Modes could not be determined from the available data, but it seems that at least two modes are present, because the magnetic and electric signals are nearly uncorrelated. Rather complex dependences on fast versus slow plasma velocities, heliographic latitude, etc. were found (Lin *et al.*, 1998, 1997).

5.2 RADIO BURSTS CAUSED BY SOLAR FLARES

Solar radio bursts were first catalogued based on observations using the radiospectrograph at Penrith, Australia (Wild *et al.*, 1963 and references therein). The nomenclature used to describe these bursts was type I, type II, type III, type IV and type V. Type II and type III bursts occur in the solar wind, whereas the other categories seem to be limited to the solar corona. Type II bursts are caused by electrons generally thought to be accelerated at coronal and IP shocks; they are characterized by a slow drift in frequency corresponding to the motion of a shock wave through the corona and solar wind (see Section 5.3). The type III bursts, of which examples are shown in Figure 5.1, are caused by electron beams escaping from solar-flare regions. Their faster frequency drift, as compared with the type II bursts, results from the velocities of the electron beam of $c/10$ to $c/3$, where c is the speed of light. For both types of bursts, the frequency drifts result because the emission frequency is directly proportional to f_{pe}, which varies approximately as r^{-1} in the solar wind, where r is the distance from the Sun.

Because type III bursts are directly associated with flares, their occurrence is correlated with solar activity. In Figure 5.2 type III burst occurrence is plotted as a function of time; overall, the correlation with solar activity (as indicated by sunspot number) is very good. Even small inflections in the sunspot-number plot can be associated with similar features in the type III occurrence probability.

5.2.1 Type III radio burst theory and observations

The hot electron distribution generated at a flare site causes electron beams to form far from the site as the electrons flow outward on open field lines. The distribution

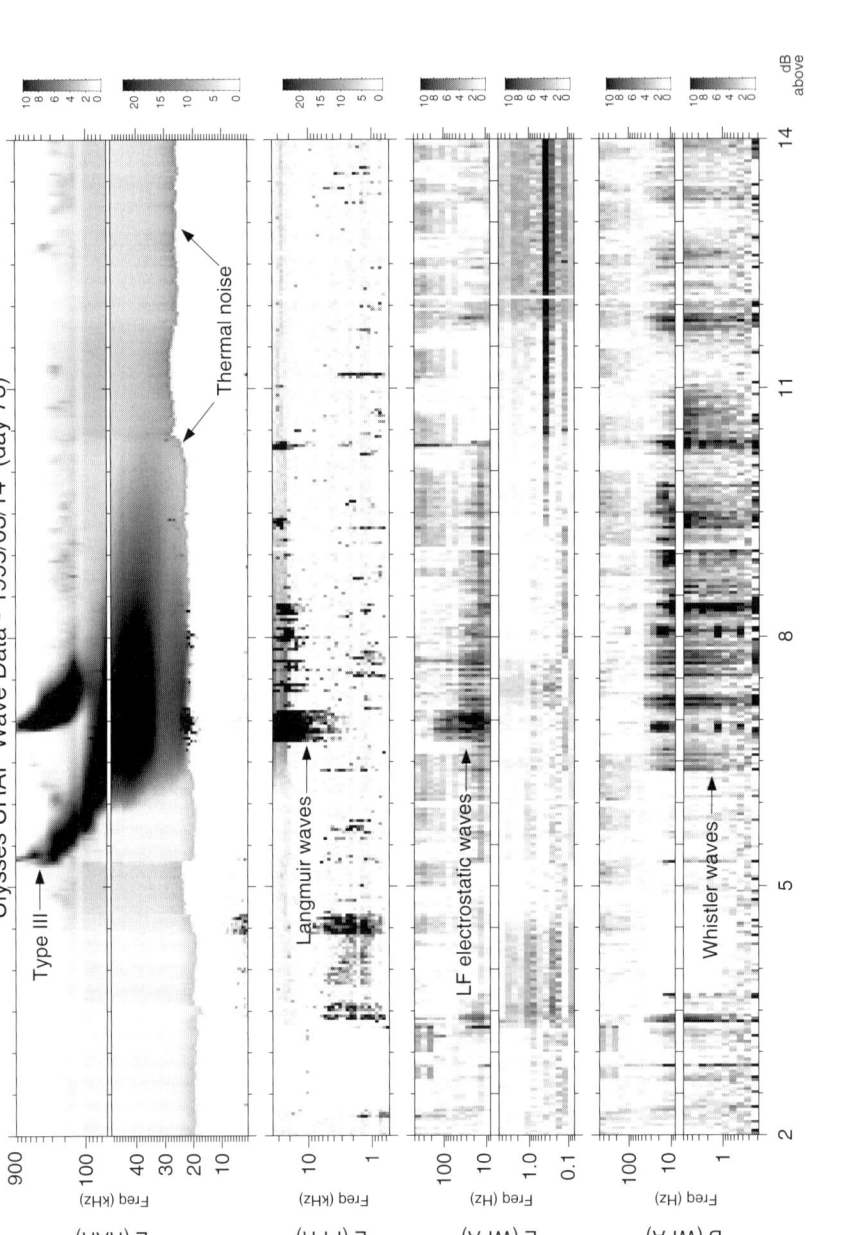

Figure 5.1. Example of an intense IP type III radio burst and associated *in-situ* waves. The panels cover the full frequency range of the *Ulysses* URAP instruments. The wave power is indicated by the colour scale (in units of dB above background.) In the top panel, two type III radio bursts are seen; the earlier, more intense event shows radio emission extending to the local electron plasma frequency (f_{pe}), occurring at approximately 22 kHz as indicated by the electron thermal noise. Langmuir, low frequency (LF) electrostatic, and whistler waves are detected in association with the arrival of the electron beam at the spacecraft. The apparent finite bandwidth of the Langmuir waves results from the filter bandwidths (14 per cent full width at half power) and from the waves' intensity relative to the dynamic range that is used for the colour

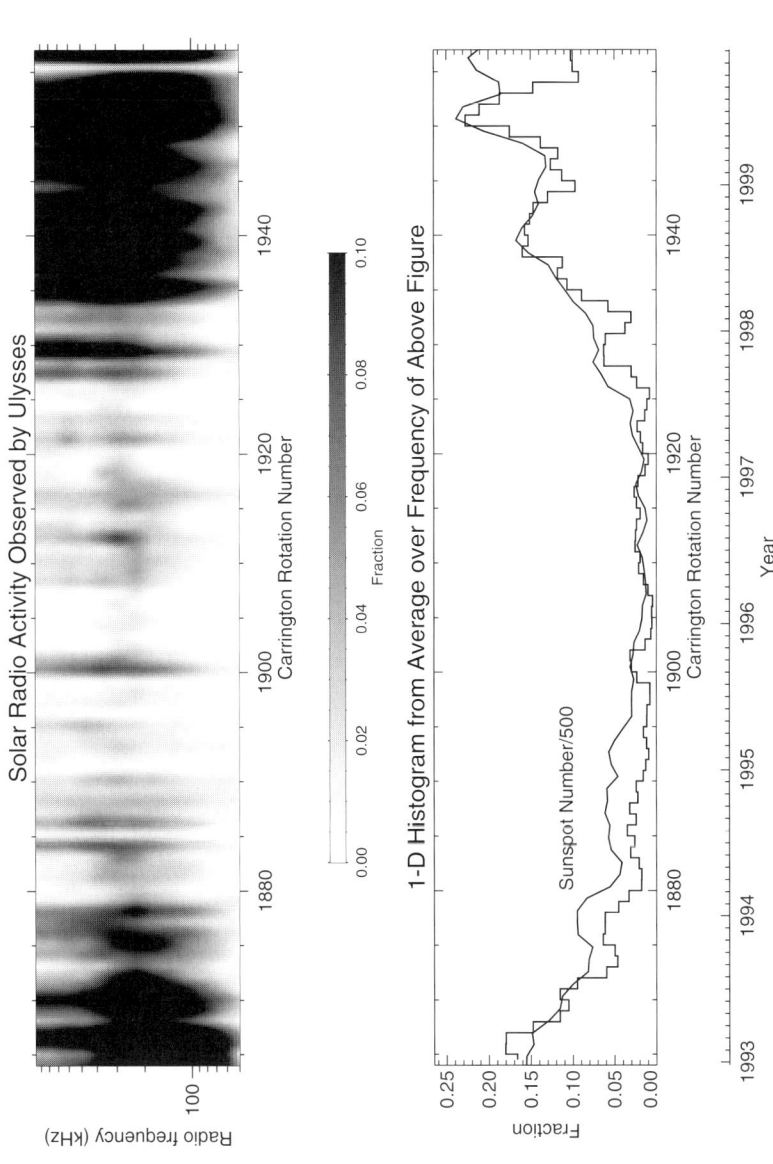

Figure 5.2. Type III burst occurrence as a function of the solar cycle. Upper panel shows the frequency-dependent occurrence over 80 Carrington rotations, as observed by *Ulysses*. Data prior to 1993 are excluded because of the interference of Jovian radio emissions when *Ulysses* was close to Jupiter. Bottom panel is the fractional occurrence averaged over frequency. The scaled sunspot number for the same interval is overplotted.

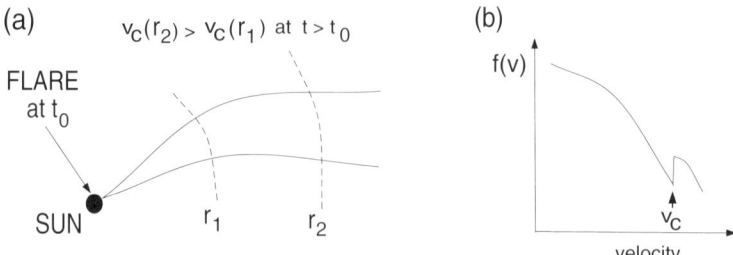

Figure 5.3. After the onset of a solar flare at time t_0, electrons escaping along open field lines (panel a) will form a cutoff (or bump-on-tail) distribution function $f(v)$ at large distances from the flare (panel b). The velocity cut-off is established by time of flight, so at a given time, it occurs at higher velocities v_c farther from the source ($r_2 > r_1$).

that would form far from the source, if there were no compensating effects, is a cut-off distribution (Figure 5.3). Such distributions arise when the source turns on abruptly. Then, only electrons with velocities faster than a cutoff velocity $v_c = r/\Delta t$ can reach a given distance r from the source in a time Δt. Such a beam-generation mechanism is important in other emission regions, such as upstream of planetary bow shocks (Filbert and Kellogg, 1979). A more gradual onset or scattering of the electrons along the path to the spacecraft would smoothe the beam distribution to form what is commonly called a bump-on-tail distribution, which has been observed for type III bursts (Lin et al., 1981, 1986).

The Langmuir waves excited by such a distribution are, in their simplest form, monochromatic electron-plasma waves at the electron plasma frequency f_{pe}. For the type III event in Figure 5.1 these waves are seen as intense bursty activity occurring at ~22 kHz from 6:40 to 8:30 UT. Their detection at the spacecraft corresponds to the time interval during which an unstable beam of electrons is flowing past the spacecraft. At much higher time resolution, these Langmuir waves are seen to occur in packets as shown in Figure 5.4. These data, from the *Ulysses* URAP FES, have 17.8-ms resolution. The wave period is significantly shorter than this, so the instrument is designed to detect the rectified, peak signal during each time resolution element. The peak electric field is approximately $10\,\text{mV}\,\text{m}^{-1}$ (in the bandwidth observed, which extends from 2 to 20 kHz).

The peak-electric-field levels achieved are an important factor in determining how the electron beam is decoupled from the waves that it produces. Such a decoupling is required to prevent the wave excitation from extracting all of the energy from the beam within the solar corona (Sturrock, 1964). If so-called strong turbulence theory applies, then the electric field is sufficient to generate underdense regions by expelling plasma through the ponderomotive force. Waves trapped in these underdense regions can self-focus and collapse, leading to an evolution to small-wave scales (large **k**) that do not interact with the beam (Goldstein et al., 1979). Conversely, if weak turbulence processes, such as induced scattering or electrostatic decay, apply, then the waves evolve towards larger scales (small **k**), which would also decouple them from the beam (Kaplan and Tsytovich, 1973; Cairns and Robinson,

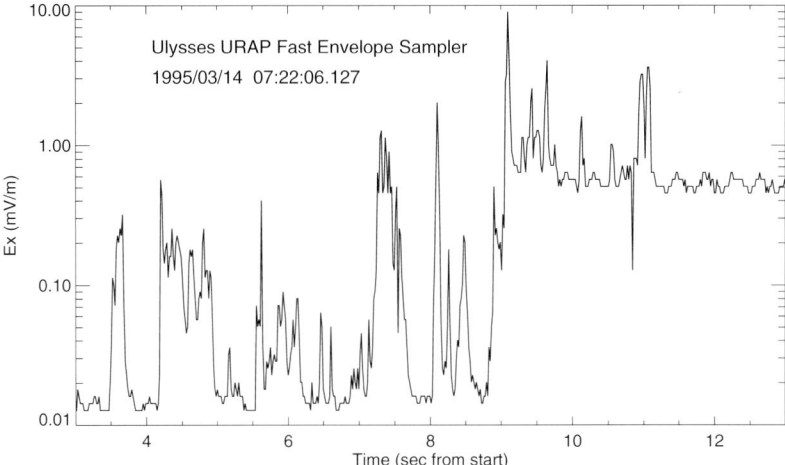

Figure 5.4. High time resolution envelope of Langmuir waves during the event in Figure 5.1, detected by the *Ulysses* URAP FES. At approximately 9 s, the large electric field ($\sim 10\,\mathrm{mV\,m^{-1}}$) caused the attenuator to be switched in, with the resulting increase in background noise after than time.

1998). It may be the case that both processes occur in different regions of the type III burst, as has been suggested for the event shown in Figure 5.1 (Thejappa amd MacDowell, 1998). A third process that may play a role in the production of radio emission is linear-mode conversion (Yin and Ashour-Abdalla, 1999).

The radio emission produced by these mechanisms is emitted at the fundamental of the electron-plasma frequency and its harmonic ($2f_{\mathrm{pe}}$). *Ulysses* was the first spacecraft with sufficient frequency resolution to show this clearly for IP type III bursts (Reiner *et al.*, 1992), although the emission at two frequencies was anticipated because many coronal type III bursts show two bands of emission, separated by a factor of approximately 2 (Kundu, 1965). Several spectral cuts through the type III burst in Figure 5.1 are shown in Figure 5.5. The enhancement in emission immediately above f_{pe} is evident; such emission is termed 'local' type III emission because it is emitted close to the spacecraft. The electromagnetic emission cannot propagate at frequencies below f_{pe}; consequently, the spectrum cuts off in the vicinity of that frequency. The peak at $2f_{\mathrm{pe}}$ is broader and likely comes from a larger region of space.

As the type III electrons radiate and spread into larger and larger regions, the capability to sustain a beam distribution above the background electrons diminishes. The question as to how far from the Sun type III burst electrons can continue to generate radio emission which has been answered by joint *Ulysses-WIND* observations (Leblanc *et al.*, 1996), which found that the lower limit for intense type III bursts is typically 10 kHz, as seen in Figure 5.6. For typical solar-wind densities, 10 kHz corresponds to the plasma frequency at approximately 3 AU. An example of such an event is a type III burst occurring on 14 July, 1991, described by Kellogg *et al.* (1996), when *Ulysses* was 3.6 AU from the Sun. The high-inclination orbit of

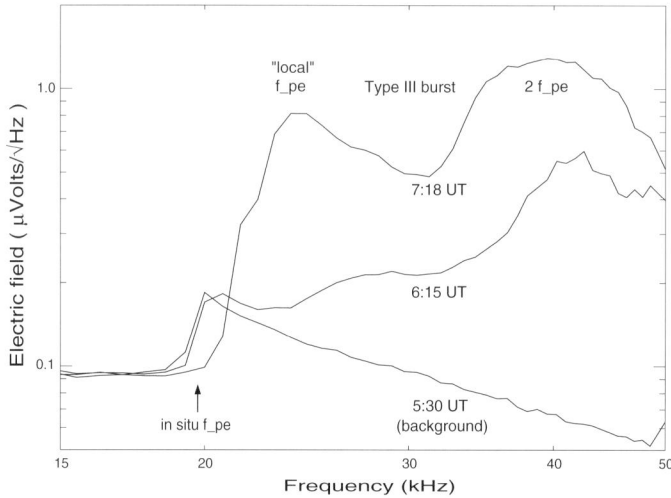

Figure 5.5. Spectral cuts through the type III burst shown in Figure 5.1. Enhanced emission occurs at frequencies near the *in-situ* electron plasma frequency and its harmonic ($2f_{pe}$).

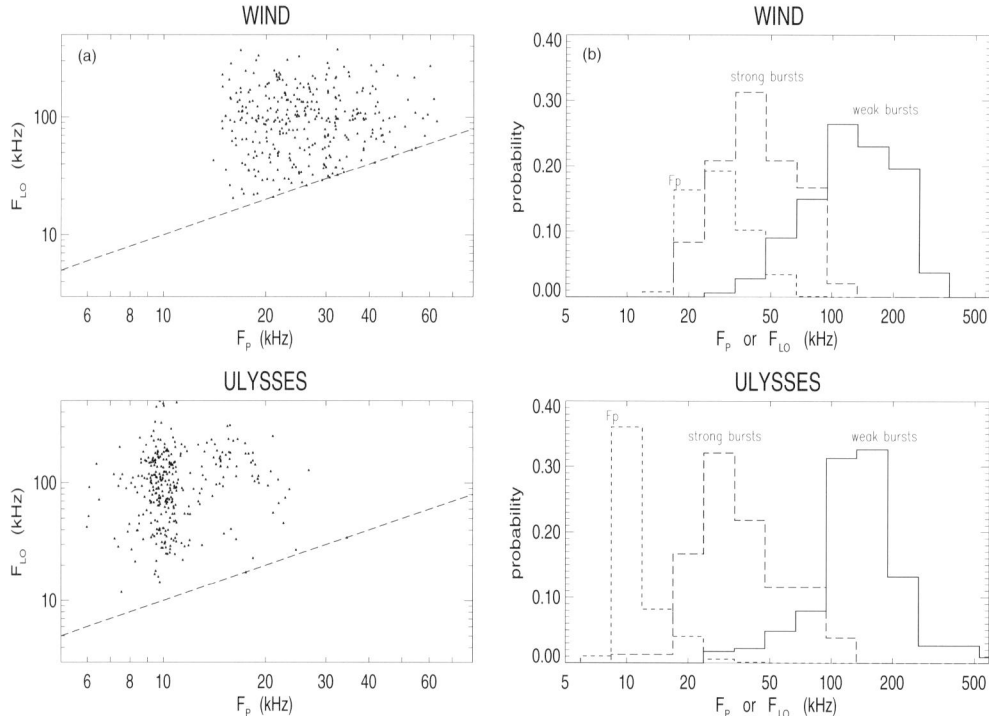

Figure 5.6. (a) Low frequency limit of type III radio bursts versus the local plasma frequency at *WIND* and *Ulysses*. (b) Histograms comparing the low frequency cutoff of strong and weak type III bursts to the local plasma frequency at *WIND* and *Ulysses*. These data suggest that the low frequency cutoffs of type III emission observed by *WIND* (at 1 AU) are the physical cutoffs. They are not observed to extend to lower frequencies by *Ulysses*. (From Leblanc *et al.*, 1996).

Ulysses is critical for this determination, because it places the typical plasma frequency at *Ulysses* at a fraction of 10 kHz, so that type III emission at frequencies lower than f_{pe} at *WIND* could be observed, if it existed. This lower limit might be different for observations at solar maximum, when the suprathermal electron background in the solar wind is enhanced.

5.2.2 Applications in remote studies of the IPM

Both the Langmuir waves and the radio waves produced by flare-accelerated electrons can be applied to the tracing of magnetic-field lines. The detection of Langmuir waves at the spacecraft is a robust proxy for the presence of electron beams. If these waves are associated in time with the occurrence of a type III burst, then the spacecraft is magnetically connected to the flare site. Such observations were used by Pick *et al.* (1995a, 1995b) to demonstrate that magnetic-field lines originating at low solar latitudes could sometimes reach high latitudes in the solar wind. These observations were made during solar minimum, when flares providing such electron beams are rare. Similar high-latitude observations during solar maximum should provide many more events for study. On a cautionary note, *Ulysses* observations (discussed in Section 5.5) have permitted the discovery that Langmuir waves are often found in so-called magnetic holes, apparently unrelated to flare-accelerated electrons. Therefore, times of Langmuir wave occurrence should be checked for magnetic-hole occurrence in the magnetic-field data before a flare association is considered reliable.

Tracing of magnetic-field lines using the *remote* observations of the type III radio emission was pioneered by Fainberg *et al.* (1972) using data from the *IMP-6* spacecraft. The technique requires that the direction to the radio source be determined at a number of frequencies using the spinning spacecraft method (Manning and Fainberg, 1980). The signal from a remote radio source is modulated sinusoidally as the antenna rotates relative to the electric field of the radio emission. *Ulysses* is among the first spacecraft to combine the signals from spin-plane and spin-axis antennas to determine the direction to the radio emission source as a function of both azimuth and elevation. Given these vector directions from the spacecraft, it is necessary to intersect them with a surface to determine the location of the source. Typically, this is accomplished by using a spherically symmetric density–distance scale (of the form $n_e \sim r^{-\alpha}$, where n_e is the electron density) and assuming that the emission is at the fundamental or harmonic of the plasma frequency. An example of such a density–distance scale measured *in situ* would be that derived for *Helios* by Bougeret *et al.*, (1984). Then the field line is drawn from the Sun through the source locations determined by the intersections.

In a modification of this approach, Reiner *et al.* (1995a) used source directions derived for bursts when *Ulysses* was at high heliographic latitudes, plus the assumption that the sources were near the ecliptic plane to derive trajectories as shown in Figure 5.7. These trajectories are remarkably close to Archimedean spirals with appropriate solar-wind velocities. During solar minimum few opportunities for such determination of type III trajectories presented themselves. More opportunities

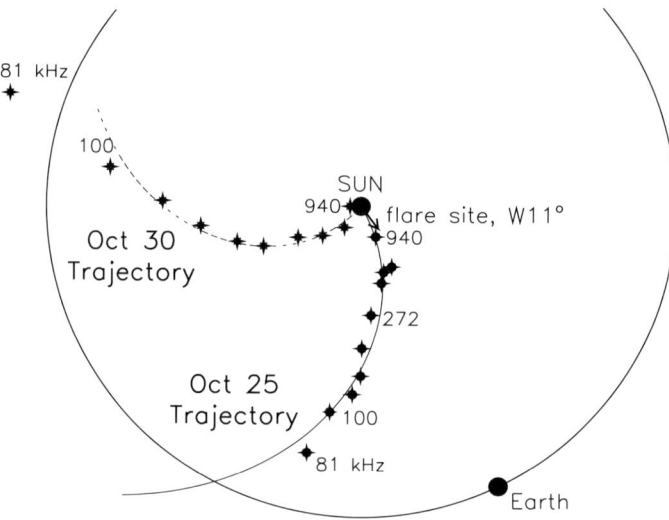

Figure 5.7. Type III radio burst trajectories observed remotely from *Ulysses* at high heliographic latitudes. (From Reiner *et al.*, 1995a).

should be forthcoming during the next *Ulysses* fast latitude scan (FLS). Some of these trajectories may show deviations in the magnetic-field geometry due to coronal mass ejections, corotating streams, and other large-scale structures in the solar wind.

Another alternative for determining the source locations of type III bursts as a function of frequency is to use triangulation from multiple spacecraft. The first triangulation of a type III burst using data from two spacecraft with three-dimensional-direction finding capabilities (*WIND* and *Ulysses*) was performed by Reiner *et al.*, (1998). Two-spacecraft triangulation of type III burst trajectories was a major goal of the original two-spacecraft Out of Ecliptic/Solar Polar (OESP) mission that led to *Ulysses*. With the arrival of the next *Ulysses* fast latitude scan during solar maximum, this goal should finally be realized for a large number of bursts.

5.3 WAVES ASSOCIATED WITH IP SHOCKS

IP shock waves perturb the distribution functions of ions and electrons and accelerate particles, all of which are potential sources of free energy for the generation of plasma waves. The details of the shock structure and of its effects on particles are determined by the criticality of the shock and by its geometry. Shock geometry (i.e. the angle of the shock normal relative to the direction of the magnetic field) plays a key role in determining which waves are found near the shock, because electrons accelerated at the shock are constrained to follow the magnetic field. This effect is particularly evident for a shock with a small radius of curvature, like Earth s bow shock. There, the magnetic-field lines tangent to the surface of the bow shock define the surface beyond which no shock-accelerated, magnetized particles can be found. Faster particles will be found farther from the point of acceleration, establishing regions known as the electron and ion foreshocks. In the case of a typical IP shock, the global radius of curvature is so large that this concept probably breaks down. Instead, we would expect that local variation in the magnetic field and in the shock normal would create a patchwork of electron and ion foreshocks, with varying wave populations. The study of such regions is somewhat limited by the high speed with which they pass the spacecraft, which means that only limited plasma data are available in the relevant time intervals. Plasma-wave data, typically available at higher time resolutions, can serve to identify such regions.

In Figure 5.8 a forward–reverse shock pair with typical wave activity is shown. Of particular note are: Langmuir waves upstream of the forward shock, ion acoustic waves with maximum intensity near the shock, and enhanced magnetic waves downstream of the shocks. Each of these waves is discussed in a following section; they correspond to the classical description of waves in the vicinity of IP shocks (Kennel et al., 1982. Some shocks do not follow this scenario; Thejappa et al., (1995) identified electrostatic wave activity extending upstream of several shocks for intervals of hours. These waves were suggested to be Doppler-shifted lower hybrid waves, generated locally by anisotropic particle distributions from the shock.

5.3.1 Low-frequency electromagnetic waves

The shock in Figure 5.8 presents a typical example of low-frequency waves in the downstream region of a shock. These waves, which are observed at frequencies as high as a fraction of f_{ce}, are typically interpreted to be whistler-mode waves (Kennel et al., 1982; Gurnett et al., 1979b). The likely source of these waves is considered to be an electron-cyclotron instability due to anisotropic electron distributions in the vicinity of the shock (Tokar et al., 1984; Pierre et al., 1997). Electromagnetic lower hybrid waves have been suggested as an alternative wave mode (Thejappa et al., 1995). These two wave modes are on the same branch of the dispersion relation and differ primarily in that they are typically called whistlers when they are nearly parallel to B and lower hybrid waves when they are highly oblique. The B/E ratios observed for these waves indicate that they propagate at angles of 50–70° from the direction of the magnetic field (Lengyel-Frey et al., 1994). The waves are then resonant with electrons from tens of eV to 2 keV for the cyclotron resonances.

5.3.2 Ion acoustic waves

As discussed earlier, the waves identified as ion acoustic waves have short wavelengths in the solar wind; consequently, the frequency at which they are observed is almost entirely due to the Doppler shift by the solar-wind velocity. The waves frequently occur when the temperature ratio $T_e/T_p < 3$–5, which presents difficulties for explaining why they are not damped. For this reason, a number of authors refer to these observed waves as Ion Acoustic-Like (IAL) waves. Hess et al. (1998) studied 124 shocks and found that the occurrence of IAL waves peaked in the vicinity of the

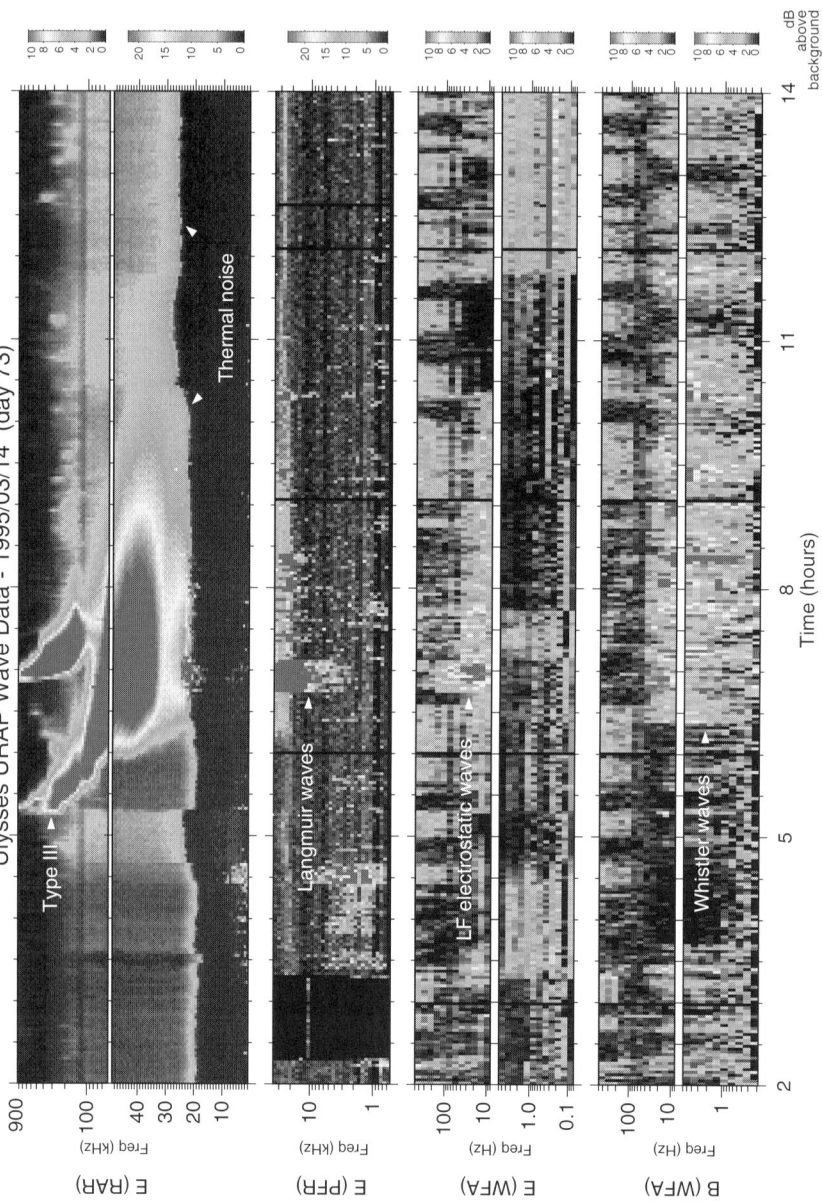

Figure 5.8. A forward-reverse shock pair from the perspective of the *Ulysses* wave data. Langmuir waves (L) are observed upstream of both shocks. Electrostatic emission is observed for more than one hour after the passage of the forward shock. Whistler wave activity is observed by both electric and magnetic sensors during the entire interval of elevated magnetic field between the shocks. The 'event' from 6:00 to 8:00 seen in the E(WFA) data is a signal caused by spacecraft manoeuvres.

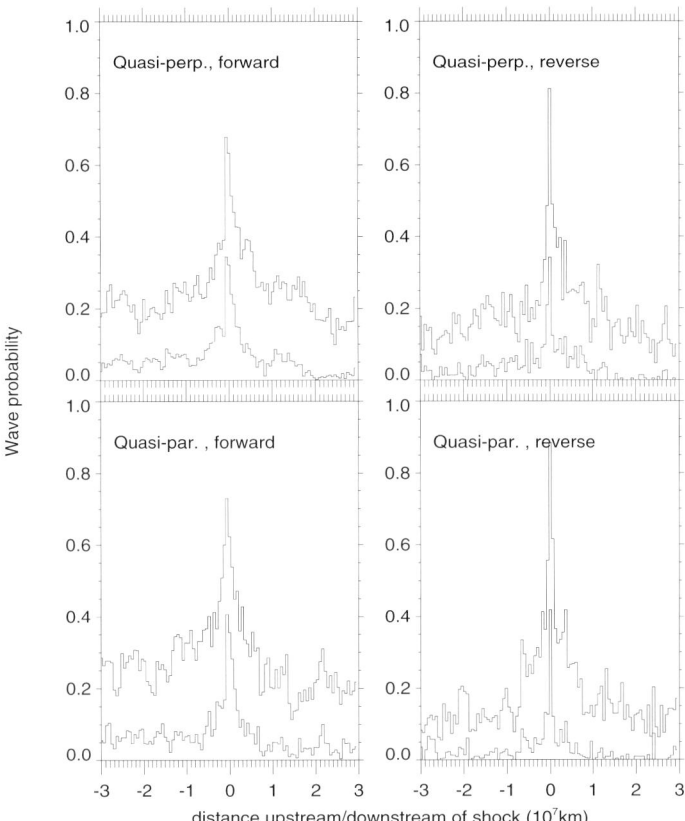

Figure 5.9. Occurrence probability of ion acoustic wave activity as a function of distance from forward and reverse, quasiparallel and quasiperpendicular shocks. The two histograms per panel show the probabilities for waves exceeding 4 and 60 nV m^{-1} Hz$^{-1/2}$. (From Hess et al., 1998). Negative distance is upstream of the shock for both forward and reverse shocks.

IP shocks (Figure 5.9). Note that IAL waves with electric fields exceeding 4 mV m^{-1} Hz$^{-1/2}$ are found for more than 40 per cent of the events – both forward and reverse shocks, independent of shock geometry. As expected from previous work (Gurnett et al., 1979a), Hess et al., (1998) demonstrated that the IAL wave occurrence was correlated with the temperature ratio T_e/T_p (correlation coefficient 0.44), but not with several other dimensionless plasma parameters. Hess et al., (1998) note that IAL wave occurrence is reduced in fast solar wind, probably due to the reduced temperature ratio.

It is plausible that a particular instability can grow faster than the damping rate for T_e/T_p less than 5. Gary and Omidi (1987) proposed that the ion–ion acoustic instability could occur for $T_e/T_p < 1$ if the ion-beam temperature is sufficiently low relative to the ion-core temperature. Another possible explanation is that the IAL waves are the electron-acoustic mode, which can be excited by ion or electron beams

(Marsch, 1985; Gary and Tokar, 1985). Hess *et al.*, (1998) identify a variety of perturbations in the proton distributions and the plasma parameters for several shocks, which suggest that more than one mechanism may play a role in exciting these waves. Given the very bursty nature of these waves (Kurth *et al.*, 1979), it is equally likely that the relevant changes in distribution functions and plasma parameters occur on timescales too short to be observed by typical spacecraft instrumentation.

5.3.3 Langmuir waves

Langmuir waves are observed upstream of a minority of IP shocks (Lengyel-Frey *et al.*, 1997). By analogy with Earth's bow shock, it is expected that electron beams or cut-off distributions will occur in regions upstream of and magnetically connected to the shock (Filbert *et al.*, 1979). Waves with greater intensity and larger bandwidths may be observed in the immediate vicinity of the shock, similar to waves seen at Earth's bow shock, although the shock strengths observed for IP shocks are typically less than those of the terrestrial bow shock.

The Langmuir waves, as well as the ion acoustic waves discussed in the previous section, are also of interest because they are expected to be the waves that merge to produce type II radio bursts. In this theory the fundamental radio emission is produced by the merging of Langmuir waves and ion acoustic waves to produce an electromagnetic wave at f_{pe} or by the decay of Langmuir waves into ion acoustic waves and electromagnetic waves. Second harmonic radio emission ($\sim 2f_{pe}$) would be produced by the merging of two oppositely directed Langmuir waves (e.g. see Melrose, 1980). For the second harmonic emission to occur, some inhomogeneity must first backscatter some Langmuir waves to permit them to merge with the Langmuir waves propagating in the direction of the beam. Ion acoustic waves may be the inhomogeneities that provide this backscatter (Cairns, 1987). Other variations for the production of type II radio emission exist, but they are almost all tied to the existence of Langmuir waves.

5.3.4 Radio waves

Ulysses was well suited to make high-frequency resolution observations of type II bursts near and beyond 1 AU (Figure 5.10). These drifts correspond to speeds from several hundreds to thousands of $km\,s^{-1}$, corresponding to the transit speeds of shocks in the corona and solar wind. A good correlation between IP shocks and type II bursts was identified by Cane (1985) for forty-eight events observed by the *ISEE-3* spacecraft during a 4-year interval. In the IP medium, these bursts are much less intense and less common than type III bursts. It has been shown (Cane, 1985; Lengyel-Frey and Stone, 1989) that there is a correlation between burst intensity and shock speed.

As indicated in Section 5.3.3, the common explanation for the production of type II emission involve Langmuir waves and lower frequency waves, known to be produced in the vicinity of IP shocks. Lengyel-Frey *et al.*, (1997) showed that these

Sec. 5.4] Waves in coronal mass ejections and magnetic clouds 245

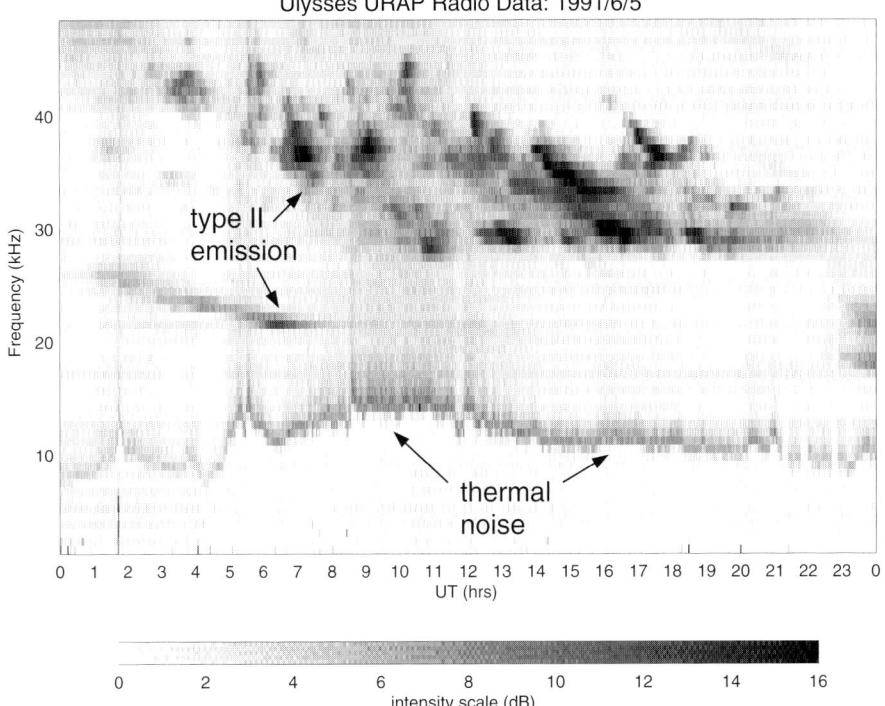

Figure 5.10. A 24-hour interval of *Ulysses* radio data with IP type II radio burst activity.

waves are most intense at distances from the shocks of less than 10^6 km. Therefore, tracking the type II emission using radio direction finding is equivalent to tracking the position of the IP shock in space. Such tracking has been done with *Ulysses* and, especially, with *WIND* radio data (Hoang et al., 1998, Reiner et al., 1998). Because the *WIND* spacecraft, at 1 AU, is several times closer to the Sun than *Ulysses*, it is able to detect weak type II emission that is not observable by *Ulysses*.

5.4 WAVES IN CORONAL MASS EJECTIONS AND MAGNETIC CLOUDS

The name Coronal Mass Ejection (CME) refers to ejecta expelled from the Sun's corona as observed by white-light coronagraphs. Solar ejecta observed by *in-situ* spacecraft are often referred to by the same name. These IP CMEs are identified by a variety of signatures including counterstreaming suprathermal electrons, helium-abundance enhancements, and low proton temperatures (e.g see Gosling, 1997). When they have sufficient translational and/or expansion velocities relative to the solar wind, the CMEs drive shocks ahead and/or behind them. Although CMEs frequently consist of complex magnetic structures, a subset of all CMEs are structures with smooth rotations of the magnetic-field, known as magnetic clouds. The definition of a magnetic cloud is a structure with enhanced magnetic field strength, large and smooth rotation in the magnetic-field direction, and low proton temperature (Burlaga et al., 1981). These signatures lead directly to the description of the magnetic cloud as a flux rope. Such a flux rope may represent an existing magnetic structure at the time the solar ejecta erupted from the corona or it may be formed by magnetic reconnection associated with the eruption.

The particle environment inside the magnetic cloud may support significant levels of plasma-wave activity. In magnetic clouds observed near and beyond 1 AU from the Sun, the electron pressure typically exceeds the ion pressure. A good example is the magnetic cloud observed by *Ulysses* from 10–12 June, 1993. This event, observed when *Ulysses* was 4.6 AU from the Sun, is characterized by $T_e/T_p > 10$ (Stone et al., 1995). Furthermore, the electron-core and halo-gas pressures are comparable in value, suggesting a highly non-Maxwellian plasma (Fainberg et al., 1996). These conditions make such magnetic clouds a preferred environment for the existence of ion acoustic waves.

As seen in Figure 5.11 extensive wave activity also occurs throughout the October 1996 magnetic cloud. These waves, believed to be ion acoustic waves, undergo reduced damping because $T_e/T_p \gg 1$. They are Doppler shifted from their rest frequency into the kHz range by the solar-wind velocity, as described in Section 5.1.1. It is notable that the waves can be Doppler shifted to frequencies above f_{pe}, as is seen for this magnetic cloud; these are frequencies where only freely propagating electromagnetic waves would typically be observed. Similar ion acoustic wave activity is reported for other magnetic clouds observed by *Ulysses* (Lin et al., 1999).

The June 1993 magnetic cloud has been interpreted by Osherovich et al., (1999) as a structure containing two flux ropes. The region where the ion acoustic wave activity is temporarily suppressed (start of 12 June, 1993) and where T_e/T_p drops to 1–2 is the separatrix between the two flux ropes. Thus, the ion acoustic waves can serve to signal boundaries in the magnetic cloud. Similar behaviour is observed for the magnetic cloud in Figure 5.11.

Other wave modes are also found in magnetic clouds and other solar ejecta. Bursts of Langmuir waves sometimes occur, perhaps excited by the bidirectional streaming of electrons. The Langmuir wave bursts often occur in small magnetic holes within the clouds (see Section 5.5.2). At frequencies below f_{ce}, electrostatic waves are often observed; the occurrence of these waves is associated with expanding solar wind (Lin et al., 1999). Intense whistler-mode waves are rarely observed in magnetic clouds, consistent with the quiet magnetic conditions found there (e.g. Figure 5.11). When the solar ejecta drive shocks, as is the case for the June 1993 event, shock-associated waves are to be expected outside the magnetic cloud.

5.5 WAVES AT IP DISCONTINUITIES AND MAGNETIC HOLES

The time resolution and completeness of the *Ulysses* database has permitted extensive statistical studies of discontinuities in the solar wind, including the

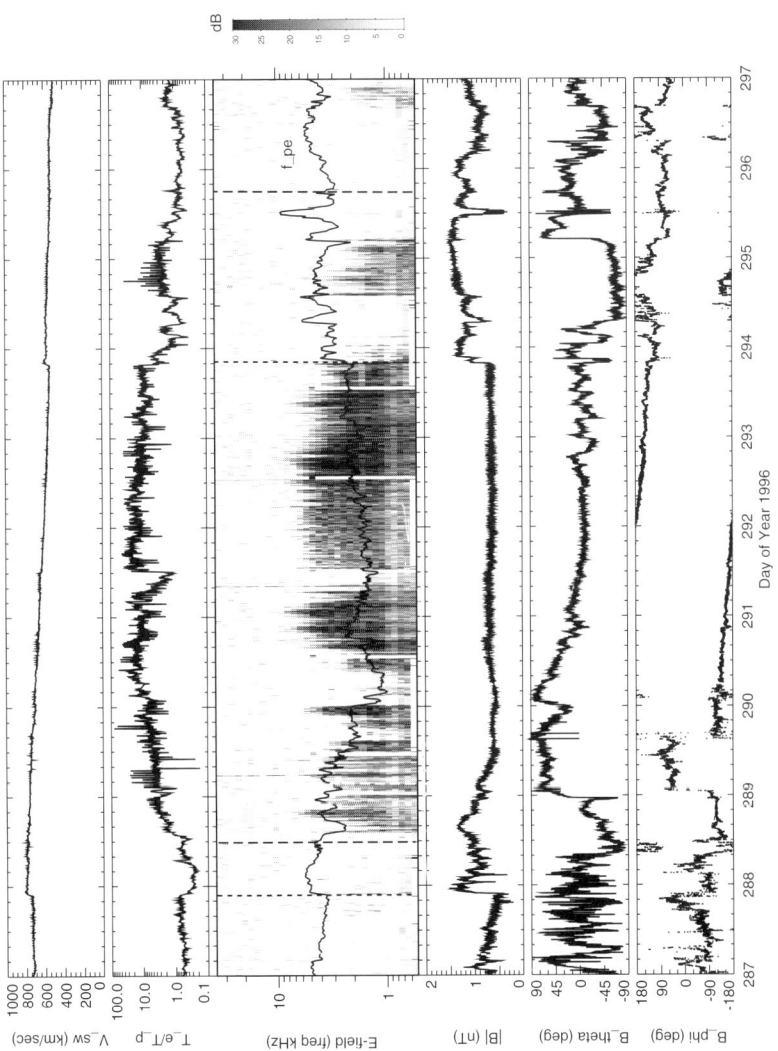

Figure 5.11. Solar wind, wave, and magnetic-field data for the magnetic cloud interval of 13–22 October 1996. Intense ion acoustic wave activity occurs when T_e/T_p is greater than 1. Note that the ion acoustic waves are frequently observed reaching frequencies greater than f_{pe} by factors of 2 to 3. Dotted lines correspond to the occurrence times of shocks; dashed lines correspond to the time interval during which bi-directional electron streaming was observed (Lin et al., 1999).

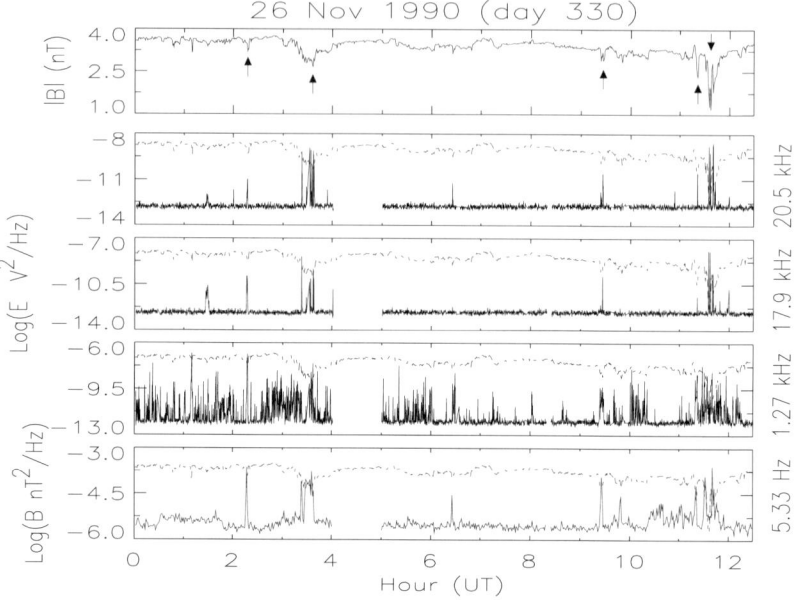

Figure 5.12. Example of Langmuir, ion acoustic, and whistler waves associated with magnetic holes, indicated by arrows. (From Lin *et al.*, 1995).

unique dataset at high heliographic latitudes. Tsurutani *et al.* (1994) have shown that directional discontinuities are more common in Alfvén-wave trains, which occur almost continuously in fast, high-latitude solar wind. The high-latitude magnetic field is dominated by large-amplitude Alfvén waves (Smith *et al.* 1995) and phase steepening of these waves leads to rotational discontinuities (Tsurutani *et al.*, 1994). Ho *et al.*, (1995) found that only 5 per cent of the directional discontinuities in a 15-day period of high-latitude magnetic data were Tangential Discontinuities (TDs). Of these TDs 19 per cent were associated with mirror-mode structures, whereas the majority of TDs were either current sheet associated or intermediate in nature.

These discontinuities play an important role in exciting certain plasma-wave activities. Lin *et al.*, (1995) show that a variety of waves are observed in the immediate vicinity of discontinuities, probably because the variation in magnetic-field intensity and direction perturbs the particles moving along the field lines. In Figure 5.12 whistler, ion acoustic, and Langmuir waves are all observed to intensify in the vicinity of a directional discontinuity associated with a significant decrease in the magnetic-field magnitude, known as a magnetic hole. The existence of such Langmuir waves in magnetic holes was discovered in the *Ulysses* data in fast solar wind at high heliographic latitudes at solar minimum. Subsequently, their existence in slow solar wind has been confirmed, albeit at a lower rate of occurrence (MacDowell *et al.*, 1996a). Lin *et al.*, (1996) conducted a statistical study of these Langmuir waves and determined that they were associated with magnetic holes in

Sec. 5.6]   The 'quiet' solar wind   249

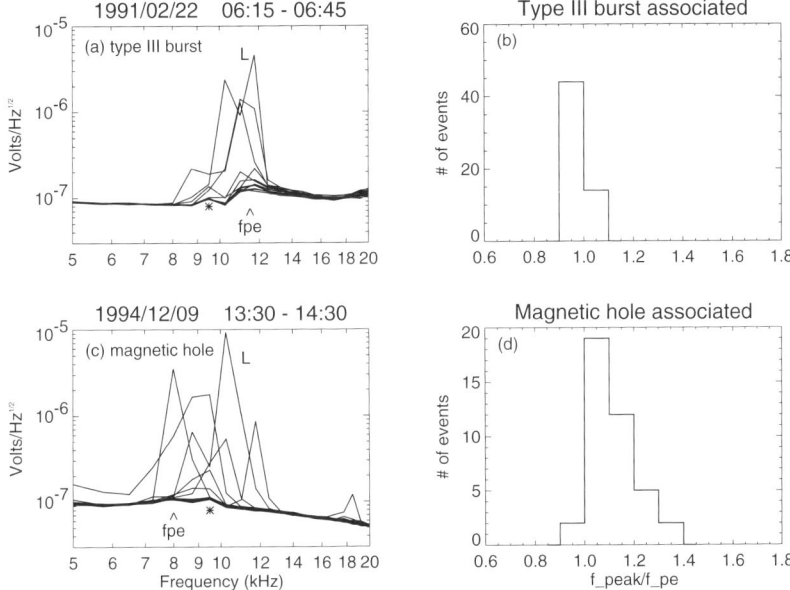

Figure 5.13. (a) Electric field spectrum of URAP RAR data on 22 February 1991, showing Langmuir waves (L) caused by energetic electrons associated with a type III radio burst. The electron plasma frequency f_{pe} obtained from the electron thermal noise spectrum is indicated. An asterisk (*) marks a weak spacecraft interference line at 9.5 kHz. (b) Histogram of f_{peak}/f_{pe} for Langmuir waves associated with several type III events observed by URAP. (c) Electric field spectra of URAP RAR data on 9 December 1994, showing waves (L) in magnetic holes. Although the spectral speak associated with f_{pe} is difficult to see on this plot, it is discernible in the original data. (d) Histogram of f_{peak}/f_{pe} for 20 intervals of wave activity in magnetic holes (September–December 1994) similar to that in (c). (From MacDowell et al., 1996b).

about 75 per cent of the cases, which suggests that the reduction in magnetic field strength is important for the excitation of the observed Langmuir waves.

The mechanism(s) whereby these waves are excited are difficult to determine, in part because of the small dimensions of the magnetic holes. Such structures, as observed by *Ulysses*, traverse the spacecraft in a median time of 22 s (Winterhalter et al., 1994). It is interesting to note that the Langmuir waves observed in these holes are frequently observed at frequencies significantly greater than f_{pe}, as shown in Figure 5.13. MacDowell et al., (1996b) attributed this to a Doppler shift enhanced by the short wavelengths of waves excited by relatively slow electron beams in the magnetic holes. Such beams have not been reported in the literature, and their detection will not be easy.

5.6 THE 'QUIET' SOLAR WIND

The preceding sections have focused on waves associated with structures in the solar wind; however, waves are also found in regions that are not perturbed by transient or corotating structures. In such cases the source of free energy may be less obvious, but there is much useful information to be derived from these less complicated wave environments.

5.6.1 Whistler waves and heat-flux regulation

Ulysses detects nearly continuous waves in the whistler-mode range ($< f_{ce}$) throughout much of its trajectory (Scime *et al.*, 1994 and Lengyel-Frey *et al.*, 1994). It is likely that such waves are always present, but their levels often decrease below the URAP search coil background at distances beyond 2 AU. Possible sources of these waves include the whistler heat-flux instability (Gary and Feldman, 1977) or the electron-temperature anisotropy (Kennel, 1996). Like the case for electromagnetic waves in this frequency range observed at shocks, it has also been suggested that these waves could be electromagnetic lower hybrid waves.

It has been suggested that whistler mode waves are responsible for electron heat-flux regulation in the solar wind (Gary *et al.*, 1994 and Bougeret *et al.*, 1984). Variations in heat flux will drive variation in the whistler-mode intensity. Data that suggests this might be occurring include those presented by Scime *et al.*, (1994).

5.6.2 VLF waves associated with expanding regions of the solar wind

Ulysses data have pointed to another relationship between heat flux and VLF waves. Lin *et al.*, (1998) discovered that enhanced electric-field noise is observed during intervals when the solar-wind velocity measured at the spacecraft is decreasing. These rarefaction regions are also associated with significant reductions in electron-heat flux. Lin *et al.*, (1998) suggest that the anti-correlation of the intensity of these waves and the heat-flux indicates that these waves have regulated the heat flux. Given that the whistler heat-flux instability has a maximum growth rate in the direction of the *B*-field and produces primarily magnetic fluctuations, these observations are surprising. Perhaps the whistler waves are propagating at large angles to the magnetic field, where the E/B ratio is large. Consequently, they are observed as electrostatic waves. Further work is required to understand the source of these waves.

5.6.3 More ion acoustic waves

Ion acoustic waves occur throughout the solar wind, not just in the vicinity of shocks and other discontinuities. Typically, the occurrence can be associated with an increase in the T_e/T_p ratio, which suggests that damping plays a significant role. On the other hand, the relatively low temperature ratios observed in the solar wind are often in the range ($T_e/T_p < 3-5$) expected to limit ion acoustic wave occurrence. In Figure 5.14 an interval of quiet, high-latitude, fast solar wind, where the temperature ratio T_e/T_p is typically less than 1.0, with ion acoustic wave activity is shown.

Given their frequency of occurrence, the limited understanding of ion acoustic waves is frustrating. Progress in this area requires particle observations at higher

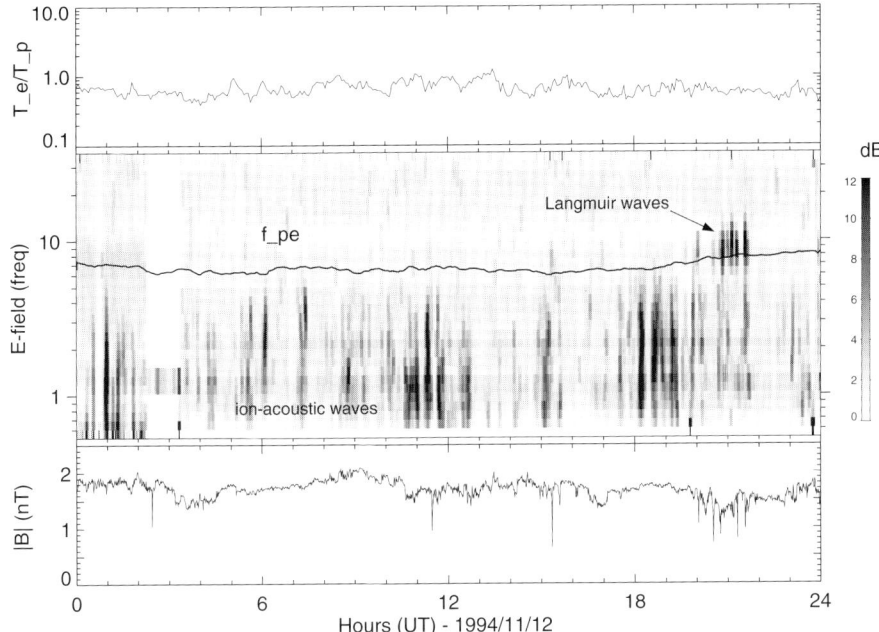

Figure 5.14. Example of ion acoustic wave activity in high-latitude, fast solar wind. The temperature ratio T_e/T_p is less than 1 throughout the entire interval. In addition to the ion acoustic wave activity, note the Langmuir wave activity at the time of magnetic hole structures.

cadences than are provided by most spacecraft spectrometers. It may be that the solar wind is marginally unstable to the growth of ion acoustic waves, so that small changes in solar-wind parameters are sufficient for the instability to develop (Hess et al., 1998).

5.6.4 Thermal noise

Another prominent feature in the wave data is the signal associated with thermal-particle populations. Shown in Figure 5.15 is a thermal noise spectrum from the *Ulysses* URAP RAR instrument (Issautier et al., 1999). The electron thermal noise is most evident when the antenna is long relative to the Debye length. Such a length permits parts of the antenna to lie outside of the volume of the local plasma in which uniform electron shielding can occur. When the antenna system is composed of a long dipole (or some other symmetric forms) whose calibration is well known, the thermal noise spectrum can be accurately related to *in-situ* solar-wind parameters. In the simplest case only the density n_e and the electron temperature T_e are required to fit the thermal noise data (Meyer-Vernet et al., 1989). More typically, the electron distribution requires additional parameters to describe it accurately (e.g. in terms of core and halo populations), each described by a density and temperature. The

252 Waves and instabilities in the three-dimensional heliosphere [Ch. 5

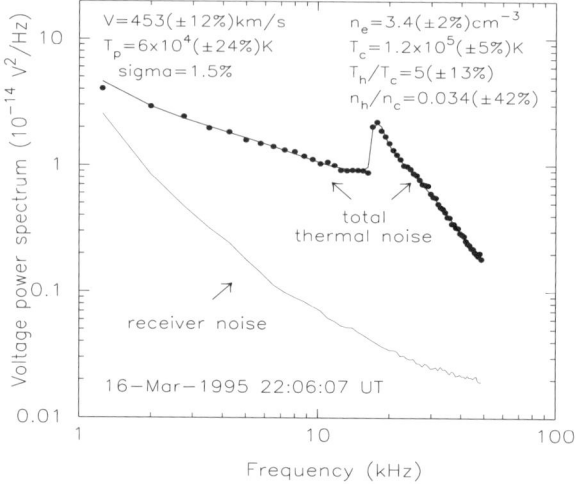

Figure 5.15. Thermal noise spectrum from the URAP with fit to derive the solar wind velocity V, proton temperature T_p, electron density n_e, electron core temperature T_c, electron temperature ration T_h/T_c, and density ratio n_h/n_c. (From Issautier *et al.*, 1999).

current understanding of the thermal noise for both electrons and protons permits calculating a synthetic thermal noise spectrum that agrees with typical data to the accuracies shown in Figure 5.15. The proton thermal noise appearing at the lower frequencies corresponds to electrical fluctuations due to proton motions that are Doppler shifted to this frequency range.

The thermal noise can be understood in the following way. Electric fields in plasma can be analysed as wave modes. In thermal equilibrium these modes are populated at energy kT per mode ($\hbar\omega \ll kT$). Small damping, i.e. an imaginary part of the mode frequency does not change this; it only determines the time to reach thermal equilibrium. Large damping spreads the spectrum. The long-wavelength Langmuir modes lie close to the plasma frequency and cause a jump in the spectrum, especially if the antennas are long enough to discriminate in favour of long wavelengths and against short-wavelength modes, which have little structure in their spectrum. If the plasma is not in thermal equilibrium, as is generally true for the solar wind, then the electric-field spectrum tends to equilibrate with electrons of the same phase velocity. This is quite fast for long-wavelength modes close to f_{pe}, so these modes sense the 'halo' electrons of the solar wind. As these have a high temperature, this effect enhances the spectrum just above f_{pe}. Farther above f_{pe}, the spectrum tends toward equilibrium with the cooler 'core' electrons.

Consequently, the routine measurement of the thermal noise spectrum by *Ulysses* permits accurate determination of the density and electron temperature, as well as determination with lower accuracy of the solar-wind velocity and proton temperature. These measurements complement those of the *Ulysses* particle spectrometers; the thermal noise measurements are insensitive to spacecraft charging and photoelectron contamination. When *Ulysses* is far from the Sun or at high heliographic latitudes, low density and temperature regions may cause the Debye length to increase beyond 100 m, which exceeds the length of the *Ulysses* dipole antenna, reducing the accuracy of the thermal noise measurements. Near 1 AU at low latitudes the Debye length is typically shorter than the *Ulysses* dipole antenna length. Other advantages of thermal noise analysis include high time resolution, ease of calibration, and long-term stability.

5.7 SUMMARY AND REMAINING QUESTIONS

The plasma wave results reviewed here depend significantly on the high latitude orbit of *Ulysses*, which has permitted comparison of waves in fast and slow solar wind. This point is evident in Figure 5.16, which presents wave data for several frequencies during the *Ulysses* fast latitude scan in 1994 and 1995. The solar-wind speed, plotted in panel (f), indicates when *Ulysses* was in fast/slow solar wind. At f_{pe}, the increased probability of Langmuir waves associated with magnetic holes in fast solar wind is evident. At $f_{pe}/3$ ion acoustic wave activity is more likely when the temperature ratio T_e/T_p is greater than 1. The occurrence of low-frequency electrostatic wave activity at 19 Hz is nearly continuous in fast solar wind. Whistler-wave activity at 3.5 Hz is associated with IP shocks and with enhanced values of the DC magnetic field in the streamer belt.

Questions still remain concerning the sources and consequences of the plasma waves discussed here. Among them:

- Are whistlers the only mediating waves for heat-flux regulation?
- What is the source of the electrostatic wave activity associated with expanding regions and fast solar wind?
- What instability leads to ion acoustic waves when $T_e/T_p \sim 1$ and why aren't they strongly damped?
- What causes Langmuir and other waves to be concentrated at magnetic holes?
- Do type III radio bursts result from multiple mechanisms?

Some of these questions can be resolved with existing data from *Ulysses* and other spacecraft making measurements in the solar wind. Others may require measurement of particle-distribution functions with time resolutions that are not presently available.

It should be noted that these *Ulysses* results focus on the state of the solar wind around solar minimum. As of this writing *Ulysses* had yet to make its out-of-ecliptic observations during the maximum in solar activity. Increased solar activity will yield more transients and more variable solar wind at high heliographic latitudes. We anticipate that the study of these events will answer some of the remaining questions about plasma waves in the solar wind and probably raise some new ones.

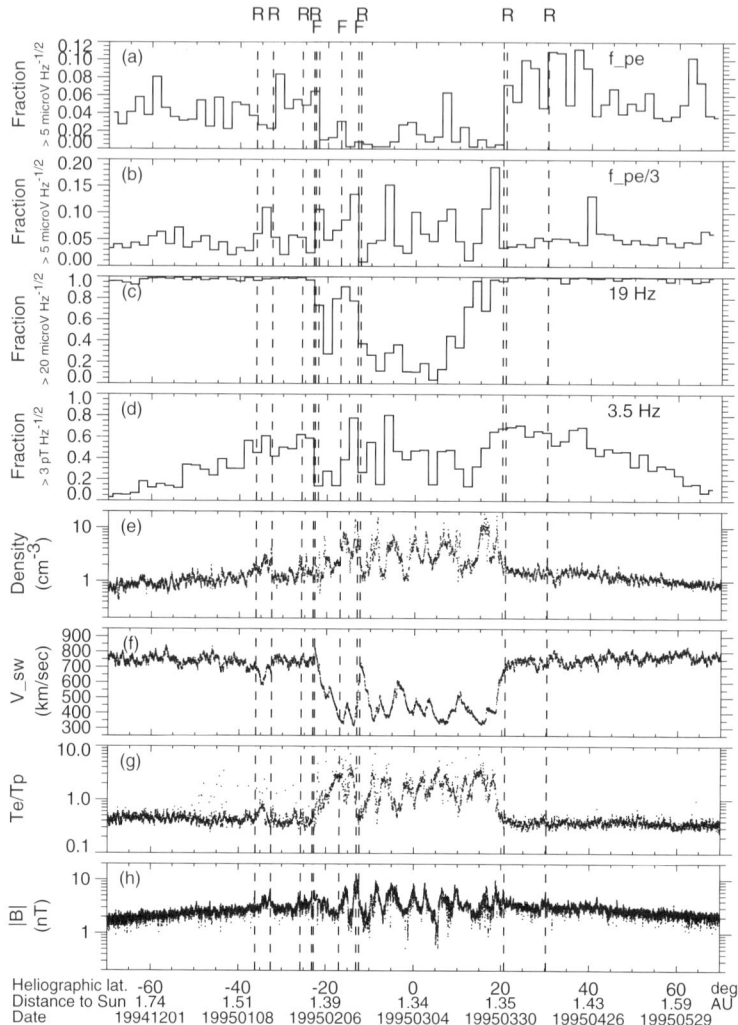

Figure 5.16. Electric and magnetic field wave activity at several frequencies along the *Ulysses* fast latitude scan in 1994–5. The histograms show the fraction of time that waves at a given frequency (f_{pe}, $f_{pe}/3$, 19 Hz, and 3.5 Hz) exceed the level indicated on the plot. Solar wind and magnetic-field data are also plotted (from Macdowell *et al.*, 1996a). Forward (F) and reverse (R) shocks are indicated by dashed lines.

5.8 REFERENCES

Beinroth, H. J. and F. M. Neubauer, (1981) Properties of whistler mode waves between .3 and 1.0 AU from Helios observations. *J. Geophys. Res.* **86**, 7755.

Bougeret, J. L., King, J. H. and Schwenn, R. (1984) Solar radio-burst and in situ determination of interplanetary electron density. *Sol. Phys.* **90**, 401.

Burlaga, L. F., Sitter, E., Mariani, F. and Schwenn, R. (1981) Magnetic loop behind the interplanetary shock: Voyager, Helios, and IMP-8 observations. *J. Geophys. Res.* **86**, 6673.

Cairns, I. H. (1987) Second harmonic plasma emission involving ion sound waves. *Plasma Phys.* **38**, 179.

Cairns, I. H. and Robinson, P. A. (1998) Constraints on nonlinear and stochastic growth theories for type III solar radio bursts from the corona to 1 AU. *Astrophys. J.* **509**, 471.

Cane, H.V. (1985) The evolution of interplanetary shocks. *J. Geophys. Res.* **90**, 191.

Fainberg, J., Evans, L. G. and Stone, R. G. (1972) Radio tracking of solar energetic particles through the interplanetary space. *Science*, **178**, 743.

Fainberg, J., Osherovich, V. A., Stone, R. G., MacDowall, R. J. and Balogh, A. (1996) Ulysses observations of electron and proton components in a magnetic cloud and related wave activity. In: Winterhalter, D. *et al.*, (ed.) *Proc. of the Eighth Intl. Solar Wind Conf.*, American Institute of Physics, pp. 554–557.

Filbert, P. C. and Kellogg, P. J. (1979) Electrostatic noise at the plasma frequency beyond the bow shock. *J. Geophys. Res.* **84**, 1369.

Gary, S. P. and Feldman, W. C. (1977) Solar wind heat flux regulation by the whistler instability. *J. Geophys. Res.* **82**, 1087.

Gary, S. P. and Omidi, N. (1985) The ion-ion acoustic instability. *J. Plasma Phys.* **37**, 45.

Gary, S. P. and Tokar, R. L. (1985) The electron-acoustic mode. *Phys. Fluids*, **28**, 2439.

Gary, S. P., Feldman, W. C., Forslund, D. W. and Montgomery, M. D. (1977) Heat flux instabilities in the solar wind. *J. Geophys. Res.* **80**, 4197.

Gary, S. P., Scime, E. E., Phillips, J. L. and Feldman, W. C. (1994) The whistler heat flux instability: Threshold conditions in the solar wind. *J. Geophys. Res.* **99**, 23,391.

Goldman, M. V., Newman, D. L., Wang, J. G. and Muschietti, L. (1996) Langmuir turbulence in space plasmas. *Physica Scripta*, **T63**, 28.

Goldstein, M. L., Smith, R. A. and Papadapoulos, K. (1979) Nonlinear instability of solar type III radio bursts, II. Application to observations near 1 AU. *Astrophys. J.* **237**, 683.

Gosling, J. T. (1997) Coronal mass ejections, an overview. In: Crooker, N. Joselyn, J. A. and Feynman, J. (eds) *Coronal Mass Ejections*, American Geophysical Union, Washington, D.C.

Gurnett, D. A. (1991) Waves and instabilities. In Schwenn R. and Marsch E. (eds) *Physics of the Inner Heliosphere*. Springer-Verlag, Berlin.

Gurnett, D. A., Marsh, E., Pilipp, W., Schwenn, R. and Rosenbauer, H. (1979a) Ion acoustic waves and related plasma observations in the solar wind. *J. Geophys. Res.* **84**, 2029.

Gurnett, D. A., Neubauer, F. M. and Schwenn, R. (1979a) Plasma wave turbulence associated with an interplanetary shock. *J. Geophys. Res.* **84**, 541.

Hess, R. A., MacDowall, R. J., Goldstein, B., Neugebauer, M. and Forsyth, R. J. (1998) Ion acoustic-like waves observed by Ulysses near interplanetary shock waves in the 3d heliosphere. *J. Geophys. Res.* **103**, 6531.

Ho, C. M., Tsurutani, B. T., Goldstein, B. E., Phillips, J. L. and Balogh, A. (1995) Tangential discontinuities at high heliographic latitudes ($\sim -80°$). *Geophys. Res. Let.* **22**, 3409.

Hoang, S., Maksimovic, M., Bougeret, J. L., Reiner, M. J. and Kaiser, M. L. (1998) Wind-Ulysses source location of radio emissions associated with the January 1997 coronal mass ejection. *Geophys. Res. Let.* **25**, 2497.

Issautier K., Meyer-Vernet, N., Moncuquet, M., Hoang, S., McComas, D. J. (1999) Quasi-thermal noise in a drifting plasma: Theory and application to solar wind diagnostic on Ulysses. *J. Geophys. Res.* **104**, 6691.
Kaplan, S. A. and V.N. Tsytovich, (1973) *Plasma Astrophysics*, Pergamon Press, Oxford.
Kellogg, P. J. (2000) Fluctuations and ion isotropy in the solar wind. *Astrophys. J.*, **528**, 480.
Kellogg, P. J., Goetz, K., Monson, S. J., Balogh, A. and Forsyth, R. J. (1996) Some remarks on waves in the solar wind. In: Winterhalter D. *et al.* (eds) *Proc. of the Eighth Intl. Solar Wind Conf.* American Institute of Physics, pp. 301–304.
Kennel, C. F. (1996) Low frequency whistler mode. *Phys. Fluids.*, **9**, 2190.
Kennel, C. F., Scarf, F. L., Coroniti, F. V., Smith, E. J. and Gurnett, D. A. (1982) Nonlocal plasma turbulence associated with interplanetary shocks. *J. Geophys. Res.* **87**, 17.
Kundu, M. R. (1965) *Solar Radio Astronomy*. Interscience, New York.
Kurth, W. S., Gurnett, D. A. and Scarf, F. L. (1979) High-resolution spectrograms of ion acoustic waves in the solar wind. *J. Geophys. Res.* **84**, 3413.
Leblanc, Y., Dulk, G. A., Hoang, S., Bougeret, J.-L. and Robinson, P.A. (1996) Type III radio bursts observed by Ulysses pole to pole, and simultaneously by Wind. *Astron. Astrophys.* **316**, 406.
Lengyel-Frey, D. and Stone, R. G. (1989) Characteristics of interplanetary type II radio emission and the relationship of shock and plasma properties. *J. Geophys. Res.* **94**, 159.
Lengyel-Frey, D., Farrell, W. M., Stone, R. G., Balogh, A. and Forsyth, R. (1994) An analysis of whistler waves at interplanetary shocks. *J. Geophys. Res.* **99**, 13325.
Lengyel-Frey, D., Thejappa, G., MacDowall, R. J., Stone, R. G. and Phillips, J. L. (1997) Ulysses observations of wave activity at interplanetary shocks and implications for type II radio bursts. *J. Geophys. Res.* **102**, 2611.
Lin, N., Kellogg, P. J. MacDowall, R. J. Scime, E. E. Phillips, J. L. Balogh, A. and Forsyth, R. J. (1997) Low frequency plasma waves in the solar wind: from ecliptic plane to the solar polar regions. *Adv. Space. Research*, **19**, 877.
Lin, N., Kellogg, P. J., Goetz, K. A., Monson, S. J. and MacDowall, R. J. (1999) Plasma waves in coronal mass ejections: Ulysses observations. In: Habbal, S. (ed.) *Solar Wind Nine*, AIP Conference Proceedings 471, pp. 673–676.
Lin, N., Kellogg, P. J., MacDowall, R. J., Balogh, A., Forsyth, R. J., Phillips, J. L., Buttighoffer, A. and Pick, M. (1995) Observations of plasma waves in magnetic holes. *Geophys. Res. Let.* **22**, 3417.
Lin, N., Kellogg, P. J., MacDowall, R. J., Scime, E. E., Balogh, A., Forsyth, R. J., McComas, D.J. and Phillips, J. L. (1998) Very low frequency waves in the heliosphere: Ulysses observations. *J. Geophys. Res.* **103**, 12023.
Lin, N., Kellogg, P. J., MacDowall, R. J., Tsurutani, B. T. and Ho, C. (1996) Langmuir waves associated with discontinuities in the solar wind: a statistical study. *Astron. Astrophys.* **316**, 425.
Lin, R. P., Levedahl, W. K., Lotko, W., Gurnett, D. A. and Scarf, F. L. (1986) Evidence for non-linear wave-wave interactions in solar type III bursts. *Astrophys. J.* **308**, 954.
Lin, R. P., Potter, D. W., Gurnett, D. A. and Scarf, F.L. (1981) Energetic electrons and plasma waves associated with a solar type III burst. *Astrophys. J.* **251**, 364.
MacDowall, R. J., Hess, R. A., Lin, N., Thejappa, G., Balogh, A. and Phillips, J. L. (1996a) Ulysses spacecraft observations of radio and plasma waves: 1991–1995. *Astron. Astrophys.* **316**, 396.
MacDowall, R. J., Lin, N., Kellogg, P. J., Balogh, A., Forsyth, R. J. and Neugebauer, M. (1996b) Langmuir waves in magnetic holes: source mechanism and consequences. *Proc. of*

the Eighth Intl. Solar Wind Conf., Winterhalter, D. et al., (ed.) American Institute of Physics, pp. 301–304.

Manning, R., and Fainberg, J. (1980) *Space Sci. Instr.* **5**, 161.

Marsch, E. (1985) Beamdriven electron acoustic waves upstream of the Earth's bow shock. *J. Geophys. Res.* **90**, 6327.

Marsch, E. and Chang, T. (1983) Lower hybrid waves in the solar wind. *J. Geophys. Res.* **88**, 6869.

Melrose, D. B. (1980) Plasma emission mechanisms for solar radio bursts. *Space Sci. Rev.* **26**, 3.

Meyer-Vernet, N. and Perche, C. (1989) Tool kit for antennae and thermal noise near the plasma frequency. *J. Geophys. Res.* **94**, 2405.

Osherovich, V., Fainberg, J. and Stone, R. G. (1999) Multi-tube model of interplanetary magnetic clouds. *Geophys. Res. Let.* **26**, 401.

Pedersen, A., Cattell, C. A., Falthammar, C. G., Formisano, V., Lindqvist, P. A., Mozer, F. and Torbert, R. (1984) Quasistatic electric-field measurements with spherical double probes on the Geos and Isee satellites. *Space Sci. Rev.* **37**, 269.

Pick, M., Buttighoffer, A., Kerdraon, A., Armstrong, T. P., Roelof, E. C., Hoang, S., Lanzerotti, L. J. Simnett, G. M. and Lemen, J. (1995a) Ulysses observations of a coronal origin particle event at 32 deg south heliographic latitude. *Space Sci. Rev.* **72**, 315.

Pick, M., Lanzerotti, L. J., Buttighoffer, A., Hoang, S. and Forsyth, R. J. (1995b) Detection of a solar electron event at an heliolatitude of 73.8 deg S. *Geophys. Res. Let.* **22**, 3377.

Pierre F., Solomon, J., Cornilleau-Wehrlin, N., Canu, P., Scime, E. E., Balogh, A. and Forsyth, R. J. (1997) Oblique emission of whistlermode waves around interplanetary shocks. *Solar Phys.* **172**, 327.

Reiner, M. J., Fainberg, J. and Stone, R. G. (1995a) Large-scale interplanetary magnetic field configuration revealed by solar radio bursts. *Science*, **270**, 461.

Reiner, M. J., Kaiser, M. L., Fainberg, J. and Stone, R. G. (1995b) Type III radio source located by Ulysses/Wind triangulation. *J. Geophys. Res.* **103**, 1923.

Reiner, M. J., Kaiser, M. L., Fainberg, J. and Stone, R. G. (1998) A new method for *studying* remote type II radio emissions from coronal mass ejection driven shocks. *J. Geophys. Res.* **103**, 29651.

Reiner, M. J., Stone, R. G. and Fainberg, J. (1992) Detection of fundamental and harmonic type III radio emission and the associated Langmuir waves at the source region. *Astrophys. J.* **394**, 340.

Scime, E. E., Bame, S. J., Feldman, W. C., Gary, S. P. and Phillips, J. L. (1994) Regulation of the solar wind heat flux from 1 to 5 AU: Ulysses observations. *J. Geophys. Res.* **99**, 23,401.

Smith, E. J., Balogh, A., Lepping, R. P., Neugebauer, M., Phillips, J. L. and Tsurutani, B. T. (1995) Ulysses observations of latitude gradients in the heliospheric magnetic field. *Adv. Space Res.* **16**, 165.

Stone, R. G., MacDowall, R. J., Fainberg, J., Hoang, S., Kaiser, M. L., Kellogg, P. J., Lin, N., Osherovich, V. A., Bougeret, J. L., Canu, P., Cornilleau-Wehrlin, N., Desch, M. D., Goetz, K., Goldstein, M. L., Harvey, C. C., Lengyel-Frey, D., Reiner, M. J., Steinberg, J. L. and Thejappa, G. (1995) Ulysses radio and plasma wave observations from the ecliptic to high southern heliographic latitudes. *Science*, **268**, 1026.

Stone, R.G., Bougeret, J. L., Caldwell, J., Canu, P., DeConchy, Y., Cornilleau-Wehrlin, N., Desch, M. D., Fainberg, J., Goetz, K., Goldstein, M. L., Harvey, C. C., Hoang, S., Howard, R., Kaiser, M. L., Kellogg, P. J., Klein, B., Knoll, R., Lecacheux, A., Lengyel-Frey, D., MacDowall, R. J., Manning, R., Meetre, C. A., Meyer, A., Monge, N.,

Monson, S., Nicol, G., Reiner, M. J., Steinberg, J. L., Torres, E., de Villedary, C., Wouters, F. and Zarka, P. (1992) The unified radio and plasma wave investigation, *Astron. Astrophys. Supp. Ser.* **92**, 291.

Sturrock, P.A. (1964) Paper presented at *Proc. of AAS-NASA Symposium on the Physics of Solar Flares*, Hess, W. N. (ed.), NASA SP-50, 357.

Thejappa, G. and MacDowall, R. J. (1998) Evidence for strong and weak turbulence processes in the source region of a local type III radio burst. *Astrophys. J.* **498**, 465.

Thejappa, G., Wentzel D. G. and Stone, R. G. (1998) Interplanetary low-frequency waves associated with high-frequency electron plasma waves. *J. Geophys. Res.* **100**, 3417.

Thejappa, G., Wentzel, D. G., MacDowall, R. J. and Stone, R. G. (1995) Unusual wave phenomena near interplanetary shocks at high latitudes. *Geophys. Res. Let.* **22**, 3421.

Tokar, R. L. and D. A. Gurnett, (1985) The propagation and growth of whistler mode waves generated by electronbeams in Earth's bow shock. *J. Geophys. Res.* **90**, 105.

Tokar, R. L., Gurnett, D. A. and Feldman, W. C. (1984) Whistler mode turbulence generated by electron beams in earth's bow shock. *J. Geophys. Res.* **89**, 105.

Tsurutani, B. T., Ho, C. M., Smith, E. J., Neugebauer, M., Goldstein, B. E., Mok, J. S., Arballo, J. K., Balogh, A., Southwood, D. J. and Feldman, W. C. (1994) The relationship between interplanetary discontinuities and Alfvén waves: Ulysses observations. *Geophys. Res. Let.* **22**, 2267.

Wild, J. P., Smerd, S. F. and Weiss, A. A. (1963) Solar bursts. *Ann. Rev. Astron. Astrophys.*, **1**, 291.

Winterhalter, D., Neugebauer, M., Goldstein, B. E., Smith, E. J., Bame, S. J. and Balogh, A. (1994) Ulysses field and plasma observations of magnetic holes in the solar wind and their relation to mirror-mode structures. *J. Geophys. Res.* **99**, 23,371.

Yin, L. and Ashour-Abdalla, M. (1999) Mode conversion in a weakly magnetized plasma with a longitudinal density gradient. *Phys. Fluids*, **6**, 449.

6

Energetic particles in the heliosphere

Louis J. Lanzerotti and Trevor R. Sanderson

6.1 INTRODUCTION AND STAGE SETTING

That the space between the Sun and the Earth is not completely a vacuum has been recognized for at least a century, ever since discussions about the observed relationships between solar activity and geomagnetic variations and the aurora borealis became more quantitative (e.g. see historical sources in Chapman and Bartels, 1940). The fact that the solar system itself was generally populated with electrons, protons, and heavier ions with a few tens of keV energy to a few tens of MeV energy (often called 'energetic particles') was not elucidated until the advent of the flight of appropriate detectors on spacecraft that travelled beyond the boundaries of the Earth's magnetosphere, beginning in the 1960's. Since that time, this energy range – intermediate between the solar wind particles and the cosmic rays that can reach Earth's surface (both galactic and solar) – has been a subject area of immense fruitfulness in the process of seeking to understand the heliosphere that encompasses the Sun, the Earth, and the other planets.

Indeed, the finding that the Interplanetary Medium (IPM) – near the ecliptic plane where all measurements prior to the *Ulysses* mission had largely been made – was filled with charged particles and electromagnetic fields was one of the earliest major achievements of the space age. The particle energy range provided by the instrumentation on the *Ulysses* spacecraft that is discussed in this chapter covers several tens of keV to several tens of MeV. This energy range is partially defined by the capabilities and design of space flight instrumentation. However, more importantly, the discussions in this chapter are based dominantly on the fundamental understanding of the IPM that has been achieved by *Ulysses* measurements of these charged particles – electrons, protons, and heavier ions – in this energy range.

After its departure from Jupiter in February 1992 near the beginning of the decline of Solar Cycle 22, instrumentation on *Ulysses* began to record the particle and field populations outside the ecliptic plane (Balogh *et al.*, 2001). A schematic overview for the entire solar-minimum orbit of *Ulysses* of the intensities of the

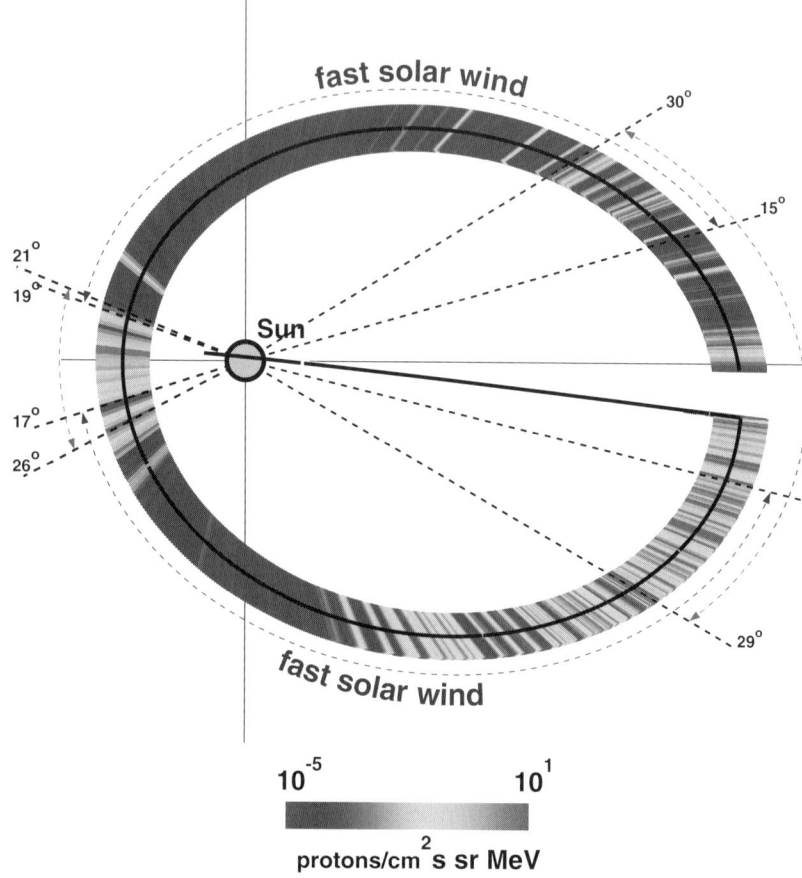

Figure 6.1. Intensity of 1.8–3.8 MeV protons as measured by the COSPIN LET instrument shown in greyscale code on a projection of the Ulysses orbit perpendicular to the ecliptic plane.

1.8–3.8 MeV protons as acquired by the Low Energy Telescope (LET) telescope of the COSPIN instrument (Simpson *et al.*, 1992) is shown in Figure 6.1. These intensities are superimposed upon a projection of the *Ulysses* orbit perpendicular to the ecliptic plane. The spacecraft left Jupiter towards the southern solar hemisphere (lower part of Figure 6.1). The proton flux intensities are greyscale-coded. Indicated are the latitudes at which slow and fast solar-wind streams were encountered, as well as the latitudes of transition between the two solar-wind states (e.g. transition regions in the south were between latitudes of 13° and 29° near 5 AU and between 26° and 17° in the fast latitude pass; see Neugebauer, 2001). The many quasiperiodic fluctuations in the proton intensities as the spacecraft transitioned from the ecliptic plane (and the slow solar-wind stream region) to the fast streams at the high latitudes are quite evident.

A more conventional overview of the solar minimum measurements by the *Ulysses* low-energy particle instrumentation on *Ulysses* is shown in Figure 6.2.

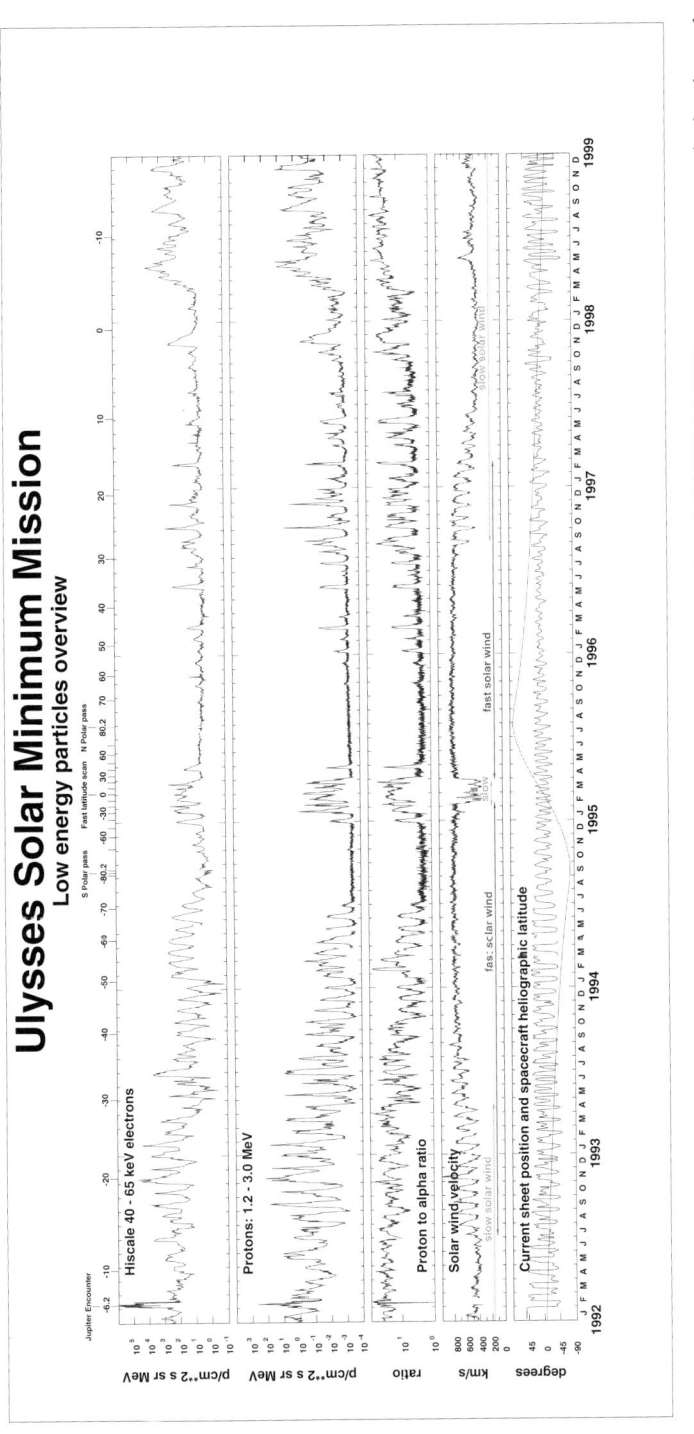

Figure 6.2. Intensities of an electron and a proton channel measured by the HI-SCALE and COSPIN LET instruments, respectively, through the Ulysses solar minimum mission. Also plotted are the proton to helium particle ratio (~1 MeV/nucl), the solar wind velocity, and, in the bottom panel, the location of the heliospheric current sheet as deduced from measurements at the Wilcox solar observatory and the heliographic latitude of Ulysses.

Here the intensities of electrons (40–65 MeV) from the HI-SCALE instrument (Lanzerotti *et al.*, 1992) and protons (1.2–3.0 MeV) from the LET telescope are plotted in the upper two panels, respectively, as a function of time throughout the solar minimum orbit. The heliolatitude is given at the top of the Figure, and thus the time and latitude can be referenced to the fluxes in Figure 6.1. The ratio of $\sim$1 MeV/nucl protons to helium particles (from the LET instrument) are shown in the centre panel, while the fourth from the top panel plots the solar-wind velocity for the entire orbit (see also Neugebauer, this volume). The bottom panel plots the location of the heliospheric current sheet (irregular solid line) for the entire *Ulysses* orbit. The current sheet determination is derived from measurements of the magnetic 'neutral line' on the Sun by the Wilcox Solar Observatory at Stanford University (Hoeksema *et al.*, 1982). The solid, smoothly-varying curve in the lowest panel of the Figure is the heliolatitude of *Ulysses*.

The overall low energy particle perspective of the *Ulysses* solar-minimum mission can be obtained by referring to both Figures 6.1 and 6.2. After departing Jupiter in February 1992, *Ulysses* began its journey out of the ecliptic plane. At the beginning of this journey, the level of solar activity was relatively high and *Ulysses* was completely immersed in a slow solar-wind regime. Moderately high fluxes of electrons and protons with no regular temporal patterns were observed. When *Ulysses* reached 13°S, the spacecraft began entering, once per solar rotation ($\sim$26 days), fast solar wind flow emanating from the southern polar coronal hole. Then, a regular sequence of electron and proton increases were measured that were produced by the presence of Corotating Interaction Regions (CIRs). Most of these CIRs had forward and reverse shocks associated with them.

When *Ulysses* reached about 29°S, the spacecraft became completely immersed in the high speed solar-wind flow. CIRs continued to be observed, propagating poleward, but only with reverse shocks associated with them (Forsyth and Gosling, 2000). Beyond $\sim$45°S, no further reverse shocks were measured. However, quasi-regular particle increases continued to be seen. These increases were seen in the protons up to $\sim$70°S and in the electrons up to 80.2°S, the highest southern latitude *Ulysses* reached. Interspersed with these quasi-regular increases were the occasional Coronal Mass Ejection (CME) and transient solar particle event.

Ulysses passed over the southern solar pole at a latitude of 80.2°S on 13 September 1994. It then began its 'Fast Latitude Scan' (FLS), moving rapidly down to, and through, the ecliptic plane and up to the northern polar regions in less than 1 year. As *Ulysses* began the FLS, irregular increases in the electron intensities were measured. However, the first proton increases were not seen until *Ulysses* reached $\sim$45°S in its descent to the ecliptic. A short sequence of CIR-related particle increases was encountered at about 22°S, just prior to entering the heliospheric current sheet-dominated region. Two major intensity peaks per rotation (compared with the one per rotation in 1992 and 1993) were seen in the proton fluxes. This effect was produced by a warp in the heliospheric current sheet (centre region of bottom panel, Figure 6.2) which gave rise to four IP magnetic sectors per solar rotation rather than the two per rotation seen during late 1992 and early 1993. A similar pattern, though not so pronounced, was seen in the electron intensities.

Ulysses departed the in-ecliptic slow solar wind at ~22°N in April 1995 as the solar cycle was approaching its minimum; one last transient event was seen on ~25 April at 40°N. No further proton increases were measured as the spacecraft passed over the northern pole during this interval around solar minimum. However, small intensity increases, just above instrument background, began to be observed in the electron intensities as the spacecraft descended past 70°N. After this time and heliolatitude location, increases were then measured more frequently in the electrons, and then in the protons, as the spacecraft again approached the near-ecliptic streamer belt. At ~30°N, slower speed solar wind flow was encountered again, but the regularity was not as pronounced, and the intensities of the increases were not as high, as those observed in 1992 and 1993 (during that interval the current sheet excursions were much greater, and the background flux of solar particles was much higher). Finally, in June 1997, *Ulysses* was once again completely immersed in the slow solar wind, remaining so until the end of the solar minimum orbit (aphelion: 5.8°S on 13 April 1998). The rise in solar activity in Solar Cycle 23, including the increased number of solar events in 1998, gave rise to elevated proton and electron fluxes during the last few months of this solar minimum mission.

As for the results from the entire *Ulysses* mission to date, measurements of low-energy charged particles have contributed importantly to fundamentally new understandings of the heliosphere in three dimensions. These new understandings, some of which were briefly addressed above, are discussed in the following sections of this chapter. These include the sources of particles, the acceleration of particles, the three-dimensional spatial and temporal distributions of particles, the contributions of the energy distributions of these particles to the plasma physics of the Interplanetary Medium (IPM), and, briefly, the role of these particles in producing detrimental effects in space-based technologies.

6.2 SOURCES OF ENERGETIC PARTICLES

The sources of the energetic particles that fill the heliosphere are major topic areas of study in solar system plasma physics. The sources of the particles are important in their own right. In addition, the particles can be used as probes and tracers of features of the IPM. Hence, measurements of the energetic particle population in the three-dimensional heliosphere, as the *Ulysses* mission has made possible, contribute to a much firmer knowledge base of these topics.

6.2.1 Solar sources

The flight of *Ulysses* during the declining phase of the solar cycle meant that few significant solar energetic particle events that resulted from active regions on the Sun were measured by the spacecraft's instrumentation. The largest series of particle events were those that occurred in the March–June, 1991, interval several months

264   Energetic particles in the heliosphere   [Ch. 6

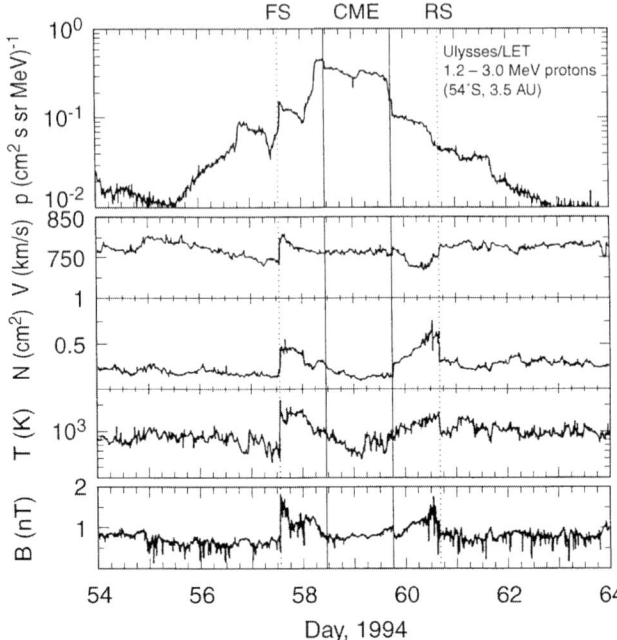

Figure 6.3. Association of interplanetary energetic protons with occurrence of a coronal mass ejection event. Times of occurrence of the forward shock (FS), the reverse shock (RS), and the CME are shown by the vertical dashed lines. Top panel: ten minute average of proton flux (1.2–3.0 MeV/nucl). The lower four panels give the time dependence of interplanetary quantities: the solar wind velocity (V), number density (n), and temperature (T), and the magnitude of the interplanetary magnetic field (B) (from Bothmer et al., 1995).

after the launch of *Ulysses* (October 1990). A special section issue of *Geophysical Research Letters* (19 June, 1992) was devoted to a set of papers that described and analysed the particle, magnetic field, and plasma wave measurements.

Nevertheless, as the spacecraft climbed to high southern heliolatitudes, several energetic particle transient events were measured by *Ulysses* instrumentation. Most of these transient events were associated with the passage over the spacecraft of CME events propagating through the IPM (e.g. Keppler et al., 1995; Lim et al., 1996; Marsden et al., 1997). These Interplanetary CME events (called ICMEs by some researchers) are largely identified from characteristic signatures in the solar wind magnetic field and plasma (Forsyth et al., 2001). Shown in Figure 6.3 is such a transient event at about 54°S and at 3.5 AU (Bothmer et al., 1995). The enhancement in the MeV-energy proton flux, as shown in the top panel, occurred several days following a CIR event (event #24 in Figure 6.5). The particle enhancement was reasonably symmetric around the interval that included strong forward and reverse shocks, with the largest fluxes occurring near the leading edge of, and within, the CME itself, and not at one of the shocks. These shocks were both identified from *Ulysses* plasma data as being driven by the overexpansion of

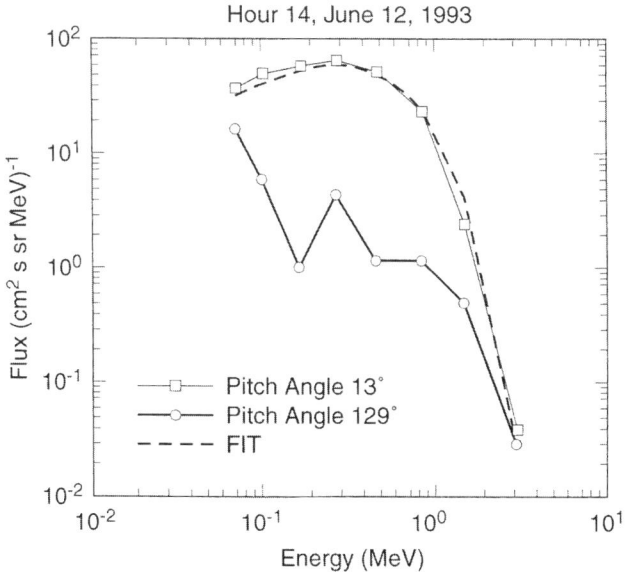

Figure 6.4. Differential energy spectra of energetic ions propagating at two different pitch angles along magnetic fields associated with a coronal mass ejection event. The ions at a pitch angle of $\sim 13°$ are propagating away from the solar corona. The dashed line represents a Maxwellian fit to the energy distribution, with a characteristic energy of 0.273 MeV (from Armstrong *et al.*, 1994).

CMEs (Gosling *et al.*, 1994). The overall particle intensities for this CME-related event are several factors of 10 less than typically seen in such events near the ecliptic plane (Bothmer *et al.*, 1995).

Not all CME-associated events are necessarily related to the acceleration of charged particles within the event and its shocks. Indeed, Armstrong *et al.*, (1994) reported a CME-associated event that was measured when *Ulysses* was 32°S and about 4.6 AU, climbing to the southern solar pole. This event (occurring between CIR events #14 and 15 Bame *et al.*, 1993) involved an interval of strong streaming of ions away from the Sun. Analysis of the energy spectra as a function of the directionality of the ions (Figure 6.4: the 13° pitch-angle spectrum corresponds to particles propagating along the magnetic field outward from the Sun; the 129° spectrum to particles propagating toward the Sun) during this event allowed Armstrong *et al.*, (1994) to conclude that the core ion beam was narrow, and that the peak flux direction of the ion beam could be characterized by a Maxwellian distribution with $kT = 270$ keV. The magnetic fields at this time were tied to the corona. A similar event, showing magnetic fields rooted in the solar corona, was analysed when *Ulysses* was at $\sim 24°$S and ~ 1.5 AU (during the FLS) (Bothmer *et al.*, 1996).

Armstrong *et al.* (1994) also concluded that the absence of any clear signatures of local shock acceleration meant that the hot particle beam could have been a long-lived hot coronal ion population propagating away from the Sun along the CME magnetic field. An analysis of solar data, including radio noise storms, at the time of this event provided additional credence to the conclusion of the hot plasma beam (Pick *et al.*, 1995b). Finally, Armstrong *et al.* (1994) concluded that in this event the weak scattering that is implied by the distribution (e.g. comparing the two pitch angle spectra in Figure 6.4) implied that the anti-sunward end of the CME magnetic flux rope was magnetically connected to the outer heliosphere rather than back to the Sun, at least during the interval of the measurement of the hot plasma beam.

In summary, the interval after about June 1991, and especially after *Ulysses* encounter with Jupiter in February 1992, was during the decline of solar cycle 22 to solar minimum. This caused there to be few 'typical' solar energetic particle events to be measured. However, there were a number of clear CME-associated events that allowed the elucidation of new features of the relevance of such events to the populating of the IMP, from the acceleration of particles at the CME shocks to the 'guidance' of particles by the CME flux rope fields. Other *Ulysses*-measured particle events, often of very low flux levels, were used to examine the latitudinal propagation of solar-originating particles, and these are discussed in Section 6.3.

6.2.2 Interplanetary sources

The acceleration of IP particles at CIRs has long been recognized as an important astrophysical process, one that has applications in a broad range of issues in space and astrophysical plasmas (e.g. McDonald *et al.*, 1976; Barnes and Simpson, 1976; Pesses *et al.*, 1978). Since it has been found that the acceleration of ions at CIRs is more intense in the outer (order few AU) solar system than in the inner (e.g. Marshall and Stone 1978; Mewaldt *et al.*, 1978; van Hollebeke *et al.*, 1978; Christon, 1981; Boufaida and Armstrong, 1985; Maclennan *et al.*, 1995), it can be deduced that the inner solar system population can be augmented by these accelerated particles (Christon, 1981; Gloecker *et al.*, 1979).

The *Ulysses* mission during solar minimum has contributed considerably to the understanding of the acceleration of IP particles by CIRs in the heliosphere within 5 AU (see Keppler (1998) for an extensive overview; *see also* Simnett *et al.*, 1998). The typical dependence of the intensities of the IP proton and alpha particle fluxes on the occurrence of CIRs is shown in Figure 6.5 as *Ulysses* climbed from 31°S to 62°S heliolatitude (Sanderson, 1995; Sanderson *et al.*, 1995). Quasi-regular, approximately 26-day period variation in the particle fluxes is evident (the numbering of the CIR-associated particle enhancements is taken from the CIR identification numbering system presented in Bame *et al.*, 1993). The vertical dashed lines in the figure show the time locations of the IP shocks (all are reverse shocks except for the two forward shocks marked F). Proton to alpha particle ($\sim$1 MeV/nucl energy) ratios of the order of 50 were observed at the lower latitudes, with the ratio decreasing with increasing latitude until the onset of larger particle enhancements after about day 20 1994, which accompanied an increase in solar activity (i.e. the occur-

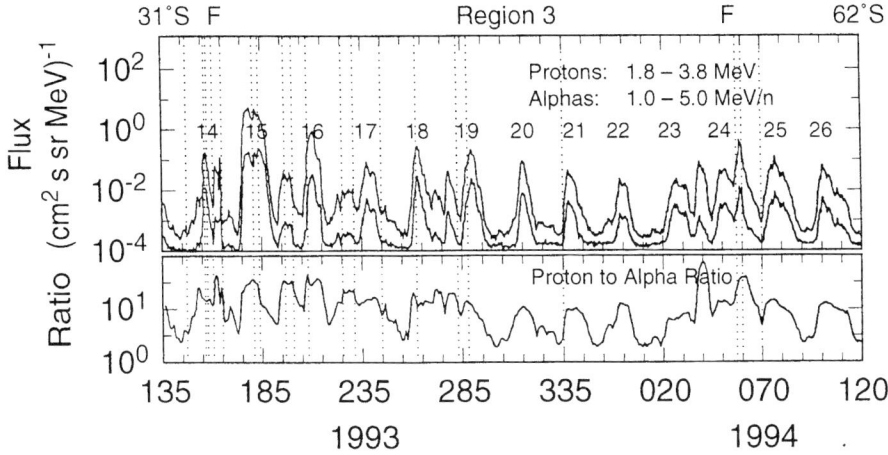

Figure 6.5. Upper panel: Proton (upper curve) and alpha particle (lower curve) fluxes as a function of time during the passage of 13 CIR events in 1993–1994. The times of reverse shock event occurrences are indicated by the vertical long dashed lines. The time of occurrence of two forward shock events (short vertical dashed lines) are labelled F. Lower panel: proton to alpha particle ration during the interval (from Sanderson, 1995 and Sanderson *et al.*, 1995).

rence of a class M4/3B flare on day 51) and the passage of a CME past *Ulysses* on day 57.

Statistical studies of proton intensity increases at corotating reverse shocks were reported from measurements that were made in the southern hemisphere (Desai *et al.*, 1997) and in both hemispheres, including the FLS (Desai *et al.*, 1998). These authors found that in the region between $\sim 29°$S and $\sim 40°$S, the 1 Mev proton intensities were well correlated with the shock strength, with the shock acceleration efficiency increasing with increasing shock strength. No similar correlation was found with the shock normal angle. It was also in this southern latitude region that Maclennan *et al.*, (1995) reported that the maximum effect of CIRs on heavy ion acceleration occurred. These authors noted that prior results (van Hollebeke *et al.*, 1978; Boufaida and Armstrong, 1985) can be explained by not only an increase in CIR acceleration efficiency between approximately the orbits of Mars and Jupiter, but also that this increase in acceleration appears to occur at heliolatitudes that correspond approximately to the edge of the heliospheric current sheet at those distances.

The Desai *et al.*, (1998) study also examined proton increases associated with recurrent compression regions, where increases are observed even when no shocks are present (as well as increases associated with forward–reverse shock pairs). The authors concluded that the Fermi shock acceleration mechanism was the dominant acceleration mechanism for producing the proton enhancements.

Using data from two instruments that covered the energy range 50 keV–20 MeV, Desai *et al.*, (1999) examined the Fisk and Lee (1980) and the Jones and Ellison (1991) models of shock diffusive theory. Comparing their predictions for the spectral indices with observations, Desai *et al.* (1999) found significant departures from the

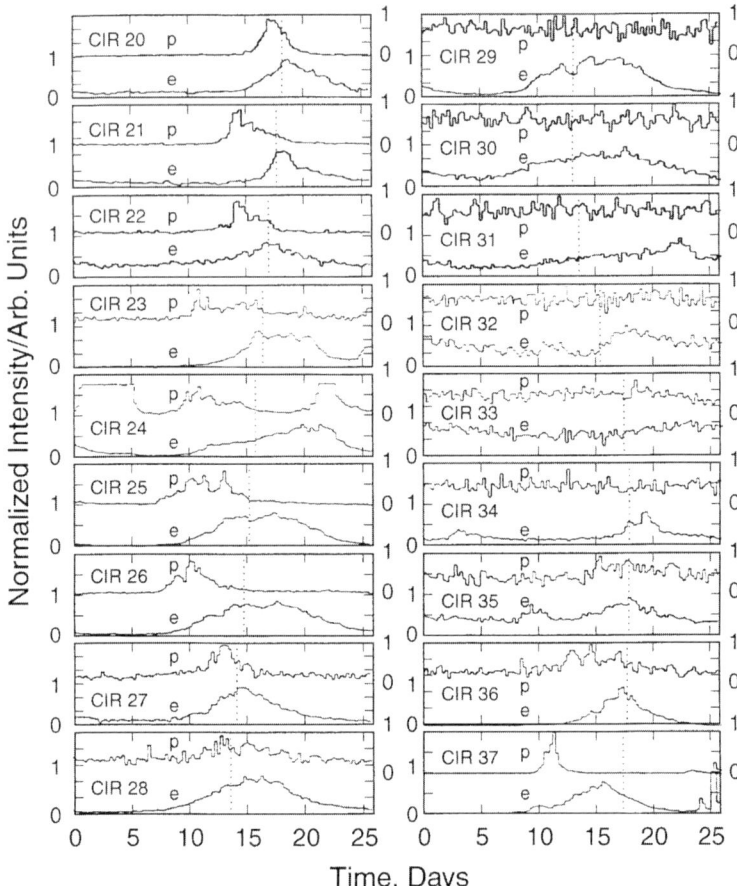

Figure 6.6. Relative timing of electron (50 keV) and proton (0.5 MeV) fluxes during 18 CIR-associated events during the southern heliosphere pass of Ulysses (from Roelof et al., 1996).

predictions of each model: the observed spectra were harder than predicted. They concluded that this was caused by either a more energetic seed population than used in the models, or an enhanced level of magnetic-field fluctuations near the trailing edges of the CIRs.

The relative timing between the CIR-associated enhancements of 50 keV electron fluxes and ~0.5 MeV proton fluxes is shown in Figure 6.6 for many of the CIRs of Figure 6.5, as well as for particle increases that occurred at higher heliolatitudes (Roelof et al., 1996). For *Ulysses* heliolatitudes below about 30°S, electron and proton enhancements occurred approximately simultaneously with the passage of a shock (e.g. CIRs #20, 21). Once *Ulysses* had climbed beyond the streamer belt in the southern hemisphere, the electron fluxes were considerably delayed with respect to the onset of the proton enhancement (e.g. CIRs #26, 31).

Similarly, Figure 6.7 shows, for CIR's 9 to 20 in 1993 (identified in Bame et al., 1993), the relative timing of the CIR-associated enhancements of ~1 MeV protons

Sec. 6.2] Sources of energetic particles 269

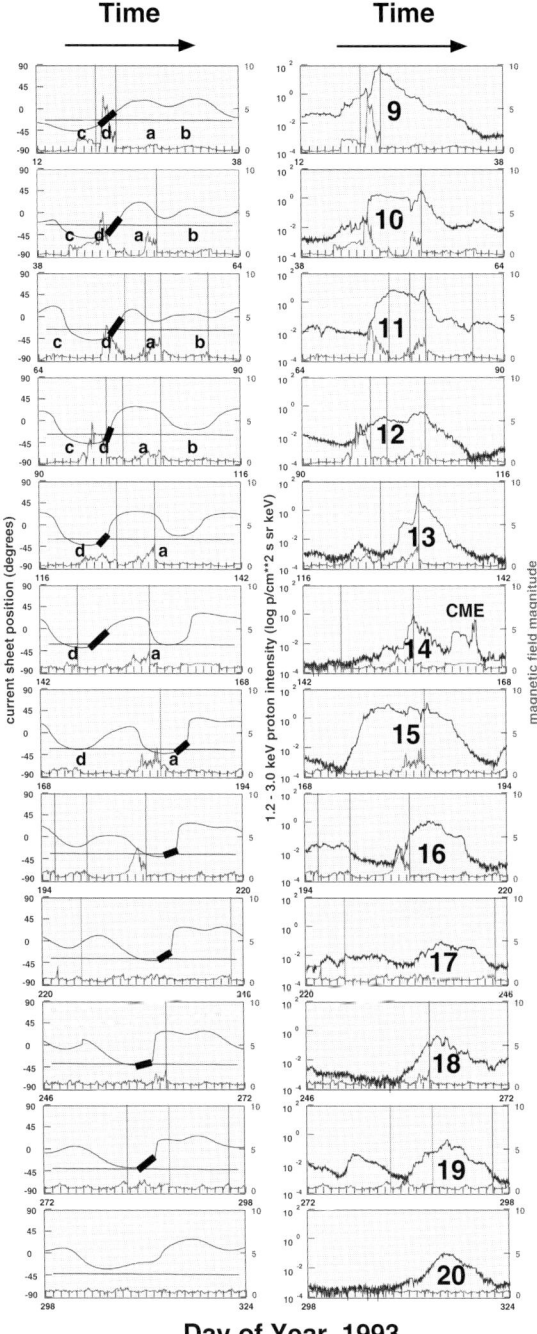

Day of Year, 1993

Figure 6.7. Twelve solar rotations (each of 26 days) beginning on 12 January 1993 during the southern heliosphere pass. Left column of panels: blue traces indicate the heliosphere current sheet at the location of *Ulysses*; the black, nearly horizontal, lines indicate the heliolatitude of *Ulysses* in each rotation. Red traces in all panels indicate the magnitude of the local interplanetary magnetic field measured at *Ulysses*. The right column of panels plots the intensities of 1.8–3.8 MeV protons as blue traces. Vertical lines in the panels show the observed times of reverse shocks. Solid bars on the heliospheric current sheet traces in the left hand column show the expected positions of the interaction regions using the model of Gosling *et al.*, (1993) (from Sanderson *et al.*, 1999).

measured at *Ulysses* (right hand set of panels), and the current sheet as observed on the Sun but ballistically projected outward to the location of *Ulysses* (blue trace, left hand set of panels (Sanderson *et al.*, 1999). The lower case letter designations in several of the panels in the left hand column are different shock compression regions as per the identification scheme of Smith *et al.* (1993). The local *Ulysses*-measured magnetic field magnitude is shown as the red traces in both sets of panels. The latitude of *Ulysses* during each solar rotation is indicated by the nearly horizontal black line in each of the left hand panels. There is an excellent association between the projected position of the current sheet at *Ulysses* and the occurrence of a measured CIR with its associated proton increase. The Sanderson *et al.* (1999) study showed how the active regions on the Sun that are responsible for the warps of the current sheet were responsible for the changes in the time of occurrence of the proton increases measured at *Ulysses*.

Indeed, at the highest heliolatitudes, where the CIRs were not measurable *in situ* by the plasma and magnetic field instruments on *Ulysses*, electron enhancements were evident to the highest latitudes that *Ulysses* measured (Lanzerotti *et al.*, 1995a). Interpretations of this feature of CIR-induced enhancements of IP particle fluxes have been made in terms of the connection of the measurement location to the poleward-moving reverse shock in the CIR (Roelof *et al.*, 1996) and to the motions of the footprints, at the Sun, of the interplanetary magnetic field lines that connect to the CIRs (Fisk, 1996; Kóta and Jokipii, 1998). As *Ulysses* descended from the north polar heliosphere (during the decline of the solar cycle to the minimum solar activity), the CIR-associated enhancements were in general far less evident than during the ascent of the spacecraft to the south polar region (Lanzerotti *et al.*, 1997).

One of the reasons that the CIR-associated increases were not so pronounced during this descent from the northern solar polar regions to the equator was that the heliosphere current sheet was much flatter than during the *Ulysses* southern hemisphere trip. Figure 6.8 (Sanderson *et al.*, 1999) clearly shows this current sheet effect, when compared with the southern hemisphere measurements in Figure 6.7. In Figure 6.8, the numbering scheme again follows that introduced in Bame *et al.*, 1993. In the panels of the left hand column, the black bars indicate local interaction regions (as in Figure 6.7), while the red bars indicate possible interaction regions remote from the spacecraft. The remainder of the format of Figure 6.8 is similar to that of Figure 6.7. The northern hemisphere trajectory depicted in Figure 6.8 occurred almost entirely during the solar minimum conditions of cycle 22.

The origins of the particles ('seed particles') that are accelerated to the IP enhancements such as those in Figure 6.6–8 are still very much debated. It seems clear that there are likely several origins, including solar-wind particles, few tens of keV solar particles, ions whose origins are interplanetary grains, and ions whose origins are interstellar atoms. This latter source will be discussed further in the next section. The importance of the solar-wind origin was explored explicitly in studies of a particle shock event measured at about 4.5 AU on *Ulysses* in the southern hemisphere (Gloeckler *et al.*, 1994). The particle phase space distribution function for the measurements made behind a forward shock of a CIR is shown in Figure 6.9 (Gloeckler *et al.*, 1994). There is a smooth distribution from thermal energies to

Sec. 6.2] Sources of energetic particles 271

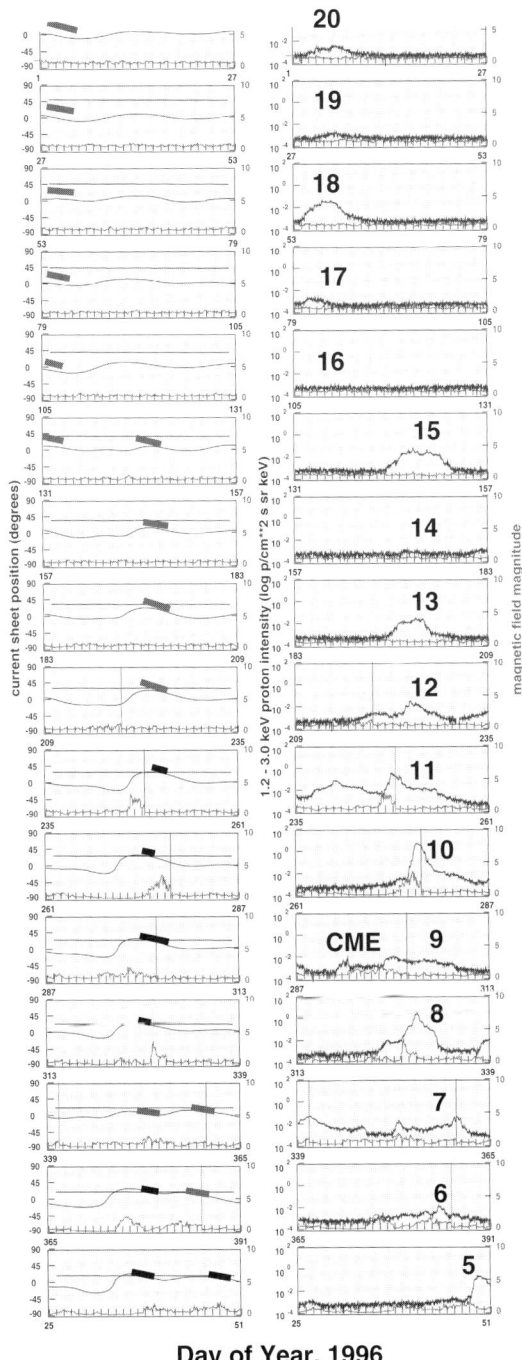

Figure 6.8. Sixteen solar rotations beginning 1 January 1996 during the northern heliosphere pass. Similar format as Figure 6.2.5 (from Sanderson *et al.*, 1999).

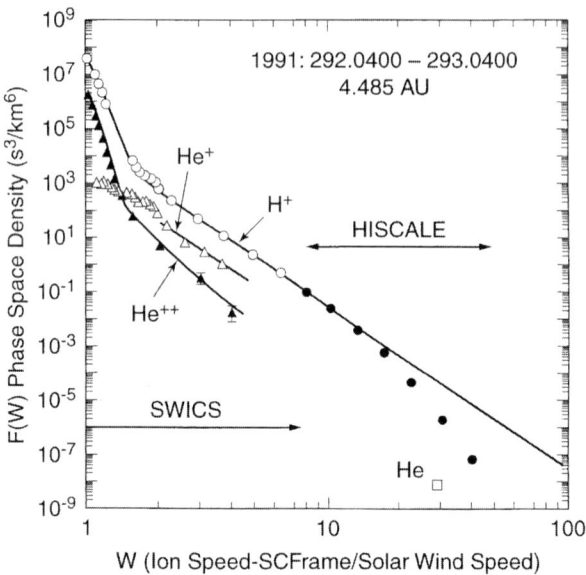

Figure 6.9. Velocity distribution functions measured by two instruments on *Ulysses* (SWICS and HI-SCALE) in the spacecraft frame of reference following a forward shock. Pickup ions from interstellar H and He are shown as H+ (open circles), He+ (open triangles), and He++ (solid triangles) (from Gloeckler *et al.*, 1994)

MeV/nucl energies. The interstellar pickup H^+ and He^+ ions are key contributors to the seed population that is accelerated to the MeV/nucl range. A two step process for the acceleration was suggested (Gloeckler *et al.*, 1994), where the pickup ions are first accelerated to higher energies within a CIR and then reach the higher energies by an acceleration process at the shock that bounds the CIR.

The composition in CIR events has been discussed by many authors, and a useful table of references in this regard, concentrating on *Ulysses* contributions, is included in Keppler, 1998. Very often the composition of CIR-accelerated ions, including ions that are not associated with interstellar atoms, can be reasonably well organized by a single characteristic velocity of the phase space distributions (e.g. Gloeckler *et al.*, 1979). This is shown in Figure 6.10 for two distribution functions during an event measured by Ulysses near the ecliptic plane at 5.2 AU (Lanzerotti and Maclennan, 1995). The particle distributions have essentially the same characteristic exponential velocity ($\sim 3000 \text{ km s}^{-1}$) over the ion velocity range from $\sim 1-\sim 3 \times 10^4 \text{ km s}^{-1}$.

6.2.3 Low energy anomalous cosmic rays

The higher energy (20 MeV/nucl) Anomalous Cosmic Rays (ACRs) are generally considered to be produced by the acceleration of the ionized interstellar components at the heliopause (Fisk *et al.*, 1974; Pesse *et al.*, 1981). However, there is no reason not to expect that lower energy ACRs that are measured in the heliosphere would

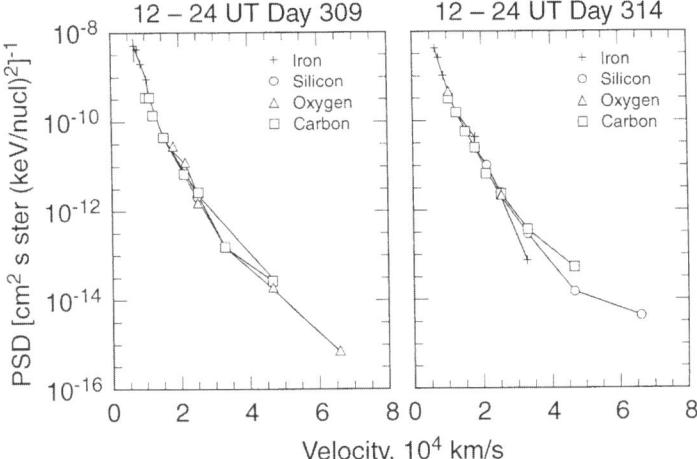

Figure 6.10. Velocity distribution functions for four ion species measured during two different CIR events, showing essentially the same characteristic exponential velocity over a wide velocity range (from Lanzerotti and Maclennan 1995)

not be energized by acceleration processes in association with such interplanetary phenomena as CIRs and travelling shocks. *Ulysses* was the first IP spacecraft to carry instrumentation that could examine in detail the time dependence of the ACR fluxes at energies in the hundreds of keV/nucl. to several MeV/nucl range, an energy region of considerable importance for understanding the physics of such cosmic rays in the heliosphere.

Shown in the top and bottom panels, respectively, of Figure 6.11 are ACR oxygen fluxes (3.5–6.8 MeV/nucl) and CIR-enhanced proton fluxes (0.43–0.83 MeV) for an interval that spans nine of the CIR events shown in Figures 6.5 and 6.6 (Reuss *et al.*, 1996). The centre panel plots the galactic cosmic ray protons measured on *Ulysses* during this interval. The horizontal bars in the top panel connect maxima of the recurrent ACR variations with succeeding minima. The fluxes are drawn with thin lines during those time intervals when the recurrent ACR variations are absent. There is a distinct anti-association between the galactic cosmic rays and the ACR variations, and the CIR-produced low energy proton enhancements (Reuss *et al.*, 1996). This anti-association is attributed to the exclusion of these energy ACR fluxes by the CIR magnetic structures (McKibben, 2001). For ions with energies below ~ 1 MeV/nucl, however, there is some evidence for the influence of CIR acceleration on ACR O fluxes (Lanzerotti *et al.*, 1995b).

Ulysses, at high south and north heliolatitudes beyond the latitudes where CIR protons were observed, provided the first measurements of the fluxes of low energy ACR ions. Energy spectra for ACR fluxes of O, N, and Ne for the energy range 0.5–8 MeV/nucl are shown in Figure 6.12 (Maclennan and Lanzerotti, 1998). These spectra, for measurements made at latitudes greater than 60°, are basically flat over this energy range. Also of significance is the fact that the fluxes of ACR O, N, and Ne

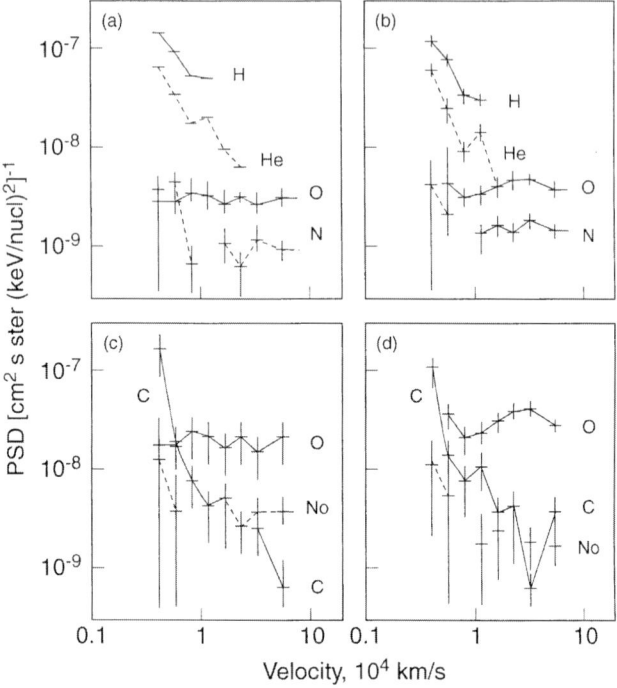

Figure 6.11. Relationship between the intensities of anomalous cosmic ray oxygen fluxes (3.5–6.8 MeV/nucl; top panel) and variations in galactic cosmic ray intensities (middle panel) during an interval of nine CIR events (indicated by the intensities of 0.43–8.1 MeV/nucl He fluxes; bottom panel) (from Reuss et al., 1996)

in this energy range are about a factor of two larger in the northern high latitudes than during the southern high latitude pass of Ulysses. This has been attributed to the easier access of ACR ions to the heliosphere inside 5 AU during the lower solar activity conditions that prevailed during the Ulysses northern hemisphere excursion than during the southern high latitude pass (Maclennan and Lanzerotti, 1998). This finding indicates that even though it is likely that CIR events can accelerate ACR ions (see above), substantial fluxes of the ACRs can come from outside 5 AU during intervals of lower lower solar activity (consistent with the interpretation of the associations in Figure 6.12).

While Maclennan and Lanzerotti (1998) concluded that their results were consistent with no heliolatitude gradient in ACR fluxes during the northern high latitude pass, Trattner et al. (1997) reported a positive latitudinal gradient of $\sim$2 per cent/degree for ions in the energy range 4–8 MeV/nucl at latitudes between 20° and 60° in the northern hemisphere. Marsden et al., (1999) also reported a positive gradient in the northern hemisphere. This latitudinal gradient was derived by making appropriate corrections for the radial gradient over the AU range that Ulysses covered. It is possible, given the smallness of both of these gradients, as well as the uncertainties

Sec. 6.3] Particle propagation and transport 275

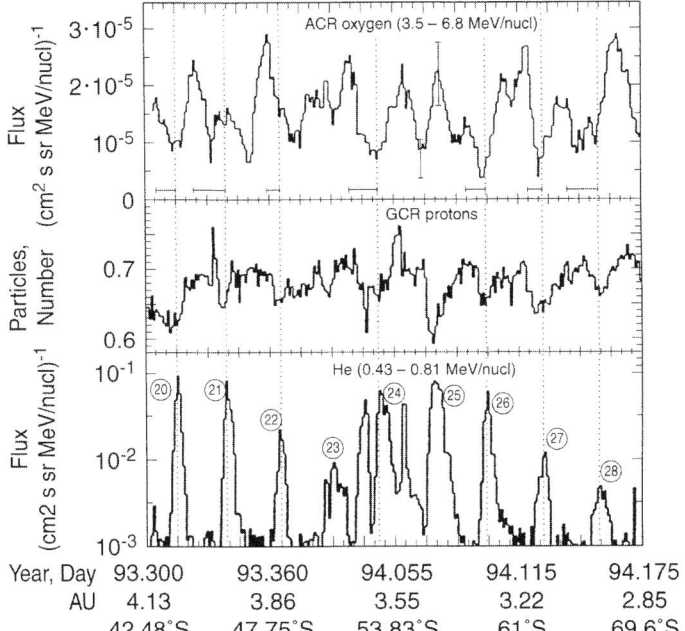

Figure 6.12. Anomalous cosmic ray oxygen, nitrogen, and neon fluxes for heliographic latitudes $>60°$S (panels (a) and (c)) and $>60°$N (panels (b) and (d)). Also shown are the energy spectra for protons, alpha particles, and carbon at these heliolatitudes (from Maclennan and Lanzerotti 1998).

that enter into the magnitude of the radial gradient, that the conclusion of Maclennan and Lanzerotti (1998) could be adjusted to agree with that of Trattner et al., (1997). The presence of a stable, positive latitudinal gradient is further supported by observations made in the southern hemisphere during the FLS (Trattner et al., 1996).

6.3 PARTICLE PROPAGATION AND TRANSPORT

The propagation of solar origin charged particles throughout the solar system in the magnetic field-filled interplanetary medium has been one of the central problems of heliospheric physics since the classic study of the large solar particle event of February 1956 (Meyer et al, 1956). Occurring prior to the space age, this event was energetic enough to be measurable with radiation detection instruments on the surface of Earth. The discussions in this paper, and the classic theoretical description of the interplanetary medium (Parker, 1963), provided a strong basis for the subsequent multi-decade examination of the nature of the interplanetary medium, including the fluctuations and the structure of the magnetic fields that influence charged particle motion. The theoretical treatments of the propagation of solar particles in the solar system are intimately linked to the problems associated with the entry of galactic cosmic rays into the heliosphere and their subsequent propagation (that is, the heliospheric modulation of galactic cosmic rays). This subject is addressed in (McKibben, 2001).

Prior to the *Ulysses* mission, studies of solar particle propagation from the Sun, including the solar corona, into the heliosphere had been carried out by instrumentation within about 30° of ecliptic plane. The *Ulysses* mission thus has provided an opportunity to study solar-origin particles that might be observed over all heliolatitudes and to derive new understanding of the propagation of these particles throughout the three-dimensional heliosphere. This section addresses several specific topics to which *Ulysses* has contributed.

6.3.1 Latitudinal particle transport

As was noted in Section 6.2, there was little solar activity that produced significant interplanetary particle events during the solar-minimum interval of the first *Ulysses* orbit of the Sun. Nevertheless, during the ascent of the spacecraft to the southern solar pole, three transient proton (~1 MeV energy) events were reported (Bothmer *et al.*, 1995; Sanderson *et al.*, 1996). The third of these events was measured at a heliolatitude of ~60°S. As discussed in Section 6.2.1, all three of these events were associated with the passage of a CME event over the spacecraft.

Solar-origin electrons were reported detected at latitudes above 50°S (Pick *et al.*, 1995a), and at latitudes as high as ~74°S (Pick *et al.*, 1995b). The event at 74°S was shown to be associated with an interplanetary Type III radio emission as measured by *Ulysses*. A careful analysis of Earth-based solar radio data (from the Nancay Radioheliograph) demonstrated that the origin was at 6°S on the Sun. Figure 6.13 plots the electron time-intensity profile at *Ulysses* for this event. The pitch-angle distribution of the electrons (inset in figure 6.13) shows the strong streaming away from the solar direction. In this inset the vertical scale is the sum of the electron fluxes in two detectors normalized to the fluxes measured at 180° pitch angle.

Pick *et al.*, (1995b) state that the event at 74°S, which was of quite low intensities, was detectable because the interplanetary background was not obscured by electrons that were produced in CIRs at the time of the solar activity. We favour an electron acceleration process in the low or middle corona for the event, followed by coronal propagation near the solar pole. Until the solar-maximum orbit of *Ulysses* (which is just beginning), this will remain the best example of the large heliolatitude range that is possible for particle propagation in the solar corona.

6.3.2 Propagation of charged particles in interplanetary structures

Theoretical research related to particle transport in the heliosphere has largely concentrated on treating the interplanetary medium as a totally stochastic environment. Random fluctuations of the interplanetary magnetic field and random fluctuations of the footpoints of magnetic fields tied to the solar surface are generally

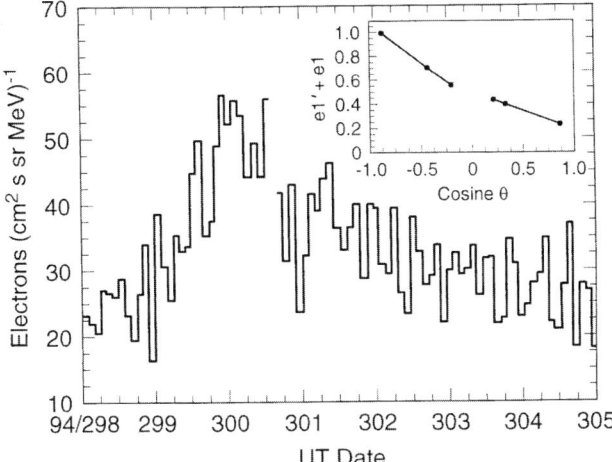

Figure 6.13. Solar electron (42–65 keV) event observed at ~74°S heliolatitude during days 298–305, 1994. The inset shows the cosine of the pitch angle of the electrons near the peak (normalized) in intensity of the event. The maximum in this distribution at cosine = −1 indicates electron propagation away from the sun at this time (from Pick *et al.*, 1995a).

invoked to explain the observations. Nevertheless, it has been recognized for some time that there are often distinct structures in the medium that seem to provide preferential particle propagation paths (e.g. Lin *et al.*, 1981; Anderson and Dougherty, 1986). These plasma structures are rooted in the solar corona and appear to 'channel' the propagation of solar-origin electrons. Indeed, electrons that propagate in these channels generally have quite long mean free paths. Such structures must be reasonably long lived, or the 'channeling' would not be observed. Such long-lived structures are not consistent with contemporary propagation theory.

Until the flight of *Ulysses*, such propagation channels had only been studied in the vicinity of 1 AU. Using *Ulysses* data, the features of several electron-propagation channels were examined to distances of nearly 5 AU (Buttighoffer *et al.*, 1995; Buttighoffer, 1998). Maia *et al.*, (1998) presented observations from 4.7 AU in the ecliptic plane of both electron and ion propagation inside a channelling structure that was itself embedded within a CIR. These observations show, most importantly, that apparently 'stable' interplanetary structures, rooted in the solar corona, can persist at least to these distances. For a solar wind velocity of the order of 500 km s^{-1}, this implies that such channels have life times of at least several days. As noted above, such long-lived structures are certainly not consistent with propagation theory.

Shown in Figure 6.14 (Buttighoffer, 1998) are order 50-keV electrons (panel e) that are detected in a propagation channel in the ecliptic plane after the *Ulysses* spacecraft has crossed into it at a distance of 4.3 AU. The geometry of the measurements (including, in other Panels, solar wind and interplanetary magnetic field

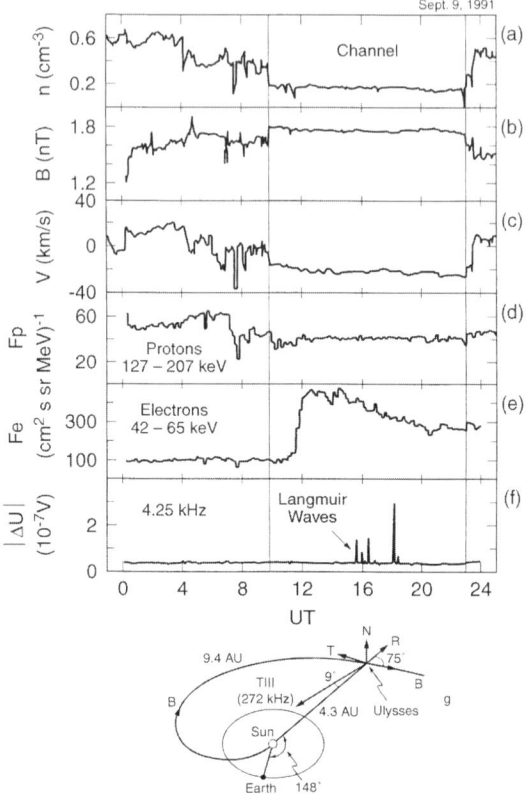

Figure 6.14. Propagation of electrons (42–65 keV; panel (d)) in an interplanetary propagation channel denoted by the shading in the figure. The solar wind number density (panel (a)) and velocity (panel (c)) and interplanetary magnetic field (panel (b)) are also shown. No enhancement in low energy protons (panel (d)) were seen. Langmuir waves (4.25 kHz; panel (f)) accompanied the electrons. The sketch below the data panels shows the geometry of the interplanetary medium at the time of the channel event (from Buttighoffer 1998).

parameters) is shown in the sketch at the bottom of the figure. Langmuir waves (panel f) were associated with the streaming electrons.

It is evident visually in panel b of Figure 6.14 that the interplanetary magnetic field has less fluctuations during the interval in which the channel is measured than outside of the channel. A quantitative examination of this feature for nine selected cases showed that low magnetic-field fluctuations are the rule for the interior of channels, consistent with the conclusion of nearly scatter-free propagation (Buttighoffer, 1998).

6.4 PERIODIC STRUCTURES IN PARTICLE FLUX TIME SERIES

As discussed in Sections 6.1 and 6.2.2, the most striking feature of the energetic-particle fluxes that were measured during the solar-minimum phase of the *Ulysses* mission is the quasi-periodic variation in particle intensities. The quasi-period is of the order of the solar rotation period, ~26 days. Indeed, during the ascent of *Ulysses* to the southern solar pole, periodic variation in the particle fluxes was shown to be clearly related to the periodic magnetic structure of the solar corona as the Sun rotates (Bai *et al.*, 1997). Variation with periods similar to the energetic particles were seen as well in galactic cosmic ray data on *Ulysses* (e.g. Simpson *et al.*, 1995; Reuss *et al.*, 1996).

However, the variation in particle fluxes appears to be more complex than the visual examinations would suggest. For example, Blake *et al.* (1997) provide evidence that the period of variation of the galactic cosmic rays as measured on *Ulysses* can change with time. Indeed, they find a dominant period of about 29 days (rather than about 27 days) for a 5-month interval in the beginning of 1994 (CIR intervals #23–27 in Figure 6.5) when *Ulysses* was at 50–60°S heliolatitude. This was also an interval when there was a slowing, or even hiatus, in the increase of the galactic cosmic ray flux as the particle modulation decreased with the decline of the solar cycle.

Thomson *et al.*, (1997) reported the results of a time-series analysis of the solar-rotation period and its first few 'harmonics' using *Ulysses* low-energy ion data for approximately the first 240 days of 1993 (see Figure 6.2). At this time, *Ulysses* traversed the latitude range from ~23–40°S. As in the Blake *et al.*, (1997) study, Thomson *et al.*, (1997) reported 'anomalous' results from their analysis. In particular, they reported that in the frequency interval 0–0.5 cycles day^{-1} there were several spectral enhancements, but that these did not occur at the frequencies expected for harmonics of the fundamental (which was at 26.5 days). As well, some of the 'harmonics' were not evident in some of the spectra.

The time interval that was studied by Blake *et al.* (1997) was included in the several sets of data (from the *Voyager* spacecraft as well as from *Ulysses*) that was used in a time-series study of particle fluxes (and some interplanetary magnetic-field data from *IMP-8* and *ISEE-3* spacecraft) (Thomson *et al.*, 1995). These authors reported finding, very surprisingly, numerous periodic components in the frequency range 1–140 mHz. They noted that these spectral components are consistent with those estimated (but not confirmed) from gravity-mode oscillations at the sun. They also reported many frequencies in the solar pressure mode range (1,000–4,000 mHz) that matched well reported p-mode frequencies.

Shown in Figure 6.15 is a section of a power spectrum calculated by Thomson of the time series of 38–53-keV electron fluxes from the HI-SCALE instrument for the southern-hemisphere time interval just after Jupiter encounter to day 175 1994 (Ladbury, 1995). Numerous spectral peaks are evident in the frequency band shown, 60–90 μHz. The spectrum was determined from analysis of three overlapped sections of data in the entire time interval. The total number of degrees of freedom was ~40. Given the background level in the spectrum (~6 flux^2 Hz^{-1}), the six peaks above ~30 flux^2 Hz^{-1} have a statistical significance of about 22σ.

Needless to say, this report (Thomson *et al.*, 1995) using predominantly *Ulysses* energetic particle data caused considerable debate in the literature. The findings have been challenged on the basis of the physics of the interplanetary medium (i.e. the result in Thomson *et al.*, 1995 is inconsistent with the medium being a totally

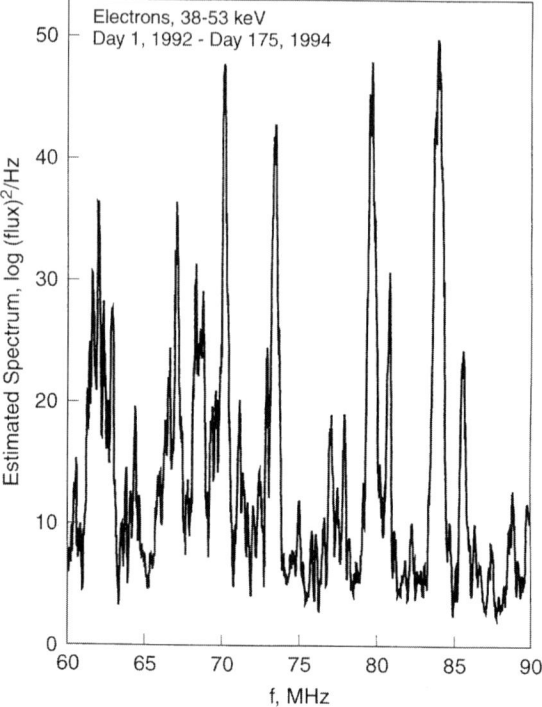

Figure 6.15. Portion of a power spectrum of 38–53 keV electron fluxes calculated with eight prolate spheroidal data windows and a time-band width product of 5.0.

stochastic environment; see e.g. Roberts *et al.*, 1996; Hoogeveen and Riley, 1998) and on the physics of solar g-mode oscillations at the solar surface (Kumar *et al.*, 1996 conclude, theoretically, that the amplitudes of solar-surface oscillations derived from the results in Thomson *et al.*, 1995 are too large for existing understanding). It is not the purpose of this chapter to enter into these debates, but just to make evident that *Ulysses* measurements have contributed to unique investigations of the heliosphere (and possibly the Sun) in ways that were not expected at the conception of the mission. Indeed, the conclusions on g-modes are now cited at times in connection with, for example, discussions of the solar interior (e.g. Fritts *et al.*, 1998) It can be expected that these types of studies, using various *Ulysses* (and other interplanetary) datasets will continue into the future.

6.5 ULYSSES AND SOLAR–TERRESTRIAL CONNECTIONS

In the excitement that has been generated by the new scientific understandings that have arisen from the data acquired by the *Ulysses* mission as it has circled the polar regions of the Sun, another aspect of the mission is frequently overlooked. This aspect comprises the contributions that *Ulysses* makes to understanding Sun–Earth connections. The end goal of much research in Sun–Earth connections is for rather practical purposes, especially in understanding the 'weather' in the space environment around Earth, and the effects of this weather on technological systems (e.g. see Lanzerotti *et al.*, 1999). The contribution of *Ulysses* to this subject is basically forward-looking research that is critical for obtaining fundamental knowledge about Sun–Earth connections.

If *Ulysses* were not present as an element of a present-day fleet of spacecraft to study the Sun, the heliosphere, and Earth during both solar-minimum and solar-maximum conditions, a 'solar polar' mission would have to be developed and flown. Without *Ulysses*, understanding of the impacts of the global heliosphere on Earth would be only partial knowledge, a kind of 'flat-land' knowledge of the heliosphere. This partial knowledge base would be totally insufficient to ultimately build complete and usable models of the Sun–Earth system for practical purposes.

Essentially all of *Ulysses* investigations contribute to some extent to the understanding of the three-dimensional heliosphere for practical purposes. These contributions have been briefly outlined in (Lanzerotti *et al.*, 1999). Energetic particles as measured on *Ulysses* (the subject of this chapter) are important elements in numerous practical applications. For example, the energetic electron and ion fluxes that are produced by solar activity can fill the heliosphere. Those that propagate to Earth can access the geosynchronous orbital location of communication satellites and can produce damage to the spacecraft electronics and solar-cell arrays.

The finding that low-energy electrons produced by solar processes associated with activity near the Sun's equatorial regions can, with little time delay, travel to high solar latitudes was a surprise (see Section 6.3.1). Thus, near-equatorial solar activity contributes importantly to establishing the interplanetary charged-particle populations, even at very high solar latitudes. This has important implications for the *Ulysses* mission during solar maximum, when solar activity can occur at high to mid-latitudes. Energetic particle observations that will be made by *Ulysses* at high heliolatitudes at these times, complemented by in-ecliptic measurements, will be critical to gain an understanding of the propagation of these particles to the ecliptic plane, where they can impact Earth's space environment and the technological systems contained therein.

A surprising finding during the recent solar-minimum interval was the effect of CIRs in producing magnetosphere disturbances that 'pumped up' the relativistic electron populations of Earth's outer magnetosphere. These electron enhancements could persist for days at a time, during which several serious anomalies occurred on communication satellites (e.g. Baker *et al.*, 1994; 1998). While CIR-related phenomena may be considerably reduced in importance during the solar maximum phase of the *Ulysses* mission, the effects on the Earth's space environment of solar transient phenomena, such as fast/slow streams, travelling shocks with accompanying particle enhancements, and CMEs, may be similar to CIRs in some circumstances. Thus, *Ulysses* measurements will be critical in providing the three-dimensional structure of such interplanetary disturbances, especially for applications to new models that can be used for predictions and for elucidation of space-weather phenomena.

6.6 SUMMARY

The foregoing sections have outlined the many new findings about the three-dimensional heliosphere that have resulted from low-energy particle investigations on the *Ulysses* spacecraft during its 360° out-of-the-ecliptic circuit of the Sun during solar-minimum conditions. These findings are numerous. They range from discoveries of the particle populations of the heliosphere at the highest heliolatitudes to understandings of the acceleration of charged particles in CIRs at all heliolatitudes to understandings of the relationships of these acceleration regions to the latitudes touched by coronal-hole boundaries. They include understandings of ACRs to the lowest energies measured to date, and the identification in the heliosphere of discrete spectral structures in the time series of particle fluxes. *Ulysses* low-energy particle (and other instrument) measurements have even contributed importantly to achieving new understandings of solar–terrestrial interactions through the new insights that have been obtained of the structure of the high-latitude heliosphere and its influence on the Earth's environment. The *Ulysses* mission has been an outstanding success in revolutionizing our understanding of the solar system.

6.7 ACKNOWLEDGEMENTS

We thank our many colleagues on the *Ulysses* COSPIN-LET, HI-SCALE and EPAC instrument teams, as well as all the *Ulysses* investigators, for their many stimulating discussions and debates over the years about the exciting new scientific insights that our instruments, and the mission, are producing.

6.8 REFERENCES

Anderson, K. A. and Dougherty, W. M. (1986) A spatially confined long-lived stream of solar particles. *Solar Phys.* **103**, 165.

Armstrong, T. P., Haggerty, D., Lanzerotti, L. J., Maclennan, C. G., Roelof, C., Pick, M., Gimnett, G. M., Gold, R. E., Krimigis, S. M., Anderson, K. A., Lin, R. P., Sarris, E. T., Forsyth, R. and Balogh, A. (1994) Observations by Ulysses of hot ($\sim$270 keV) coronal particles at 32° south heliolatitude and 4.6 AU. *Geophys. Res. Lett.* **21**, 1747.

Bai, T., Hoeksema, J. T., Weber, M. and Acton, L. W. (1997) Solar origin of the 26-day periodicity observed by Ulysses. *J. Geophys. Res.* **102**, 9793.

Baker, D. N., Allen, J. H., Kanekal, S. G. and Reeves, G. D. (1998) Disturbed space environment may have been related to pager satellite failure. *Eos, Trans. Am. Geophys. Union*, **79**, 477.

Baker, D. N., Kanekal, S., Blake, J. B., Klecker, B. and Rostoker, G. (1994) Satellite anomalies linked to electron increase in the magnetosphere. *Eos, Trans. Am. Geophys. Union*. **75**, 401.

Balogh, A., Marsden, R. G. and Smith, E. J. (2001) Introduction. In *The Heliosphere Near Solar Minimum: The Ulysses Perspective*, Chapter 1.

Bame, S. J., Goldstein, B. E., Gosling, J. T., Harvey, J. W., McComas, D. J., Neugebauer, M. and Phillips, J. L. (1993) Ulysses observations of a recurrent high speed solar wind stream and the heliomagnetic streamer belt. *Geophys. Res. Lett.* **20**, 2323.

Barnes, C. A. and Simpson, J. A. (1976) Evidence for interplanetary acceleration of nucleons in corotating interacting regions. *Ap. J. Lett.* **210**, L91.

Blake, J. B., Looper, M. D., Keppler, E., Heber, B., Kunow, H. and Quenby, J. J. (1997) Ulysses observations of short-period ($\leq$30 days) modulation of the galactic cosmic rays. *Geophys. Res. Lett.* **24**, 671.

Bothmer, V., Desai, M. I., Marsden, R. G., Sanderson, T. R., Trattner, K. J., Wenzel, K.-P., Gosling, J. T., Balogh, A., Forsyth, R. J. and Goldstein, B. E. (1996) Ulysses observations of open and closed magnetic field lines within a coronal mass ejection. *Astron. Astrophys.* **316**, 493.

Bothmer, V., Marsden, R. G., Sanderson, T. R., Trattner, K. J., Wenzel, K.-P., Balogh, A., Forsyth, R. J. and Goldstein, B. E. (1995) The Ulysses south polar pass: Transient fluxes of energetic ions. *Geophys. Res. Lett.* **22**, 3369.

Boufaida, M., and Armstrong, T. P. (1985) Observations of 0.3 to 2.0 MeV protons in the 1 to 5 AU region with Voyager 1 and Ulysses compared to IMP8. *Proceedings 23rd ICRC*, **3**, 109.

Buttighoffer, A. (1998) Solar electron beams associated with radio type III bursts: Propagation channels observed by Ulysses between 1 and 4 AU. *Astron. Astrophys.* **335**, 295.

Buttighoffer, A., Pick, M., Roelof, E. C., Hoang, S., Mangeney, A., Lanzerotti, L. J., Forsyth, R. J. and Phillips, J. L. (1995) *J. Geophys. Res.* **100**, 3369.

Chapman S. and Bartels, J. (1940) *Geomagnetism*, Vol. 2., Oxford Univ. Press.

Christon, S. P. (1981) On the origin of the MeV energy nucleon flux associated with CIRs. *J. Geophys. Res.*, **86**, 8852.

Desai, M. I., Marsden, R. G., Sanderson, T. R., Balogh, A., Forsyth, R. J. and Gosling, J. T. (1997) Particle acceleration at corotating reverse shocks in the southern heliosphere: Ulysses results. *Geophys. Res. Lett.* **24**, 1155.

Desai, M. I., Marsden, R. G., Sanderson, T. R., Balogh, A., Forsyth, R. J., Gosling, J. T., (1998) Particle acceleration at corotating interaction regions in the three-dimensional heliosphere. *J. Geophys. Res.* **103**, 2003.

Desai, M. I., Marsden, R. G., Sanderson, T. R., Lario, D., Roelof, E. C., Simnett, G. M., Balogh, A., Forsyth, R. J. and Gosling, J. T. (1999) Energy spectra of 50 keV to 20-MeV protons accelerated at Corotating Interaction Regions at Ulysses. *J. Geophys. Res.* **104**, 6705.

Fisk L. A., and Lee, M. A. (1980) Shock acceleration of energetic particles in corotating interaction regions in the solar wind. *Astrophys. J.* **237**, 620.

Fisk, L. A. (1996) Motion of the footpoints of heliospheric magnetic field lines at the Sun: Implications for recurrent energetic particle events at high heliolatitudes. *J. Geophys. Res.* **101**, 15547.

Fisk, L. A., Kozlovsky, B. and Ramaty, R. (1974) An interpretation of the observed oxygen and nitrogen enhancements in low energy cosmic rays. *Ap. J. Lett.* **190**, L35.

Forsyth, R. J., and Gosling, J. T. (2000) Corotating and transient structures in the heliosphere. In: *The Heliosphere at Solar Maximum: The Ulysses Perspective*. Chapter 3.

Fritts, D. C., Vadas, S. L. and Andreassen, O. (1998) Gravity wave excitation and momentum transport in the solar interior: Implications for a residual circulation and lithium depletion. *Astron. Astrophys.* **333**, 343.

Gloeckler, G., Geiss, J., Roelof, E. C., Fisk, L. A., Ipavich, F. M., Ogilvie, K. W., Lanzerotti, L. J., von Steiger, R. and Wilken, B. (1994) Acceleration of interstellar pickup ions in the disturbed solar wind observed on Ulysses. *J. Geophys. Res.* **99**, 17637.

Gloeckler, G., Hovestadt, D. and Fisk, L. A. (1979) Observed distribution functions of H, He, C, O, and Fe in corotating energetic particle streams: Implications for interplanetary acceleration and propagation. *Ap. J.* **230**, L191.

Gosling, J. T. (1994) A forward-reverse shock pair in the solar wind driven by over-expansion of a coronal mass ejection: Ulysses observations. *Geophys. Res. Lett.* **21**, 237.

Hoeksema, J. T., Wilcox, J. M. and Scherrer, P. H. (1982) Structure of the heliospheric current sheet in the early portion of sunspot cycle 21. *J. Geophys. Res.* **87**, 10331.

Hoogeveen, G. W. and Riley, P. (1998) The search for solar gravity-mode oscillations in the solar wind using Ulysses plasma data. *Solar Phys.* **179**, 167.

Jones, F. C. and Ellison, D. C. (1991) The plasma physics of shock acceleration. *Space Sci. Rev.* **58**, 259.

Keppler, E. (1998) The acceleration of charged particles in corotating interaction regions (CIR) – A review with particular emphasis on the Ulysses mission. *Surveys in Geophys.* **19**, 211.

Keppler, E., Fraenz, M., Krupp, N. and Reuss, M. K. (1995) Energetic particle observations in the high heliographic latitudes. *Nucl. Phys. B (Proc. Suppl.)*, **39A**, 87.

Kóta, J. and Jokipii, J. R. (1998) Modelling of 3-D corotating cosmic ray structures in the heliosphere. *Space Sci. Rev.* **83**, 137.

Kumar, P., Quataert, E. J. and Bahcall, J. N. (1996) Observational searches for solar g-modes: Some theoretical considerations. *Ap. J.* **458**, L83.

Ladbury, R. (1995) Is the answer blowing in the solar wind? *Physics Today*. **48**, 17, September.

Lanzerotti, L. J. (1998) Ulysses and Sun-Earth connections. *Ulysses Mission Bulletin No. 8*, NASA Jet Propulsion Laboratory, Pasadena, CA, pg. 2, March.

Lanzerotti, L. J., and Maclennan, C. G. (1995) Study of distribution functions of interplanetary particles accelerated at co-rotating interaction region at ∼5 AU. *Space Sci. Rev.* **72**, 335.

Lanzerotti, L. J., Gold, R. E., Anderson, K. A., Armstrong, T. P., Lin, R. P., Krimigis, S. M., Pick, M. Roelof, E. C., Sarris, E. T., Simnett, G. M. and Frain, W. E. (1992) Heliosphere instrument for spectra, composition, and anisotropy at low energies. *Astron. Astrophys. Suppl. Ser.* **92**, 349.

Lanzerotti, L. J., Armstrong, T. P., Gold, R. E., Maclennan, C. G., Roelof, E. C., Simnett, D. J., Thomson, G. M., Anderson, K. A., Hawkins, S. E., Lin, R. P., Pick, M., Sarris, E. T. and Tappin, S. J. (1995a) Over the solar pole: Low energy interplanetary charged particles. *Science*, **268**, 1010.

Lanzerotti, L. J., Maclennan, C. G., Gold, R. E., Armstrong, T. P., Roelof, E. C., Krimigis, S. M., Simnett, G. M., Sarris, E. T., Anderson, K. A., Pick, M. and Lin, R. P. (1995b) Measurements of anomalous cosmic ray oxygen at heliolatitudes ∼25 to ∼64. *Geophys. Res. Lett.* **22**, 333.

Lanzerotti, L. J., Maclennan, C. G., Armstrong, T. P., Roelof, E. C., Gold, R. E. and Decker, R. B. (1997) Low energy charged particles in the high latitude heliosphere. *Adv. Space Res.* **19**, 851.

Lanzerotti, L. J., Thomson, D. J. and Maclennan, C. G. (1999) Engineering issues in space weather, in *Modern Radio Science 1999*, ed. M. Stuchly, Oxford Univ. Press.

Lim, T. L., Quenby, J. J., Reuss, M. K., Keppler, E., Kunow, H., Heber, B. and Forsyth, R. J. (1996) Ulysses observations of energetic particle acceleration and the superposed CME and CIR events of November 1992. *Ann. Geophysicae* **14**, 400.

Lin, R. P., Potter, D. W., Gurnett, D. A. and Scarf, F. L. (1981) Energetic electrons and plasma waves associated with a solar type III radio burst. *Ap. J.* **251**, 364.

Maclennan, C. G. and Lanzerotti, L. J. (1998) Low energy anomalous ions at northern heliolatitudes. *Geophys. Res. Lett.* **25**, 3473, 1998.

Maclennan, C. G., Lanzerotti, L. J., Simnett, G. M. and Sayles, K. A. (1995) Heliolatitude dependence of interplanetrary ions. *Geophys. Res. Lett.* **22**, 3361.

Maia, D., Malandraki, O. Pick, M. Sarris, E. T. Kasotakis, G. Lanzerotti, L. J. Maclennan, C. G. and Trochoutsos, P. C. (1998) Particle propagation channel detected at 4.7 AU inside a corotating interaction region. *J. Geophys. Res.* **103**, 9545.

Marsden, R. G., Desai, M. I., Sanderson, T. R., Forsyth, R. J. and Gosling, J. T. (1997) Effects of the 5 October 1996 CME at 4.4 AU: Ulysses observations. *Proc. 25th Int. Cosmic Ray Conference*, **1**, 337.

Marsden, R.G., Sanderson, T. R., Tranquille, C., Trattner, K. J., Anttila A. and Torsti, J. (1999) On the Gradients of ACR Oxygen at Intermediate Heliocentric Distances: Ulysses/SOHO Results. *Adv. Space Res.*, **23**(3), 531.

Marshall, F. and Stone, E. C. (1978) Characteristics of sunward flowing protons and alpha particle fluxes of moderate intensity. *J. Geophys. Res.* **83**, 3289.

McDonald, F. B., Teegarden, B. J., Trainor, J. H., von Rosenvinge, T. T. and Webber, W. R. (1976) The interplanetary acceleration of energetic nucleons. *Ap. J. Lett.* **203**, L149.

McKibben, R. B. (2001) Cosmic rays at all latitudes in the inner heliosphere. In: *The Heliosphere Near Solar Maximum: The Ulysses Perspective*, Chapter 8.

Mewaldt, R. A., Stone, E. C. and Vogt, R. E. (1978) The radial diffusion coefficient of 1.3–2.3 MeV protons in recurrent proton streams. *Geophys. Res. Lett.* **5**, 965.

Meyer, P., Parker, E. N. and Simpson, J. A. (1956) Solar cosmic rays of February, 1956, and their propagation through interplanetary space. *Phys. Rev.* **104**, 768.

Neugebauer, M. (2001) The solar wind and interplanetary magnetic field in three dimensions. In: *The Heliosphere near Solar Minimum: The Ulysses Perspective*, Chapter 2.

Parker, E. N. (1963) Interplanetary Dynamical Processes. *Interscience*.

Pesses, M. E., Jokipii, J. R. and Eichler, D. (1981) Cosmic ray drift, shock wave acceleration, and the anomalous component of cosmic rays. *Ap. J.*, **246**, L85.

Pesses, M. E., Van Allen, J. A. and Goertz, C. K. (1978) Energetic protons associated with interplanetary active regions 1–5 AU from the Sun. *J. Geophys. Res.* **83**, 553.

Pick, M., Lanzerotti, L. J., Buttighoffer, A., Sarris, E. T., Armstrong, T. P., Simnett, G. M., Roelof, E. C. and Kerdraon, A. (1995a) The propagation of sub-MeV solar electrons to heliolatitudes above 50 S. *Geophys. Res. Lett.* **22**, 3373.

Pick, M., Lanzerotti, L. J., Buttighoffer, A., Hoang, S. and Forsyth, R. J. (1995b) Detection of a solar particle event at an heliolatitude of 73.8 S. *Geophys. Res. Lett.* **22**, 3377.

Pick, M., Buttighoffer, A., Kerdraon, A., Armstrong, T. P., Roelof, E. C., Hoang, S., Lanzerotti, L. J., Simnett, G. M. and Lemen, J. (1995c) Ulysses observations of a coronal origin particle event at 32 south heliographic latitude. *Space Sci. Rev.* **72**, 315.

Reuss, M. K., Fraenz, M. and Keppler, E. (1996) Recurrent variations of anomalous oxygen in association with corotating interaction regions. *Ann. Geophysicae.* **14**, 585.

Roberts, D. A., Ogilvie, K. W. and Goldstein, M. L. (1996) The nature of the solar wind. *Nature.* **381**, 31.

Roelof, E. C., Simnett, G. M. and Tappin, S. J. (1996) The regular structure of shock-accelerated 40–100 keV electrons in the high latitude heliosphere. *Astron. Astrophys.* **316**, 481.

Roelof, E. C., Simnett, G. M., Decker, R. B., Lanzerotti, L. J., Maclennan, C. G., Armstrong, T. P., and Gold, R. E. (1997) Reappearance of recurrent low-energy particle events at Ulysses/HI-SCALE in the northern heliopshere. *J. Geophys. Res.* **102**, 11251.

Sanderson, T. R. (1995) Ulysses energetic ion observations during the declining phase of solar cycle 22. *Adv. Space Res.* **16**, 267.

Sanderson, T. R., Bothmer, V., Marsden, R. G., Trattner, K. J., Wenzel, K.-P., balogh, A., Forsyth, R. J. and Goldstein, B. E. (1996) Ulysses observations of energetic ions over the south pole of the Sun. *Proc. Solar Wind Eight, AIP Conf. Proc.*, ed. D. Winterhalter *et al.*, **414**.

Sanderson, T. R., Marsden, R. G., Tranquille, C., Balogh, A., Forsyth, R. J., Goldstein B. E. and Hoeksema, J. T. (1999) Current sheet control of recurrent particle increases at 4–5 AU. *Geophys. Res. Lett.* **26**(14), 1785–1788.

Sanderson, T. R., Marsden, R. G., Wenzel, K.-P., Balogh, A., Forsyth, R. J. and Goldstein, B. E. (1995) High-latitude observations of energetic ions during the first Ulysses polar pass. *Space Sci. Rev.* **72**, 291.

Simnett, G. M., Kunow, H., Flueckiger, E., Huber, B., Horbury, T., Kota, J., Lazarus, A., Roelof, E. C., Simpson, J. A., Zhang, M. and Decker, R. B. (1998) Corotating particle events. *Space Sci. Rev.* **83**, 215.

Simpson, J. A., Anglin, J. D., Balogh, A., Berkovitch, M., Bouman, J. M., Budzinski, E. E., Burrows, J. R., Carvell, R., Connell, J. J., Ducros, R., Ferrand, P., Firth, J., Garcia-Munoz, M., Henrion, J., Hynds, R. J., Iwers, B., Jacquet, R., Kunow, H., Lentz, G., Marsden, R. G., Mckibben, R. B., Mueller-Mellin, R., Page, D. E., Perkins, M., Raviart, A., Sanderson, T. R., Sierks, H., Treguer, L., Tuzzolino, A., J.Wenzel, K.-P., and Wibberenz, G. (1992) The Ulysses cosmic ray and solar particle investigation. *Astron. Astrophys. Suppl. Ser.*, **92**, 365.

Simpson, J. A., Anglin, J. D., Bothmer, V., Connell, J. J., Ferrando, P., Heber, B., Kunow, H., Lopate, C., Marsden, R. G., Mckibben, R. B., Müller-Mellin, R., Paizis, C., Rastoin, C., Raviart, A., Sanderson, T.R., Sierks, H. Trattner, K. J., Wenzel, K.-P., Wibberenz, G. and Zhang, M. (1995) Cosmic ray and solar particle investigations over the south polar regions of the Sun. *Science*, **268**, 1019.

Smith, E. J., Neugebauer, M., Balogh, A., Bame, S. J., Erdös, G., Forsyth, R. J., Goldstein, B. E., Phillips, J. L. and Tsurutani, B. T. (1993) Disappearance of the heliospheric sector sector structure at Ulysses. *Geophys. Res. Lett.*, **20**, 2327.

Thomson, D. J., Maclennan, C. G. and Lanzerotti, L. J. (1995) Propagation of solar oscillations through the interplanetary medium. *Nature*, **376**, 139.

Thomson, D. J., Maclennan, C. G. and Lanzerotti, L. J. (1996) The nature of the solar wind. *Nature*. **381**, 32.

Thomson, D. J., Maclennan, C. G. and Lanzerotti, L. J. (1997) Recurrences of interplanetary interaction regions at southern solar latitudes and approximate harmonics. *Adv. Space Res.* **20**, 103.

Trattner, K. J., Marsden, R. G. Bothmer, V. Sanderson, T. R. Wenzel, K.-P. Klecker, B. and Hovestadt, D. (1996) Ulysses COSPIN/LET: Latitudinal Gradients of Anomalous Cosmic Ray O, N and Ne. *Astron. Astrophys.* **316**, 519.

Trattner, K. J., Marsden, R. G. Sanderson, T. R. (1997) The Ulysses North Polar Pass: Latitudinal gradients of anomalous cosmic ray O, N and Ne. *Geophys. Res. Lett.* **24**, 1719.

Van Hollebeke, M. A. I., McDonald, F. B. Trainor, J. H. and von Rosenvinge, T. T. (1978) The radial variation of corotating energetic particle streams in the inner and outer solar system. *J. Geophys. Res.* **83**, 4723.

7

Heliospheric and interstellar phenomena revealed from observations of pickup ions

George Gloeckler, Johannes Geiss and Lennard A. Fisk

7.1 INTRODUCTION

Pickup ions are produced when neutral particles from the interstellar gas, or interplanetary dust, or local bodies such as planets, comets, and asteroids are ionized in the solar wind. These pickup ions provide us with a new tool to explore both nearby local interstellar space and regions close to the Sun. Interstellar pickup ions are an important component of the heliospheric ion population and play a significant role in the dynamics of the outer heliosphere. Observations of pickup ions tell us much about the interactions of suprathermal ions with a magnetized plasma, and about their acceleration and transport in the solar wind. Pickup ions are the source of Anomalous Cosmic Rays (ACRs), and a significant source of energetic particles accelerated by heliospheric shocks inside the heliosphere.

The importance of pickup ions is mainly due to the information they carry regarding their sources, and the dynamic processes governing their interaction with the solar wind. It is notable that nature has provided a convenient mechanism in the solar wind of converting very low-density, low-speed neutrals into ions with energies which are readily observable with modern plasma composition spectrometers.

The presence of interstellar hydrogen atoms inside the heliosphere was established from analyses of Ly-α sky background maps almost 30 years ago (Bertaux and Blamont, 1971; Thomas and Krasse, 1971). Interstellar pickup ions were discovered much later with the advent of new ion mass spectrometers: He^+ at 1 AU (Möbius *et al.*, 1985), and H^+ (Gloeckler *et al.*, 1993), $^3He^{++}$ (Gloecker and Geiss, 1996), $^4He^{++}$ (Gloeckler *et al.* 2000a) and heavy pickup ions (Geiss et al., 1994) at several AU.

In this chapter we present a synopsis of *Ulysses* pickup ion observations and results which provide insights regarding their sources and transport in the heliosphere. These observations were made with the Solar Wind Ion Composition Spectrometer (SWICS), designed primarily for measurements of the composition of the solar wind (Gloeckler *et al.*, 1992). SWICS measures the intensity of solar wind and suprathermal ions as a function of their energy (E), mass (m), and charge state (q), in the E/q range from 0.6 to 60 keV/e. To determine the ion parameters, E, m, and q, SWICS uses a combination of an energy per charge analysis with energy resolution $\Delta E/E \sim 0.04$ followed by a post-acceleration of 23 kV, a time-of-flight analysis and an energy measurement using solid-state detectors. The ion distributions are sampled once every $\sim$13 minutes. With the double and triple coincidence techniques used, the background levels are extremely low allowing measurements of very low-flux levels of pickup ions. The viewing angle of SWICS is such that in one spin period ($\sim$12 sec) a π steradian cone is sampled. The cone is centred on the spin axis of the spacecraft which falls within 20° of the Sun direction.

We draw heavily on published work by Gloeckler *et al.* (1991a), Geiss *et al.* (1994a, 1995, 1996), Gloeckler and Geiss (1996, 1998), and Fisk *et al.* (1997), and provide a review of observational findings regarding the sources and transport of pickup ions in the heliosphere, and physical processes in the heliosphere–interstellar cloud interaction region beyond the termination shock.

We begin by reviewing the basic physics of pickup-ion production, followed by a description of observations and resulting implications of the light pickup-ion triad H^+, He^+ and He^{++}. Next we review information provided by pickup-ion observations regarding the interface region beyond the heliospheric termination shock, which allows us to deduce the ionization fraction of H and He in the Local Interstellar Cloud (LIC). We then present measurements of the heavy interstellar pickup ions, O, N, and Ne and discuss how the interstellar abundance of the gas in the (LIC) is deduced. Finally, we describe observation of pickup ions from an extended inner source, and review our knowledge of the interaction of pickup ions with the solar wind. We conclude with a summary for the chapter and a brief discussion of future work and prospects.

7.2 PRODUCTION AND MEASUREMENTS OF PICKUP IONS

Ions in the expanding corona are virtually all of solar origin. With increasing distance from the Sun, however, non-solar ions are gradually mixed in, so that when the solar wind approaches the termination shock (estimated to be between $\sim$85 AU and 110 AU in the direction of motion of the Sun through the interstellar cloud), their relative abundance is estimated to have reached the order of 10 per cent. Pickup ions in the heliosphere originate from extended sources such as the interstellar gas, interstellar grains and interplanetary grains, as well as from local sources such as comets or planetary atmospheres as is illustrated in Figure 7.1.

At relatively unperturbed solar-wind conditions, two distinct types of ion velocity distributions are observed in interplanetary space. (1) The dominant solar-wind ions are found in a characteristically narrow Mach angle that corresponds to Mach numbers of typically 10 to 20. (2) Far less abundant are the

Sec. 7.2] Production and measurements of pickup ions 289

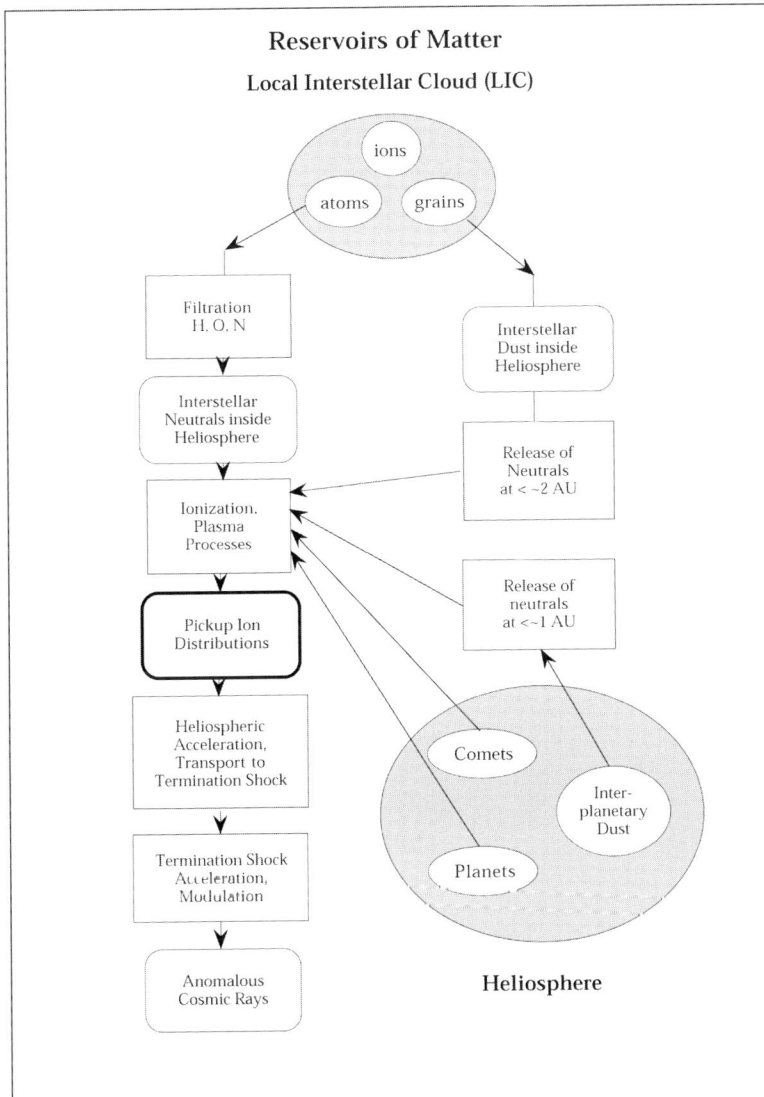

Figure 7.1. Schematic representation of sources and processes leading to the creation of pickup ions in the heliosphere and their subsequent transport and acceleration to form the ACRs. In addition to the interstellar source, a new extended inner source (interstellar and interplanetary dust) inside the heliosphere has been identified using **SWICS**/*Ulysses* data. Inside the large ovals (reservoirs of matter) specific sources of pickup ions are shown in small ovals. Physical processes are indicated in rectangular boxes. These processes produce the particle populations shown in boxes with rounded corners.

pickup ions which have broad suprathermal distribution functions that in the undisturbed high-speed, high-latitude solar wind show a sharp decrease at the upper velocity limit of twice the solar-wind speed (Möbius et al., 1985; Gloeckler et al., 1993). At heliocentric distances of several AU and sufficiently far from local sources, their number densities are of the order of 10^{-3} or less relative to the number densities of the solar wind ions.

For presenting and comparing different pickup ion populations, it is advantageous to relate V_{ion}, the speed of an incoming ion to V_{sw}, the locally measured solar wind bulk speed. We will use:
$$W = V_{\text{ion}}/V_{\text{sw}} \qquad (7.1)$$
for the relative speed of an ion in the reference frame of the Sun (or the spacecraft). Then:
$$\mathbf{w} = (\mathbf{V}_{\text{ion}} - \mathbf{V}_{\text{sw}})/V_{\text{sw}} \qquad (7.2)$$
relates the relative velocity $\mathbf{w}$ in the reference frame of the solar wind to the ion velocity $\mathbf{V}_{\text{ion}}$ in the spacecraft frame.

Three criteria are available for identifying pickup ion populations and determining their sources of origin:

(1) *Charge states.* Solar-wind ions of heavy elements have high charges reflecting the typical coronal temperature of $\sim 10^6$ K. Pickup ions, on the other hand, are predominately singly charged. The reason for this is simple. Pickup ions, created as singly charged, are rapidly convected by the solar wind from regions of high EUV and solar-wind flux that caused their ionization to distances where further ionization becomes negligible. (The only important exception is He which also has a relatively large cross-section for double ionization.) Therefore, the vast majority of pickup ions produced from atoms or molecules outside magnetospheres remains singly charged. Measurements of ionization states of heavy ions is one of the easiest ways to separate and identify low-charge state, heavy pickup ions from the far more abundant, highly charged solar-wind ions.

(2) *Spatial distributions in the heliosphere.* Ions produced from the interstellar gas have a very characteristic spatial distribution in the heliosphere. Their number density depends on both the angle between the interstellar wind flow direction and the observer, as well as on the ionization rate which determines the distance from the Sun of their maximum abundance (e.g. Thomas, 1978; Holzer, 1989). The density of ions of local origin (e.g. comets, Jupiter, etc.), on the other hand, decreases with distance from the source (cf. Neugebauer et al., 1987; Geiss et al., 1994c; Ogilvie et al., 1995). However, local sources of neutrals, located between the Sun and the point of observation, can be detected even at large distances from the source as was recently demonstrated by Grünwaldt et al. (1997) who, with an instrument near Earth, recorded pickup ions produced from neutrals originating from Venus, and Gloeckler et al. (2000) who detected pickup ions in the ion tail of comet Hyakutake at a distance of 3.75 AU from its nucleus.

(3) *Velocity distributions.* In the solar wind, pickup ions are produced with $W \ll 1$ from slowly moving atoms or molecules, such as the interstellar gas atoms,

cometary constituents, and atoms and ions liberated from grains (e.g. by evaporation or sputtering). These ions are immediately picked up by the electromagnetic field of the solar wind, and rapidly convected outward. Pitch-angle scattering and/or variation in the magnetic field (**B**) direction are then required to spread the ions into shell-like distributions in **w** and, in particular, change W of some of the ions from less than to greater than 1. When the average direction of **B** is nearly perpendicular to V_{sw} (i.e. for heliocentric distances greater than about a few AU at low latitude) distributions showing little anisotropy between $W < 1$ and $W > 1$ are expected and are indeed observed (Gloeckler et al., 1993, 1997). On the other hand, at high latitudes or very close to the Sun where the magnetic field is oriented more radially, the velocity distribution of pickup ions is found to be highly anisotropic with little scattering into the $W > 1$ phase-space hemisphere (Fisk et al., 1997; Gloeckler et al., 1995). In either case, pickup ions are adiabatically cooled in the expanding solar wind. Thus, when moving away from the region where they are created, the shell-like distribution of these pickup ions shrinks to lower w values. All these processes tend to produce a pickup ion velocity distribution in the solar wind frame that is relatively flat to $w = w_c \cdot 1$ with a sharp cut-off beyond w_c. These highly suprathermal pickup-ion distributions are distinctly different from the thermal spectra of solar-wind ions. For protons and $^4He^{++}$ this difference in the velocity distributions provides the only means of separating pickup hydrogen and $^4He^{++}$ from solar wind protons and alpha particles, respectively.

Using one or more of the three criteria given above, the following pickup ions of interstellar origin have been discovered: $^1H^+$ (Gloeckler et al., 1993), $^4He^+$ (Möbius et al., 1985; Gloeckler et al., 1993), $^3He^+$ (Gloeckler and Geiss, 1996), $^4He^{++}$ (Gloeckler et al., 1997), N^+, O^+, and Ne^+ (Geiss et al., 1994c). The dependence of their abundances on heliocentric distance and latitude, as well as the shape of their velocity distributions show that the majority of the pickup ions of these species in the heliosphere as a whole is produced from the neutral interstellar gas as it flows through the solar system.

C^+ ions are also found at all solar latitudes and distances visited by *Ulysses* (see Section 7.7). However, their distribution in space and their phase-space density spectra differ radically from pickup ions produced from the interstellar gas. Geiss et al. (1995, 1996) have shown that the majority of the C^+ ions as well as a fraction of O^+ and N^+ are produced by an 'extended inner source' that is located at solar distances below a few AU. In Section 7.7 the origin of this newly found source of pickup ions is discussed in some detail.

7.3 HYDROGEN AND HELIUM PICKUP IONS

The orbit of *Ulysses* and the low background capabilities of the SWICS instrument (Gloeckler et al., 1992) made it possible to discover pickup hydrogen (Gloeckler et al., 1993) whose existence was postulated in the early 1970s (e.g. by Blum and Fahr,

1970; Axford, 1972 and references therein; and Vasyliunas and Siscoe, 1976), and whose detection was anticipated with the launches of the deep-space missions *Pioneers 10* and *11* and *Voyagers 1* and *2*. The exceptionally low background of SWICS and the long time periods that *Ulysses* spent in the fast, steady and quiet solar wind from the polar coronal holes enabled Gloeckler *et al.* (1997) to identify the uncommon ^{4}He^{++} pickup ions and from these measurements deduce the abundance of He in the local interstellar cloud. With SWICS Gloeckler and Geiss (1996) also discovered the extremely rare pickup ^{3}He$^+$ and used its measured abundance to place a new lower limit on the amount of missing matter in the universe.

7.3.1 Pickup-ion velocity distributions observed in the high-speed solar wind of the polar coronal hole

7.3.1.1 Protons

Interstellar pickup hydrogen ions are distinguishable from the far more abundant solar-wind protons by their distinctly different velocity distributions as displayed in Figure 7.2. The proton phase-space density shown was accumulated over a 100-day period (10 April to 19 July 1994) when *Ulysses* was at an average heliocentric distance of 3.0 AU and at a high latitude ($-66°$) in the undisturbed and fast solar wind from the south-polar coronal hole. The solar-wind proton distribution around $W = 1$ is narrow and well represented by a kappa function in the solar wind frame:

$$f_{sw}(w) = f_o[1 + (w/\theta)^2/\kappa]^{-(\kappa+1)} \tag{7.3}$$

$$\theta = V_{th}[1 - (1.5/\kappa)]^{1/2} \quad \text{for } \kappa > 1.5 \tag{7.4}$$

The fit (curve labelled Solar wind) to the solar-wind peak around $W = 1$ is the kappa distribution (equation 7.3) transformed to the spacecraft frame using equation (7.2), and integrated over the instrument view angles (see Gloeckler, 1996 for details). Best values for the density, thermal speed V_{th} and κ were 0.30 cm^{-3}, 32.3 km s^{-1} and 6.3, respectively. We note in passing that a maxwellian distribution ($\kappa \gg 20$) fails to fit the solar wind proton distribution we observe.

The speed distribution of interstellar pickup hydrogen, on the other hand, is broad and extends to $W \approx 2$ at which point the phase-space density drops by about three orders of magnitude. The fact that the two distributions are quite distinct implies that the interaction between the solar wind and pickup ions is sufficiently weak so that no indication of thermalization of the pickup ion populations is apparent. The very sharp drop in phase-space density at $W \approx 2$ indicates little energy diffusion, in the quiet high-speed solar wind at high latitudes.

The dotted curve labelled $\xi = 0$ represents the predicted phase-space density of pickup protons using the 'hot model' of Thomas (1978) for the spatial distribution of hydrogen atoms in the heliosphere. We use the Vasyliunas and Siscoe (1976) equation $f_{iso}(R, \Theta, w)$ derived under the assumption of rapid pitch-angle scattering

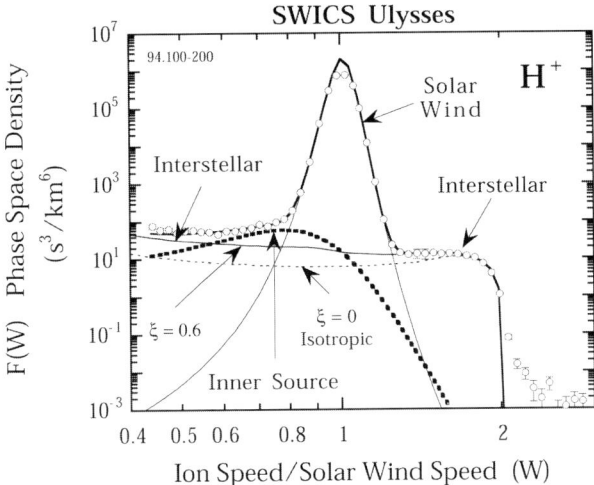

Figure 7.2. Phase-space density of H$^+$ versus W, the ion speed in the spacecraft frame of reference divided by the solar-wind speed. This time-averaged spectrum was observed with SWICS during a 100-day period in 1994 when *Ulysses* was in the steady high-speed (785 km s^{-1} average speed) stream of the south-polar coronal hole. The average direction of the magnetic field was almost radial (165°). The two highest density points closest to $W = 1$ fall below the kappa-curve fit (labelled Solar wind) because instrument dead-time saturated the proton count rate used here. We note that the solar-wind proton-number density is computed from the start detector rate (see Gloeckler *et al.*, 1992) which does not suffer from these dead-time effects. In these and subsequent plots the phase-space density $F_{m,m/q}(W)$ is computed from the differential energy per charge flux of the ion with mass (m) and mass/charge (m/q) in the standard fashion using a calculated value for the isotropic geometrical factor. Model distribution functions are then computed, first in the solar-wind frame of reference and then transformed to the spacecraft frame, integrating over the SWICS view angles using the measured directional geometrical factor to obtain the predicted count rate for each of the sixty-four W steps. These count rates are then converted to a predicted phase-space density for each W step using the calculated isotropic geometrical factor of the instrument (see Gloeckler, 1996 for further details).

and hence isotropy (see also Gloeckler, 1996) for computing the phase-space density in the solar-wind frame of reference of the resulting pickup protons at the location (R, Θ) of *Ulysses*, with R the heliocentric radial distance in AU and Θ the angle between the direction of motion of the Sun relative to the interstellar cloud and the Sun–*Ulysses* line.

Finally, we make the co-ordinate transformation to the spacecraft frame and integrate over the view directions of SWICS. The parameter values used for the density, temperature, and flow speed of atomic hydrogen at the termination shock were 0.115 cm^{-3} (Gloeckler *et al.*, 1997), 7,000 K (Witte *et al.*, 1996; Geiss and Witte, 1996), and 20 km s^{-1} (Lallement, 1996; Geiss and Witte, 1996), respectively. The loss rate ($\beta_{\text{Hloss}} = 5.5 \times 10^{-7}$ s^{-1}) and the ratio of radiation pressure to the gravitational force ($\mu = 1$) were taken from the best fit to the spatial distribution of the pickup H$^+$

measured between 1.5 and 3.3 AU (see Figure 7.6). The total production rate of pickup protons from atomic hydrogen required to match the observations for W above ~ 1.7 was $\beta_{\text{Hprod}} = 2.25 \times 10^{-7}\,\text{s}^{-1}$. This value is considerably below the measured value of the production rate $(3.5 \times 10^{-7}\,\text{s}^{-1})$ due to charge-exchange with solar wind protons alone. Furthermore, it is clear that the model distribution, based on the assumption of rapid pitch-angle scattering, fails to account for the observed spectral shape. In fact, no isotropic (in the solar wind frame) velocity distribution of any kind would fit the observed spectrum both below and above $W = 1$. Based on this result, Gloeckler et al. (1995) and Fisk et al. (1997) concluded that fast pitch-angle scattering leading to isotropic phase-space density distributions in the solar-wind frame does not occur during the time period and solar-wind conditions of these observations.

The solid curve labelled $\xi = 0.6$ uses a modified form of the Vasyliunas and Siscoe 1976) equation. Instead of fast scattering at all pitch angles we assume a strong suppression of scattering at $90°$ pitch angle, but allow rapid scattering for all pitch angles less than as well as greater than $90°$ (Fisk et al., 1997). The modified form of the Vasyliunas and Siscoe (1976) equation we use is $f_+(R, \Theta, w, \theta) = f_{\text{iso}}(R, \Theta, w) \cdot (1 + \Delta(w))$ for $0 < \theta < \pi/2$ and $f_-(R, \Theta, w, \theta) = f_{\text{iso}}(R, \Theta, w) \cdot (1 - \Delta(w))$ for $\pi/2 < \theta < \pi$, where θ is the angle between the velocity $\mathbf{w}$ of the pickup ion and the direction of the average magnetic field $\mathbf{B}$ taken to point inward. In the present case of nearly radial ($165°$) average magnetic field, pickup ions created at $W \ll 1$ stay at speeds below $W = 1$ with only a small fraction scattering past $90°$ into the speed range above $W = 1$. For simplicity we assume here $\Delta(w) = \xi$ to be independent of w (but see equations 7.12 and 7.13). The degree of anisotropy ξ is related to the probability of scattering through $90°$, and is given in terms of the density ratio in the $<90°$ ($W < 1$) to that in the $>90°$ pitch angle ($W > 1$) hemisphere of phase-space $f_+/f_- = (1 + \xi)/(1 - \xi)$.

The $\xi = 0.6$ model distribution fits the observed pickup proton speed spectrum above $W = 1.25$ quite well. The parameters used to compute this model distribution are the same as those for the $\xi = 0$ model curve, except for β_{Hprod}, the total production rate and μ. To obtain the best match to the observed distribution above $W = 1.25$ required $\mu = 1.17$ and $\beta_{\text{Hprod}} = 4.95 \times 10^{-7}\,\text{s}^{-1}$. The contributions to the total production rate are primarily from charge exchange with solar wind protons (measured to be $3.5 \times 10^{-7}\,\text{s}^{-1}$) with the rest coming from ionization by solar UV and electron impact (Rucinski et al., 1996). For the anisotropy parameters we chose $\xi = 0.6$. Using eq. (2) of Fisk et al. (1997) an anisotropy $\xi = 0.6$ indicates a large ($\sim 1.2\,\text{AU}$) mean free path.

While the $\xi = 0.6$ fit matches the observed spectrum above $W = 1.25$ extremely well it falls below the measured distribution for $W < 0.85$. When data from the $45°$ Sun sector is excluded (as was done in Gloeckler et al., 1995) the observed phase-space density below $W \sim 0.62$ falls right on the $\xi = 0.6$ curve. To account for the relatively large contributions to the total density below $W = 0.8$, especially in the Sun sector, Gloeckler and Geiss (1998b) postulated the existence of an extended inner source of pickup protons as indicated by the dashed curve. The shape of this 'inner source' proton spectrum is similar to that of the inner source carbon and oxygen distribution discussed in Section 7.7 and shown in Figures 7.13. 7.14, and 7.15.

7.3.1.2 Doubly charged helium-4

In Figure 7.3 we show the ^{4}He^{++} speed distribution averaged over the same 100-day period in 1994 used for protons of Figure 7.2. The distribution of alpha particles is very similar to that of H$^+$, showing the clear separation of the solar wind alphas from interstellar pickup ^{4}He^{++}, which is produced almost entirely by double charge exchange of atomic helium with solar-wind alpha particles (Gloeckler et al., 1997; Gloeckler and Geiss, 1998a; Rucinski, et al., 1996). The distribution function of pickup ^{4}He^{++} (solid curve) is again anisotropic with $\xi = 0.4$, which corresponds to a mean free path of ~ 1 AU using eq. (2) of Fisk et al., 1997.

To obtain the model pickup ^{4}He^{++} distribution function we used the hot model of Thomas (1978) with a neutral interstellar helium density of 0.0153 cm^{-3} (Gloeckler et al., 1997), a temperature of 7,000 K (Witte et al., 1996; Geiss and Witte, 1996; Witte et al., 1993) and a loss rate of 0.55×10^{-7} s^{-1} (Rucinski et al., 1996) (also see Figure 7.4). A total of 90 per cent of the total production rate comes from double charge exchange of solar-wind alphas with interstellar atomic He and the rest from charge exchange with pickup ^{4}He$^+$ and photoionization (Gloeckler et al., 1997; Rucinski et al., 1998). Thus, the dominant portion of the total production rate of 0.0215×10^{-7} s^{-1} used here is known reasonably well from the product of the

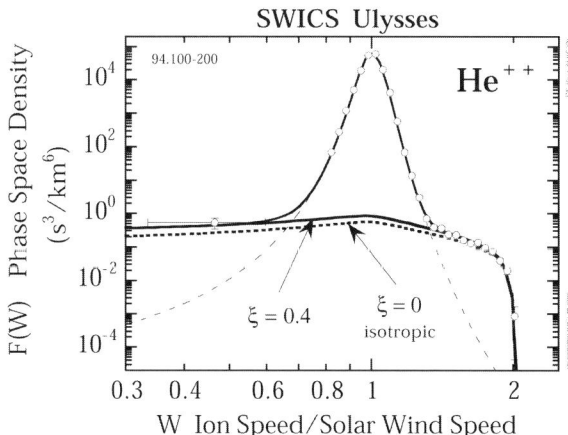

Figure 7.3. Same as Figure 7.2 for He^{++}. This time-averaged spectrum was taken during the same time period as the proton distribution of Figure 7.2. Pickup He^{++} phase-space densities were used to derive the absolute interstellar helium number density of 0.0153 cm^{-3} because its production rate (almost entirely by double charge exchange with solar wind alpha particles) was measured by SWICS. This method eliminates instrumental systematic error because the interstellar He density is then essentially proportional to the ratio of the pickup He^{++} flux (at W close to 2) and the solar wind He^{++} flux (at $W = 1$). Because of the lower count rate no saturation of the rate near $W = 1$ occurs.

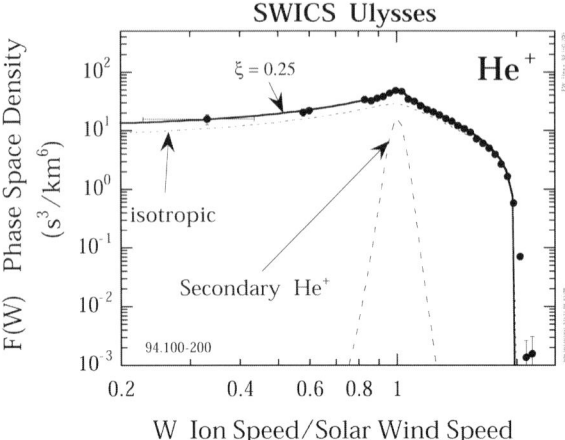

Figure 7.4. Same as Figure 7.3 for He^+. Although pickup He^+ is about 30 times more abundant than He^{++} it could not be used to compute the interstellar helium density because its rate of production, almost entirely by solar UV was not measured during the time of these observations. See text for further details and explanation of the various model curves.

measured solar-wind alpha-particle flux (adjusted to 1 AU) and the double charge exchange cross-section of $1.9 \times 10^{-16}\,cm^2$ (Rucinski et al., 1998) at an energy corresponding to the $\sim 800\,km\,s^{-1}$ wind. Best values for the solar wind alpha particle density, thermal speed V_{th}, and κ were $0.0145\,cm^{-3}$, $36\,km\,s^{-1}$ and 5.2, respectively.

7.3.1.3 Singly charged helium-4

Interstellar $^4He^+$ pickup ions are the most abundant suprathermal ions at 1 AU and comparable in density with pickup protons at ~ 5 AU. They were the first of the interstellar pickup ions to be discovered (Möbius et al., 1985), although the detailed spectrum of these ions over a wide velocity range was only observed with the SWICS/*Ulysses* instrument (Gloeckler et al., 1993). Figure 7.4 shows the $^4He^+$ phase-space density as a function of W. The time period for this average spectrum was the same as used for the two previous figures. SWICS samples almost all of the relevant phase space in the case of pickup $^4He^+$. We show the $^4He^+$ phase-space density over the entire interval of speeds between $\sim 0.2 < W <\sim 2$. The small gaps around $W = 0.5$ and $W = 0.7$ result from spill-over of solar-wind protons and alpha particles for which no corrections have as yet been made. To obtain the $^4He^+$ densities in the important speed interval around $W = 1$ we carefully subtracted known contributions from the measured fluxes of solar wind Si^{+7}, Mg^{+6}, Fe^{+14}, as well as O^{+6} and Fe^{+12}. The solid and dotted curves are the predicted spectra using the same parameters that were used for the computation of the $^4He^{++}$ pickup-ion distribution, except for the production rate. The dotted curve, which assumes an isotropic distribution, does not fit the measured distribution. The solid curve, which fits the data over the entire speed range, uses an anisotropy factor $\xi = 0.25$, corresponding to a mean free path of ~ 0.8 AU. The total production rate required to match the data was 0.8×10^{-7} s^{-1}. We note that the major part of the production rate of ^{4}He$^+$ is due to photoionization. The EUV photon flux was not measured during this observation period.

In the region around $W = 1$ significant contributions to the production rate of pickup ^{4}He$^+$ can also come from electron-impact ionization close (<1 AU) to the Sun (Rucinski et al., 1996; Rucinski et al., 1998). In this same W interval there are also contributions from secondary ^{4}He$^+$, produced primarily in the charge-exchange reaction ($\sigma = 8.5 \times 10^{-16}$ cm^2, Rucinski et al., 1998) of solar-wind alpha particles with atomic hydrogen. In this reaction ^{4}He$^+$ is produced with the same velocity distribution as solar wind ^{4}He^{++}, as well as a small fraction of pickup hydrogen. For the time period used in this analysis we calculate the production rate and flux of secondary ^{4}He$^+$ to be 0.078×10^{-7} s^{-1} and 300 cm^{-2} s^{-1} respectively using the measured solar wind alpha parameters and the atomic hydrogen spatial distribution deduced from our pickup proton measurements. The dashed curve in Figure 7.4 represents the calculated secondary ^{4}He$^+$ velocity distribution whose shape is taken to be the same as that of the measured solar wind alpha particles of Figure 7.3. Adding the secondary ^{4}He$^+$ to our model fit did not entirely account for the small peak in the ^{4}He$^+$ phase-space density around $W = 1$. To obtain the excellent fit (solid curve) to the observed spectrum over the entire range of speeds required additional production of ^{4}He$^+$ by electron-impact ionization. The production rate we used was $\beta_{el} = (0.16 \times 10^{-7}) \times R^{-2.3}$ s^{-1}, with R the heliocentric distance in AU.

7.3.1.4 Singly charged helium-3

The rarest of the light-mass pickup ions observed is ^{3}He$^+$. Long accumulation times were required to positively identify this uncommon pickup ion and measure its abundance relative to ^{4}He$^+$ (Gloeckler and Geiss, 1996). The mass/charge distribution of ions with masses between 2 and 6 amu and $1.6 < W < 2.0$ measured by SWICS during the 40-month period (July 1992 to November 1995) of high-speed solar wind is given in the left-hand panel of Figure 7.5. The triple coincidence requirement reduced residual background to about 10 per cent of the ^{3}He$^+$ counts (less than one count in 3 years). While limited by counting statistics, the peak due to ^{3}He$^+$ is clearly visible at mass per charge around 3, between the peaks at mass/charge 2 and 4 from the far more abundant pickup ^{4}He^{++} and ^{4}He$^+$, respectively.

The phase-space density of ^{3}He$^+$ (divided by 2.48×10^{-4}) is compared with that of ^{4}He$^+$ in the right-hand panel of Figure 7.5. Within the statistical errors the two spectra are comparable as expected, further strengthening the case for the correct identification of interstellar ^{3}He. We also note that ^{3}He$^+$ was not detected above $W = 2$ indicating the characteristic cut-off of a pickup-ion spectrum. Based on the results shown in Figure 7.5, Gloeckler and Geiss (1996, 1998a) were able to place a new lower limit on the amount of missing matter in the early universe, and concluded that the amount of ^{3}He production by stars was less than predicted by some models of stellar chemical evolution.

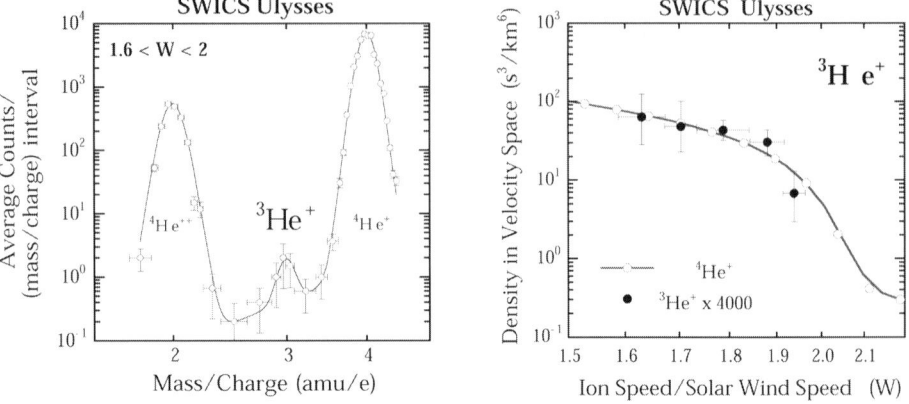

Figure 7.5. (left) Plot of the average triple-coincidence counts of ions with masses between 2 and 6 amu and with normalized speed W between 1.6 and 2 per mass/charge (m/q) intervals versus the mean m/q of each interval. Triple coincidence is required to determine both the mass and the mass/charge of the ions (Gloeckler et al., 1992) and to reduce the background sufficiently to detect these extremely rare ions. In regions of low count rate the m/q intervals were increased to contain at least one count. The peak at m/q of $\sim$3 is due to $^3\text{He}^+$. Note that $^3\text{He}^+$ is well resolved from far more abundant pickup $^4\text{He}^+$ to the right and pickup $^4\text{He}^{++}$ to the left. The time interval between the detection of adjacent $^3\text{He}^+$ followed a Poisson distribution with a mean time interval of 121 days. Because of the extremely low fluxes of $^3\text{He}^+$, data were taken during a 40-month period when the solar-wind speed exceeded 750 km s^{-1} for maximum detection efficiency. (right) Phase-space density of $^4\text{He}^+$ and $^3\text{He}^+$ (divided by 2.48×10^{-4}) versus W (ion speed/solar-wind speed). Despite the large statistical error for $^3\text{He}^+$ the similarity of the two spectral shapes is obvious. No $^3\text{He}^+$ counts were detected above $W = 2$ consistent with the expected cut-off in density. Below $W \sim 1.5$ the probability of $^3\text{He}^+$ producing triple coincidence becomes extremely small since such ions have insufficient energy to trigger the 40-keV threshold solid-state detector (from Gloeckler and Geiss, 1996, 1998a).

7.3.1.5 Spatial distribution of atomic hydrogen

The spatial distribution of an atomic species from which the corresponding pickup ions are created depends on the loss rate β_{loss} for removing neutrals through ionization and on μ, the ratio of the force on the atom due to radiation pressure to the gravitational force. The loss rate and μ can be obtained from the shape of the distribution function of the pickup-ion species between $W \sim 1.3$ and 2, as was discussed above, or from measurements of the gradual change (gradient) of the pickup-ion fluxes with heliocentric distance R at a given angle Θ with respect to the direction of the interstellar wind flow. The gradient of the ratio of pickup hydrogen to pickup $^4\text{He}^+$ is shown in Figure 7.6 which includes data for the time periods of Figures 7.2–7.4. Assuming $\mu = 1$ for protons and $\mu = 0$ for He, the best fit to the observed gradient is with $\beta_{\text{loss}}(\text{hydrogen}) = (5.5 \pm 1.0) \times 10^{-7}\,\text{s}^{-1}$. This same

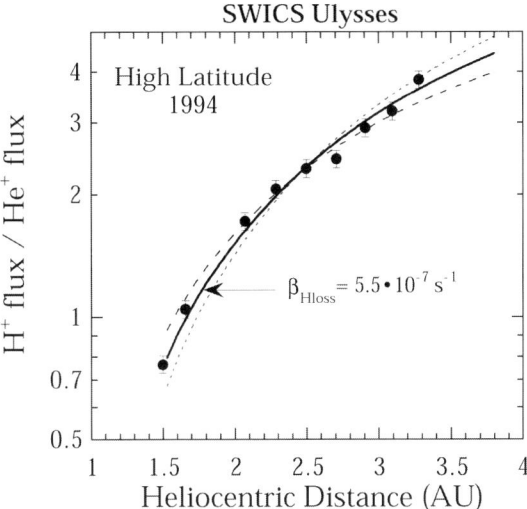

Figure 7.6. Ratio of H^+ to $^4He^+$ pickup-ion fluxes versus heliocentric distance. The flux ratio is obtained from the measured phase-space densities of H^+ and He^+ for W between 1.75 and 2. Each data point is a 1-month average. During the entire time period of these observations *Ulysses* was in the high-speed stream of the south-polar coronal hole. The best fit to the data is the solid curve which was computed assuming μ, the ratio of radiation to gravitational forces on hydrogen to be 1, using a hydrogen loss rate of $5.5 \times 10^{-7}\,s^{-1}$, a He loss rate (Rucinski *et al.*, 1996) of $0.55 \times 10^{-7}\,s^{-1}$, and a H/He density ratio at the termination shock of 7.5 (see Table 7.1). The dashed and dotted curves were calculated using H loss rates of $4.5 \times 10^{-7}\,s^{-1}$ and $6.5 \times 10^{-7}\,s^{-1}$, respectively, but keeping the other parameter values the same.

value for the loss rate was used to derive the model pickup proton distribution shown in Figure 7.2.

7.3.2 Pickup-ion velocity distributions observed in the in-ecliptic, low-speed solar wind

7.3.2.1 *Pickup-ion distributions in the unperturbed slow solar wind*

In the slow, in-ecliptic, unperturbed solar wind, strong high-speed tails are observed in the velocity distributions of H^+, He^+, and He^{++} even in the absence of travelling shocks, waves or Corotating Interaction Regions (CIRs). These tails indicate the presence of a baseline population of energized ions, with speeds W as high as 5–10, at all times. Figure 7.7 shows the spectra measured in the baseline solar wind, when no CIRs, waves, or shocks were locally detected. In contrast to this, distributions in the unperturbed, quiet high-speed, high-latitude solar wind show little evidence of suprathermal tails (see Figures 7.2–7.4) (i.e. few ions above twice the solar-wind speed exist there). Thus, the presence of substantial high-velocity tails in the ion velocity distributions in the in-ecliptic solar wind is one of many characteristics

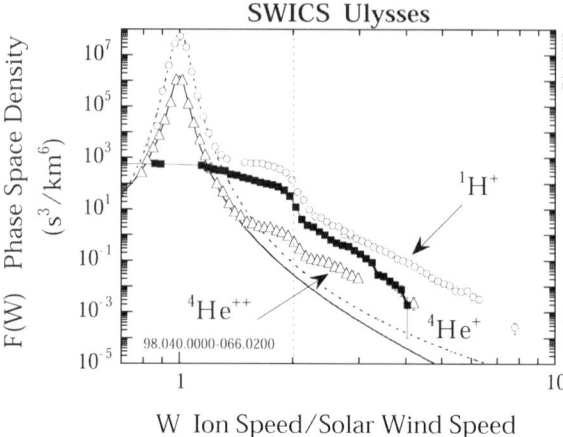

Figure 7.7. Phase-space density versus W of H^+, He^+, and He^{++} ions in the slow (375 km s^{-1} average speed), in-ecliptic solar wind. High-velocity tails are clearly visible in all distributions indicating the presence of pre-accelerated ions in the absence of shocks. Suprathermal interstellar pickup H^+ and He^{++} rise above the solar-wind distributions in the W range between ~1.5 and 2. For $W > 2$ ten times more He^+ than He^+ is observed, implying preferential acceleration of the suprathermal pickup ions even in the absence of local shocks or waves. Dashed and solid curves are kappa-function fits to the solar wind H^+ ($\kappa = 4.5$) and He^{++} ($\kappa = 3.0$), respectively (from Gloeckler, 1999).

that distinguishes the low-speed from the high-speed wind. One of the open questions then is how these ubiquitous ions, with speeds several times that of the solar wind, are created in the slow wind (and not the fast wind) at times when there are no shocks observed locally. Because of their slow mobility, it is unlikely that these ions were accelerated by distant shocks and then transported to the shock-free regions of the low-latitude heliosphere. In addition, we note that in the baseline, in-ecliptic solar wind ten times more He^+ than He^{++} is observed in the tails ($W > 2$), indicating that pickup ions are preferentially accelerated compared with the bulk solar wind (Gloeckler *et al.*, 1994).

7.3.2.2 Pickup-ion distributions in the turbulent slow solar wind

The high-velocity tails observed in the slow solar wind become even more pronounced in the more turbulent wind, for example inside CIRs. In the left panel of Figure 7.8 we compare the spectral shapes of H^+, He^+, and He^{++} measured downstream of the forward shock (FS) with those observed in the downstream region of the reverse shock (RS) in the late December 1992 CIR. The normalization factors applied to the reverse shock data are shown in the figure as fractions in parentheses. There are several remarkable features apparent from these data. First, we note that more He^+ than He^{++} is accelerated even though solar-wind alpha particles are at

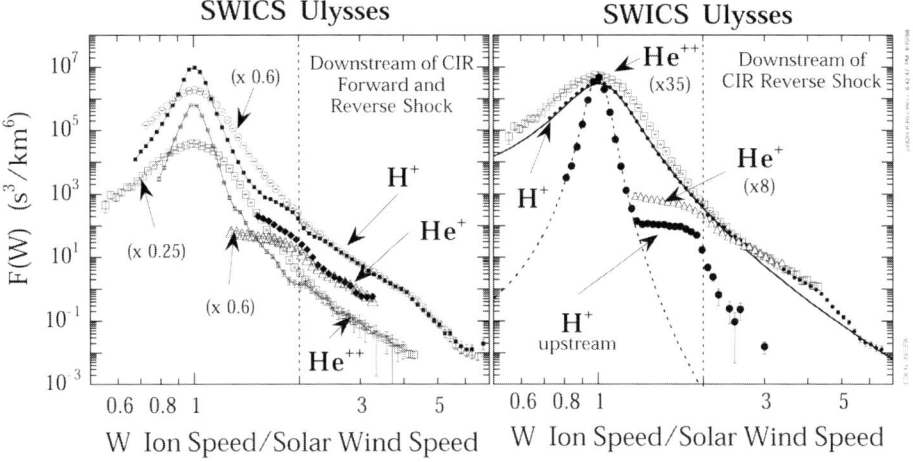

Figure 7.8. (left) Distribution functions of H^+, He^+, and He^{++} downstream of the forward shock (small filled squares, filled diamonds, and small open squares, respectively) and downstream of the reverse shock (open circles, open triangles, and open squares, respectively) of the late December 1992 CIR. The time period used for averaging the downstream FS spectra was from 26 December 19:41 to 28 December 07:47. For the downstream RS distributions the time interval was 28 December 07:47 to 29 December 23:47. The downstream RS distributions have been multiplied by the fractions shown in parentheses in order to match the respective spectra at the FS in the high-velocity region above $W = 2.4$. Notice the remarkable similarity in the spectral shapes of all three ion species in the high-velocity tail region. (right) Distribution functions of H^+ (small filled circles), He^+ (open triangles) and He^{++} (open squares) in the region downstream of the RS. The accumulation times were 28 December 07:47 to 29 December 23:47. The He^+ and He^{++} phase-space densities were multiplied by 8 and 35 respectively in order to match the proton spectrum in the high-velocity tail above $W = 2.5$. For comparison we also show the proton distribution (filled circles) in the upstream RS region (29 December 23:47 to 1 January 1993 00:59). We note that (a) the tails (above $W = 2.5$) of all three ion species have the same shape in the downstream region of the shock, and that the shape is not a simple power law as predicted by standard shock acceleration models, (b) the He^+/He^{++} ratio in the tail is ~ 4.4, and (c) the $H^+/(He^+ + He^{++})$ ratio in the tail is about 6.5.

least a factor of 10^3 more abundant than pickup He^+. Second, the spectral shapes in the high-speed ($W > \sim 2.5$) tail region behind the RS are identical (within experimental uncertainties) to those behind the FS. This is even more remarkable because the two shocks are so different. The FS is weaker ($M_s = 1.55$) than the RS ($M_s = 2.38$) and one is quasi-perpendicular while the other is quasi-parallel (Balogh *et al.*, 1995). It is evident from the velocity distributions that the solar wind behind the reverse shock is heated more than the wind downstream of the weaker forward shock. In this hotter solar wind downstream of the RS more suprathermal solar wind ions are available for injection into the acceleration process than in the colder solar wind behind the FS. The fact that four times more He^{++} and only 1.7 times more H^+ and He^+ is accelerated behind the RS compared with the FS implies that downstream of the RS the heated, suprathermal solar wind ions are competitively accelerated along with pickup ions. Behind the forward shock, on the other hand, where the heating is less, fewer suprathermal solar wind ions are available. Thus, downstream of this relatively weak forward shock the more abundant pickup ions are the principal components available for injection and pre-acceleration.

In the right panel of Figure 7.8 we contrast the phase-space distribution of protons (small filled circles) in the downstream region of the CIR RS with that of He$^+$ (open triangles), multiplied by a factor of 8, and He^{++} (open squares), multiplied by 35. These normalization factors were chosen so that all three velocity distributions overlapped in the tail region above $W \sim 2.5$. For reference we also show the averaged distribution function of protons (filled circles) measured in the region upstream of the RS (see caption to Figure 7.8 for time interval). In comparing the upstream with the downstream H$^+$ distribution we note that the solar-wind proton distribution in the downstream region behind the RS is very broad and non-maxwellian. The solid curve is a kappa-function fit to protons with $V_{th} = 67.2$ km s^{-1} and an extremely low value of $\kappa = 3.1$. This is to be contrasted with the upstream solar-wind proton distribution (dotted curve) which is much colder ($V_{th} = 29$ km s^{-1}) and has a weaker tail ($\kappa = 4.9$). In the downstream region the pickup hydrogen, quite obvious in the upstream spectrum, is completely obscured by the over ten times more abundant heated solar-wind protons above $W \sim 1.5$. We therefore conclude that in the case of strong heating illustrated here, suprathermal solar wind protons are predominantly injected for further acceleration to higher energies. On the other hand, in instances where less heating occurs (as was shown to be the case downstream of the FS in this CIR and as is often the case behind FSs of CIRs in general) the protons in the high-speed tail will be a mixture of suprathermal solar wind and pickup protons (Gloeckler *et al.*, 1994).

The speed distributions of all three species in the tail region have identical shapes within experimental uncertainties) above $W \sim 2.5$. This, combined with the fact that the FS and RS spectra also have the same shapes (see left panel of Figure 7.8), implies that the acceleration mechanism depends primarily on the ion speed. In addition, the spectral shapes are not simple power laws, as would be predicted by standard shock-acceleration models, providing further evidence that the high-velocity tail distributions we observe are not produced by a simple shock-acceleration mechanism, but rather by the turbulence downstream of shocks as is discussed further below.

Downstream of both the shocks and the strong waves that proceed the formation of shocks (Gloeckler, 1999), suprathermal tails increase in strength by one to two orders of magnitude. But, as was shown by Gloeckler (1999), the turbulence associated with the wave seemed to strengthen the tails as much if not more than that associated with the shocks. It is thus not the shock itself but rather the strong turbulence (that also heats the solar wind) associated with the shock or the wave that is required for the formation of strong tails (Gloeckler *et al.*, 1994; Gloeckler, 1999; Schwadron *et al.*, 1996).

7.3.2.3 Implications for sources and acceleration processes

In summary, in the disturbed solar wind, as is the case in CIRs between FSs and RSs, the distribution functions of solar wind ions become broader and highly pronounced tails extending to high energies are observed for both pickup as well as solar wind ions (cf. Figure 7.8). Clearly, pickup He^+ (and H^+) are easily accelerated in these regions, explaining the once-puzzling ubiquitous presence of energetic (0.4–0.6 MeV amu^{-1}) $^4He^+$ measured during a $\sim$1.5-year period in 1978–1979 at 1 AU by Hovestadt et al. (1984), to be interstellar $^4He^+$ pickup ions that were most likely accelerated in CIRs at $\sim$3–6 AU and then transported back to 1 AU. The fact that the acceleration process produces spectra of identical (but often complex) shapes despite the large difference in the shock parameters reinforces the idea that the acceleration responsible for the formation of tails is not due to the shocks themselves but rather to the turbulence in the downstream regions of the shocks as also reported by Gloeckler et al. (1994), Schwadron et al. (1996), and Gloeckler (1999).

We now come to the question of what material is injected and thus further accelerated to MeV energies. We have shown that pickup ions (in particular He^+) are clearly pre-accelerated and thus are available for further acceleration to MeV energies at shocks. We also provided observational evidence that the heated, suprathermal solar wind is also available for injection and further acceleration. Just how much of each of these two populations is injected in CIRs will very likely depend on the relative fraction of these populations in the suprathermal range, perhaps around $W = 2$. Since the solar wind generally is heated more behind the RS than it is behind the FS (Wimmer-Schweingruber, 1997) we would expect that ions accelerated at the RS are predominantly solar wind ions. On the other hand, when the solar wind speed is high, pickup-ion distributions extend to higher speeds, and at twice the solar wind speed they may again have densities comparable with or higher than those of solar-wind ions heated to these higher speeds. It seems to be the case that in CIRs at $\sim$3–5 AU, the heated solar wind distributions for H and the pickup-ion H distributions have roughly comparable densities near $W \sim 2$ and therefore relatively small changes in either the solar wind thermal speed or its bulk speed will affect the proportion of solar wind H versus pickup-ion H accelerated in CIRs. Observations of the larger abundance of energetic (MeV) H and He particles relative to oxygen in CIRs compared with their relative abundance in the solar wind (Gloeckler et al., 1979; Scholer et al., 1979; Reames et al., 1991) support our conclusion that pickup H and He form a significant part of the seed population of energetic particles accelerated in these turbulent regions.

The inner-source pickup ions (see Section 7.7) could also contribute substantially to the energetic particles accelerated in CIRs (Gloeckler et al., 2000). The inner source (Geiss et al., 1995; Geiss et al., 1996) has a carbon-to-oxygen ratio $\sim$1. Indeed the average C/O ratio of CIR-accelerated particles is also $\sim$1 and significantly higher than the solar-wind C/O ratio (Gloeckler et al., 1979; Scholer et al., 1979; Reames et al., 1991; Fränz et al., 1995). The composition of inner-source pickup ions also includes elements such as Mg and Si with an abundance comparable with that of the solar wind. In fact, Gloeckler et al. (2000) argue that a significant fraction of inner-source pickup ions originate from solar wind absorbed by interplanetary dust and eventually released as slow moving (compared with the solar wind) atoms (Gruntman, 1996). Again, the composition of MeV particles heavier than oxygen in CIRs resembles that of the inner-source pickup ions (which themselves reflect the composition of the solar wind).

At heliocentric distances of $\sim$2–6 AU, where CIRs predominate in the lower latitude regions of the heliosphere, it appears that all three sources – interstellar pickup ions, solar wind, and inner-source pickup ions – inject more or less comparable amounts of material for further acceleration by CIR shocks. Depending on solar wind conditions one of these sources may predominate. However, at larger heliocentric distances ($>\sim$10 AU), the contributions from interstellar pickup ions will grow, while those from both the solar wind and the inner source will diminish. This is consistent with the measured composition of the $\sim$10–100-MeV ACRs which reflect that of interstellar pickup ions. While most of the acceleration to hundreds of MeV is likely to take place at the heliospheric termination shock, some pre-acceleration of pickup ions inside the heliosphere is required in order to compensate for adiabatic cooling and to provide for easy injection of the pickup ions into diffusive shock acceleration at the termination shock. The same mechanisms that produce the ever-present suprathermal tails in the in-ecliptic slow solar wind in the absence of shocks could also pre-accelerate interstellar pickup ions in the in-ecliptic distant heliosphere right up to the termination shock. Another possibility, suggested by Fisk (1996) is acceleration at distances far beyond $\sim$10 AU at the interface between the low-speed flow in the ecliptic and the high-speed flow from the polar coronal holes at higher latitudes. Presumably, pickup ions in the turbulence associated with this interface would be pre-accelerated as efficiently as in CIRs.

7.4 INTERSTELLAR PICKUP IONS WITH MASSES HEAVIER THAN HELIUM

Heavy pickup ions (O^+, N^+, and Ne^+) of interstellar origin were discovered by Geiss et al. (1994a) using SWICS/Ulysses data at distances beyond several AU. Presence of singly charged heavy ions of non-solar origin is clearly evident from Figure 7.9 which shows a mass/charge (m/q) histogram of ions with speeds above 1.4 times the solar-wind speed ($W > 1.4$) in the mass/charge range above $10\,\mathrm{amu\,e^{-1}}$. The prominent peak at $m/q = 16$ is due to singly charged interstellar pickup oxygen, with smaller peaks at m/q of 14 and 20 corresponding, respectively, to pickup N^+ and Ne^+. In particular, C^+ ($m/q = 12$) is not observed in this velocity range, placing a rough upper limit on C^+/O^+ of $\sim$0.05. These observations are consistent with an interstellar origin for these three heavy pickup ion species. Atomic O, N, and Ne penetrate deep into the solar system before they become pickup ions. Carbon, on

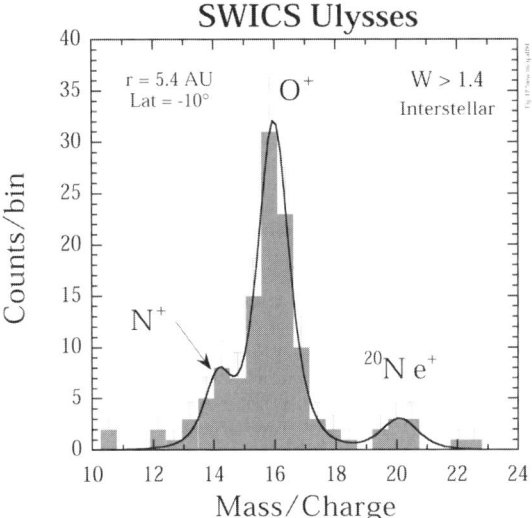

Figure 7.9. Plot of the double coincidence counts of ions with normalized speed W above 1.4 per mass/charge (m/q) intervals versus the mean m/q of each interval. These SWICS data were accumulated during a 135-day period in 1992 at an average distance of 5.4 AU and average heliolatitude of $-10°$ (also see Geiss et al., 1994). The average solar-wind speed was 442 km s^{-1}. In addition to the prominent peak at mass/charge 16 (^{16}O$^+$), ^{20}Ne$^+$ and ^{15}N$^+$ are also clearly observed.

the other hand, is mostly ionized in the local interstellar cloud (Frisch, 1995, 1997). Thus, a large fraction of interstellar C is excluded from the heliosphere.

Conclusive evidence for interstellar pickup oxygen comes from the velocity distribution function of O$^+$ shown in Figure 7.10. These SWICS data (and those shown in Figure 7.9) were accumulated during a 135-day period in 1992 at a distance of 5.4 AU and an average heliolatitude of $-10°$. During this time period the solar wind speed remained relatively steady at 442 ± 36 km s^{-1}. The distribution has the characteristic cut-off at $W = 2$, and the flat shape of the spectrum is similar to that of pickup hydrogen. Both the sharp cut-off and the flatness of the distribution function indicate a source location beyond 5.4 AU, the heliocentric distance of *Ulysses* when these data were obtained. A Jovian origin was ruled out by Geiss et al. (1994c). The solid curve is a model fit for the interstellar oxygen using 12×10^{-7} s^{-1} for the loss and production rates of O (see Rucinski et al., 1996) and the same anisotropy factor as for hydrogen ($\xi = 0.6$). The agreement with the observed spectrum above $W = 1.3$ is excellent. We find the atomic oxygen number density at the termination shock to be $(6.9 \pm 1.7) \times 10^{-5}$ cm^{-3}.

The velocity distribution functions of N$^+$ and Ne$^+$ are similar in shape to that of O$^+$ shown in Figure 7.10. By fitting model curves of interstellar N and Ne using loss and production rates given by Rucinski et al. (1996) we deduce the abundance of these atoms near the termination shock. The results along with the abundance of H, ^{4}He, ^{3}He, and O are given in columns 2 and 3 of Table 7.1.

306 Heliospheric and interstellar phenomena revealed from observations of pickup ions [Ch. 7

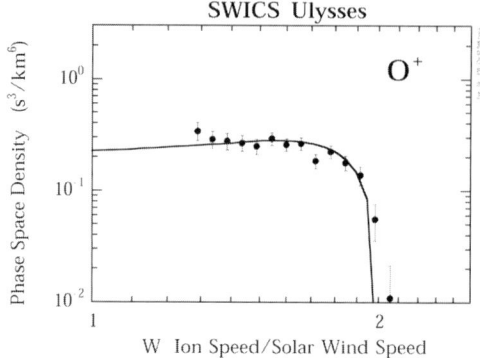

Figure 7.10. Phase-space density of O^+ versus W averaged over 135 days from 17 February to 1 July 1992 when *Ulysses* was still in the ecliptic plane. The average solar wind speed was relatively steady at 442 ± 36 km s^{-1} and the distance from the Sun remained close to 5.4 AU during the entire time period of these observations. The solid curve is the model calculation of interstellar pickup O^+ assuming an anisotropy factor $\xi = 0.6$ (see text). However, because the average magnetic field was nearly perpendicular (112°) to the solar wind flow direction the distribution function would appear to be isotropic in the spacecraft frame with nearly equal densities above and below $W = 1$.

Table 7.1. Densities of atoms at the heliospheric termination shock (~85–110 AU) and densities of atoms and ions in the local interstellar cloud (LIC) of elements and isotopes measured as pickup ions with SWICS on *Ulysses*.

	Termination shock		LIC				Solar system abundance ratio†
			Density (cm^{-3})				
Isotope	Ratio	Density (cm^{-3})	Atoms	Ions*	Total	Ratio	
^{1}H	7.5	0.115 ± 0.015	0.20	0.04	0.24	10	10
^{4}He	1.000	0.0153 ± 0.002	0.0153	0.009	0.024	1	1
^{3}He	2.48 ± 10^{-4}	$(3.8 \pm 1.0) \times 10^{-6}$	3.8×10^{-6}	2.2×10^{-6}	6.0×10^{-6}	2.5×10^{-4}	–
^{14}N	0.6×10^{-3}	$(9.2 \pm 3.0) \times 10^{-6}$	1.1×10^{-5}	2.3×10^{-6}	1.3×10^{-5}	0.54×10^{-3}	1.12×10^{-3}
^{16}O	4.5×10^{-3}	$(6.9 \pm 1.7) \times 10^{-5}$	9.6×10^{-5}	2.1×10^{-5}	1.2×10-4	5.0×10^{-3}	8.51×10^{-3}
^{20}Ne	1.0×10^{-3}	$(1.5 \pm 0.5) \times 10^{-5}$	1.5×10^{-5}	1.5×10^{-5}	3.0×10^{-5}	1.25×10^{-3}	1.23×10^{-3}

* For nitrogen and oxygen we assumed the same ionization fraction as measured for hydrogen (0.18).
† Anders and Grevesse (1988). The solar-system abundance of Ne is not well established.

7.5 IONIZATION STATE OF THE LOCAL INTERSTELLAR CLOUD

The density of interstellar neutral hydrogen (and to a lesser extent oxygen and nitrogen) reaching the inner heliosphere is significantly reduced by charge-exchange in the so-called filtration region in the local interstellar cloud (LIC). This roughly 100–200-AU wide region (e.g. Holzer, 1989; Baranov and Malama, 1995; Steinolfson and Gurnett, 1995) lies just beyond the heliopause, a boundary separating solar from interstellar plasma. There the LIC plasma, forced to move around the heliopause, slows down and in a process known as filtration (e.g. Baranov and Malama, 1995), converts some fraction of the H atoms to fast protons by proton–hydrogen charge exchange. This process also affects other atoms with a reasonably large charge exchange cross-sections with protons. The ram and thermal pressure of the newly created protons and the original LIC plasma, combined with the pressure of the LIC magnetic field draped around the heliopause, fix the size of the heliosphere. The amount of reduction of atomic hydrogen passing through the filtration region is roughly proportional to the average proton density in that region.

Combining the values for the densities of atomic H and He at the termination shock (from Table 7.1) with measured parameters in the LIC listed in Table 7.2, we evaluate (see Gloeckler et al., 1997 for details) the atomic H density in the LIC, $n(\text{H I})$, and the average proton density in the filtration region, $\langle n(\text{p}) \rangle$. The computational steps and results are given in Table 7.2. We find that the LIC neutral hydrogen density, $n(\text{H I}) = 0.20 \pm 0.03 \, \text{cm}^{-3}$, and that about 40 per cent of the LIC atomic hydrogen is excluded from the heliosphere by charge exchange in the filtration region.

It is possible to make an estimate of the strength of the LIC magnetic field B based on pressure-balance arguments and using the information from Table 7.2. We note that a value of $B < 3.7 \, \mu\text{G}$, combined with parameter values from Table 7.2, implies that the Sun moves supersonically (and super-Alfvénically) relative to the LIC. Assuming $B < 3.7 \, \mu\text{G}$ there will then be a bow shock at the outer boundary of the filtration region. The magnetic field orientation inferred to be nearly parallel (Frisch, 1995) to the shock surface will make it a perpendicular MHD shock. First, we find the dependence of the total LIC pressure P_{LIC} at the heliopause on the magnetic field and densities of atomic hydrogen $n(\text{H I})$, protons $n(\text{H II})$, and ionized helium $n(\text{He II})$, in the LIC just beyond the bow shock. The total stagnation pressure at the heliopause from the interstellar medium is given by Gloeckler et al., 1997 and Holzer, 1989 as:

$$P_{\text{LIC}} = \tfrac{7}{8}(\rho V_o^2) + \tfrac{3}{8}(2n_e kT) + 3(B^2/8\pi) + P_{\text{H}} \quad (7.5)$$

where $V_o = 25.3 \, \text{km s}^{-1}$ is the relative speed of the Sun through the local cloud, $\rho = m_{\text{H}}[n(\text{H II}) + 4n(\text{He II})]$, $n(\text{H II}) = \langle n(\text{p}) \rangle / r$, r is the bow shock compression ratio, $n(\text{He II}) = [n(\text{H I}) + n(\text{H II})]/R_2 - n(\text{He I})$, $n_e = n(\text{H II}) + n(\text{He II})$, and $P_{\text{H}} = f \cdot n(\text{H I}) \cdot [m_{\text{H}} V_o^2 + kT]$ is the pressure from that fraction $f = [1 - \exp(-\langle n(\text{p}) \rangle L\sigma)]$ of hydrogen atoms converted to $\sim 25 \, \text{km s}^{-1}$ protons in the filtration region by charge exchange. Using the standard jump condition (Siscoe, 1983) at the bow shock, the upstream magnetic field pressure is

$$B^2/8\pi = \rho V_o^2[(4/r) - 1 - 5r(2n_e kT)]/(r+5) \quad (7.6)$$

We have neglected contributions to the pressure from cosmic rays, since the observed gradients and thus the resulting force should be small, as well as from dust. Since the

Table 7.2. Measured and derived average parameters in the LIC and in the very local neighbourhood of the solar system.

Parameter	Units	Local* interstellar medium	Very local interstellar cloud ($<\sim 2,000$ AU)
$n(H_{TS})$	cm^{-3}		0.115 ± 0.025**
$n(He\,\textsc{i}) = n(He_{TS})$	cm^{-3}		0.0153 ± 0.0018**
V_o	km s^{-1}		25.3 ± 0.4†
T	K	$7,000 \pm 600$†	
$R_1 = n(H\,\textsc{i})/n(He\,\textsc{i})$		13 ± 1††	
$R_2 = n(H)/n(He)$		10‡	
$n(H\,\textsc{i}) = R_1 \cdot n(He\,\textsc{i})$	cm^{-3}		0.20 ± 0.03
$F = n(H_{TS})/n(H\,\textsc{i}) =$ filtration factor			0.58 ± 0.15
$\langle n(p) \rangle = (-1/L\sigma) \cdot \ln(F/1.46)$	cm^{-3}		0.073 ± 0.031§
$n(H\,\textsc{ii}) = n(p)/r$	cm^{-3}		0.040 ± 0.017¶
$n(H) = n(H\,\textsc{i}) + n(H\,\textsc{ii})$	cm^{-3}		0.240 ± 0.034¶
$n(He\,\textsc{ii}) = [n(H)]/R_2 - n(He\,\textsc{i})$	cm^{-3}		0.009 ± 0.004¶
$n_e = n(H\,\textsc{ii}) + n(He\,\textsc{ii})$	cm^{-3}		0.0515 ± 0.0165¶
$X_H = n(H\,\textsc{ii})/n(H)$		$0.17\| < X_H < 0.4$††	0.18 ± 0.01¶#
$X_{He} = n(He\,\textsc{ii}) \cdot R_2/n(H)$		$0.3\| < X_{He} < 0.5$††	0.36 ± 0.02¶#

* Average values measured or calculated over distances of a few to ~ 100 pc and assumed to be also valid in the immediate neighbourhood of the heliosphere.
** Densities near the termination shock from column 3 of Table 7.1.
† From Witte et al. (1993, 1996) and Geiss and Witte (1995).
†† From Dupuis et al. (1995).
‡ From Anders and Grevesse (1988).
§ Proton density in the filtration region: the width of the filtration region, $L = 160 \pm 40$ AU (Holzer, 1989; Baranov and Malama, 1995; Steinolfson and Gurnett, 1995), and $\sigma = 5.3 \times 10^{-15}$ cm^2 is the proton–hydrogen charge-exchange cross-section (Fite et al., 1962) at $V_{rel} = 19$ km s^{-1}, the average relative speed between protons and H atoms in the filtration region. The factor of $1.46 = V_{rel}/(V_{rel} - 6\,\text{km s}^{-1})$ accounts for that fraction of slow-moving H atoms, produced by charge exchange, that enter the heliosphere, contribute to $n(H_{TS})$, and reduce the average speed of interstellar H inside the heliosphere (Lallement et al., 1993) to 20 km s^{-1}.
¶ For compression ratio, $r = 1.8$.
\| From Cheng and Bruhweiler (1990).
\# For $80 < R_{TS} < 85$ AU see caption to lower panel of Figure 7.11 for error limits.

values of ρ, n_e, and B all depend on the shock compression ratio r, P_{LIC} also depends only on r, since all other parameters are known (see Table 7.2). We note that magnetic field pressure and P_H contribute significantly to P_{LIC}.

The location R_{TS} of the termination shock (a boundary where the supersonic solar wind becomes subsonic) is found from pressure balance between the solar wind and the LIC plasmas at the heliopause. The solar-wind pressure at the heliopause P_{SW} depends on the location of the termination shock R_{TS} on the solar-wind density and speed just inside the termination shock and decreases rapidly with increasing R_{TS}. The stagnation pressure at the heliopause from the solar wind P_{SW} is determined using the usual gas dynamic calculation, relating conditions just inside the termination shock to those measured by Burlaga et al. (1996) at 49.1 AU with

Voyager 2. We assume that in the distant heliosphere the solar wind expands uniformly but is decelerated slightly in the process of picking up interstellar atoms, neglects magnetic-field pressure but includes thermal pressure from pickup ions (Gloeckler *et al.*, 1997):

$$P_{SW} = \tfrac{7}{8} m_H \cdot [1.16 n_{sw}(p)] \cdot [49.1 U_{sw}/R_{TS}]^2$$
$$+ \tfrac{3}{8}(m_H/7) \cdot (U_{sw}/R_{TS}) \cdot \beta_{prod} n(H_{TS}) \qquad (7.7)$$

where $U_{sw} = V_{sw}[1 - \tfrac{6}{7} \cdot \sigma_p \cdot n(H_{TS}) \cdot R_{TS}]/[1 - (6/7) \cdot \sigma_p \cdot n(H_{TS}) \cdot 49.1]$ is the reduced solar-wind speed due to momentum loss to pickup hydrogen, $n_{sw}(p) = 0.0017\,\mathrm{cm}^{-3}$ and $V_{sw} = 534\,\mathrm{km\,s}^{-1}$ is the 1996 averaged *Voyager 2* solar-wind proton density and speed at 49.1 AU (Burlaga *et al.*, 1996), the factor of 1.16 accounts for the solar wind He/H of 4 per cent, and $\beta_{prod} = 7 \times 10^{-7}\,\mathrm{s}^{-1}$. Assuming that there is no flow across or along the heliopause in the direction of motion of the Sun, the stagnation pressure on either side of the heliopause must be equal (Holzer, 1989). Thus all LIC parameters, such as B and ρ, calculated as a function of the compression ratio may also be expressed in terms of T_{TS} by combining equations (7.5), (7.6), and (7.7). For a given R_{TS}, the compression ratio r is determined and vice versa. Given either R_{TS} or r, the magnetic-field strength B and ρ in the LIC are also determined. The dependence of B on R_{TS} is shown in the upper panel of Figure 7.11.

The shaded vertical region in Figure 7.11 indicates the range of R_{TS} values derived from observations of anomalous cosmic rays with *Voyagers 1* and *2* (Stone *et al.*, 1996), putting the minimum distance to the termination shock at 80 AU. This places an upper limit on B of $3\,\mu\mathrm{G}$ (upper dashed curve), justifying our assumption of supersonic flow. Our most probable calculated field (solid curve) agrees remarkably well with other estimates of the interstellar magnetic-field strength of $1.4 \pm 0.15\,\mu\mathrm{G}$ (Frisch, 1995, 1997) represented by the shaded horizontal region.

The variation with R_{TS} of the fractional ionization of hydrogen $X(H)$ and helium $X(He)$, and the electron density n_e in the very local interstellar cloud is shown in lower panel of Figure 7.11. Our values for both $X(H)$ and $X(He)$ are well below the measured (Dupuis *et al.*, 1995) upper limits given in Table 7.2. However, imposing the calculated lower limit of 0.17 on $X(H)$ (Cheng and Bruhweiler, 1990) places an upper limit on R_{TS} of ~ 85 AU. Thus, the most probable value of B is between ~ 1.3 and $\sim 2\,\mu\mathrm{G}$ for the case of supersonic flow. The corresponding bow shock compression ratios range from 1.5 to 1.9, making it a weak shock.

7.6 COMPOSITION OF THE LOCAL INTERSTELLAR CLOUD

The absolute number densities of atomic H, He, N, O, and Ne of the interstellar gas inside the heliosphere can be obtained without much ambiguity from measurements of the phase-space densities of the corresponding pickup ions. We have used our measurements of the velocity distributions of pickup ions (such as shown in Figures 7.2, 7.3, 7.4, 7.5, 7.10 and 7.16) to derive the atomic-number densities of ^{1}H, ^{4}He, ^{3}He, ^{14}N, ^{16}O and ^{20}Ne at the termination shock (~ 100 AU). The ratios (relative to

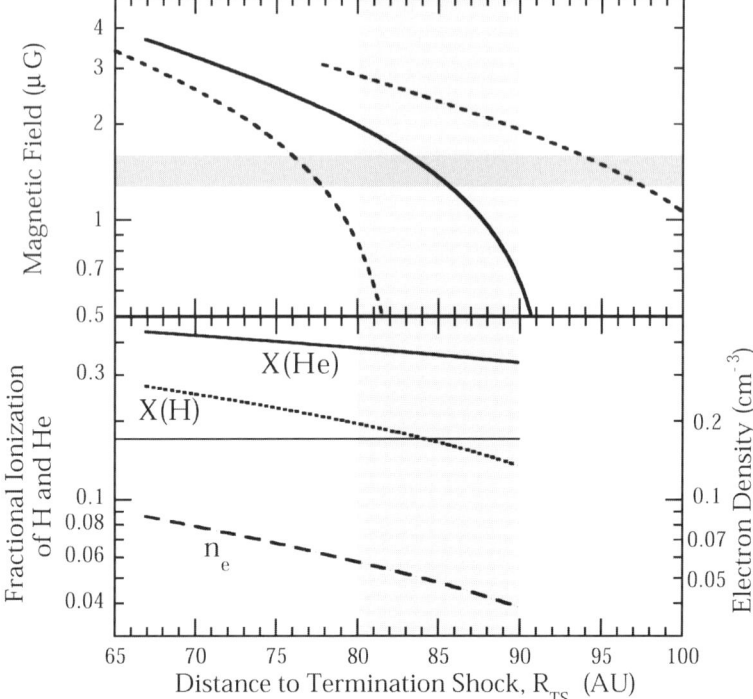

Figure 7.11. (top) Variation of the calculated magnetic-field strength B just beyond the interstellar bow shock with distance to the termination shock R_{TS}. Mean (solid curve) and 1σ limits (dotted curves) are shown. (bottom) Average values of the electron density and ionization fractions of H and He calculated as a function of R_{TS} using values from Table 7.2 and computational steps given in the text. The shaded vertical region indicates the range of locations of the termination shock determined by Stone *et al.* (1996). The horizontal line is the lower limit of $X(H)$ calculated by Cheng and Bruhweiler (1990) (from Gloeckler *et al.*, 1997).

^{4}He) and the number densities are given in columns 2 and 3, respectively, of Table 7.1. The ^{1}H and ^{4}He values are from Gloeckler *et al.* (1997), and the ^{3}He abundance is taken from Gloeckler and Geiss (1998b). The interstellar heavy-element abundances are derived from SWICS measurements of O^+ (cf. Figure 7.10) and from similar distribution functions of N^+ and Ne^+ (e.g. Figure 7.16). The errors given next to the values include estimates of systematic errors in the ionization rates of the respective species. These systematic errors are in general the largest sources of uncertainty.

Matter in the local interstellar cloud exists in the form of atoms, ions, and grains (or dust). Only atoms and very low-charged grains can enter the heliosphere as indicated in Figure 7.1. For H, O, and N there is a reduction in the density of atoms reaching the termination shock compared with their abundance in the LIC due to charge-exchange ionization with the slow moving (compared with neutrals) interstellar protons in the filtration region (roughly 100–200 AU wide) just beyond the heliopause (see Section 7.5 and Gloeckler et al., 1997; Holzer, 1989; Baranov and Malama, 1995; Zank and Pavis, 1996). Because of the small charge-exchange cross-section for noble gases, He and Ne are virtually unaffected by this process. Using SWICS measurements of pickup ions (H^+ and He^{++}), Gloeckler et al. (1997) have found the amount of reduction (filtration) for hydrogen and from this computed the fractional ionization for H and He in the LIC (see Section 7.5 and Gloeckler et al., 1997). The derived densities in the LIC of hydrogen and helium atoms and ions are given in columns 4 and 5 of Table 7.1.

The depletion of atomic oxygen in the heliosphere caused by filtration in the interaction region beyond the heliopause can be estimated by comparing our oxygen density at the termination shock (6.9×10^{-5} cm^{-3}) with the atomic oxygen density in the LIC. The LIC density of O is derived by combining the interstellar $n(H\,\textsc{i})/n(O\,\textsc{i})$ ratio, determined by Linsky et al. (1995) to be 2090, with our estimate of the atomic hydrogen density of 0.20 cm^{-3} in the local cloud (column 4 of Table 7.1). We arrive at an ~28 per cent loss of atomic oxygen by filtration using the estimated LIC atomic oxygen density of 9.6×10^{-5} cm^{-3} and our measured oxygen density at the termination shock. Another way to estimate the filtration of oxygen is to use the simple relationship $F = \exp(-\langle n(p)\rangle L\sigma_o)$ between the filtration factor F and charge-exchange cross-section for oxygen σ_o (see Table 7.2). The correction factor of 1.46 used for H does not apply here because negligible amounts of charge-exchanged neutral oxygen enters the heliosphere. Since the charge-exchange cross-section of oxygen with protons is about one-third the value of the hydrogen with protons cross-section (Rucinski et al., 1996) the oxygen filtration is calculated to be ~0.71, giving a density of 9.7×10^{-5} cm^{-3} for atomic oxygen in the LIC. Thus, both methods yield about the same amount of filtration for atomic oxygen. About 28–29 per cent of neutral oxygen is lost due to filtration. The loss of H by filtration is about 42 per cent (Gloeckler et al., 1997). Thus, the observed filtration losses of H and O are consistent with the theoretical predictions of Geiss et al. (1994b) who showed that under all conditions the filtration loss caused by charge exchange is smaller for O than for H. The charge-exchange cross-section between nitrogen and protons is estimated to be about a five times smaller than between hydrogen and protons. Using the expression for the filtration factor given above and the ratio of cross-sections we find the nitrogen filtration to be 0.84 giving a LIC atomic nitrogen density of 1.1×10^{-5} cm^{-3}. Calculations by Frisch and Slavin (1996) of atomic abundance ratios (relative to He) of H, N, O and Ne in the solar neighborhood are in good agreement with the values given in column 4 of Table 7.1.

The ionization fractions of the high FIP (first ionization potential) N, O, and Ne are not well known, but believed to be far less than for low FIP elements (e.g. C, Mg, Fe) which are highly ionized (e.g. Frisch, 1997). We have assumed the same ionization fraction for N and O as for hydrogen (0.18) to compute the ion densities and total densities of N and O in the LIC. For Ne we combined the solar-system abundance with our atom density value to deduce an ionization fraction of 0.5. This is justified because Ne is unlikely to be trapped in interstellar grains. In comparing the abundance ratios (relative to ^{4}He) of elements measured in the LIC (column 7 of Table 7.1) with those in the protosolar nebula given in column 8

(Anders and Grevesse, 1988), we find that nitrogen and oxygen are underabundant in the LIC by about a factor of roughly 2. The abundance of N, O, and other heavy elements in B-stars is about 70 per cent lower than in the Sun (e.g. Frisch, 1998). Assuming that our estimate of the ionization fractions of O and N is correct, our results suggest that the composition in the LIC may be better represented by B-star rather than solar abundances. Using B-star rather than solar abundances our results further suggest that about 20 per cent of the total oxygen abundance in the LIC may be in the form of grains.

7.7 A NEW PICKUP-ION POPULATION FROM AN EXTENDED INNER SOURCE

A new population of pickup ions was discovered by Geiss et al. (1995) who found unusually large fluxes of singly charged carbon at speeds below $W = 1.2$. This was surprising because interstellar carbon is highly ionized and thus excluded from the heliosphere. Measurement of the radial gradient and other evidence ruled out singly charged ions with solar-wind kinetic properties as the source. Geiss et al. (1995) concluded that the C^+ they detected as well as some of the N^+ and O^+ were produced from an inner source, located at a distance below a few AU from the Sun and extending over all heliolatitudes.

To illustrate the presence of C^+ we show in Figure 7.12 the mass/charge histogram of heavy, singly-charged ions with W between 0.8 and 1.0 observed during 1994, when *Ulysses* was in the high-speed solar wind and at high latitudes. In contrast to what was seen at low latitude and high W (cf. Figure 7.9), C^+ is now present at a level comparable with O^+. A peak at $m/q = 14$ is also clearly visible. This peak is most likely due to N^+ and not to solar wind Fe^{+4} because no significant amount of Fe^{+5}, which is likely to be more abundant than Fe^{+4}, is observed at $m/q = 11.2$.

The distribution functions of C^+ and O^+ for this same 1-year period, when *Ulysses* was at high latitudes, are shown in Figure 7.13. Because of the high solar-wind speed ($\langle V_{sw} \rangle = 780\,\mathrm{km\,s^{-1}}$) the SWICS energy per charge upper limit of $60\,\mathrm{keV\,e^{-1}}$ prevented detection of C^+ and O^+ beyond W of 1.26 and 1.1, respectively (dotted and dashed vertical lines in Figure 7.13). This instrumental limit excludes interstellar O^+ which otherwise would be visible beyond $W \sim 1.1$ as is indicated by the heavy dotted, nearly horizontal model curve. The velocity distributions of C^+ and O^+ in the observable W range are identical within experimental errors but may have rather complicated shapes. The observed phase-space densities decrease modestly with decreasing W below ~ 0.95 but fall more rapidly above $W \sim 0.95$. Concentration of density around W close to 1 suggests strong adiabatic cooling, placing the source of these heavy pickup ions well below the average radial distance of *Ulysses* of 2.8 AU during 1994 when these data were taken.

We have begun to investigate the characteristics of this inner source (or inner sources), located between *Ulysses* and the Sun, by measuring the velocity distributions of heavy pickup ions at different positions of the spacecraft. The rapid pole to

Sec. 7.7] A new pickup-ion population from an extended inner source 313

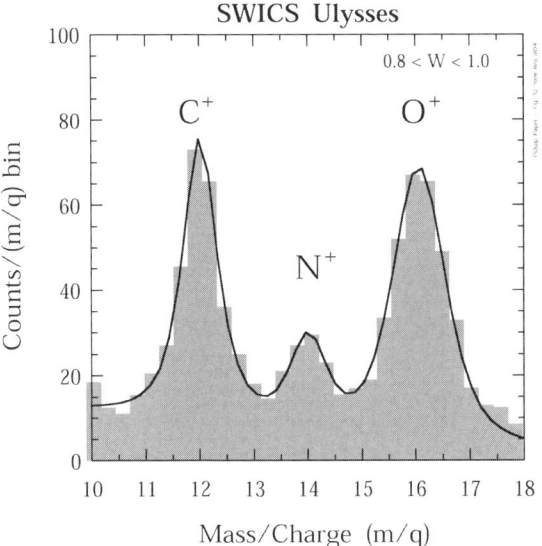

Figure 7.12. Plot of the double coincidence counts per mass/charge (m/q) interval of ions with normalized speed W (ion speed/solar-wind speed) between 0.8 and 1.0 versus the mean m/q of each interval. These SWICS data were accumulated during all of 1994 at an average distance of 2.8 AU from the Sun and average heliolatitude of $-65°$. The average solar-wind helium speed was 784 km s^{-1}. In addition to the ^{16}O$^+$ and ^{14}N$^+$, ^{12}C$^+$ is now observed at a level comparable with ^{16}O$^+$.

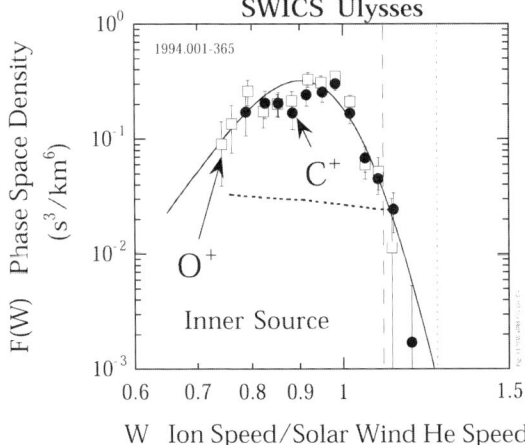

Figure 7.13. Phase-space density of C$^+$ (filled circles) and O$^+$ (open squares) versus W during all of 1994. The vertical dashed and dotted lines indicate the average upper W cut-off for O$^+$ and C$^+$, respectively, above which the SWICS instrument upper energy per charge limit of 60 keV e^{-1} prevents detection of these singly charged heavy ions when the solar-wind speed is high. The dotted, nearly horizontal curve is the expected contribution of interstellar O$^+$. It is evident that interstellar O$^+$ is entirely obscured by the far more abundant O$^+$ from the inner source. The solid curve is drawn to guide the eye.

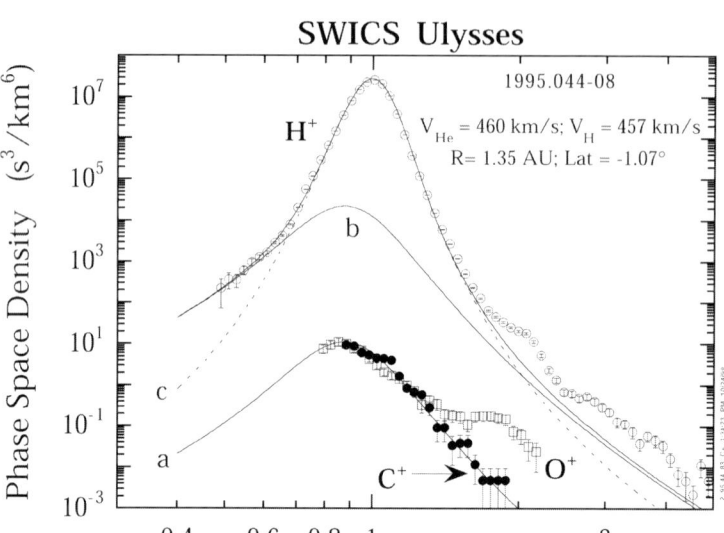

Figure 7.14. Velocity distributions of pickup H^+, C^+, and O^+. These time-averaged distributions were measured with SWICS during the time period from day 44 to day 83 in 1995, when *Ulysses* rapidly crossed the ecliptic plane at 1.35 AU. The average solar-wind speed was 460 km s^{-1}. The entire C^+ distribution consists of inner-source pickup ions since little if any interstellar pickup carbon enters the heliosphere. The O^+ distribution has a small interstellar component above $W = 1.4$, but below 1.4 is predominantly of inner-source origin. Curve (a) is an empirical fit to the carbon distribution representing the inner-source spectrum. Curve (b) is curve (a) multiplied by 2,000. The H^+ distribution below $W = 0.65$ is attributed to inner-source pickup hydrogen, with densities well above curve (c) which is a kappa-function fit to solar-wind protons in the W range between 0.7 and 1.7. Interstellar pickup hydrogen (around $W = 2$) and a high-velocity proton tail are also observed.

pole passage of *Ulysses* in 1995 provided the opportunity to observe pickup-ion distributions when *Ulysses* was again in the ecliptic plane but now only 1.4 AU away from the Sun. Because much of this time the solar-wind speed was well below ~600 km s^{-1} the distribution functions of C^+ and O^+ over the entire speed range up to $W = 2$ could be measured. The C^+ and O^+ spectra we observed are shown in Figure 7.14 along with the velocity distribution of protons for the same time period. The distribution functions of C^+ and O^+ shown in Figure 7.14 are again nearly identical, reaching a maximum of ~10 s^3/km^6 near $W = 0.85$, and, as those shown in Figure 7.13, decrease rapidly with increasing W, indicating once more strong adiabatic cooling, and placing the location of the inner source well below 1.4 AU. For O^+ the spectrum becomes flat for $W > \sim 1.5$ with a cut-off at $W = 2$. The phase-space density in this high W interval is entirely consistent with oxygen of interstellar origin and an atomic oxygen density of 6.9×10^{-5} cm^{-3} at the termination shock. Thus, as close as 1.4 AU observations of interstellar pickup oxygen should be possible at least at solar minimum, since it appears to be not entirely obscured by the O^+ from the inner source.

The inner-source contribution to the proton spectrum is also quite evident. The kappa fit (curve c) to the solar-wind proton distribution leaves an excess density below $W = 0.65$. Curve (b), which is the fit to the C^+ spectrum multiplied by 2,000, fits the proton data for $W < 0.65$ extremely well. Because the density ratio of inner source H : C : O is close to the solar-wind ratio of $2,000 : 0.7 : 1$, these observations are most consistent with an origin for these inner-source ions being the solar wind that was first absorbed by interplanetary dust grains close to the Sun, and then released as slow-moving neutrals or ions (Gloeckler *et al.*, 2000; Schwadron *et al.*, 2000). Inner-source protons are also observed at high latitudes in the fast wind as is evident from Figure 7.2. However, at high latitudes the pickup hydrogen density from the inner source is much less (Schwadron *et al.*, 2000). The mass/charge histogram for heavy ions with W between 0.85 and 1.4 accumulated during roughly the same time period is given in Figure 7.15. In addition to C^+ and O^+, smaller amounts of N^+ and Ne^+ are observed as indicated by the dotted curve. Furthermore, there is a hint that some molecular species, such as CH^+ (mass 13) and H_2O^+ (mass 18) may also be present.

In Figure 7.16 we show the distribution functions of O^+, N^+, and Ne^+ averaged for a 1-year period when *Ulysses* was again near its 5-AU aphelion in the ecliptic plane. In all the spectra the typical shape of interstellar pickup ions is observed for W above $\sim$1.3. The model curves for the interstellar portion of the distributions (relatively flat solid curves) were calculated using interstellar atomic densities given in column 3 of Table 7.1. However, for W below $\sim$1.3 additional contributions, beyond that expected from the interstellar gas source are clearly evident in each of the spectra. This extra component of pickup ions (dotted curves) has most likely the same origin as the pickup C^+ discovered by Geiss *et al.* (1995). Thus, in addition to carbon, the inner source also produces pickup hydrogen, nitrogen, oxygen, and neon. The presence of substantial amounts of inner-source pickup Ne^+ and H^+ suggests strongly and perhaps conclusively that these particles originate predominantly as solar-wind ions which become embedded in dust grains, are eventually released as molecules and atoms, which are then ionized to form the pickup ions we observe. Ne, H, and to a lesser degree N should be depleted in grains, making it difficult to explain the composition we observe solely as the result of evaporation or sputtering of the original material contained in grains. We have extended our composition measurements of the inner-source pickup ions beyond Ne by looking at the mass per charge range above 20. Above m/q of $\sim$20 the phase space sampled by SWICS becomes progressively more restrictive with increasing ion mass and solar-wind flow speed, resulting in very small counting statistics for these heavy ions. Figure 7.17 (from Gloeckler *et al.*, 2000) shows the mass per charge histogram of inner-source pickup ions ($0.8 < W < 1.2$) averaged over the period from 21 November 1994 to 30 May 1995. Despite the poor statistics, peaks around mass/charge 24 and 28 (Mg^+ and Si^+, respectively) are evident. While our observations indicate roughly a solar-wind-type composition for the inner-source heavy pickup ions, the Mg and Si observed is probably a combination of the release of captured solar wind and sputtering of the original dust material.

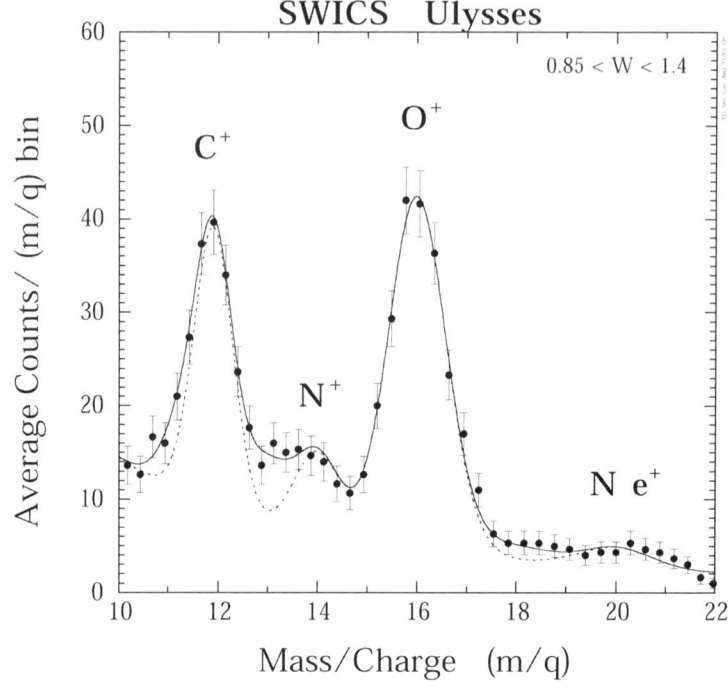

Figure 7.15. Plot of the double coincidence counts per mass/charge (m/q) interval of ions with W between 0.85 and 1.4 versus the mean m/q of each interval. These SWICS data were accumulated from $-44°$ to $+41°$ latitude during the fast latitude pass of *Ulysses*. The dotted curve is a fit to the data assuming an inner-source composition of only C^+, N^+, O^+, and Ne^+. The solid curve includes molecular ions CH^+ and H_2O^+ and solar wind Fe^{+5} in addition to C^+, N^+, O^+, and Ne^+.

The distribution functions of the inner-source pickup ions are complicated and not too well defined at present. The shape of the inner-source distributions viewed from difference is most likely due to different spatial source profiles at different latitudes. Peaked distributions would be expected to be the result of adiabatic cooling of ions picked up far away from where they are detected (Gloeckler *et al.*, 1993; Fisk, 1997; Vasyliunas and Siscoe, 1976; Gloeckler, 1996). If we accept adiabatic energy losses then the peaked distributions observed indicate a source located predominately upstream of *Ulysses*, very likely well within 1 AU from the Sun. The fact that the phase-space densities of both O^+ and C^+ always peak somewhat below $W = 1$ is consistent with very weak pitch-angle scattering and a large scattering mean free path (Schwadron *et al.*, 2000).

Schwadron *et al.* (2000) have used a simple transport model to provide an important consistency check for the postulated production mechanism for the inner-source pickup ions through release of atoms and molecules from grains saturated with solar-wind material. With this model they have also placed firm constraints on the total dust geometric cross-section. The model assumes a density

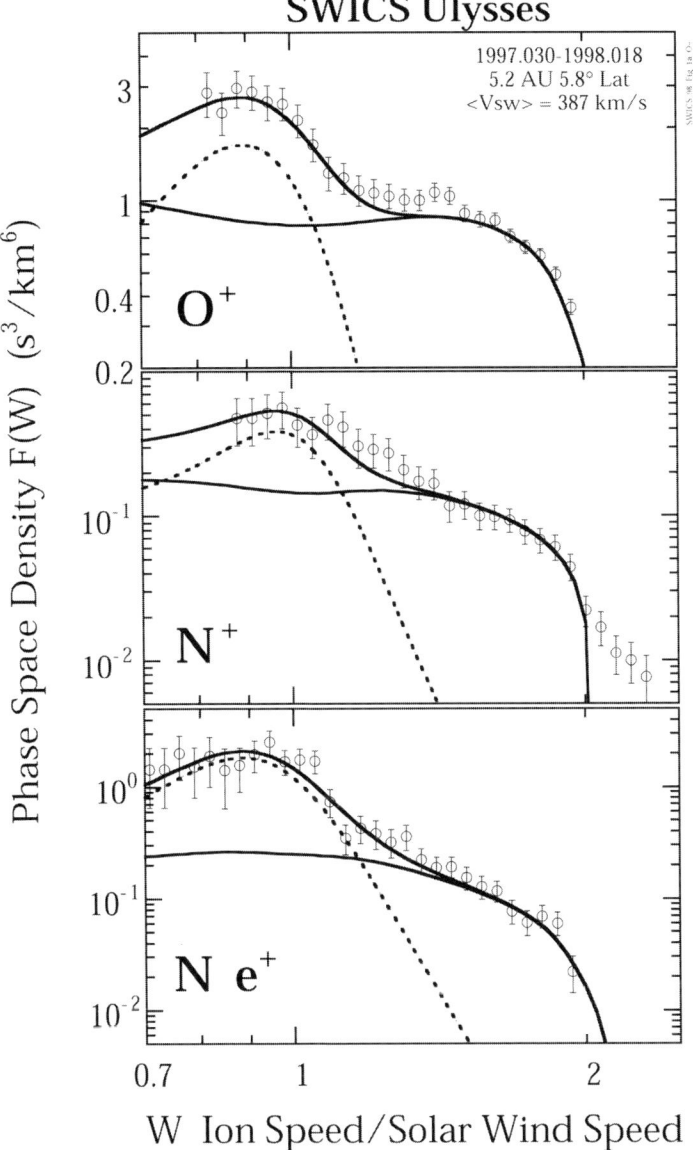

Figure 7.16. One-year averaged (30 January 1997–18 January 1998) velocity distributions of O^+, N^+, and Ne^+ when *Ulysses* was again in the ecliptic plane but at $\sim$5.2 AU. The long averaging interval was required in order to reveal the contributions from the inner source (dotted curves). In each of the panels the solid curve passing through the data points is the sum of the inner-source contribution (dotted curve) and the interstellar spectrum (relatively flat solid curve). A strong inner-source contribution for Ne indicates that the dominant source of inner-source pickup ions is the 'recycled solar wind'. O^+, N^+, and Ne^+ all have interstellar components that are seen above $W \sim 1.3$.

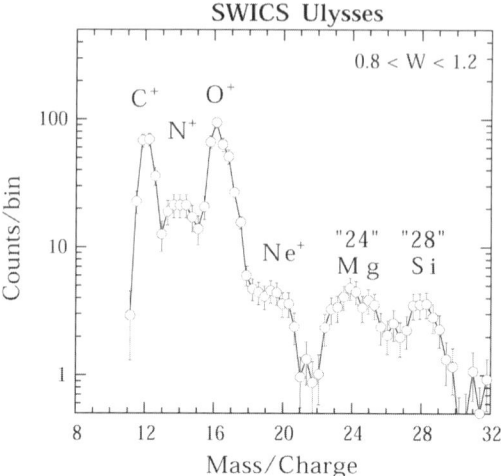

Figure 7.17. Mass per charge distributions of heavy pickup ions in the inner source. These data were taken when *Ulysses* was relatively close to the Sun at latitudes less than ~60°. A 5-point running average of counts in mass/charge bins greater than 17 was used to smoothe the data. Peaks at mass/charge 24 and 28 are clearly visible indicating the presence of singly charged Mg and Si (as well as C, N, O, and Ne) in the inner source (from Gloeckler *et al.*, 2000).

profile for the dust n_d given by Schwadron *et al.* (2000) and references therein:

$$n_d(R) = n_o(R_o/R)^p \exp\{-(L/R_o)[(R_o/R) - 1]\} \qquad (7.8)$$

where n_o is the total number of grains per unit volume at $R_o = 1$ AU, and p and L are free parameters. Schwadron *et al.* (2000) also assumed a solar-wind density given by $n_{sw}(R) = n_1(R_o/R)^2$ (n_1 is the solar-wind density at 1 AU), a constant solar-wind speed V_{sw}, and solar-wind abundance ratios χ_X of an element X relative to protons. Then the rate per unit volume at which an element X is absorbed by the grains is:

$$S_{aX}(R) = \chi_X n_{sw}(R) V_{sw} [n_d(R)/n_o] \Gamma \qquad (7.9)$$

where Γ is the total dust cross-section per unit volume. Presumably under equilibrium conditions the rate at which element X is released is equal to the rate at which it is absorbed. However, the particles released are likely to be in molecular form. Eventually these molecules break up and end up as ionized atoms in a complicated process (e.g. Gruntman, 1996). These complications are folded into a single factor ξ_X, taken to be a constant, which is the fraction of inner-source element X that is detected. Then the source rate per unit volume for inner-source pickup ions is given by:

$$S_{rX^+}(R) = \beta_{X^+}[R_o/R]^2 n_d(R) \qquad (7.10)$$
$$\beta_{X^+} = \xi_X \chi_X V_{sw} n_1 \Gamma / n_o \qquad (7.11)$$

where the β_{X^+} is the production rate of inner-source pickup ion X^+ at 1 AU.

The full transport equation for pickup ions is solved assuming weak scattering through 90° pitch angle (the angle between the magnetic field direction and the instantaneous ion velocity). Weak pitch-angle scattering of this type was shown by Fisk et al. (1997) and Gloeckler et al. (1995) to explain the large anisotropies observed in the distributions of interstellar pickup ions (Gloeckler et al., 1995) (see also Figures 7.2, 7.3, and 7.4). The solution to the transport equation then consists of an isotropic part $f_{\text{iso}} = (f_+ + f_-)/2$ and an anisotropic part $\Delta = (f_+ - f_-)/2$, where f_+ and f_- are the respective phase-space densities in the anti-sunward and sunward hemispheres. Given a constant scattering mean free path λ (expressed in units of AU), the following analytic solution for the pickup-ion distribution function in the solar-wind frame is obtained:

$$f_{\text{isoX}^+}(w, R) = \tfrac{3}{8}\pi V_{\text{sw}}^{-4} \beta_{\text{X}^+}[R_o^2/R] w^{-3/2} n_d(Rw^{3/2}) \quad (7.12)$$

$$\Delta_+(w, R) = -f_{\text{isoX}} \exp\{-(6Rw/\lambda)(1 - w^{1/2})\} \quad (7.13)$$

where R is the distance from the Sun where the distribution is observed, $w = |\mathbf{w}|$ is defined by equation (7.2), β_{X^+}, the production rate of inner-source pickup ion X^+, is given by equation (7.11), and $N_d(Rw^{3/2})$ is the density profile of the dust spatial distribution (equation 7.8) computed at the reduced distance $Rw^{3/2}$.

Using equations (7.12) and (7.13) Schwadron et al. (2000) have obtained fits (after transformation to the spacecraft reference frame, see Gloeckler, 1996) to the distribution functions of C^+ and O^+ shown in Figure 7.13 and H^+ (not shown) measured during all of 1994 when *Ulysses* was in the high-speed, high-latitude solar wind. From the best fits to the data they found the following values for the model parameters: $\lambda > 2$ AU, $n_o \beta_{H^+} = 6.2 \times 10^{-10}$ cm^{-3} s^{-1}, $L = 0.05$ AU, $p = 1.2$. They also found that the observed phase-space density ratio of inner-source pickup hydrogen to carbon was $= 310$, much below that ratio in the in-ecliptic wind pickup ions as well as the corresponding ratio in the typical solar wind. With these values of the model parameters they were able to place a lower limit for Γ of 1.3×10^{-17} cm^{-1}.

7.8 INTERACTION OF PICKUP IONS WITH THE SOLAR WIND

The large mean free path of pickup ions in the inner heliosphere has been remarked on in previous sections. Its cause, however, is not clear. Nor is the general lack of scattering of low-energy particles in the heliosphere well understood. One of the enduring problems in heliospheric physics is the long mean free paths of low energy, and low-rigidity particles in the solar wind, despite the apparent presence of substantial turbulence which should result in substantially smaller mean free paths (e.g. Bieber et al., 1994; Palmer, 1982).

In Fisk et al. (1997) the causes of the long mean free path of pickup ions was explored. Presumably, three potential causes should be considered, which in turn will have consequences for the distribution function of pickup ions which may be discernible:

1. Scattering could be small at all pitch angles, in which case we might expect that the distribution function of particles propagating outward in the solar-wind frame is peaked near pitch angles of 90°, or the distribution function in the solar-wind frame could be described as being proportional to $\sin(\alpha)$, where α is the pitch angle.
2. Scattering could be small only near 90° pitch angles, but particles experience scattering at other pitch angles. In this case we would expect that the distribution function of particles propagating outward relative to the solar wind would be nearly isotropic, although at a smaller value than the distribution function of inward-propagating particles.
3. Scattering could be small at 90° pitch angles and small at other pitch angles in the direction away from the Sun. In this case, particles would find that their pitch angles are reduced by the expansion of the magnetic field in the solar-wind frame, and the particles become focused around small values of the pitch angle. The distribution function of outward-propagating particles could then be described as proportional to $\cos(\alpha)$.

Distribution functions that are proportional to $\sin(\alpha)$ or $\cos(\alpha)$ are centred on the magnetic-field direction. Thus, if these are the forms of the distribution functions, the ratio of the distribution functions in the speed range $1.8 = W = 2.0$ to those in the range $1.6 = W = 2.0$ as seen by SWICS, should vary with the magnetic-field direction. Conversely, if the distribution function is isotropic there should be no correlation with the magnetic-field direction.

In Fisk et al. (1997) a careful analysis of the magnetic field and its fluctuations was made with the observed ratio of the distribution functions in the speed range $1.8 = W = 2.0$ to those in the range $1.6 = W = 2.0$, and with the predicted ratios for the $\sin(\alpha)$, $\cos(\alpha)$, and isotropic distribution functions. The results are shown in Figure 7.18. Clearly, the data are most consistent with a distribution of outward-propagating particles that is isotropic. A distribution function that is proportional to $\sin(\alpha)$ lies noticeably below the observed values; a distribution function proportional to $\cos(\alpha)$, noticeably above. Thus, the data are most consistent with the cause of the long mean free path to be the inability of the particles to scatter through pitch angles of 90°.

In quasilinear theory particles can have difficulty in scattering through 90° pitch angles because the required resonant wave number goes to infinity, where there is no power in the fluctuations. It would be perhaps surprising if this was the only reason for particles not scattering through 90° pitch angles since the heliospheric magnetic field does contain some magnitude fluctuations which can cause particles to scatter through 90°, and resonate broadening effects that may be important. There may be some other feature of turbulence in the solar wind which restricts scattering near 90° pitch angles.

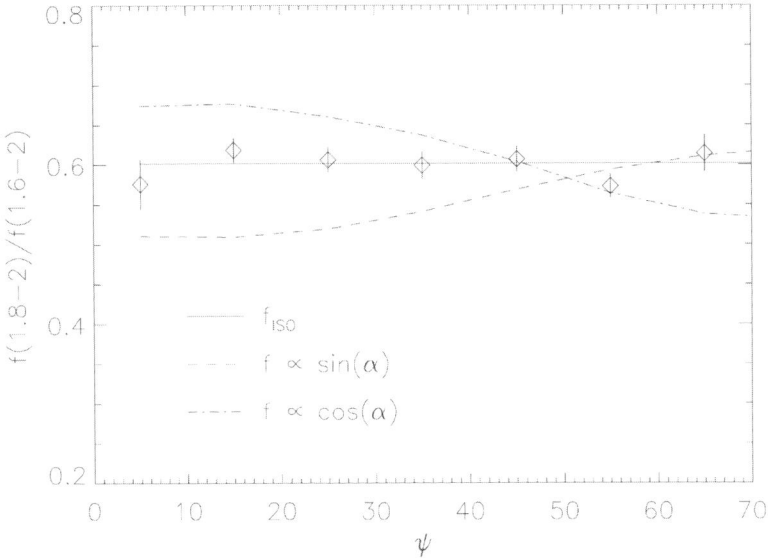

Figure 7.18. The dependence of the ratio of the average distribution function observed between W of 1.8–2.0 to that observed between W of 1.6–2.0 on the magnetic field angle ψ is compared with theoretical estimates (from Fisk et. al., 1997).

7.9 SUMMARY AND CONCLUSIONS

In this chapter we have provided an overview of pickup-ion observations with the SWICS instrument on *Ulysses*. While SWICS was designed and optimized for the first measurements of the composition of the solar wind under all conceivable flow conditions, the same multi-coincidence techniques used in SWICS and the favourable orbit of *Ulysses* made it possible to investigate six pickup-ion species of interstellar origin and to discover the inner source of pickup ions. Following the discovery of interstellar He^+ at 1 AU in 1985, we have now, with these *Ulysses* observations, increased our information on interstellar pickup ions substantially, and in the process extended significantly our knowledge of conditions in the local interstellar cloud and in the interface region between it and the heliosphere. Thus, from pickup-ion observations we were able to deduce the amounts of filtration of interstellar hydrogen and oxygen, find the ionization fractions of H and He in the local interstellar cloud, and put limits on the strength of the LIC magnetic field. Once *Voyager 1* locates the position of the termination shock it will be possible to deduce a firm value for the LIC magnetic field from pickup-ion measurements.

From pickup-ion observations we have now established the most comprehensive list of elemental abundances in the local interstellar cloud. The first measurement of $^3He/^4He$ in the present-day galaxy came from pickup He observed with *Ulysses*. The accuracy of many of these results will be improved with additional observations and analysis.

There were several results that were contrary to expectation. Observations of pickup ions forced us to conclude that pitch-angle scattering is inhibited at 90° pitch angle, and that pre-acceleration of ions, other than by shocks, is likely to occur. The existence of the ubiquitous high-velocity tails in the velocity distributions of pickup ions in the slow, in-ecliptic solar wind presents an interesting theoretical challenge.

A particularly noteworthy achievement was the discovery of a new population of pickup ions from an extended inner source, and the identification of the likely origin (recycled solar wind) for these inner-source pickup ions. Pickup-ion observations give us a powerful new tool to explore processes in the inner-source region which is not easily reached by spacecraft. Pickup-ion measurements from spacecraft at 1 AU, such as *WIND*, *SOHO*, and *ACE*, will add significantly to the information provided by *Ulysses* observations. Being closer to the inner source than *Ulysses* pickup-ion instruments on these spacecraft will give us a more accurate determination of the composition from this source, and a more detailed density profile of the dust distribution in the ecliptic plane within 1 AU. However, *Ulysses* observations of inner-source pickup ions remain unique because they give us the only chance to map the three-dimensional spatial structure of this source.

The *Ulysses* orbit does not allow us to find the kinetic temperature of the interstellar gas from observations of pickup ions. The temperature of interstellar helium has been measured with the GAS instrument on *Ulysses* by observing directly He atoms reaching the orbit of *Ulysses*. Pickup He ions measured in the gravitational focusing cone have recently provided a comparable value of $\sim 8,000$ K for the helium temperature. We anticipate that pickup-ion observations on *Cassini* will give us the temperatures of heavy interstellar atoms such as N, O, Ne, and Ar.

To achieve the next level of advances, beyond what is currently possible, will require pickup-ion instruments with a sensitivity several hundred times that of SWICS and comparable instruments now in use. This would improve dramatically the statistical accuracy of the composition measurements of the interstellar gas and allow for the first time precise determinations of isotopic abundance ratios of H, He, O, Ne, and Ar. These measurements have important implications for studies of Big Bang nucleosynthesis and galactic chemical evolution. With such instruments the inner source could be probed in far more detail. We have the technical ability to build such instruments now and place them into an appropriate elliptical orbit around the Sun, to sample both the interstellar gas at 3–4 AU and at the same time be close to the inner source at 1 AU or less.

7.10 ACKNOWLEDGMENTS

We are very grateful to the many individuals (see Gloeckler *et al.*, 1992) at the University of Maryland, the University of Bern, the Max-Planck-Institute für Aeronomie and the Technical University of Braunschweig for their numerous outstanding contributions to the success of the SWICS experiment on *Ulysses* during the many years of its development phase. We wish to thank especially Ed Tums, the SWICS project engineer, who in addition to designing the 30-kV and other high voltage supplies, supervised the overall design, fabrication and testing of SWICS, and Fred Ipavich, Fritz Gliem and Wolfgang Rieck for developing and implementing the onboard data-reduction algorithms which enabled high data recovery. Of particular benefit have been the many illuminating discussions with Nathan Schwadron and Thomas Zurbuchen concerning evolution of the distribution functions of pickup ions, with Daniel Rucinski regarding ionization rates of interstellar atoms in the heliosphere and with Priscilla Frisch and Vlad Izmodenov on abundances in the LIC. We thank Christine Gloeckler for her help with data reduction. This work was supported in part by NASA/JPL contract 955460 (G.G. and L.F.) and by the International Space Science Institute and the Swiss National Science Foundation (J.G.).

7.11 REFERENCES

Anders, E. and Grevesse, N. (1988) *Geochim. Cosmochim. Acta.* **53**, 197.
Axford, W. I. (1972) The interaction of the solar wind with the interstellar medium. *Proceedings of the Solar Wind Conference, Asilomar, CA, 1971*, edited by C. P. Sonett, P. J. Coleman, Jr and J. M. Wilcox. NASA SP–308, 609.
Balogh, A., Gonzalez-Esparza, J. A., Forsyth, R. J., Burton, M. E., Goldstein, B. E., Smith, E. J. and Bame, S. J. (1995) Interplanetary shock waves: Ulysses observations in and out of the ecliptic plane. *Space Sci. Rev.* **72**, 171.
Baranov, V. B. and Malama, Y. G. (1995) *J. Geophys. Res.* **100**, 755.
Bertaux, J. L. and Blamont, J. E. (1971) Evidence for a source of extraterrestrial Lyman alpha emission: The interstellar wind. *Astron. Astrophys.* **11**, 200.
Bieber, J. W., Matthaeus, W. H. and Smith, C. W. (1994) Proton and electron mean free path: the Palmer consensus revisited. *Astrophys, J.* **420**, 294.
Blum, P. W. and Fahr, H.-J. (1970) Interaction between interstellar hydrogen and the solar wind. *Astron. Astrophys.* **4**, 280–290.
Burlaga, L. F., Ness, N. F., Belcher, J. W., Lazarus, A. J. and Richardson, J. D. (1996) Voyager observations of the magnetic field, interstellar pick-up ions and solar wind in the distant heliosphere.. In: von Steiger, R., Lallement, R. and Lee, M. A. (eds), *The Heliosphere in the Local Interstellar Medium*, Space Sciences Series of ISSI, Vol. 1. *Space Sci. Rev.* **78**, 33.
Cheng, K.-P. and Bruhweiler, F. C. (1990) Ionization processes in the local interstellar medium: Effects of the coronal substrate. *Astrophys. J.* **364**, 573.
Dupuis, J., Vennes, S., Bowyer, S., Pradhan, A. K. and Thejll, P. (1995) Hot white dwarfs in the local interstellar medium: hydrogen and helium interstellar column densities and stellar effective temperatures from Extreme–Ultraviolet Explorer spectroscopy. *Astrophys. J.* **455**, 574.
Fisk, L. A. (1996) Implications of a weak termination shock. In: von Steiger, R., Lallement, R. and Lee, M. A. (eds), *The Heliosphere in the Local Interstellar Medium*, Space Sciences Series of ISSI, Vol. 1. *Space Sci. Rev.* **78**, 129.
Fisk, L. A., Schwadron, N. A. and Gloeckler, G. (1997) Implications of fluctuations in the distribution functions of interstellar pickup ions for the scattering of low rigidity particles. *Geophys. Res. Lett.* **24**, 93.
Fite, W. L., Smith, A. C. H. and Stebbings, R. F. (1993) *Proc. R. Soc. London A* **268**, 527.

Fränz, M., Keppler, E., Krupp, N., Rouss, M. K. and Blake, J. B. (1995) *Space Sci. Rev.* **72**, 339.

Frisch, P. C. (1995) *Space Sci. Rev.* **72**, 499.

Frisch, P. C. (1997) The interstellar cloud surrounding the solar system. In: Jokipii, J. R., Sonett, C. P. and Giampapa, M. S. (eds), *Cosmic Winds and the Heliosphere*. University of Arizona Press, Tucson, p. 733.

Frisch, P. C. (1998) Interstellar matter and the boundary conditions of the heliosphere. *Space Sci. Rev.* **86**, 107.

Frisch, P. C. and Slavin, J. D. (1996) Relative ionization in the nearest interstellar gas. In: von Steiger, R., Lallement, R. and Lee, M. A. (eds), *The Heliosphere in the Local Interstellar Medium*, Space Sciences Series of ISSI, Vol. 1. *Space Sci. Rev.* **78**, 223.

Geiss, J. and Witte, M. (1996) Properties of the interstellar gas inside the heliosphere. In: von Steiger, R., Lallement, R. and Lee, M. A. (eds), *The Heliosphere in the Local Interstellar Medium*, Space Sciences Series of ISSI, Vol. 1. *Space Sci. Rev.* **78**, 229–238, 1996.

Geiss, J., Gloeckler, G., Mall, U., von Steiger, R., Galvin, A. B. and Ogilvie, K. W. (1994a) Interstellar oxygen, nitrogen and neon in the heliosphere. *Astron. Astrophys.* **282**, 924–933.

Geiss, J., Gloeckler, G. and von Steiger, R. (1994b) Solar and heliospheric processes from solar wind composition measurements. *Phil. Trans. R. Soc. London Ser. A.* **349**, 213–226.

Geiss, J., Gloeckler, G. and Mall, U. (1994c) Origin of the O^+ pickup ions in the heliosphere. *Astron. Astrophys.* **289**, 933.

Geiss, J., Gloeckler, G., Fisk, L. A. and von Steiger, R. (1995) C^+ pickup ions in the heliosphere and their origin. *J. Geophys. Res.* **100**, 23373.

Geiss, J., Gloeckler, G. and von Steiger, R. (1996) Origin of C^+ ions in the heliosphere. In: von Steiger, R., Lallement, R. and Lee, M. A. (eds), *The Heliosphere in the Local Interstellar Medium*, Space Sciences Series of ISSI, Vol. 1. *Space Sci. Rev.* **78**, 43.

Gloeckler, G. (1996) The abundance of atomic ^1H, ^4He and ^3He in the local interstellar cloud from pickup ion observations with SWICS on Ulysses. In: von Steiger, R., Lallement, R. and Lee, M. A. (eds), *The Heliosphere in the Local Interstellar Medium*, Space Sciences Series of ISSI, Vol. 1. *Space Sci. Rev.* **78**, 335.

Gloeckler, G. (1999) Observation of injection and pre–acceleration processes in the slow solar wind. In: *Corotating Interaction Regions*, Space Sciences Series of ISSI. *Space Sci. Rev.* **89**, 91–104.

Gloeckler, G. and Geiss, J. (1996) Abundance of ^3He in the local interstellar cloud. *Nature* **381**, 210.

Gloeckler, G. and Geiss, J. (1998a) Measurement of the abundance of helium-3 in the Sun and in the local interstellar cloud with SWICS on Ulysses. In: *Primordial Nuclei and Their Galactic Evolution*, Space Sciences Series of ISSI, Vol. 4. *Space Sci. Rev.* **84**, 275.

Gloeckler, G and Geiss, J. (1998b) Interstellar and inner source pickup ions observed with SWICS on Ulysses. *Space Sci. Rev.* **86**, 127.

Gloeckler, G., Hovestadt, D. and Fisk, L. A. (1979) Observed distribution functions of H, He, C, O and Fe in corotating energetic particle streams: implications for interplanetary acceleration and propagation. *Astrophys. J.* **230**, L191.

Gloeckler, G., Geiss, J., Balsiger, H., Bedini, P., Cain, J. C., Fischer, J., Fisk, L. A., Galvin, A. B., Gliem, F., Hamilton, D. C., Hollweg, J. V., Ipavich, F. M., Joss, R., Livi, S., Lundgren, R., Mall, U., McKenzie, J. F., Ogilvie, K. W., Ottens, F., Rieck, W., Tums, E. O., von Steiger, R., Weiss, W. and Wilken, B. (1992) The solar wind ion composition spectrometer. *Astron. Astrophys. Suppl. Ser.* **92**, 267–289.

Gloeckler, G., Geiss, J., Balsiger, H., Fisk, L. A., Galvin, A. B., Ipavich, F. M., Ogilvie, K. W., von Steiger, R. and Wilken, B. (1993) Detection of interstellar pickup hydrogen in the solar system. *Science* **261**, 70–73.

Gloeckler, G., Geiss, J., Roelof, E. C., Fisk, L. A., Ipavich, F. M., Ogilvie, K. W., Lanzerotti, L. J., von Steiger, R. and Wilken, B. (1994) Acceleration of interstellar pickup ions in the disturbed solar wind observed on Ulysses. *J. Geophys. Res.* **99**, 17637.

Gloeckler, G., Schwadron, N. A., Fisk, L. A. and Geiss, J. (1995) Weak pitch angle scattering of few MV rigidity ions from measurements of anisotropies in the distribution function of interstellar pickup H^+. *Geophys. Res. Lett.* **22**, 2665.

Gloeckler, G., Fisk, L. A. and Geiss, J. (1997) Anomalously small magnetic field in the local interstellar cloud. *Nature* **386**, 374.

Gloeckler, G., Fisk, L. A., Geiss, J., Schwadron, N. A. and T. H. Zurbuchen, T. H. (2000a) The elemental composition of the inner source pickup ions. *J. Geophys. Res.* **105**, 7459–7463.

Gloeckler, G., Geiss, J., Schwadron, N. A., Fisk, L. A., Zurbuchen, T. H., Ipavich, F. M., von Steiger, R., Balsiger, H. and Wilken, B. (2000b) Interception of comet Hyakutake's tail at a distance of 500 million kilometres. *Nature* **404**, 576.

Grünwaldt, H., Neugebauer, M., Hilchenbach, M., Bochsler, P., Hovestadt, D., Bürgi, A., Ipavich, F. M., Reiche, K.-U., Axford, W. I., Balsiger, H., Galvin, A. B., Geiss, J., Gliem, F., Gloeckler, G., Hsieh, K. C., Kallenbach, R., Klecker, B., Livi, S., Lee, M. A., Managadze, G. G., Marsch, E., Möbius, E., Scholer, M., Verigin, M. I., Wilken, B. and Wurz, P. (1997) Venus tail ray observation near Earth. *Geophys. Res. Lett.* **24**, 1163.

Gruntman, M. (1996) H^{2+} pickup ions in the solar wind: outgassing of interplanetary dust. *J. Geophys. Res.* **101**, 15555.

Holzer, T. E. (1989) Interaction between the solar wind and the interstellar medium. *Ann. Rev. Astron. Astrophys.* **27**, 199.

Hovestadt, D., Klecker, B., Gloeckler, G., Ipavich, F. M. and Scholer, M. (1984) Survey of He^+/He^{2+} abundance ratios in energetic particle events. *Astrophys. J. Lett.* **282**, L39.

Lallement, R. (1996) Relations between ISM inside and outside the heliosphere. In: von Steiger, R., Lallement, R. and Lee, M. A. (eds), *The Heliosphere in the Local Interstellar Medium*, Space Sciences Series of ISSI, Vol. 1. *Space Sci. Rev.* **78**, 361.

Lallement, R., Bertaux, J.-L. and Clarke, J. T. (1993) *Science* **260**, 1095.

Linsky, J. L., Diplas, A., Wood, B. E., Brown, A., Ayres, T. R. and Savage, B. D. (1995) Deuterium and the local interstellar medium properties for the Procyon and Capella lines of sight. *Astrophys. J.* **451**, 335.

Möbius, E., Hovestadt, D., Klecker, B., Scholer, M., Gloeckler, G. and Ipavich, F. M. (1985) Direct observation of He^+ pick–up ions of interstellar origin in the solar wind. *Nature* **318**, 426–429.

Neugebauer, M., Lazarus, A. J., Altwegg, K., Balsiger, H., Goldstein, B. E., Goldstein, R., Neubauer, F. M., Rosenbauer, H., Schwenn, R., Shelley, E. G. and Ungstrup, E. (1987) The pick–up of cometary protons by the solar wind. *Astron. Astrophys.* **187**, 21–32.

Ogilvie, K. W., Gloeckler, G. and Geiss, J. (1995) Neutral oxygen near Jupiter. *Astron. and Astrophys.* **299**, 925.

Palmer, I. D. (1982) Transport coefficients of low-energy cosmic rays in interplanetary space. *Rev. Geophys Space Phys.* **20**, 335.

Reames, D. V., Richardson, I. G. and Barbier, L. M. (1991) *Astrophys. J. Lett.* **382**, L43.

Rucinski, D., Bzowski, M. and Fahr, H.-J. (1998) Minor helium components co–moving with the solar wind. *Astron. Astrophys.* **334**, 337–354.

Rucinski, D., Cummings, A. C., Gloeckler, G., Lazarus, A. J., Möbius, E. and Witte, M. (1996) Ionization processes in the heliosphere – rates and methods of their determination. In: von Steiger, R., Lallement, R. and Lee, M. A. (eds), *The Heliosphere in the Local Interstellar Medium*, Space Sciences Series of ISSI, Vol. 1. *Space Sci. Rev.* **78**, 73.

Scholer, M., Hovestadt, D., Klecker, B. and Gloeckler, G. (1979) The composition of energetic particles in corotating events. *Astrophys. J.* **227**, 323.

Schwadron, N. A., Fisk, L. A. and Gloeckler, G. (1996) Statistical acceleration of interstellar pick–up ions in co–rotating interaction regions. *Geophys. Res. Lett.* **23**, 2871.

Schwadron, N. A., Geiss, J., Fisk, L. A., Gloeckler, G., Zurbuchen, T. H. and von Steiger, R. (2000) Inner source distributions: theoretical interpretation, implications and evidence for inner source protons. *J. Geophys. Res.*, **105**, 7465–7472.

Siscoe, G. L. (1983) In: Carovillano, R. L. and Forbes, J. M. (eds), *Solar System Magnetohydrodynamics: Concepts and Basic Equations in Solar Terrestrial Physics*, p. 74. D. Reidel Publishing Co., Dordrecht, Holland.

Steinolfson, R. S. and Gurnett, D. A. (1995) Distance to the termination shock and heliopause from a simulation analysis of the 1992–1993 heliospheric radio emission event. *Geophys Res. Lett.* **22**, 651.

Stone, E. C., Cummings, A. C. and Webber, W. R. (1996) The distance to the solar wind termination shock in 1993 and 1994 from observations of anomalous cosmic rays. *J. Geophys. Res.* **101**, 11017.

Thomas, G. E. (1978) The interstellar wind and its influence on the interplanetary environment. *Ann. Rev. Earth Planet. Sci.* **6**, 173.

Thomas, G. E. and Krasse, R. F. (1971) Ogo–5 measurements of the Lyman alpha sky background. *Astron. Astrophys.* **11**, 218.

Vasyliunas, V. M. and. Siscoe, G. L. (1976) On the flux and the energy spectrum of interstellar ions in the solar system. *J. Geophys. Res.* **81**, 1247.

Wimmer-Schweingruber, R. F., von Steiger, R. and Paerli, R. (1997) Solar wind stream interfaces in corotating interaction regions: SWICS/Ulysses results. *J. Geophys. Res.* **102**, 17407.

Witte, M., Rosenbauer, H., Banaszkiewicz, M. and Fahr, H.-J. (1993) *Adv. Space Res.* **13**(6), 121.

Witte, M., Banaszkiewicz, M. and Rosenbauer, H.(1996) Recent results on the parameters of the interstellar helium from the Ulysses/GAS experiment. In: von Steiger, R., Lallement, R. and Lee, M. A. (eds), *The Heliosphere in the Local Interstellar Medium*, Space Sciences Series of ISSI, Vol. 1. *Space Sci. Revs.* **78**, 289.

Zank, G. P. and Pauls, H. L. (1996) Modeling the heliosphere. In: von Steiger, R., Lallement, R. and Lee, M. A. (eds), *The Heliosphere in the Local Interstellar Medium*, Space Sciences Series of ISSI, Vol. 1. *Space Sci. Rev.* **78**, 95.

8

Cosmic rays at all latitudes in the inner heliosphere

R. Bruce McKibben

8.1 INTRODUCTION

As *Ulysses* continues its exploration of the heliosphere during its second orbit of the Sun, our picture of the heliosphere is quite different from the heliosphere as we imagined it when *Ulysses* was launched in 1990. Results from *Ulysses*' first polar orbit of the Sun are primarily responsible for the change in our perceptions. For cosmic rays, *Ulysses*' first exploration of the high solar latitude regions discovered a heliosphere which, at solar minimum, was far more homogeneous and allowed far easier propagation in three dimensions than had been anticipated from either theory or observations available before the launch of *Ulysses*. On the other hand, *Ulysses* also found that the high-latitude inner heliosphere was much less accessible to cosmic radiation than the perhaps naive ideas current during the development of the *Ulysses* mission had suggested. Rather than being able to observe the nearly unmodulated cosmic-ray flux over the poles, as had been hoped, *Ulysses* found that modulation at high latitudes was only slightly weaker than near the equator, even though the solar wind and magnetic structure changed dramatically from the equator to the poles. The effort to assimilate and understand the new information is ongoing, and is providing for stimulating re-imaginings of the modes of propagation of cosmic rays through the heliosphere. The first part of this chapter (Section 8.2) will review these observations and the ideas that they have stimulated during the first polar orbit of *Ulysses* about the Sun.

While *Ulysses*' primary goal may have been exploration of the heliosphere, it also carried experiments that look beyond the heliosphere to questions more closely related to galactic astrophysics than to heliospheric physics. For cosmic rays, the expectation that the low-energy ($\sim 10^2$ MeV/nucl) cosmic-ray flux over the Sun's poles might be subject to much weaker modulation than in the ecliptic justified the inclusion of very high resolution instrumentation to study the elemental and isotopic composition of the cosmic radiation at low energies for elements from hydrogen through nickel. In the expected high intensities over the poles, it was hoped that in its few months of observations at polar latitudes *Ulysses* would be able to make critical measurements bearing on the propagation of low-energy cosmic rays through the Galaxy, the nucleosynthetic history of cosmic rays, and on the flux and spectrum of cosmic rays in interstellar space down to energies inaccessible from observations in the equatorial zone. While hopes for a high-latitude window to the unmodulated low-energy interstellar spectrum were disappointed, the composition measurements made by *Ulysses* using data accumulated throughout the mission provided the first clear separation of isotopes of elements up through the iron peak. These measurements have already had significant impact on models of the origin, acceleration, and galactic propagation of cosmic rays. Section 8.3 will review the progress and impact of *Ulysses* composition measurements on our understanding of cosmic rays in the Galaxy.

Finally, in Section 8.4 we will look forward to the discoveries yet to be made as *Ulysses* continues its exploration of the latitudinal structure of the heliosphere during the solar-maximum phase of its periodic variation in response to the solar activity cycle.

8.1.1 Motivation for the *Ulysses* investigations

The overall motivation for the *Ulysses* mission was succinctly expressed by Page (1985) in his opening remarks for the *19th ESLAB Symposium*, which was dedicated to defining what was then known of the three-dimensional heliosphere as a background to studies to be performed with *Ulysses* after its expected launch in 1986:

> *Since the beginning of the space age there have been those who realized that since the narrow slice of the heliosphere in which the earth finds itself is probably so unrepresentative of the heliosphere as a whole, a mission out of the ecliptic had to be organized.*

For cosmic rays, the difficulty of in-ecliptic observations is that the cosmic rays are constrained to first order to follow the Archimedean spiral interplanetary field lines. If they were strictly to follow the field lines in from the boundary at a presumed distance of ~100 AU to reach Earth, they would have to traverse path lengths as long as 5,000 AU along the field. As a result, the entry of cosmic rays at low heliographic latitudes is very sensitive to diffusion perpendicular to the magnetic field, which, for resonant scattering, is generally very much slower than diffusion along the field. Thus, solar modulation is very effective in the ecliptic, completely excluding particles with interstellar energies less than several hundred MeV/nucl from the inner heliosphere (e.g. Urch and Gleeson, 1971). Because of the geometry of the Archimedean spiral as a function of latitude, it was presumed that much shorter path lengths along the field lines would allow more cosmic rays to reach the inner heliosphere at high latitudes, leading, possibly, to much weaker modulation. Since interplanetary scintillation measurements (e.g. Rickett and Coles, 1983) indicated a faster solar wind at high latitudes, the Archimedean spiral at high latitudes was expected to be even more loosely wound, shortening the path still further. In the ideal case, the field line emanating from the rotational pole of the Sun would be a straight line with direct connection to the interstellar medium. Thus, with the near-ecliptic *Pioneer 10* spacecraft still deeply embedded in the modulation region at a distance of ~35 AU from the Sun during *Ulysses'* development (e.g. McKibben, 1985), the best near-term hope for observing the nearly unmodulated cosmic-ray spectrum seemed to lie at high latitudes. Whether or not the unmodulated cosmic-ray beam would be observed at polar latitudes, by the early 1980s it had begun to be realized that, whereas modulation had so far been treated primarily with spherically symmetric models, these models could not be accurate physical descriptions of the modulation process. Not only was there direct evidence of departures from spherical symmetry in the heliosphere, as, for example, in the scintillation measurements of high-latitude solar-wind speeds referred to above, but there were significant effects in modulation, primarily gradient and curvature drifts in the non-uniform interplanetary magnetic field, that were intrinsically three dimensional and could not be described or incorporated in spherically symmetric models. Thus, to test predictions of the newly developed non-spherically symmetric models (e.g. Kóta and Jokipii, 1983; Burger *et al.*, 1987), an out-of-the-ecliptic mission was required.

8.1.2 Status of modulation and cosmic-ray composition studies prior to *Ulysses*

8.1.2.1 Modulation studies

The general status of observations, theory, and modelling of modulation just prior to the launch of *Ulysses* was summarized by McKibben (1987), Stone (1987), and by several authors in a collection of papers at the *First COSPAR Colloquium* (McKibben, 1989a; Le Roux and Potgieter, 1989). At the time of *Ulysses* launch, the necessity of including gradient and curvature drifts in models of modulation was generally agreed as a result of confirmation of three striking predictions of drift models for which no other explanation was readily at hand.

First, as shown by numerous observations (e.g. McKibben, 1989a), the profile of the cosmic-ray intensity maximum in the solar minimum of 1987 was sharply peaked, as predicted by drift theories for the negative sign of the solar magnetic dipole then in effect (Kóta and Jokipii, 1983). The intensity profiles of previous minima in 1965 (peaked, negative dipole) and 1975-1976 (broad, positive dipole) had also been consistent with drift model predictions.

Second, clear changes in the electron/nucleon ratio consistent with predictions of drift models (e.g. Potgieter and Moraal, 1985) were observed with the change in sign of the solar dipole during a 22-year solar magnetic cycle (Garcia-Munoz *et al.*, 1986).

Third, following its fly-by of Saturn, *Voyager 1* had risen to ~20°N latitude in 1986 when the sign of the solar dipole was negative, and had detected a clear negative latitudinal gradient in the fluxes of cosmic-ray nuclei and anomalous components (Christon *et al.*, 1986a; Cummings *et al.*, 1987). A positive latitudinal gradient had been found when *Pioneer 11* rose to 17°N latitude in 1975–1976 (Bastian *et al.*, 1979; McKibben, 1989b), when the sign of the solar dipole was positive. Both observations were consistent with predictions of theories of modulation incorporating drifts (e.g. Kóta and Jokipii, 1983). With these confirmations, *Ulysses*' observations were eagerly anticipated to provide a crucial test of both the general and detailed predictions of drift modulation models.

8.1.2.2 Composition studies

Cosmic rays constitute the only sample of matter to which we have direct access that comes from remote regions of the Galaxy. Their elemental and isotopic composition contains information about their origins, acceleration, and propagation through the Galaxy obtainable in no other way. As a result, one of the long-term goals of cosmic-ray astrophysics has been to develop a complete catalogue of the elemental and isotopic abundances of cosmic-ray nuclei. This goal has been addressed using instruments of steadily increasing geometric factor and resolution. This continuing development has culminated today in the *Ulysses* COSPIN High Energy Telescope (HET), which has been making measurements with isotopic resolution from hydrogen through nickel since 1990, and in the Cosmic Ray Isotope Spectrometer (CRIS) on the *Advanced Composition Explorer* (*ACE*), which is now collecting data with a geometric factor in excess of $100\,\text{cm}^2\,\text{sr}^{-1}$ and isotopic resolution for all elements from hydrogen through zinc.

At the time of *Ulysses*' launch, the state of abundance measurements was much less developed. A summary of the state of measurements as of 1983 (the originally scheduled launch date of *Ulysses*) is to be found in reviews by Simpson (1983) and by Mewaldt (1989), which provide comprehensive lists of references on which the following summary is based.

At energies below $\sim 1\,\text{GeV/nucl}$ the elemental abundances had been well measured for elements up through nickel, and less well measured, primarily because of much poorer available statistics due to their low abundances, through the actinide elements (Binns *et al.*, 1989). Comparisons with solar-system abundances (Cameron, 1982) showed clear evidence of the effects of propagation through the Galaxy, establishing a mean path length of $\sim 8\,\text{gm}\,\text{cm}^{-2}$ at these energies. After correction for spallation of nuclei during propagation through the interstellar medium, elemental abundances at the cosmic-ray source were found to be remarkably similar to solar-system abundances if allowances were made for preferential injection and acceleration of elements with low first-ionization potential or low volatility (e.g. Meyer *et al.*, 1997 and references therein).

Isotopic measurements had been for the most part restricted to the lighter elements. Measurements of the radioactive isotope ^{10}Be had determined the cosmic-ray mean age to be approximately 10–15 million years and the mean density through which the cosmic rays had travelled to be ~ 0.1–$0.3\,\text{atoms}\,\text{cm}^{-3}$. It was not yet possible to measure with sufficient precision other radioactive clock isotopes (^{26}Al, ^{36}Cl, and ^{54}Mn) that could confirm this conclusion. However, the relatively young age established for the cosmic rays compared with the age of solar-system materials ($\sim 4.5 \times 10^9$ yr) offered the possibility, with sufficiently accurate measurements, of performing significant studies of galactic chemical and isotopic evolution.

As *Ulysses* was being prepared for launch, some isotopic abundance results with useful accuracy had been reported for elements up through Si, based primarily on measurements from instruments on *IMP-8*, *ISEE-3*, and *Voyager 1* and *2*, and a few results of low accuracy were available up through Fe and Ni. Of the elements measured, only Ne showed significant deviations from solar-system abundances after correction back to the source, a situation that has been confirmed by more recent measurements as discussed below. Of the instruments reporting isotopic measurements, only the *ISEE-3* instrument employed trajectory sensing detectors, which gave the potential for extending isotopic measurements through the FeNi range of elements. Such measurements were reported by Leske (1993).

The COSPIN High Energy Telescope (HET) on *Ulysses* (Simpson *et al.*, 1992), together with a nearly identical telescope flown almost simultaneously on the short-lived Earth-orbiting mission *CRRES* (*Chemical Release and Radiation Effects Satellite*) (DuVernois *et al.*, 1996a) were the first instruments in space to provide stable, large geometrical factor trajectory sensing that permitted accurate measurements of the isotopic composition of elements from hydrogen through nickel. Principal objectives of the *Ulysses* measurements were to search for isotopic variations from solar-system abundances, to examine other radioactive clocks for determining the cosmic-ray age, to look for evidence of an enhanced contribution from explosive nucleosynthesis as evidence for direct injection of supernova ejecta into the cosmic-ray source, and to look for evidence of delayed acceleration and re-acceleration from examination of electron-capture isotopes. All of these measurements required the high isotopic resolution that *Ulysses* HET could provide.

8.1.3 Summary of available *Ulysses* measurements

For study of the cosmic radiation, defined to consist of electrons and nuclei above about 10 MeV/nucl, almost all measurements available from *Ulysses* were provided by the instruments of the Cosmic and Solar Particle Investigations (COSPIN) consortium, described in detail by Simpson *et al.* (1992). As shown in Figure 8.1, for electrons, protons, and helium nuclei, measurements distributed among the five telescope systems of the COSPIN experiment package provided clean identification of these particles with good energy resolution over an energy range from a few tenths of an MeV/nucl, through several hundred MeV/nucl. Pulse height analysis with the Kiel Electron Telescope (KET) extended this range to more than 1,000 MeV/nucl for protons and helium. Integral fluxes were measured above the prime analysis ranges. For composition studies, as shown in Figure 8.2, the Low Energy Telescope (LET) and High Energy Telescope (HET) provided PHA-identified elemental resolution from hydrogen through Fe (Ni for the HET) over a nearly continuous energy range from a few MeV/nucl through a few hundred MeV/nucl, depending upon the nuclear charge. In addition, using its power to determine and correct for the angle of incidence using trajectory sensing detectors, the HET provided isotopic

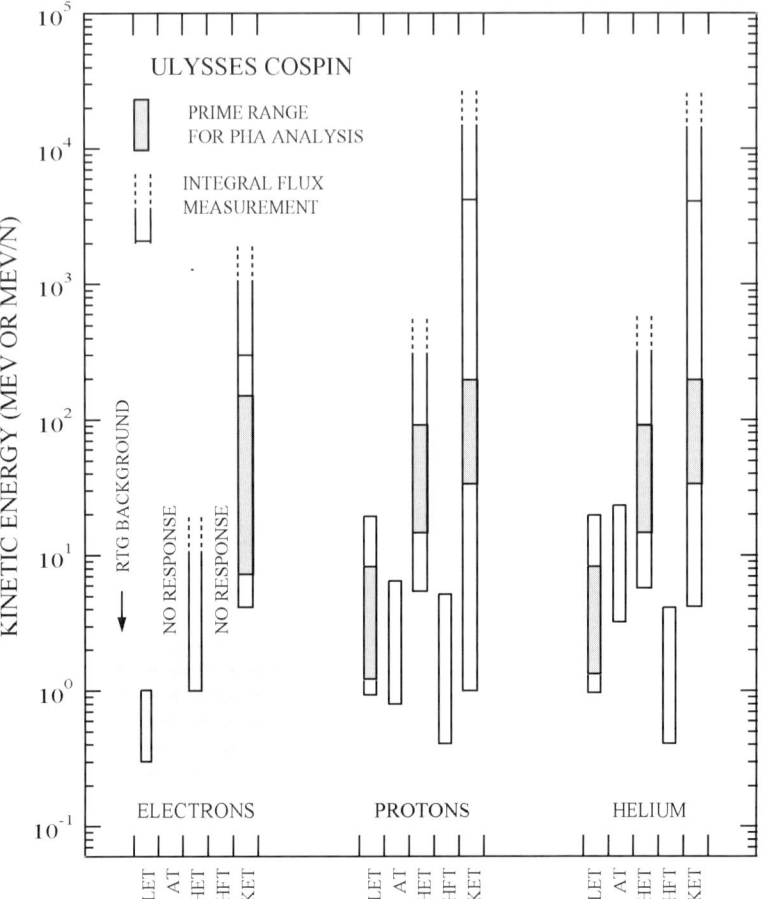

Figure 8.1. Energy ranges for electrons, protons, and helium over which COSPIN sensors provide measurements. Energies >10 MeV are of principal interest for cosmic-ray studies (from Simpson *et al.*, 1992).

resolution from ~200–400 MeV/nucl at iron, and at somewhat lower energies for lighter nuclei.

The measurement capabilities of the COSPIN experiment were oriented both towards the requirements of an exploratory mission, where wide dynamic range and broad capabilities are required to provide useful information from a range of environments that could not be fully anticipated, and towards the need to make definitive measurements during the short periods at high latitudes where the opportunity to make further measurements was unlikely to be repeated for decades after the end of the *Ulysses* mission. A particular driving goal was to achieve the precision required to make definitive measurements of the cosmic-ray composition during the few months spent in the expected low-modulation region over the poles, and to

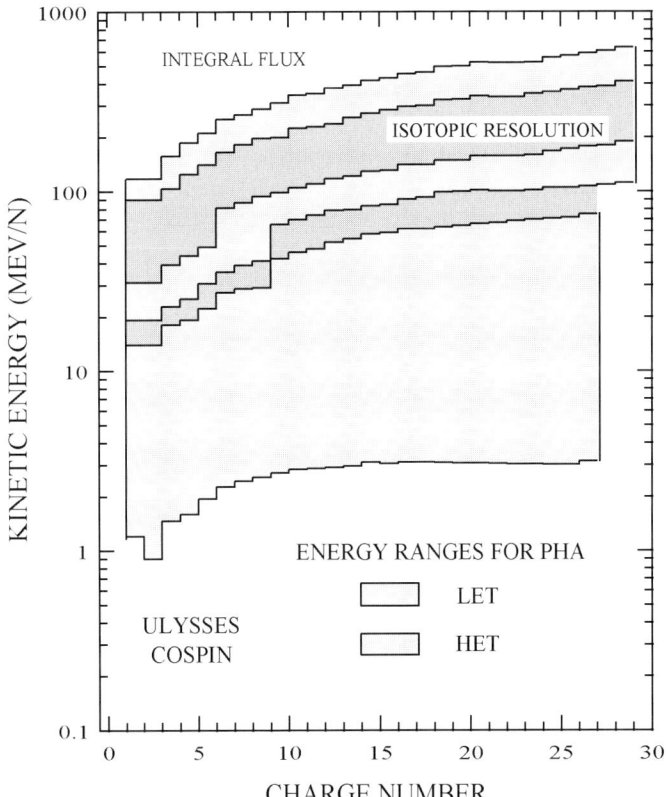

Figure 8.2. Energy and nuclear charge ranges over which COSPIN sensors provide measurements for study of the elemental and isotropic composition of cosmic rays (from Simpson *et al.*, 1992).

provide the necessary energy resolution, particularly for electrons, to study the evolution of the modulated spectra of electrons and nucleons as a function of location in the heliosphere. Thus the measurements were driven both by questions of astrophysical interest, and by questions concerning the structure of the heliosphere.

While the instrumentation was not designed to provide definitive measurements in Jupiter's magnetosphere during the fly-by, it nevertheless provided useful information concerning the radiation in and escaping from the magnetosphere. In making the first traversal of the dusk side of the Jovian magnetosphere, it discovered intense activity, including large-scale organized particle flows and quasi-periodic bursts of energetic nucleons and electrons with intensities sufficient to make significant contributions to the energetic-particle population filling Jupiter's outer magnetosphere (Zhang *et al.*, 1995). In particular, studies of Jovian electrons in the solar wind provided unique information for studies of particle propagation in the heliosphere and will be discussed below.

8.2 HELIOSPHERIC MODULATION OF COSMIC RAYS, ELECTRONS AND ATYPICAL COMPONENTS

8.2.1 Summary of modulation theory and available modelling tools.

While there had been clear observational evidence for the influence of solar activity on the cosmic radiation since the work of Forbush (1938a, b, 1954), the basis for our current understanding of modulation was laid out by Parker (1965), who first wrote down the equation (8.1) that describes the variation in the density U (particles per cm^3 per MeV) of cosmic rays in the heliosphere as a result of their interaction with the interplanetary magnetic field carried by the solar wind:

$$\nabla \cdot \kappa \cdot \nabla U - \nabla \cdot \mathbf{V} U + \tfrac{1}{3} \nabla \cdot \mathbf{V} \frac{\partial}{\partial T}(\alpha T U) - \mathbf{v}_\mathrm{D} \cdot \nabla U = \frac{\partial U}{\partial T} = 0 \qquad (8.1)$$

The interplanetary field is frozen into the solar wind in the upper corona and then carried outward throughout the heliosphere, forming an Archimedean spiral as a result of the rotation of the Sun at the base of the solar wind once every 26 days. Setting $\frac{\partial U}{\partial T} = 0$ in equation (8.1) assumes that the cosmic-ray modulation is in a steady state (i.e. that galactic cosmic rays are sufficiently mobile so that they come into equilibrium with heliospheric conditions in a time short compared with the timescale for significant changes in these conditions). We now know that this is seldom the case, but the basic physics of the process is unaffected if $\frac{\partial U}{\partial T} \neq 0$.

The terms in equation (8.1) correspond, in order, to:

- *diffusion* of the cosmic rays through the irregular magnetic field, governed by a diffusion tensor κ which describes the scattering properties of the field for transport both along and perpendicular to the field direction;
- *convection* by the solar wind with a velocity $\mathbf{V}$;
- *adiabatic deceleration* of the cosmic rays as a result of their being embedded in the radially expanding solar wind (T is the kinetic energy of the particles, and $\alpha = (T + 2T_0)/(T + T_0)$ where T_0 is the particle rest mass); and
- *gradient and curvature drifts* arising from the gradients of the interplanetary field as a result of its radial expansion and the curvature corresponding to the spiral structure of the field. The drift is characterized by a velocity $\mathbf{v}_\mathrm{D}$, given by:

$$\mathbf{v}_\mathrm{D} = \frac{cvp}{3q} \nabla \times \frac{\mathbf{B}}{B^2} \qquad (8.2)$$

where c is the speed of light, v, p, and q are the particle speed, momentum, and charge, respectively, and $\mathbf{B}$ is the magnetic field. In Parker's original equation, the drifts were expressed as off-diagonal elements in the diffusion tensor. In early studies of modulation, the role of drifts was neglected. However, a series of papers by Jokipii and his co-workers beginning with Jokipii *et al.* (1977) demonstrated that in the classical Parker Archimedean spiral field the drift velocities for most cosmic rays could be comparable with the solar-wind velocity. This realization sparked development of increasingly sophisticated computational models to incorporate this intrinsically 3-dimensional effect into theoretical descriptions of the modulation.

Drifts take on special significance for *Ulysses* measurements because, of all the processes that are included in the modulation equation, only drifts produce radically different motions through the heliosphere for electrons and nucleons. Because the drift velocities are normal to the mean magnetic field and to the gradient and radius of curvature of the field, for the classical Archimedean spiral model field, the drifts transport particles primarily in the latitudinal direction throughout most of the heliosphere, with the velocities oppositely directed for electrons and protons. Thus significantly different latitudinal profiles of modulation might be expected for electrons and nucleons. The Sun's magnetic polarity reverses at solar maximum every 11 years, and for almost the entire first solar orbit of *Ulysses* the north pole of the Sun had positive magnetic polarity. In such a field, positively charged nucleonic cosmic rays drift in over the poles and downward towards the equator, and positive latitudinal gradients would be expected for these particles. Electrons drift in the opposite sense, inwards along the equatorial current sheet and up toward the poles, leading possibly to negative latitude gradients. However, Potgieter *et al.* (1977) find that realistic values of modulation parameters can produce positive latitude gradients for low-rigidity electrons, even in the presence of drifts. Nevertheless, there is a firm expectation that spatial and temporal profiles of electrons and nucleons will differ significantly as a result of the influence of drifts.

Because of the complexity of the modulation equation (equation 8.1), analytic solutions have not been found, even for simple and idealized representations of the heliosphere. Thus approximations and numerical methods must be used to solve the equation and to investigate the implications of various model choices for the observed modulation. The complexity of the software required to obtain useful results has limited the number of fully developed codes for study of modulation. By the early 1990s, only two groups, one centred at the University of Arizona in the USA and the other at Potchefstroom University in South Africa, had achieved computational codes that treat more than one spatial dimension in modulation. At the time of the launch of *Ulysses*, computational models had advanced to the point of allowing three-dimensional simulation (either two spatial dimensions and time [Le Roux and Potgieter, 1989] or three spatial dimensions [Kóta and Jokipii, 1983]) of the modulation, which allowed full inclusion of the effects of drifts into the analysis.

Based on these models, and the best available knowledge prior to the high-latitude phase of the *Ulysses* mission, a prediction of the intensity profile as a function of latitude was made by Potgieter and Haasbroek (1993). Their model predicted a factor of ~ 10 increase in the intensity of ~ 1-GeV protons from the ecliptic to $80°$ heliographic latitude, and a factor of ~ 400 increase for ~ 200-MeV protons. That no such dramatic changes were observed is one of the major

336 Cosmic rays at all latitudes in the inner heliosphere [Ch. 8

discoveries of *Ulysses* and has posed significant challenges for modulation models and our understanding of cosmic ray propagation in the heliosphere.

8.2.2 Observations of solar cycle modulation

Figure 8.3 contains a summary and overview of the intensity of cosmic-ray nucleons at Earth and *Ulysses*, and of the solar-wind speed and magnetic field at *Ulysses* from the time *Ulysses* left the ecliptic following Jupiter fly-by in 1992 through the end of

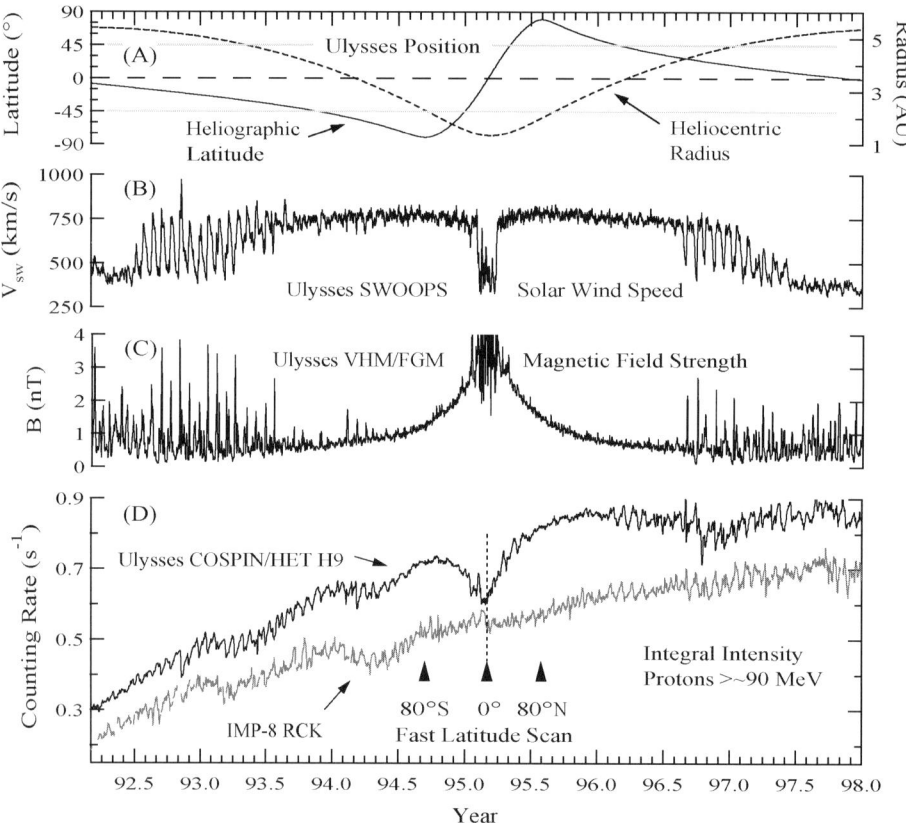

Figure 8.3. (A) *Ulysses* heliographic longitude and latitude versus time. To guide the eye, grey horizontal lines mark latitudes of ±45°. (B) 6-hour average solar-wind speeds from the *Ulysses* SWOOPS instrument. (C) Daily average magnetic-field strength from the *Ulysses* VHM/FGM instrument. (D) Daily average counting rates of protons >~90-100 MeV from the *Ulysses* COSPIN HET and the University of Chicago cosmic-ray experiment on the Earth satellite *IMP-8*. Normalization of the two counting rates is arbitrary. The mean energy of response is ~2 GeV.

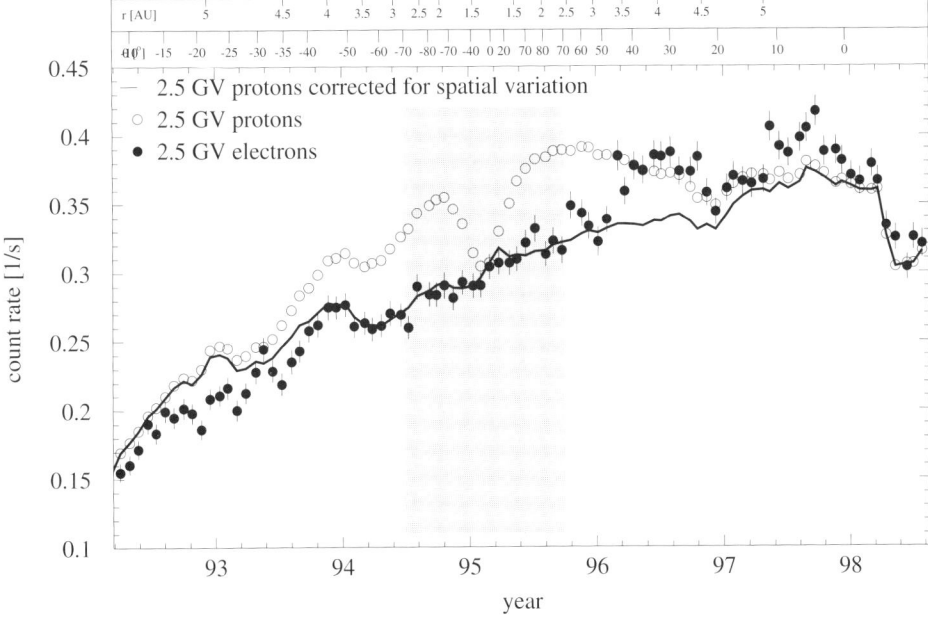

Figure 8.4. Average intensities (26-day) of 2.5-GV electrons (filled circles) and protons (open circles). The solid line is the proton intensity corrected to the equator using measured radial and latitude gradients (from Heber et al., 1999a).

solar minimum in late 1997. In this figure the modulation is represented by the integral intensity of cosmic rays $>\sim 90$ MeV, for which the mean energy of response is ~ 2 GeV. Essentially all of the significant conclusions relating to the three-dimensional solar modulation of nucleonic cosmic rays can be derived from inspection of this figure. At lower energies for both galactic cosmic rays and anomalous components, the magnitude of changes varies, but the qualitative picture is the same. For electrons, less comprehensive observations exist, primarily because of the great intrinsic difficulty of making clean, well-resolved measurements of the cosmic-ray electron flux (see Ferrando et al., 1996 and Heber et al., 1999b for a discussion of the electron response of the KET). Furthermore, no measurements directly comparable with those on *Ulysses* exist from instruments at Earth. A summary of the *Ulysses* electron measurements, compared with protons of similar energy, is shown in Figure 8.4.

8.2.2.1 Long-term temporal variations during the Ulysses mission

The most significant change in intensity both at *Ulysses* and at Earth is the overall increase in intensity due to the decline of solar activity, and consequently, of the strength of solar modulation. In fact, the first conclusion that can be drawn from Figure 8.3 is that, aside from the clearly latitude-related changes observed during the

Fast Latitude Scan (FLS) in 1994–1995, the profile of modulation was essentially the same at Earth and *Ulysses* throughout the period shown.

Ulysses was launched in 1990 near the time of solar maximum, and minimum solar activity was not observed until 1997. Thus, underlying all intensity variation resulting from *Ulysses*' changing position in the heliosphere was a temporal trend of increasing intensity that was much larger than any spatial effects. For the relativistic cosmic rays shown in Figure 8.3, the increase at Earth was more than a factor of 3. For lower energy cosmic rays and especially for anomalous components, the increase was much greater, for example exceeding a factor of 10 for $\sim$30–70 MeV/nucl helium, which is a mix of galactic and anomalous component helium.

Because of the magnitude and rate of the changes in the solar-cycle modulation, it becomes especially important to have a monitor of purely temporal changes at 1 AU. For almost all cosmic ray and anomalous component fluxes measured by *Ulysses*, *IMP-8* has provided the necessary reference measurements. The exception is the electron measurements provided by the KET, for which no satisfactory reference monitor exists. This raises special difficulties for interpretation of the electron measurements.

Aside from predicted spatial effects for the electron versus proton fluxes, drift models also predict different modulation for electrons and protons in alternate solar cycles, depending on the sign of the solar magnetic dipole. During the solar minimum of 1972–1977, when the sign of the north solar pole magnetic field was positive, the electron-to-nucleon ratio at the same rigidity was observed to be higher than in the solar minimum of 1987, where the magnetic polarity was reversed (e.g. Evenson, 1998). During the *Ulysses* mission, the north-polar polarity has been consistently positive, so drift models would predict a high electron-to-proton ratio at solar minimum.

As shown in Figure 8.4, *Ulysses* observations at least qualitatively support this prediction. The solid line in the figure represents the proton intensity measured by the KET corrected to the ecliptic in a manner to be discussed below, and normalized to the electron flux during *Ulysses*' crossing of the ecliptic in 1995. Throughout the period shown, the electron-to-proton ratio was gradually increasing, and the highest electron-to-proton ratios were obtained in 1996 and 1997, very near solar minimum. Heber *et al.* (1999a) discuss these observations in relation to a drift-dominated modulation model with a wavy equatorial current sheet, and demonstrate an anti-correlation between the tilt of the current sheet and the electron-to-proton ratio, in qualitative agreement with expectations from modulation models in which drifts play an important role.

8.2.2.2 *Radial gradients*

The *Ulysses* mission trajectory is not well suited for measurement of radial gradients. Except for the initial Earth–Jupiter transfer portion of the trajectory, radial motion of the spacecraft has in all cases been coupled with latitudinal motion, leading to ambiguities as to whether observed changes are dominated by latitudinal or radial gradients. During the Earth–Jupiter transfer (October 1990–February 1992), solar

Table 8.1. *Ulysses* radial gradient measurements (1–5 AU).

Species	Energy range (MeV or MeV/nucl)	Gradient (%/AU)	Reference
Protons	34–69	~0	McDonald et al., 1997
	>~90	6.8 ± 0.6	Binns et al., 1989
	320–2,100	4.9 ± 1.3	Cameron, 1982
	>2,100	2.4 ± 0.4	Cameron, 1982
	~2,000	3–3.5	Paizis et al., 1995
Helium	34–50	10 ± 0.5	McDonald et al., 1997
	320–2100	4.3 ± 2.0	Cameron, 1982
	>2,100	2.5 ± 0.4	Cameron, 1982

activity was so high that there were few periods when fluxes were not contaminated with solar energetic particles. As a result, there have been few reports of radial gradient measurements from *Ulysses*. Table 8.1 contains a few representative radial gradient measurements that have been reported over the radial range 1-5 AU when Ulysses was in or near the ecliptic. The measurements can be summarized by the statement that all are consistent with gradients of a few (i.e. <10) per cent AU^{-1}.

In general, the measurements are in line with measurements made previously, although as McDonald et al. (1997) point out, precise comparisons are difficult because of the differing analysis techniques and spacecraft pairs that have been used over the last few solar cycles.

8.2.2.3 Latitudinal gradients

The principal objective of the *Ulysses* mission was, of course, to explore latitudinal variation in the heliosphere. As shown in Figure 8.3A, during the first solar orbit two scans of the latitude range 0–80° were made in both the northern and southern hemispheres of the heliosphere. Following Jupiter fly-by in early 1992, the spacecraft climbed to 80°S latitude while moving inward from 5.4 AU in the ecliptic to 2.3 AU over the south pole. Maximum south latitude was reached in September 1994. Then, in a period of about 1 year the spacecraft made a rapid scan of latitude from 80°S to 80°N, reaching a perihelion of 1.3 AU near the heliographic equator and returning to a radius of 2.0 AU at a latitude of 80° over the north pole in early August 1995. Following the north-polar pass, the spacecraft returned to the ecliptic near 5.4 AU, crossing the heliographic equator in December 1997. Near aphelion, the spacecraft remained within 25° of the equator and at radii between 4.4 and 5.4 AU from September of 1996 until April of 1999.

For study of latitude effects, by far the easiest period to interpret is the so-called FLS in 1994–1995, both because of its short duration with respect to the timescale of changes in the solar-cycle modulation and because of the restricted radial range over which the spacecraft moved. As a result, latitude effects were almost cleanly separated from temporal and radial effects.

The existence of a latitude effect is clearly visible in the relativistic protons shown in Figures 8.3D and 8.4. During the FLS the intensity decreased by about 20 per cent from the south pole to the equator, and then increased again to a slightly higher level over the north pole. The first conclusion to be drawn is that the latitude gradients for cosmic-ray nuclei are positive, consistent with expectations of drift models for the current sign of the solar magnetic dipole. With the sequence of positive latitude gradients observed by *Pioneer 11* in 1974–1975 (Bastian *et al.*, 1979; McKibben, 1989b), negative gradients by *Voyager 1/2* in 1986–1987 (Christon *et al.*, 1986a; Cummings *et al.*, 1987), and now again positive gradients in 1994–1995, this prediction of drift models must be considered to be fully confirmed.

On the other hand, the surprise in the *Ulysses* observations is that the total effect is so small. As mentioned above, prior to *Ulysses*, the expectation (Potgieter *et al.*, 1997) had been for an increase of a factor of 10. For other particle species, gradients similarly much smaller than expected were observed, and for low-energy protons ($<\sim100$ MeV) and for electrons of all energies (e.g. Ferrando *et al.*, 1996; Ferrando, 1997), essentially no latitude gradient at all was observed.

While the latitude gradients were easiest to measure during the FLS, Paizis *et al.* (1995) have developed a clever analysis technique, based on the assumption that the radial and latitudinal variation is separable, which allows gradients to be deduced from the slow latitude scans as well. In this case the intensity at *Ulysses* can be represented as:

$$I_U = I_E \exp[G_r(r-1) + G_\theta \Delta\theta] \tag{8.3}$$

where I_E is the intensity at Earth, r is the radial position of *Ulysses*, $\Delta\theta$ is the latitudinal difference between Earth and *Ulysses*, and G_r and G_θ are the radial and latitudinal gradients, respectively. From this it follows that, for $Y = \ln(I_U/I_E)/(r-1)$ and $X = |\Delta\theta|/(r-1)$:

$$Y = G_r + G_\theta X \tag{8.4}$$

and G_ρ and G_θ can be derived graphically from a run of observations even if the spacecraft is simultaneously moving in radius and latitude. Values from Paizis *et al.* included in Tables 8.1 and 8.2 are based on this technique.

Table 8.2 contains a representative sample, from among a bewildering variety, of latitude gradients that have been reported in the literature based on the *Ulysses* measurements. At various phases of the mission, gradients have been reported for various subsets of the observations and for various species. While the values vary somewhat from report to report, the values listed in Table 8.2 are consistent with the range of values reported. Errors have been omitted in most cases as they result from formal errors on a fitting procedure and thus are unrealistically small. All reported values except for the very low energy anomalous helium fluxes reported by Trattner *et al.* (1996) are less than ~1 per cent per degree, and result in a pole/equator flux ratio of less than a factor 2. This is the principal result of the *Ulysses* measurements.

Table 8.2. *Ulysses* latitudinal gradient measurements (0°–±60°).

Species	Energy range (MeV/nucl)	Latitude range (deg)	Latitude gradient (% deg^{-1})	Reference
Protons	35–70	70S–10S	0.10	Simpson et al. (1996)
		10S–50N	0.14	Simpson et al. (1996)
	34–69	30S–60S	0.06	McDonald et al. (1997)
		30N–60N	0.32	McDonald et al. (1997)
	75–90	70S–50N	0.25	Simpson et al. (1996)
	>~100	70S–50N	0.37	Simpson et al. (1996)
	~2,000		0.3–0.4	Paizis et al. (1995)
	~15,500		0.2–0.3	Paizis et al. (1995)
Helium	35–70	70S–10S	0.77	Simpson et al. (1996)
		10S–50N	0.89	Simpson et al. (1996)
	34–50	30S–60S	0.76	McDonald et al. (1997)
		30N–60N	0.83	McDonald et al. (1997)
	70–95	70S–50N	0.55	Simpson et al. (1996)
	145–255	30S–60S	0.27	McDonald et al. (1997)
		30N–60N	0.18	McDonald et al. (1997)
Nitrogen	4–7	80S–80N	1.2 ± 0.3	Trattner et al. (1996)
	8–20		1.3 ± 0.2	
Oxygen	4–8	80S–80N	1.8 ± 0.3	Trattner et al. (1996)
	16–20		2.0 ± 0.3	
Neon	4–8	80S–80N	2.8 ± 0.5	Trattner et al. (1996)
	9–30		2.0 ± 0.4	

However, the reduction of the observations to a single parameter, 'the gradient', by a fitting procedure inevitably misses some significant features in the observations.

The detailed observations for three measured particle fluxes are graphically summarized in Figure 8.5, where latitudinal intensity profiles relative to the intensity measured simultaneously at Earth are shown for $\gtrsim 100$ MeV protons, ~35–70 MeV protons, and ~35–70 MeV/nucl helium. The helium contains a strong contribution from the anomalous helium component. Inspection of Figure 8.5 demonstrates several significant features of the latitudinal profile of modulation as observed by *Ulysses* that are not adequately represented by a simple latitude gradient.

North–south asymmetry

As first reported by Simpson et al. (1996) and Heber et al. (1996a), and as is clearly visible in Figure 8.5, intensities relative to the near-Earth intensities were somewhat higher in the north polar zone than in the south polar zone, by amounts ranging from 6 to 15 per cent. The point of minimum intensity relative to the near-Earth

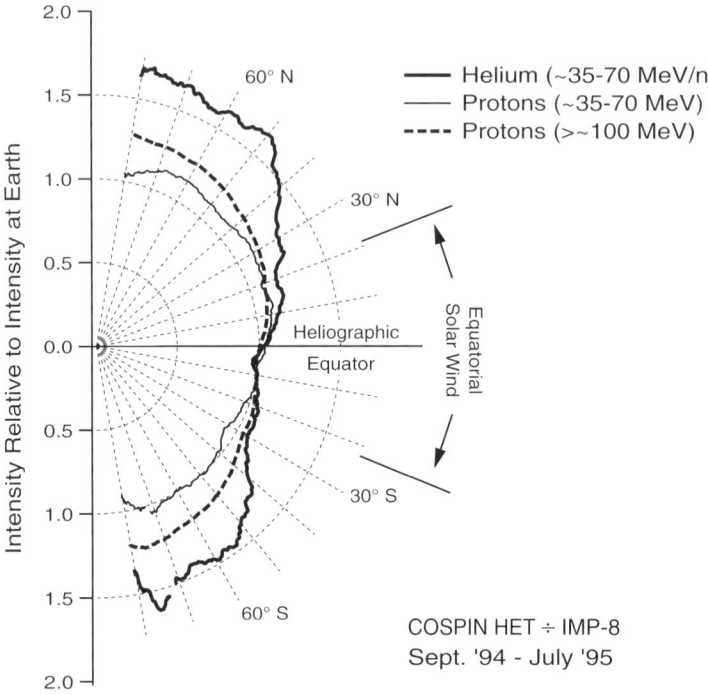

Figure 8.5. Polar plot of the ratio of intensities measured at *Ulysses* to those measured at *IMP-8* at Earth as a function of *Ulysses* latitude for relativistic and low-energy galactic cosmic-ray protons and low-energy anomalous helium measured during *Ulysses*' FLS. Boundaries of the equatorial solar wind as determined from the SWOOPS instrument are indicated (from McKibben et al., 1996).

intensity is also offset about 10° south of the heliographic equator, and Simpson et al. (1996) showed that with respect to 10°S latitude, the latitudinal variation in the northern and southern hemispheres was remarkably symmetrical. This led to the suggestion that the north-polar excess was the result simply of the action of a nearly identical latitudinal gradient in the two hemispheres over a larger latitudinal range in the north than in the south. However, McKibben et al. (1996) noted that comparing the excess for several different energy ranges and particle species showed no strong correlation between the size of the latitudinal gradient and the size of the excess. Simpson et al. (1996) also suggested that the observed difference in winding of the interplanetary magnetic field, with the south-polar fields exhibiting slightly tighter winding than the north-polar fields (Forsyth et al., 1996) might be partly responsible for the asymmetry. Heber and Burger (1999) have summarized the current status of attempts to model this asymmetry, and conclude that no clear and compelling explanation for the north-polar excess has yet been put forth.

An obvious explanation for offset of the plane of symmetry from the heliographic equator would be a southward displacement of the mean location of the heliospheric current sheet dividing north and south magnetic polarities. Indeed, as reported by McDonald *et al.* (1997), the source surface location of the current sheet calculated by J. T. Hoeksema from the Wilcox Observatory solar magnetograms is consistent with a southward offset of the current sheet of about the right magnitude. Since the current sheet is frequently assumed to be the surface of symmetry for cosmic-ray modulation, as drift velocities are directed either towards or away from the current sheet in the two hemispheres, this would seem to be a plausible explanation.

However, to preserve $\nabla \cdot \mathbf{B} = 0$, such an offset would require a significantly stronger radial component of the magnetic field in the south-polar regions than in the north. In contradiction to this, Smith and Balogh (1995) and Suess *et al.* (1996) report that the radial component is essentially constant from pole to pole, and Erdös and Balogh (1998) have explicitly integrated the observed radial component from pole to pole and find the observations inconsistent with offsets even as small as $5°$. Very recently, Smith *et al.* (2000) have pointed out that these arguments implicitly assume that the magnitude of the radial magnetic field was invariant with time during *Ulysses*' FLS. Using observations from *WIND* as well as *Ulysses*, they find that a gradual change in B_r in the positive and negative sectors took place during the FLS in such a way as to allow a southward offset of the mean current sheet and at the same time preserve the appearance of a latitude-independent B_r at *Ulysses*. This suggestion appears very plausible, and may well provide the explanation for the southward offset of the point of symmetry for the modulation. It also points up the dangers and difficulties of attempting to untangle spatial from temporal variation in a dynamically evolving environment.

Latitudinal structure in the gradient

In addition to the north–south asymmetry, there is clear structure in the latitude dependence of the modulation. Most obvious is the decrease in the gradient at high latitudes, leading to regions above about $60°$S and $50°$N where there is little dependence of the intensity on latitude. This high-latitude decrease in the latitudinal gradient is also seen in observations reported by Heber *et al.* (1996b) and by McDonald *et al.* (1997). McDonald *et al.* report evidence that the latitude gradient may vary continuously with latitude, with a maximum near $50°$ in both the south and north hemispheres. As shown in Figure 8.3, there is no obvious gross change in solar-wind or magnetic-field measurements at latitudes above the current sheet that correlates with these changes in the gradient. McDonald *et al.* suggest that the decrease in the latitude gradient may be related to enhanced turbulence found over the solar poles, but a clear explanation for this effect has not been found.

Somewhat more interpretable if less obvious is the observation by Simpson *et al.* (1995) and Paizis *et al.* (1995) that the latitudinal gradient in the low-speed region of the solar wind swept by the current sheet is nearly zero. The effect was most obvious in the period just after Jupiter fly-by when *Ulysses* was just beginning its climb to high southern latitudes, although Heber *et al.* (1998) report observation of a similar but more latitude-restricted effect when *Ulysses* returned to the equatorial zone near 5 AU in 1997.

Suppression of the measured latitude gradients in the current-sheet region may in part be a measurement difficulty since arguably the relevant parameter is distance from the current sheet (e.g. Christon *et al.*, 1986b). In the region swept by the current sheet, this is very difficult to determine. However, once well above the equatorial zone, distance to the current sheet is well correlated with heliographic latitude, at least in near-solar-minimum conditions where the current sheet is confined to a narrow, near-equatorial zone. On the other hand, there may be physical causes as well due to the enhanced turbulence in the relatively more disturbed solar wind of the current-sheet region and the rapid drifts of particles along the current sheet tending to equalize the flux over the latitude range swept by the sheet.

Insensitivity to the solar-wind velocity

Another unanticipated feature of the latitudinal variation that is apparent from Figure 8.5, and also from Figure 8.3, is the apparent insensitivity of the modulation level to the rapid and dramatic change in solar-wind velocity observed as *Ulysses* passed from the fast polar solar wind to the slow equatorial wind and back during the FLS. Especially on the northern-hemisphere crossing of the boundary, despite a change in solar-wind velocity of more than a factor of 2 in a matter of a few days, almost no change in the profile of the modulated intensity was observed at *Ulysses*. Since the solar-wind velocity figures directly or indirectly into every term except the diffusion term of the modulation equation (equation 8.1), this suggests that latitudinal transport of particles by diffusion or other means (see below) must be sufficiently rapid to smoothe out the effects of this strong latitudinal dependence of the solar-wind speed.

Rigidity dependence of the gradient

The rigidity dependence of the latitude gradient has been addressed by Paizis *et al.* (1997), McKibben *et al.* (1998), and Heber and Potgieter (1999). Figure 8.6 summarizes observations of the latitude gradients of galactic cosmic rays (Figure 8.6A) and anomalous components (Figure 8.6B), and compares them with similar measurements in the outer heliosphere from the *Pioneer* and *Voyager* spacecraft. The gradients for galactic cosmic rays show a broad maximum at rigidities of a few GV, with a very gradual decline to higher rigidities, and a sharp drop to near zero for rigidities below ~ 1 GV. Gradients compiled by Heber and Potgieter (1999) and Paizis *et al.* (1997) show similar behaviour. Paizis *et al.* (1997) have interpreted the decrease at low energies in terms of the Compton–Getting factor, which derives from the shape of the cosmic-ray spectrum. At low rigidities, the slope of the spectrum is positive, so that the flux increases with increasing rigidity, and the effects of adiabatic deceleration on the intensity at a given rigidity are reduced. As a result, gradients resulting from adiabatic deceleration in the solar wind are also reduced. However, this conclusion is strictly valid only within the context of the force-field approxima-

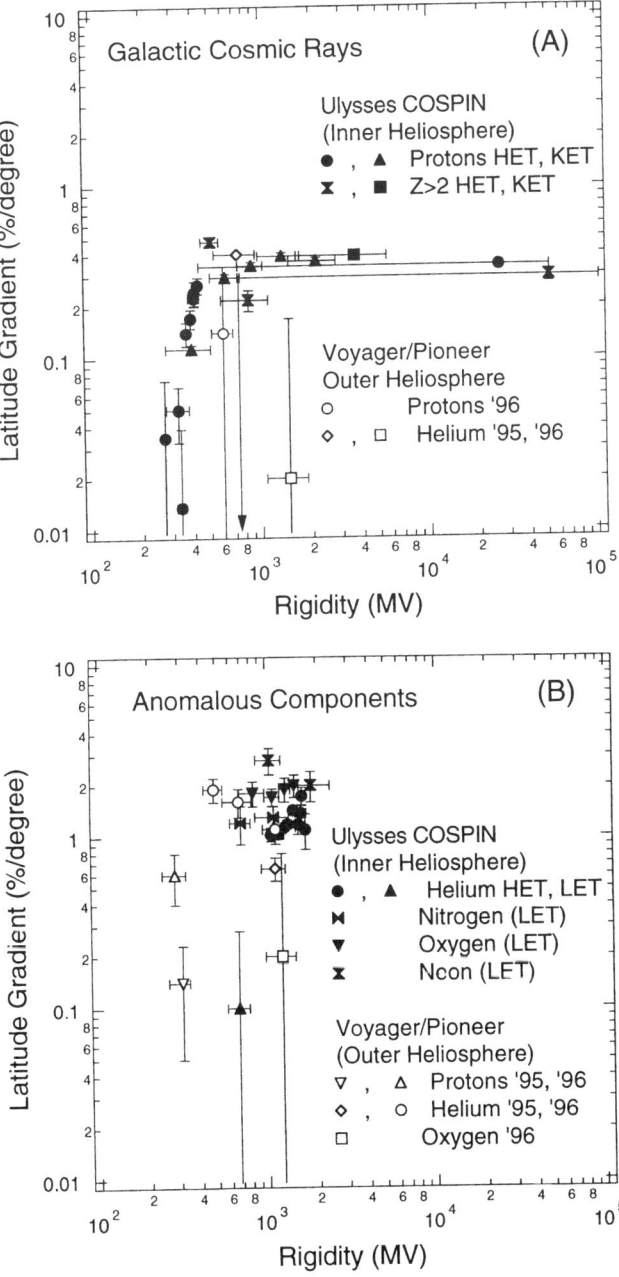

Figure 8.6. Compilation of latitude gradients measured for galactic cosmic rays and for anomalous components by *Ulysses* and, for comparison, by *Voyager 1.2* and *Pioneer 10/11* instruments, plotted as a function of particle rigidity. Anomalous components are assumed to be singly ionized. Measurements are compiled from Trattner *et al.*, 1996, 1995; Heber *et al.*, 1996a; McKibben *et al.*, 1996 and McDonald *et al.*, 1997) (from McKibben *et al.*, 1998).

tion for the modulation equation (Gleeson and Axford, 1968), and must be validated by full numerical solutions of the equation.

Using a much more sophisticated and complete computational model, including anisotropic perpendicular diffusion enhanced in the poleward direction, a transition to fast solar wind at high latitudes, and enhanced turbulence in the polar fields, Potgieter *et al.* (1997) have attempted to reproduce the observed rigidity dependence of the gradients for protons and electrons, and have achieved qualitative agreement with the observations for both protons and electrons. Burger *et al.* (1999) have pushed even further, allowing the rigidity dependences of the perpendicular diffusion coefficients in the polar and ecliptic directions to differ, and have achieved quantitative agreement for the proton gradients. Further explorations of the effects of variation in the heliospheric configuration and propagation parameters allowed by these models will be especially useful in understanding the effects of changes associated with the solar cycle during *Ulysses* second solar orbit. For anomalous components, the rigidity range over which measurements exist is not sufficient to show any clear rigidity dependence, but there is a suggestion that the gradients may decrease towards low rigidity below a few hundred MV.

8.2.3 Observations of short-term, 26-day recurrent modulations

As is apparent from Figure 8.3, one of the more striking features of the cosmic-ray intensity as observed by *Ulysses* is the prolonged sequence of quasi-periodic variations observed as *Ulysses* rose from the equator to high southern latitudes, and, again, although somewhat less regular in their period, as *Ulysses* returned to the equator after the FLS. At moderate latitudes this variation was clearly associated with large variation in the solar-wind speed and consequent compressions of the interplanetary magnetic field as *Ulysses* was immersed alternately in the fast solar wind from the polar coronal hole and slow equatorial solar wind in the course of each solar rotation. For the rise to the south pole, Bai *et al.* (1997) have shown that the equatorward extension of the fast solar wind giving rise to this variation was the result of the development of a stable lobe of the south-polar coronal hole in 1992 which persisted at least through mid-1995. Up to latitudes of $\sim 30°$, the spacecraft was in the region swept by the current sheet (Balogh *et al.*, 1995). Beyond 30° the magnetic-sector structure was no longer observed, but solar-wind speed variation and reverse shocks originating from the interactions between the fast and slow streams continued to be observed regularly up to $\sim 40°$S latitude (González-Esparza *et al.*, 1996). Above $\sim 40°$S latitude, little velocity structure was seen in the solar wind, and few shocks were observed, although Balogh *et al.* (1995) report isolated recurrent reverse shocks observed in March–April 1994 at latitudes of 55–58°S. Variation in cosmic-ray intensity in association with CIRs produced by fast solar-wind streams are a well-known phenomenon in the equatorial zone. A comprehensive collection of studies of the CIR effect has recently been published by Balogh *et al.* (1999) summarizing the work of a series of workshops devoted to the CIR phenomenon. From approximately mid-1992 through mid-1993, the observations at *Ulysses* were entirely consistent with what was known of CIR-induced variations from earlier missions.

The significant new result from *Ulysses* is that intensity variation continued to high latitude in the *Ulysses* observations even in the absence of significant structure in the solar wind. Figure 8.7 shows a detrended history of the *Ulysses* variation in relativistic cosmic rays together with the solar-wind speed as measured by the *Ulysses* SWOOPS experiment (McComas, 1999) for the period from Jupiter fly-by through the FLS. Variation (26-day) are clearly apparent throughout the particle-intensity record. The maximum amplitude was observed near the maximum latitude reached by the current sheet. This is consistent with the prediction by Pizzo (1994) for maximum compression of the solar wind in this region, based on three-dimensional MHD simulations of the stream interactions resulting from a tilted dipole solar magnetic structure with high-speed solar wind emanating from the coronal holes poleward the streamer belt.

While the amplitude of the variation is diminished at higher latitudes, comparison of the intensities with the 26-day tics shown in Figure 8.7 through the polar pass provides evidence that 26-day intensity variation in relativistic cosmic rays continued at least to 70°S latitude, and possibly all the way to the maximum latitude of 80°S. Similar observations were reported from the COSPIN KET observations (Kunow et al., 1998). Note that the persistence of the variation with a period of 26 days to high latitude is consistent with rigid rotation of the magnetic structure of the heliosphere, independent of latitude, at the observed photospheric equatorial rotation rate.

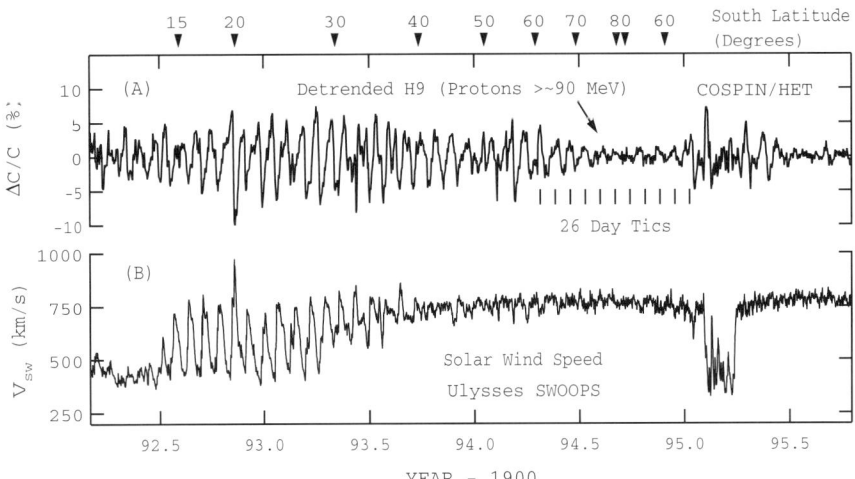

Figure 8.7. (A) Daily average intensity of $>\sim 90\,\text{MeV}$ protons at *Ulysses*, detrended by subtracting running 27-day average intensities from the daily averages. (B) Average solar-wind velocities (6-hour) for the same period measured by the *Ulysses* SWOOPS instrument.

Confirming the continued influence of near-equatorial CIRs at polar latitudes, at lower energies ~1-MeV protons accelerated by the CIR shocks were also observed to latitudes of ~50°S, and ~50 keV electrons from the CIRs continued to be observed all the way to 80°S latitude (e.g. Simnett and Roelof, 1998).

The rigidity dependence of the 26-day variation has been addressed by McKibben *et al.* (1995), and, more recently, by Paizis *et al.* (1999). Similar to the case for latitude gradients, they find a maximum in the amplitude for a rigidity of ~1 GV, with a rapid decrease towards zero amplitude at lower rigidities and a somewhat more gradual decrease at higher energies.

Figure 8.8, taken from Zhang (1997), suggests that, more than a simple similarity in rigidity dependence, there may be a fundamental relationship between the

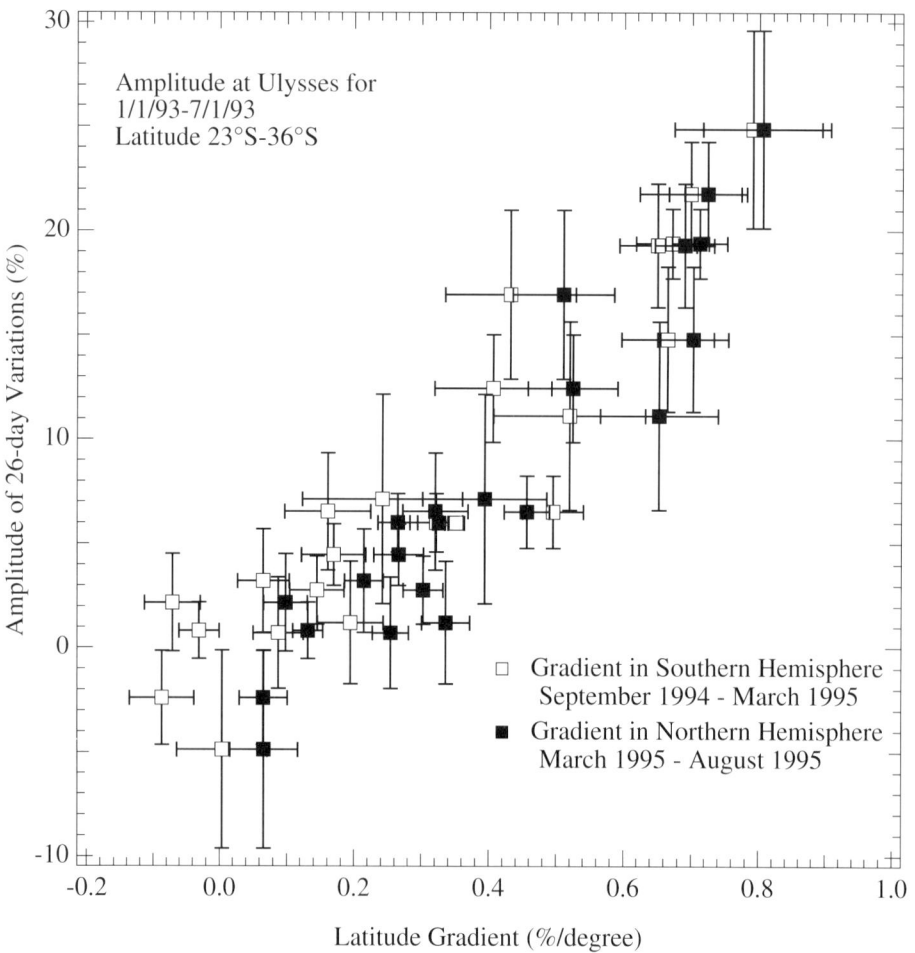

Figure 8.8. Correlation between the amplitude of 26-day variation and the size of measured latitude gradients for a variety of energy intervals for protons, helium, carbon, and oxygen (from Zhang, 1997).

processes governing the 26-day variations, which by their nature must be a local phenomenon to retain the imprint of a local structure, and the global heliospheric modulation as measured by the latitude gradients determined by *Ulysses*. Over the complete rigidity range for which measurements are available, Zhang finds a tightly defined linear correlation between the amplitudes of the 26-day variation and the size of the latitude gradients. Such a relationship was unexpected, and its implications are not yet clear.

8.2.4 Implications of measurements for modulation theory and models

The challenges posed by *Ulysses* measurements for modulation theory and models may be summarized by two points:

(1) The inner heliosphere is remarkably homogeneous as a function of latitude for cosmic rays, with less than a factor of 2 difference between the pole and equator even for the anomalous components, which are most sensitive to changes in parameters governing modulation.
(2) There appears to be rapid communication between the equatorial and polar regions. Not only is variation in modulation imposed by low-latitude corotating structures observed all the way to the polar regions, but also ~1-MeV protons and ~50-keV electrons presumably accelerated by these structures also find their way to the interplanetary field lines encountered over the poles. This last observation was completely unexpected.

The relatively homogeneous nature of the inner heliosphere had already been suggested at the time of launch of *Ulysses* when Jokipii and Kóta (1989) recognized that any transverse components resulting from even small irregularities in the polar magnetic field will vary with radius as $1/r$, whereas radial components will fall off as $1/r^2$. Therefore, even if the polar fields are nearly radial in the inner heliosphere, any transverse fluctuations in the field direction, such as are almost certain to be imposed by the convective motions of the supergranulation in the solar atmosphere, will become dominant outside radii of a few AU. As a result, rather than experiencing a free ride to the inner heliosphere along nearly radial fields, the cosmic rays see predominantly transverse field at all latitudes, and modulation in the polar regions could be nearly as effective as in the equatorial region.

This importance of transverse fluctuations for the polar magnetic structure has been fully verified by the *Ulysses* observations (Balogh, 1998). The magnetic field in the polar regions was found to contain high levels of fluctuations deriving from inertial length turbulence, transverse Alfvénic fluctuations, and convected transverse variation of the sort that might be imposed by the supergranulation convection. Thus access of particles to the polar regions is significantly impeded, both as a result of diffusive scattering by the irregularities and, particularly for solar cycles where the solar magnetic dipole has a positive sign, by disruption of the smooth idealized Parker spiral field and consequent reduction in average drift velocities.

Aside from the modifications to the polar field, the *Ulysses* observations of small latitudinal gradients, and particularly the high-latitude observations of 26-day variation in modulation have required other changes in models. Potgieter (1998) has given a brief summary of modifications to the standard modulation models that are required for consistency with the *Ulysses* measurements. The principal modification is the requirement for enhanced latitudinal transport. In the context of traditional models for the modulation and heliospheric magnetic-field structure, this is accomplished by preferentially increasing the perpendicular diffusion coefficient in the latitudinal direction, possibly to levels as high as 30 per cent of the parallel diffusion coefficient (Potgieter, 1998). Jokipii *et al.* (1995) have shown that an enhancement of the latitudinal diffusion coefficient arises naturally from the radial development of irregularities imposed by the supergranulation that lead to the classical random walk of fields (Jokipii and Parker, 1968). They have also shown that even modest enhancements in the latitudinal diffusion can result in persistence of the 26-day recurrent modulation to the polar regions (e.g. (Kóta and Jokipii, 1995).

However, models that depend exclusively on latitudinal diffusion have difficulty accounting for the observation of very low energy CIR-accelerated particles in the polar regions. Furthermore, a difficulty only recently pointed out (McKibben, 1999) is that earlier studies of the latitudinal propagation of Jovian electrons with *Pioneer 11* (Hamilton and Simpson, 1979) and *Ulysses* (Simpson *et al.*, 1993; Ferrando *et al.*, 1993) found cross-field diffusion coefficients for Jovian electrons that were smaller in the latitudinal direction than in the ecliptic, in direct contradiction to the conclusions above. Thus it is not clear that a simple enhancement of latitudinal diffusion is all that is needed to reconcile the models with the observations.

A much more radical suggestion has been made by Fisk (1996), who investigated the implications of the observation that while both the tilted magnetic axis of the solar dipole field and the coronal magnetic structure rotate rigidly, the photosphere, in which the interplanetary magnetic-field lines are ultimately rooted, rotates differentially, with the polar regions rotating significantly more slowly than the equatorial. As a result of this complex interaction, and noting that the interplanetary magnetic field is drawn primarily from super-radial expansion of field lines in the polar coronal hole (e.g. Smith and Balogh, 1995), he proposed that field lines from the tilted solar dipole passing from the differentially rotating solar photosphere and out into the rigidly rotating coronal-hole structure will be subject to large-scale mixing of field lines in latitude over several solar rotations. As a result, field lines that may have originated at the solar poles may be found to connect directly to regions near the ecliptic at radii of 10–15 AU from the Sun. This large-scale mixing would yield direct field-line connections between polar and equatorial zones at large radii, and even at smaller radii could enhance latitudinal transport by adding a systematic latitudinal component to the magnetic field. However, since the predicted latitudinal component is small, localized latitudinal transport, such as is measured in studies of Jovian electrons in the vicinity of Jupiter, may not be much enhanced.

This suggestion represents a major departure from the traditional Parker Archimedean spiral field. The field geometry is so complex that, as of yet, it has not been possible to incorporate it into existing models for the modulation. Thus it is not yet possible to make quantitative studies of the implications for modulation. The current status of this question is summarized by Fisk and Jokipii (1999).

8.3 COMPOSITION STUDIES

In contrast to studies of the intensity variation of cosmic rays, which relate primarily to the dynamic heliospheric structure that controls modulation, studies of the composition of cosmic rays relate mainly to questions that look beyond the heliosphere. For example, studies of the anomalous component provide insight into the composition of the very local interstellar medium and the nature of the heliospheric termination shock. Studies of the stable isotopes of the galactic cosmic rays provide information about the nature of the process by which the nuclei were produced and accelerated and the propagation history of the cosmic rays through the Galaxy. Finally, studies of the radioactive isotopes provide a measure of the age of the cosmic rays and also information concerning their acceleration history. With its large geometric factor, high-mass resolution instrumentation, *Ulysses* provided the first chance to obtain a nearly complete catalogue of the isotopic composition of the cosmic rays from hydrogen through nickel.

8.3.1 Instrumentation for isotopic studies

The standard technique for measuring the composition of cosmic rays is the so-called dE/dx versus residual-energy analysis whereby the rate of energy loss is measured in a thin detector, and the total remaining energy E of the nucleus is measured in a detector thick enough to bring the particle to rest. When dE/dx is plotted versus residual energy E particles of a given charge and mass are distributed along a roughly hyperbolic track with velocity increasing monotonically along the E axis. Since for a given E, dE/dx varies with nuclear charge, Z as Z^2 elemental abundances are easily determined with this technique. Separation by mass is a much weaker effect, produced by the slightly lower energy per nucleon, and thus velocity, corresponding to a given signal E in the residual energy detector for a heavier isotope compared with a lighter one. For iron, addition of one nucleon results in less than a 2 per cent increment to the E signal, and to achieve accurate separation of the isotopes, measurements with accuracy of order 0.1 per cent are required.

The primary obstacle to identification of individual isotopes in the cosmic rays is that they are isotropic in their directions of arrival, so that particles impinge on the detector system at a range of angles determined by the acceptance cone of the detector system. Consequently, nuclei of the same charge, mass, and velocity, leave a range of energy deposits in the dE/dx detector depending on their angle of incidence. This problem can be reduced by designing a very narrow cone of acceptance, but then the geometrical factor becomes so small that it is impossible to collect enough events to provide the statistical accuracy necessary for composition studies on any but the most abundant nuclear species. Thus, for a practical experiment it is necessary not only to measure accurately the energy losses of individual particles in

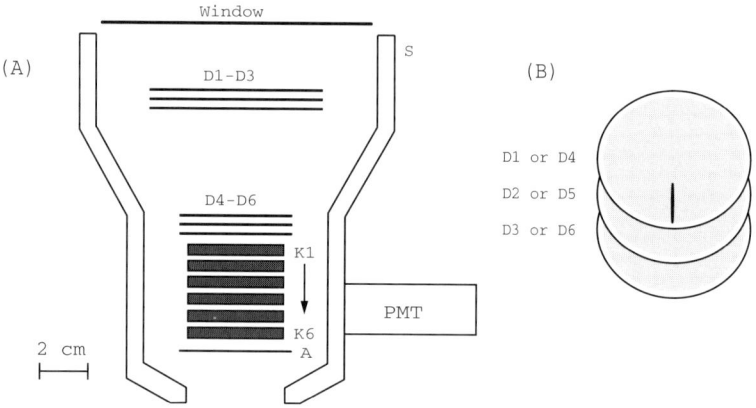

Figure 8.9. (A) Schematic of the COSPIN HET detector arrangement. D1–D6 are Si multi-strip position-sensing detectors, each of which provides location of the penetration of a particle to an accuracy of ~150 microns in one dimension. K1–K6 are 0.5 cm thick Si detectors, and A and S are, respectively, Si and plastic scintillator anicoincidence detectors. (B) Orientation of the strips on the position-sensing detector to form a hodoscope. In each set of D1–3 and D4–6, the strips which measure the position of the particle are rotation by 60° relative to strips in the adjacent detectors (from Connell and Simpson, 1997a).

the detector system, but also to measure the trajectory of the particle so that corrections for the angle of incidence can be applied.

The high-mass resolution measurements made by the COSPIN HET on *Ulysses* were made possible by the development of large-area position-sensitive semiconductor detectors (Lamport et al., 1976). As described by Simpson et al. (1992) and shown in Figure 8.9, two sets of three position-sensing detectors (D1–D3 and D4–D6), are used to determine trajectories of particles incident within the ~40° half-angle acceptance cone for particles which penetrate to the thick silicon residual energy detectors (K1–K6). The use of six position-sensing detectors rather than the minimum of four required for trajectory determination provides redundancy and allow for in-flight checks of consistency and determination of corrections for non-linearities and misalignments. In the energy ranges for which isotopic resolution for stopping particles is possible, the HET has a geometrical acceptance factor of ~4 to ~8 cm^2 sr, depending on energy. The energy ranges for isotopic abundance measurements as a function of Z are shown in Figure 8.2.

As a measure of the accuracy of this trajectory-determination system, based on analysis of in-flight data it has proved possible to determine the rotational orientation of the position-sensing detectors to ~0.02°, and their inclination to the axis of symmetry of the telescope to better than 1° (Connell and Simpson, 1997a). Coupled with electronics that maintain stability against thermal variation at a level of

Sec. 8.3] **Composition studies** 353

138 ppm °C^{-1} (or better than the thermal variation of the Si response to incident particles), the *Ulysses* HET, together with a similar telescope on the short-lived *CRRES* (DuVernois et al., 1996a) provided measurements for isotopic studies that are still arguably the most accurate that have been achieved. While the current measurements from the much larger geometrical factor *ACE* are of comparable accuracy and have the benefit of high statistics even for rare nuclei, many of the first definitive measurements of astrophysically important isotopic abundances have been made by *Ulysses*.

8.3.2 Measurements of stable isotopes in the galactic cosmic rays

Figures 8.10 and 8.11 contain mass distributions measured by the COSPIN HET for the isotopes of C, N, O, Ne, Mg, Si, Fe, and Ni. These mass distributions have been used to derive the measured isotope ratios listed in column 5 of Table 8.3. The

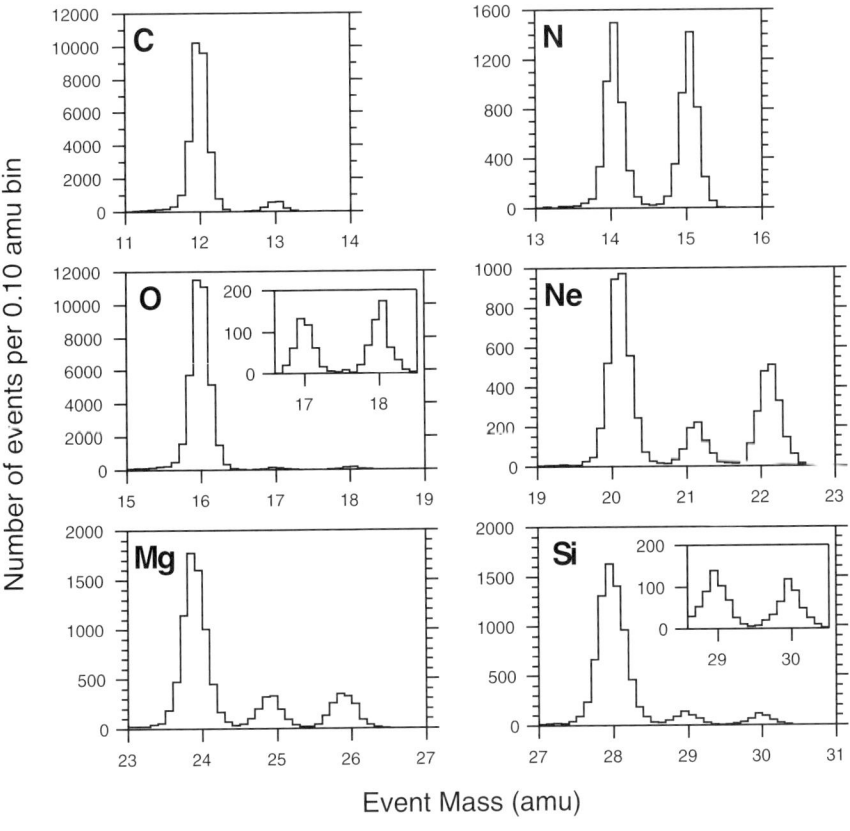

Figure 8.10. Mass histograms for C, N, O, Ne, Mg, and Si in the cosmic radiation as measured by the COSPIN HET (from Connell and Simpson, 1997b).

354 Cosmic rays at all latitudes in the inner heliosphere [Ch. 8

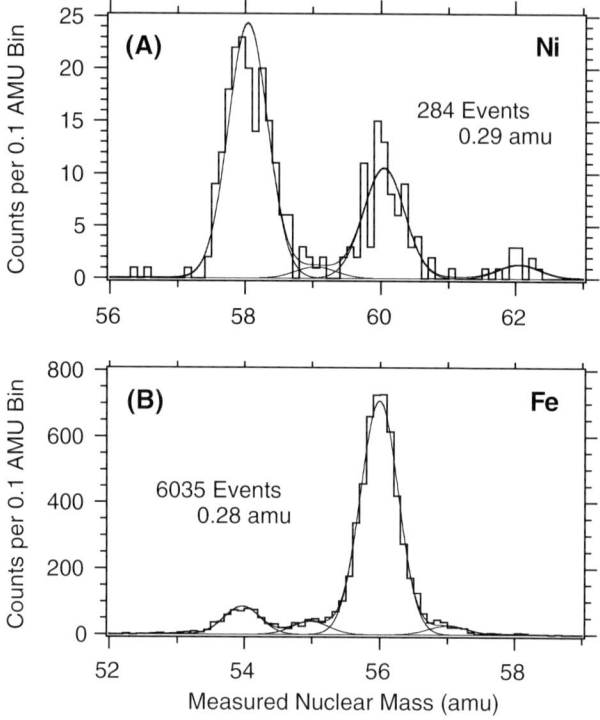

Figure 8.11. Mass histograms for Fe and Ni in the cosmic radiation as measured by the COSPIN HET (from Connell and Simpson, 1997a).

measured isotope ratios reflect the entire history of the cosmic ray particles from their injection and acceleration at the source through propagation in the interstellar medium and finally modulation by the solar wind as they enter the heliosphere. At each stage, to a greater or lesser extent, the abundance ratios may be modified from the original, at-rest abundances that would be measured at the source. Correction for the effects of modulation to arrive at the abundances that would be measured outside the heliosphere is the most straightforward. The method of detection (i.e. stopping a cosmic ray particle in a physical detector) means that the energy range measured in energy/nucleon is slightly lower for heavier isotopes than for lighter isotopes of the same chemical species, simply because at a given energy per nucleon the heavier isotopes have greater penetrating power. Furthermore, the modulation depends upon momentum per unit charge, or magnetic rigidity, and thus on mass per charge. The energy of the particles is also significantly reduced by adiabatic deceleration in penetrating the heliosphere, so that the energy in the local interstellar medium is typically several hundred MeV/nucleon higher than the energy at which the particles are measured. Fortunately, models for modulation are well developed, and computational tools have been developed that make it easy to correct the ratios observed in the solar system to those that would be observed immediately outside the

Table 8.3. Measurements of stable cosmic-ray isotopic ratios.

Isotopic ratio	N	σ (amu)	$\langle E \rangle$ MeV/amu	Measured (%)	Source (%)	GCR source/ solar system	Ref.
^{7}Be/Be	1,525	0.10	~107	56.1 ± 1.3	–	No Be in source	d
^{9}Be/Be				39.3 ± 1.3	–		d
^{13}C/^{12}C	32,374	0.11	129	6.3 ± 0.2		0.8 ± 0.1 ± 0.7	b
^{15}N/N	8,277	0.11	138	49.8 ± 0.6		0.6 ± 0.2 ± 1.3	b
^{17}O/^{16}O	39,382	0.12	152	1.21 ± 0.06		−3.1 ± 1.4 ± 4.8	b
^{18}O/^{16}O				1.47 ± 0.07		−0.2 ± 0.3 ± 1.0	b
^{21}Ne/^{20}Ne	6,416	0.14	168	23.0 ± 0.9		100 ± 20 ± 63	b
^{22}Ne/^{20}Ne				59.0 ± 3.7		2.97 ± 0.15 ± 0.33	b
^{27}Al/^{24}Mg	1,453 (^{27}Al)		~190		9.5 ± 0.6	1.04 ± 0.08	d
^{25}Mg/^{24}Mg	10,002	0.16	190	20.5 ± 0.6		1.05 ± 0.05 ± 0.09	b
^{26}Mg/^{24}Mg				23.1 ± 0.7		1.07 ± 0.05 ± 0.08	b
^{29}Si/^{28}Si	8,154	0.17	210	8.2 ± 0.4		1.06 ± 0.07 ± 0.08	b
^{30}Si/^{28}Si				6.3 ± 0.3		0.97 ± 0.09 ± 0.12	b
^{33}S/^{32}S	827	0.23		19.0 ± 2.4	2.6 ± 2.4	3.2 ± 3.0	c
^{34}S/^{32}S				24.2 ± 2.7	6.2 ± 0.6	1.4 ± 0.6	c
^{37}Cl/^{35}Cl	250	0.2	238		$49.4^{+7.3}_{-6.9}$	$1.51^{+0.22}_{-0.21}$	e
^{54}Fe/^{56}Fe	6,035	0.28	290	11.4 ± 0.6	9.3 ± 0.6	1.5 ± 0.1	a
^{55}Fe/^{56}Fe				5.4 ± 0.4	1.6 ± 0.5	...*	a
^{57}Fe/^{56}Fe				$3.9^{+0.35}_{-0.38}$	$3.7^{+0.33}_{-0.36}$	$1.6^{+0.14}_{-0.16}$	a
^{58}Fe/^{56}Fe				$0.34^{+0.10}_{-0.14}$	$0.18^{+0.10}_{-0.14}$	$0.6^{+0.3}_{-0.4}$	a
^{59}Ni/^{58}Ni	284	0.29	309	$4.6^{+2.6}_{-2.1}$	$2.6^{+2.1}_{-1.7}$	...*	a
^{60}Ni/^{58}Ni				$43.2^{+6.7}_{-6.4}$	$43.2^{+6.7}_{-6.4}$	1.1 ± 0.2	a
^{61}Ni/^{58}Ni				<1.6	<1.2	<0.7	a
^{62}Ni/^{58}Ni				$5.8^{+2.3}_{-1.9}$	$5.4^{+2.2}_{-1.8}$	$1.0^{+0.4}_{-0.3}$	a

* SS abundance = 0%.
References: (a) Connell and Simpson (1997a); (b) Connell and Simpson (1997b); (c) Thayer (1997); (d) Connell (1998); (e) Connell et al. (1998).

heliosphere in a common energy range for all isotopes of a given element. The process of making these corrections has been discussed by DuVernois (1997).

Propagation through the interstellar medium impresses more substantial changes on the abundances as a result of spallation reactions between cosmic-ray nuclei and nuclei in the interstellar medium. For example, since the abundance of Li, Be, and B in the cosmic rays is several orders of magnitude greater than in the solar system, these elements are almost entirely produced by spallation of heavier nuclei, mainly C and O, during propagation. Further constraints are provided by the abundances of sub-Fe nuclei, which are significantly enhanced in the cosmic radiation as a result of spallation of primary Fe-group nuclei. Determinations of the abundances of these and other secondary nuclei, together with laboratory measured cross-sections for their production and destruction by nuclear interactions, can be incorporated into a computational model (the weighted slab model) to derive an energy dependent path-length distribution for cosmic rays propagating through the Galaxy. Using this path-length distribution, the changes in both chemical and isotopic abundances that have occurred during propagation can then be assessed for all cosmic-ray nuclei. There is an extensive literature on this subject, and entry points are provided by Garcia-Munoz et al. (1987) and DuVernois et al. (1996b).

The measured abundances, corrected for propagation and modulation, are considered to be the source abundances for the cosmic rays. Differences in the source abundances from solar-system abundances might be expected:

(1) because of chemical evolution of the Galaxy between the time of the formation of the solar system ~4.5 billion years ago (e.g. Timmes et al., 1995) and the time of acceleration of cosmic rays, determined from radioactive nuclei in the cosmic rays (see below) to be approximately 10–20 million years ago;
(2) because of contributions from exotic sources such as Wolf–Rayet ejecta (Cassé et al., 1982), or supernova ejecta; and
(3) because of selection processes in the injection of material into the acceleration process (e.g. the now well-known apparent first ionization potential bias [Cassé and Paul, 1982] for elemental abundances).

Column 7 of Table 8.3 contains a summary of the isotopic ratios derived for the cosmic-ray source from the measured ratios compared with solar-system abundances as determined by Anders and Grevesse (1989) and Cameron (1982). Cosmic-ray abundance ratios that differ by more than 3 s from solar-system ratios are shown on a tinted background in Table 8.3. The principal conclusion to be drawn from Table 8.3 is that, with very few exceptions, the source abundances of the cosmic rays look remarkably like solar-system material. The largest deviation, already known from measurements made before *Ulysses* (e.g. Lukasiak et al., 1994) is in ^{22}Ne/^{20}Ne. To explain this anomaly, Woosley and Weaver (1981) have suggested that enhanced metallicity in the source stars might result in the needed enhancement of ^{22}Ne, and Cassé and Paul (1982) have suggested a Wolf–Rayet source. However, most investigators find that these models also predict enhancements in heavy isotopes of Mg, Si, and S, which are not consistent with the *Ulysses* measurements (Connell and Simpson, 1997b; Thayer, 1997). Thus the source of the ^{22}Ne enhancement remains unexplained.

Smaller deviations from solar-system abundances are found for the various isotopes of Fe. For isotopes lighter than ^{56}Fe, spallation from primary ^{56}Fe nuclei in the course of propagation is the principal source for the enhanced abundances observed. Thus uncertainties in production cross-sections for these isotopes may play some role in producing a higher than solar-system abundance. The modest enhancement in ^{57}Fe cannot be explained in this manner since, as shown by comparing the measured versus source ratio listed in Table 8.3, little of the ^{57}Fe is produced as secondaries in propagation. Connell and Simpson (1997a) have noted that this may be an indication of a contribution to the cosmic-ray source from freshly processed material from supernovae. Observations of SN1987a have suggested that the ratio of the radioactive isotopes ^{57}Ni/^{56}Ni, which decay to ^{57}Fe and ^{57}Fe would produce a ^{57}Fe/^{56}Fe about two times solar, and that some models of Type II supernova nucleosynthesis would produce similar excesses (Clayton et al., 1992 and references therein). On theoretical grounds, Theilemann et al. (1997) have suggested that Type Ia supenovae would overproduce ^{57}Fe, along with ^{62}Ni and ^{58}Ni, by more than a factor of 2 compared with solar.

For the Ni isotopes, Connell and Simpson (1997a) remark that the absence of deviations from solar-system abundances is not consistent with models of nucleosynthesis requiring enrichment in neutron-rich isotopes of Ni (Arnett, 1996). Furthermore, the absence of a significant source abundance of ^{59}Ni, which has an electron capture half-life of 76,000 years (Connell and Simpson, 1997a; Wiedenbeck et al., 1999) implies no significant delay between nucleosynthesis and acceleration of the newly produced nuclei.

8.3.3 Measurements of radioactive isotopes in the galactic cosmic rays

8.3.3.1 *Cosmic-ray chronometer isotopes*

Among the nuclei produced as secondaries from spallation reactions of primary cosmic rays on nuclei in the interstellar gas are radioactive nuclei, some with half-lives comparable with the average age of the cosmic rays themselves. Observations of the stable isotopes that are primarily of secondary origin in the cosmic rays (e.g. boron or many of the odd-Z sub-uron elements), constrain the amount of material the cosmic rays have passed through since their acceleration (e.g. Garcia-Munoz et al.1987; DuVernois et al., 1996b). However, the stable isotopes provide no information about how long the cosmic rays have been propagating through the Galaxy. Observations of the abundances of radionuclides with very long half-lives, coupled with knowledge of the cross-sections for their production by spallation, provide a measure of the average time that the cosmic rays have been propagating in the Galaxy. With knowledge of both the average residence time for cosmic rays in the Galaxy and the total amount of material they have passed through, the average density seen by the cosmic rays can also be determined.

The best nuclei for these studies, based on their abundance and lifetimes, are ^{10}Be, ^{26}Al, and ^{36}Cl. ^{54}Mn is also useful, but uncertainty concerning its half-life as a fully stripped nucleus limits the accuracy of conclusions that can be drawn (however, see Wuosmaa et al., 1998 and Zaerpoor et al., 1997) for recent progress on the beta-decay half-life of ^{54}Mn).

The first measurement of the cosmic-ray lifetime was made using ^{10}Be (Garcia-Munoz et al., 1977, 1981; Wiedenbeck and Greiner, 1980). Beryllium is particularly useful because it is relatively abundant in cosmic rays, and there is no concern about a possible primary component from the source. Since the abundance of Be in cosmic rays is $\sim 10^6$ times greater than the solar-system abundance of Be, all cosmic-ray

Table 8.4. Measurements of radioactive clock isotopes.

Isotopic ratio	N	$\langle E \rangle$ MeV/amu	Decay mode	$\tau_{1/2}$ (Myr)	Measured (%)	Cosmic-ray age (Myr)	I.S. $\langle \rho \rangle$ (cm^{-3})	Ref.
^{10}Be/Be	1,525	~107	β^-	1.6	4.6 ± 0.6	26^{+4}_{-5}	0.19 ± 0.03	a
^{26}Al/^{27}Al	1,545	~200	ec	0.72	6.1 ± 0.7	19 ± 3	$0.26^{+0.05}_{-0.04}$	b
			β^+	0.87				
^{36}Cl/Cl	250	238	β^-	0.308	5.2 ± 1.8	18^{+10}_{-6}	$0.28^{+0.12}_{-0.11}$	c
^{54}Mn/Mn	331	280	ec	8.5×10^{-7}	12 ± 3	~18	$0.28^{+0.13}_{-0.08}$	d
			β^-	~1–2		(1 Myr $\tau_{1/2}$)	(1 Myr $\tau_{1/2}$)	e

References: (a) Connell (1998); (b) Simpson and Connell (1998); (c) Connell et al. (1998); (d) DuVernois (1997); (e) Wuosmaa et al. (1998).

beryllium nuclei are secondary particles produced by spallation from primary cosmic rays. As fully stripped nuclei in the cosmic radiation, ^{7}Be and ^{9}Be are stable, but ^{10}Be undergoes β^- decay with a half-life of $\sim 1.6 \times 10^6$ years.

Based on analysis of 906 Be nuclei collected by experiments on the *IMP-7* and *-8* spacecraft, of which ~10 per cent were ^{10}Be, Garcia-Munoz et al. (1977, 1981) derived a confinement time for cosmic rays in the Galaxy of 14^{+13}_{-5} years and an average density of $0.23^{+0.13}_{-0.11}$ atoms cm^{-3}. Subsequent work based on smaller numbers of nuclei collected by instruments on *ISEE-3* (Wiedenbeck and Greiner, 1980) and on *Voyager 1* and *2* (Lukasiak et al., 1994) yielded results consistent with the *IMP* measurements.

With its large geometrical factor and high resolution for isotopic measurements (0.10 amu for *Ulysses* versus 0.25 amu for the *IMP* measurements [Connell, 1998]) *Ulysses* offered a significant increase in the precision of the measurement. The results, shown in Table 8.4, give a confinement time of 26^{+4}_{-5} Myr, and a mean density of 0.19 ± 0.03 atoms cm^{-3}, consistent with but much more accurate than earlier measurements.

There are two remarkable and important conclusions drawn from these measurements of ^{10}Be:

(1) The cosmic rays are young by almost any astrophysical standard, so that they must continuously be regenerated. Coupled with estimates of the total energy invested in cosmic rays in the Galaxy, this imposes severe demands on the power input for the acceleration process required to maintain the cosmic-ray flux. Only supernovae appear to be capable of providing the necessary energy.
(2) The mean density experienced by the cosmic rays is significantly less than the mean density of material in the galactic disc, usually estimated as ~1 atom cm^{-3}. This led to the suggestion that the cosmic rays must spend a significant portion of their time in a low-density region of the Galaxy, most likely a galactic halo (Simpson and Garcia-Munoz, 1988).

The importance of these conclusions for the energy balance of the Galaxy and for galactic structure required that they be confirmed by measurements of other radioactive nuclei. Unfortunately, the suitable nuclei, isotopes of Al, Cl, and Mn, are all considerably heavier than Be and less abundant in the cosmic rays. Until the launch of the *Ulysses* mission with the COSPIN HET, there was no instrument in space that had both a large enough geometric factor to make possible collection of a significant number of the less abundant nuclei, and a high enough resolution to allow identification of the radioactive isotopes of heavier nuclei, to provide the necessary bmeasurements.

Measurements of the ^{26}Al/^{27}Al ratio were first reported, based on partially resolved samples of a few nuclei, by Wiedenbeck (1983) and Lukasiak *et al.* (1998). Using data from the COSPIN HET Simpson and Connell (1998) have reported the abundance of ^{26}Al based on ninety-two fully resolved ^{26}Al nuclei and 1,453 ^{27}Al nuclei.

^{26}Al is especially interesting because, in addition to being useful for cosmic-ray chronometry, it is detectable remotely throughout the Galaxy as a γ-ray emitter. As a fully stripped nucleus, ^{26}Al has a half-life of 8.7×10^5 years against β^+ decay to ^{26}Mg, which is stable. However, an ^{26}Al nucleus that retains its electrons can also decay via electron capture, emitting a 1.809-MeV γ-ray in the process. Against both electron capture and β^+ decays, ^{26}Al has a half-life of 7.2×10^5 years. The 1.809-MeV γ-ray has been detected from the Galaxy (Mahoney *et al.*, 1984) and CGRO has found that it extends with low intensity, but with localized hot spots, throughout the Galaxy (Diehl *et al.*, 1995). Thus there is a source that continually injects ^{26}Al into the interstellar medium, most likely through supernova ejecta. However, as discussed by Simpson and Connell (1998), based on models for the steady-state amount of ^{26}Al in the interstellar medium resulting from ejecta and its dilution factor in the medium, it is unlikely that the ^{26}Al visible through the γ-ray emissions makes a significant contribution to the ^{26}Al observed in the cosmic rays. Assuming that all cosmic-ray ^{26}Al is produced by spallation of heavier cosmic rays during propagation, the cosmic ray life-time and the mean density along the cosmic-ray path, shown in Table 8.4, are quite consistent with conclusions derived from ^{10}Be.

The third of the well-characterized cosmic-ray chronometers, ^{36}Cl, has a half-life of 3.01×10^5 years, and was resolved for the first time by measurements from the COSPIN HET (Connell *et al.*, 1998). As shown in Table 8.4, measurements of this isotope provide yet further confirmation of the confinement time and low mean density experienced by the cosmic rays.

It is significant to note that since the half-lives of ^{10}Be, ^{26}Al, and ^{36}Cl differ by more than a factor of 5, the surviving nuclei have sampled propagation regions of greatly different depth, perhaps by a factor of more than 10^2 (Connell *et al.*, 1998). This is significant because the very local interstellar medium is known to be of very low density (e.g. Frisch, 1995). For a short half-life nucleus, this might suggest efficient confinement in a local region of the Galaxy. However, the consistency of the results for a broad range of half-lives gives the conclusion that the cosmic rays experience lower than average density in their propagation throughout the Galaxy more general validity.

The final chronometer for the age of the cosmic rays that has been measured is ^{54}Mn (DuVernois, 1997). As a fully stripped nucleus, ^{54}Mn has a very long beta-decay half-life, estimated as $1-2 \times 10^6$ years. However, as an atom, it decays very rapidly by electron capture so that direct laboratory measurements of the β^- lifetime are essentially impossible. Nevertheless, using the estimated lifetime, the deduced cosmic-ray confinement time and mean density, shown in Table 8.4, are consistent with results from the other chronometer isotopes. Recently, stimulated by these measurements, Wuosmaa et al. (1998) have performed laboratory measurements of the even more infrequent, but more easily identifiable, β^+ decay, and, through theoretical arguments, have deduced a life-time for ^{54}Mn against β^- decay of $\sim 6 \times 10^5$ years, close to the previously estimated range and to the value used in Table 8.4 to estimate the confinement time and mean density during propagation.

8.3.3.2 *Electron-capture isotopes*

Another application of studies of radioactive isotopes that is just beginning to be undertaken is the study of isotopes that decay rapidly via electron capture. Because the cosmic rays are fully stripped after acceleration, as are the secondaries that result from spallation reactions during propagation, electron capture is strongly suppressed during interstellar propagation. A long-standing suggestion about propagation in the Galaxy, summarized by Silberberg et al. (1998), is that cosmic-ray acceleration may not be a one-time event, but may occur episodically during propagation in the Galaxy through interactions (e.g. with shock waves generated in the interstellar medium by supernovae), or continuously as a result of interstellar turbulence. While the re-acceleration hypothesis offers explanations for some observations (e.g. the decrease in secondary to primary rations observed at low energies [Silberberg et al., 1998]), the explanation is not unique, and modifications to other propagation parameters may also produce the observed results.

Electron-capture isotopes offer a probe of this process since, if re-acceleration is significant, cosmic rays observed at an energy of several hundred MeV/nucl will have spent part of their propagation time at perhaps significantly lower energies. Electron pickup by a stripped nucleus depends strongly on the energy of the nucleus, increasing rapidly towards low energies. Thus, if an electron-capture nucleus in the cosmic radiation spends any significant part of its time at low energies prior to re-acceleration, it is more likely to pick up an electron and decay than if the cosmic-ray nucleus were to have spent all its time at typical cosmic-ray energies. This should show up as a depletion of the abundance of electron-capture nuclei in the cosmic rays, and an enhancement of the daughter products.

The electron-capture isotopes ^{49}V and ^{51}Cr are particularly well suited for these studies. First, their primary source in the cosmic rays is from spallation of ^{56}Fe (Connell and Simpson, 1999), so that multiple sources do not complicate the interpretation. As a neutral atom ^{49}V decays to ^{49}Ti with a half-life of 331 days, and ^{51}Cr decays to ^{51}V with a half-life of 28 days. Thus, if any significant re-acceleration occurs, we might expect a decrease in the ratios ^{49}V/^{49}Ti and ^{51}Cr/^{51}V compared with expectations from a cosmic-ray propagation model that otherwise satisfactorily

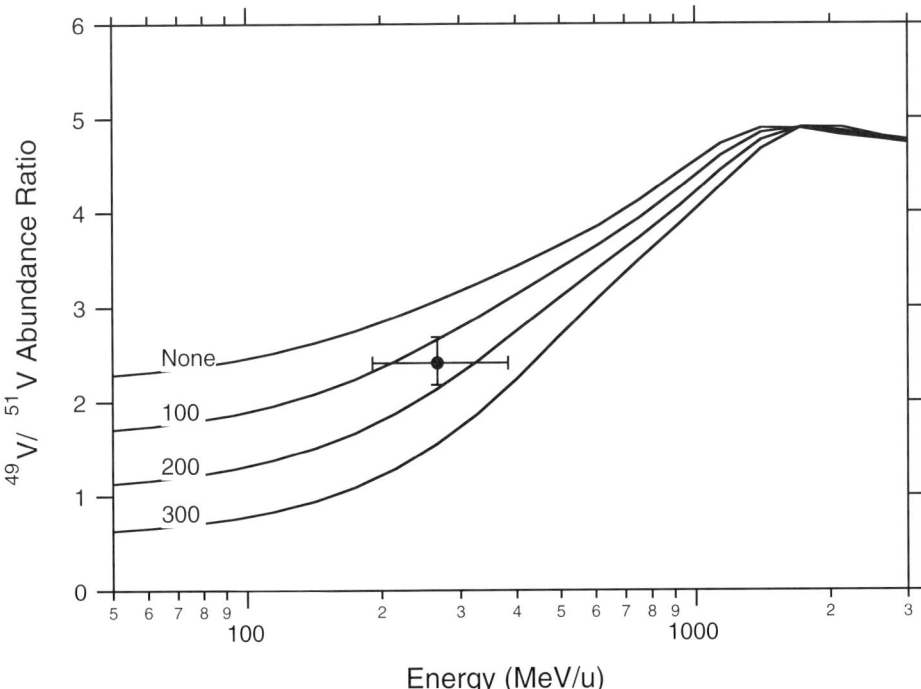

Figure 8.12. $^{49}V/^{51}V$ abundance ratio measured by the *Ulysses* HET compared with predictions for energy boosts via re-acceleration of 0 (top curve) through 300 MeV (from Connell and Simpson, 1999).

fits the observed abundances. Even more sensitive would be the $^{49}V/^{51}V$ ratio, since electron capture would lead to depletion of ^{49}V and enhancement of ^{51}V. Connell and Simpson (1999) have reported the first measurements of these isotopes using the COSPIN HET. Their result for $^{49}V/^{51}V$ is shown in Figure 8.12. Based on a very simplified model for the re-acceleration, they find an indication for re-acceleration at a level of about 100–200 MeV/nucl during propagation.

Clearly, as Connell and Simpson (1999) point out, this conclusion needs confirmation from studies of other electron capture isotopes, and some further investigation of the sensitivity of the result to the mode of re-acceleration. Continuation of the *Ulysses* mission through a second solar orbit will provide the opportunity to continue these investigations.

8.4 FUTURE CONTRIBUTIONS EXPECTED FROM THE CONTINUING ULYSSES MISSION

8.4.1 Modulation studies

As this chapter is being written, *Ulysses* is near 40° south latitude, climbing toward its second south-polar pass, where it will reach maximum latitude of 80°S in

November–December 2000. At the same time, since late 1997, solar activity, the inclination and complexity of the heliospheric current sheet and solar modulation of cosmic rays have been increasing steadily. By the time *Ulysses* reaches maximum latitude, conditions will be near those of solar maximum. Already the increasing inclination and convolution of the current sheet has resulted in extension to higher latitudes of the equator-like modulation regime, and has suppressed the strong 26-day variation that was observed during the previous climb to south-polar latitudes.

Among questions to be investigated during the solar-maximum polar passes are:

- Is the modulation more or less uniform as a function of latitude during solar-maximum conditions?
- Do quiet high-speed solar wind and quiescent modulation conditions persist at the poles even at solar maximum?
- Do propagating shocks, CMEs, and other disturbances penetrate to the high-latitude regions of the heliosphere, and how do they affect modulation if they do?
- What role, if any, do drifts play in modulation in the complex magnetic structure of the solar-maximum heliosphere?
- Do disturbances in the equatorial zone produce effects near the poles, and vice versa?
- Does the apparent enhanced latitudinal transport of cosmic rays continue even in the chaos of solar maximum?
- What can observations of solar energetic particles near the poles tell us about propagation of energetic particles through the heliosphere at solar maximum?

Answers to these questions will provide significant new information about the structure and characteristics of the heliosphere as a medium for propagation of cosmic rays.

Since the solar dipole field is expected to reverse near solar maximum, observations during the declining phase of solar activity as *Ulysses* returns to the equator will provide especially valuable information concerning the latitudinal structure of modulation for both protons and electrons as the complex heliospheric field resolves back towards a dominant dipole field with reversed polarity. Estimates of spacecraft resources give high confidence of the capability for producing useful scientific operation of the spacecraft through 2004, when *Ulysses* will be near aphelion in the equatorial zone of the heliosphere. Thus it is reasonable to hope for significant observations from *Ulysses* in a relatively quiet heliosphere with reversed magnetic polarity.

On the theoretical side, we can expect increasing sophistication of models to address the observations of modulation. In particular, efforts are already underway to generate models that can incorporate the complex Fisk (1996) magnetic-field structure (e.g. Kóta and Jokipii, 1999]). Another rethinking of computational approaches to the modulation problem, also stimulated by the need to simulate the complex magnetic-field structures defined by *Ulysses* has been reported by Zhang (1999a, b). Reducing the sophisticated ideas embodied in this work to com-

8.4.2 Composition studies

For studies of the composition of galactic cosmic rays the heliosphere is, in plain terms, a nuisance. As solar activity increases, it will become more so as intensities decrease and as interference from solar energetic particles increases. Nevertheless, with the large geometrical factor of the *Ulysses* HET, even at solar maximum useful studies of galactic cosmic-ray abundances can be done. The effects of modulation can be corrected for, and accumulation of the event statistics needed to address abundances of rare nuclei in the cosmic radiation continues even at solar maximum. While many of the principal isotopes of astrophysical importance have already been addressed, continued accumulation of statistics will provide increasingly accurate determinations of abundance ratios, and will provide an important confirmation and consistency check on analysis performed with the much larger geometrical factor instruments on *ACE*.

A particular focus of the composition studies during the remainder of the *Ulysses* mission will be study of the electron-capture isotopes, of which the vanadium studies above are only one example. Table 8.5 (Simpson, 1999) contains a list of electron-capture nuclei that can be studied with the various instruments that have isotopic resolution capability. *Ulysses* has only made a start. In addition to obtaining accurate measurements, significant effort will continue in interpretation and validation of the complex codes necessary to account for the network of nuclear reactions that can take place during cosmic-ray propagation.

Finally, exploiting the likely frequent presence of large fluxes of solar energetic particles in the heliosphere near solar maximum, the composition capabilities of the COSPIN HET and LET will be used to the maximum extent possible to study composition of the solar particles for indicators of the location and mechanism of

Table 8.5. Candidate electron-capture cosmic-ray secondary isotopes.

Parent isotope	Daughter isotope	Half-life (at rest)	Accessibility with data from		
			CRRES	Ulysses	ACE
^{44}Ti	^{44}Ca	67 year	No	No	Yes
^{55}Fe	^{55}Mn	2.7 year	Yes	Yes	Yes
^{49}V	^{49}Ti	337 day	No	Yes	Yes
^{54}Mn	^{54}Cr	312 day	No	Yes	Yes
^{57}Co	^{57}Fe	272 day	No	Yes	Yes
^{7}Be	^{7}Li	53 day	Yes	Yes	Yes
^{37}Ar	^{37}Cl	35 day	No	Yes	Yes
^{51}Cr	^{51}V	28 day	No	Yes	Yes

acceleration, and the degree to which propagation through dense regions of the solar corona may play a role in distribution of energetic particles through the heliosphere.

8.4.3 Summary outlook

Rather than being finished with its prime mission, *Ulysses* is in the middle of an ongoing mission of great excitement. The dynamic heliosphere is really four-dimensional. If the temporal dimension has as an appropriate unit of measure the 22-year solar magnetic cycle, though *Ulysses* may have spanned the three spatial dimensions once, it is less than half-way through its investigation of the temporal variation of the heliosphere in regions not accessible to in-ecliptic spacecraft. The results from *Ulysses* so far have changed our picture and the way we think about the heliosphere. We can confidently expect that continued observations by *Ulysses* out of the ecliptic in different phases of the solar cycle will continue to confound, surprise, and instruct us concerning the true nature of the heliosphere in which we live.

8.5 ACKNOWLEDGMENTS

It is a pleasure to acknowledge helpful conversations and other assistance from many people, particularly J. A. Simpson and J. J. Connell for advice concerning abundance studies and L. A. Fisk for assistance with theoretical studies concerning modulation. I am also grateful to A. Balogh for use of magnetic-field observations from the VHM/FGM experiment and to D. McComas for use of solar-wind speed observations from the SWOOPS experiment. This work was supported in part by NASA/JPL contract 955432.

8.6 REFERENCES

Anders, E. and Grevesse N. (1989) Abundances of the elements: Meteoritic and Solar. *Geochim. Cosmochim. Acta*, **53**, 197–214.
Arnett, D. (1996) Supernovae and nucleosynthesis, Princeton Univ. Press, Princeton, NJ.
Bai, T., Hoeksema, J. T., Weber, M. and Acton, L. W. (1997) Solar origin of the 26-day periodicity observed by Ulysses. *J. Geophys. Res.* **102**, 9793–9799.
Balogh, A. (1998) Magnetic fields in the inner heliosphere. *Sp. Sci. Rev.* **83**, 93–104.
Balogh, A., Smith, E. J., Tsurutani, B. T., Southwood, D. J., Forsyth, R. J. and Horbury, T. S. (1995) The heliospheric magnetic field over the south polar region of the sun, *Science*, **268**, 1007–1013.
Balogh, A., Gosling, J. T., Jokipii, J. R., Kallenbach, R. and Kunow, H. (eds.) (1999) Corotating interaction regions: Proceedings of an ISSI Workshop 6–13 June 1998, Bern, Switzerland. *Sp. Sci. Rev.* **90**, Nos. 1–2.
Bastian, T. S., McKibben, R. B., Pyle, K. R., and Simpson, J. A. (1979) Variations in the intensity of galactic cosmic rays and the anomalous helium as a function of solar latitude. Paper presented at: *Proceedings 16th International Cosmic Ray Conference*, Kyoto, Japan, **12**, 318–323.

Binns, W. R., Garrard, T. L., Israel, M. H., Klarmann, J., Stone, E. C. and Waddington, C. J. (1989) The abundances of the heavier elements in the cosmic radiation. In: Waddington, C. J. (ed.) *Cosmic Abundances of Matter*, AIP Press, New York, pp. 147–167.

Burger, R. A., Moraal, H. and Potgieter, M. S. (1987) Inclusion of a wavy nuetral sheet in two–dimensional drift models, Paper presented at: *Proceedings 20th International Cosmic Ray Conference*, Moscow, **3**, 283–286.

Burger, R. A., Potgieter, M. S. and Heber, B. (1999) Rigidity dependence of near-earth latitude gradients. Paper presented at: *Proceedings 26th International Cosmic Ray Conference, Salt Lake City, Utah*, **7**, 242–245.

Cameron, A. G. W. (1982) Elemental and nuclidic abundances in the solar system. In: Barnes, C., Clayton, R. N. and Schramm, D. N. (eds) *Essays in Nuclear Astrophysics*, Cambridge Univ. Press, pp. 23–43.

Cassé, M. and Goret, P. (1978) Ionization models of cosmic ray sources. *Astrophys. J.* **221**, 703–712.

Cassé, M. and Paul, J. A. (1982) On the Stellar Origin of the 22Ne excess in cosmic rays. *Astrophys. J.* **258**, 860–863.

Christon, S. P., Cummings, A. C., Stone, E. C., Behannon, K. W., Burlaga, L. F., Jokipii, J. R. and Kóta, J. (1986a) Differential measurement and model calculations of cosmic ray latitudinal gradient with respect to the heliospheric current sheet. *J. Geophys. Res.* **91**, 2867–2877.

Christon, S. P., Stone, E. C. and Hoeksema, J. T. (1986b) Evidence for a latitude gradient of the cosmic ray intensity associated with a change in the tilt of the heliosphere current sheet, *Geophys. Res. Lett.* **13**, 777–780.

Clayton, D. D., Leising, M. D., The, L., Johnson, W. N. and Kurfess, J. D. (1992) The ^{56}Co Abundance in SN 1987A. *Astrophys. J.* **399**, L141–144.

Connell, J. J. (1998) Galactic cosmic ray confinement time: Ulysses High Energy Telescope Measurements of the Secondary Radionuclide ^{10}Be. *Astrophys. J.* **501**, L59–L62.

Connell, J. J. and Simpson, J. A. (1997a) Isotopic abundances of Fe and Ni in Galactic cosmic ray sources. *Astrophys. J.* **475**, L61–L64.

Connell, J. J. and Simpson, J. A. (1997b) High resolution measurements of the isotopic composition of Galactic cosmic ray C, N, O, Ne, Mg, and Si from the Ulysses HET. Paper presented at: *Proceedings 25th International Cosmic Ray Conference, Durban*, **3**, 381–384.

Connell, J. J. and Simpson, J. A. (1999) Ulysses HET Measurement of electron capture secondary isotopes: Testing the Role of Cosmic Ray Re–acceleration, *Proceedings 26th International Cosmic Ray Conference, Salt Lake City*, **3**, 33–36.

Connell, J. J., DuVernois, M. A. and Simpson, J. A. (1998) The cosmic ray radioactive nuclide ^{36}Cl and its propagation in the galaxy. *Astrophys. J.* **509**, L97–L100.

Cummings, A. C., Stone, E. C. and Webber, W. R. (1987) Latitudinal and radial gradients of anomalous and galactic cosmic rays in the outer heliosphere. *Geophys. Res. Lett.* **14**, 174–177.

Diehl, R. Durpraz, C., Bennett, K., Bloemen, H., Hermsen, W., Knödlseder, J., Lichti, G., Morris, D., Ryan, J., Schönfelder, V., Steinle, H., Strong, A., Swanenburg, B., Varendorff, M. and Winckler, C. (1995) COMPTEL observations of galactic ^{26}Al emission. *Astron. Astrophys.* **298**, 445–460.

DuVernois, M. A. (1997) Galactic cosmic ray manganese: Ulysses High Energy Telescope Results. *Astrophys. J.* **481**, 241–252.

DuVernois, M. A., Garcia-Munoz, M., Pyle, K. R., Simpson, J. A. and Thayer, M. R. (1996a) The isotopic composition of galactic cosmic-ray elements from carbon to silicon:

The Combined Release and Radiation Effects Satellite Investigation. *Astrophys. J.* **466**, 457–472.

DuVernois, M. A., Simpson, J. A. and Thayer, M. R. (1996b) Interstellar propagation of cosmic rays: Analysis of the Ulysses Primary and Secondary Elemental Abundances. *Astron. Astrophys.* **316**, 555–563.

Erdös, G. and Balogh, A. (1998) The symmetry of the heliospheric current sheet as observed by Ulysses during the fast latitude scan. *Geophys. Res. Lett.* **5**, 245–248.

Evenson, P. (1998) Cosmic Ray Electrons. *Sp. Sci. Rev.* **83**, 63–73.

Ferrando, P. (1997) MeV to GeV Electron Propagation and Modulation: Results of the KET–Telescope Onboard Ulysses. *Adv. Sp. Res.* **19**(6), 905–915.

Ferrando, P., Ducros, R., Rastoin, C., raviart, A., Kunow, H., Müller-Mellin, R., Sierks, H., and Wibberenz, G. (1993) Propagation of jovian electrons in and out of the ecliptic plane. *Adv. Space Res.* **13**, 107–110.

Ferrando, P., Raviart, A., Haasbroek, L. J., Potgieter, M. S. Dröge, W., Heber, B., Kunow, H., Müller-Mellin, Sierks, H., Wibberenz, G. and Paizis, C. (1996) Latitude variations of ~7 MeV and >300 MeV cosmic ray electron fluxes in the heliosphere: Ulysses COSPIN/KET results and implications. *Astron. Atrophys.* **316**, 528–537.

Fisk, L. A. (1996) Motion of the footpoints ot heliospheric magnetic field lines at the sun: Implications for Recurrent Energetic Particle Events at High Heliographic Latitudes, *J. Geophys. Res.* **101**, 15,547–15,553.

Fisk, L. A. and Jokipii, J. R. (1999) Mechanisms for latitudinal transport of energetic particles in the heliosphere. *Space Sci. Rev.* in press.

Forbush, S. E. (1938a) On world-wide changes in cosmic ray intensity, *Phys. Rev.* **54**, 975–988.

Forbush, S. E. (1938b) On cosmic ray effects associated with magnetic storms. *Terrest. Mag. and Atmos. Elect.* **43**, 203–218.

Forbush, S. E. (1954) World-wide cosmic ray variations, 1937–1952, *J. Geophys. Res.* **39**, 525–542.

Forsyth, R. J. Balogh, A., Horbury, T. S. Erdös, E., Smith, E. J. and Burton, M. E. (1996) The heliospheric magnetic field at solar minimum: Ulysses Observations from Pole to Pole, *Astron. Astrophys.* **316**, 287–295.

Frisch, P. (1995) Characteristics of nearby interstellar matter, *Sp. Sci. Rev.* **72**, 499–592.

Garcia-Munoz, M., Mason, G. M. and Simpson, J. A. (1977) The age of the galactic cosmic rays derived from the abundance of ^{10}Be. *Astrophys. J.* **217**, 859–877.

Garcia-Munoz, M., Simpson, J. A. and Wefel, J. P. (1981) The propagation lifetime of galactic cosmic rays determined from the measurement of the beryllium isotopes. Paper presented at: *Proceedings 17th Internataional Cosmic Ray Conference, Paris*, **2**, 72–75.

Garcia-Munoz, M., Meyer, P., Pyle, K. R., Simpson, J. A. and Evenson P. (1986) The dependence of solar modulation on the sign of the cosmic ray particle charge. *J. Geophys. Res.* **91**, 2858–2866.

Garcia-Munoz, M., Guzik, T. G., Simpson, J. A., Wefel, J. P. and Margolis, S. H. (1987) Cosmic ray propagation in the galaxy and in the heliosphere: The Pathlength Distributions at Low Energy. *Astrophys. J. Suppl.* **64**, 269–304.

Gleeson, L. J. and Axford, W. I. (1968) Solar modulation of Galactic cosmic rays. *Astrophys. J.* **154**, 1011–1026.

González-Esparza, J. A., Balogh, A., Forsyth, R. J., Neugebauer, M., Smith, E. J. and Phillips, J. L. (1996) Interplanetary shock waves and large–scale structures: Ulysses' Observations In and Out of the Ecliptic Plane, *J. Geophys. Res.* **101**, 17,057–17,071.

Hamilton, D. C. and Simpson, J. A. (1979) Jovian electron propagation out of the solar equatorial plane: Pioneer 11 Observations. *Astrophys. J.* **228**, L123–L127.

Heber, B. and Burger, R. A. (1999) Modulation of galactic cosmic rays at solar minimum. *Sp. Sci. Rev.* **89**, 125–138.

Heber, B. and Potgieter, M. S. (1999) Galactic cosmic ray observations at different heliospheric latitudes. *Adv. Sp. Res.*

Heber, B., Raviart, A., Paizis, C., Bialk, M., Dröge, W., Ducros, R., Ferrando, P., Kunow, H., Müller-Mellin, R., Rastoin, C., Roehrs, K. and Wibberenz, G. (1993) Modulation of galactic cosmic ray particles observed on board the Ulysses spacecraft. Paper presented at: *Proceedings 23rd International Cosmic Ray Conference, Calgary*, **3**, 461–464.

Heber, B., Dröge, W., Ferrando, P., Haasbroek, L. J., Kunow, H., Müller-Mellin, R., Paizis, C., Potgieter, M. S., Raviart, A. and Wibberenz, G. (1996a) Spatial variation of >40 MeV/n nuclei fluxes observed during the Ulysses rapid latitude scan. *Astron. Astrophys.* **316**, 538–546.

Heber, B., Dröge, W., Kunow, H., Müller-Mellin, R., Wibberenz, G., Ferrando, P., Raviart, A. and Paizis, C. (1996b) Spatial variation of >107 MeV proton fluxes observed during the Ulysses rapid latitude scan: COSPIN/KeT Results, *Geophys. Res. Lett.* **23**, 1513–1516.

Heber, B., Bothmer, B., Dröge, W., Kunow, H., Müller-Mellin, R., Sierks, H., Wibberenz, G., Ferrando, P., Raviart, A., Paizis, C., Potgieter, M. S., Burger, R. A., Hattingh, M., Haasbroek, L. J. and McComas, D. (1998) Latitudinal distribution of >106 MeV protons and its relation to the ambient solar wind in the inner southern and northern heliosphere: Ulysses Cosmic and Solar Particle Investigation Kiel Electron Telescope Results. *J. Geophys. Res.* **103**, 4809–4816.

Heber, B. Ferrando, P., Raviart, A., Wibberenz, G., Müller-Mellin, R., Kunow, H., Sierks, H., Bothmer, V., Posner, A., Paizis, C. and Potgieter, M. S. (1999a) Differences in the temporal variations of Galactic cosmic ray electrons and protons: Implications from Ulysses at Solar Minimum. *Geophys. Res. Lett.* **26**, 2133–2136.

Heber, B. Raviart, A., Ferrando, P., Sierks, H., Paizis, C., Kunow, H., Müller-Mellin, R., Bothmer, V. and Posner, A. (1999b) Determination of 7–30 MeV electron intensities: Ulysses COSPIN KET Results. Paper presented at: Proceedings 26th International Cosmic Ray Conference, *Salt Lake City, Utah*, **7**, 186–189.

Jokipii, J. R. and Parker, E. N. (1968) Random walk of magnetic lines of force in astrophysics, *Phys. Rev. Lett.* **21**, 44–77.

Jokipii, J. R., and Kóta, J. (1989) The polar heliospheric magnetic field. *Geophys. Res. Lett.* **16**, 1–4.

Jokipii, J. R., Levy, E. H. and Hubbard, W. B. (1977) Effects of particle drift on cosmic-ray transport. I. general properties, application to solar modulation. *Astophys. J.* **213**, 861–868.

Jokipii, J. R., Kóta, J., Giacalone, J., Horbury, T. S. and Smith, E. J. (1995) Interpretation and consequences of large–scale magnetic variances observed at high heliographic latitude. *Geophys. Res. Lett.* **22**, 3385–3388.

Kóta, J. (1989) Diffusion, drifts, and modulation of galactic cosmic rays in the heliosphere. In: Grzedzielski, S. and Page, D. E. (eds). *Physics of the Outer Heliosphere*, COSPAR Colloquia Series, Vol. 1, Pergamon, pp. 119–131.

Kóta, J. and Jokipii, J. R. (1983) Effects of drift on the transport of cosmic rays VI. A 3-dimensional model including diffusion. *Astrophys. J.* **265**, 573–581.

Kóta, J. and Jokipii, J. R. (1995) Corotation variations of cosmic rays near the south heliospheric pole. *Science*, **268**, 1024–1025.

Kóta, J. and Jokipii, J. R. (1999) Cosmic ray modulation and the structure of the heliospheric magnetic field, *Proceedings 26th International Cosmic Ray Conference, Salt Lake City*, **7**, 9–12.

Kunow, H., Heber, B. and Simpson, J. A. (1998) 26–Day modulation of high rigidity particles by CIRs, pp. 225–230 in Simnett *et al.* Corotating Particle Events, Report of Working Group 2. *Sp. Sci. Rev.* **83**, 215–258.

Lamport, J. E., Mason, G. M., Perkins, M. A. and Tuzzolino, A. J. (1976) A large area circular position sensitive si detector, *Nucl. Inst. and Meth.* **134**, 71–76.

Le Roux, J. A. and Potgieter, M. S. (1989) A time–dependent drift model with a simulated wavy neutral sheet for the solar modulation of cosmic rays. In: Grzedzielski, S. and Page, D. E. (eds). *Physics of the Outer Heliosphere*, COSPAR Colloquia Series, Vol. 1, Pergamon, pp. 143–146.

Leske, R.A. (1993) The elemental and isotopic composition of galactic cosmic ray nuclei from scandium through nickel. *Astrophys. J.* **405**, 567–583.

Lopate, C., McKibben, R. B., Pyle, K. R. and Simpson, J. A. (1991) Radial Gradients of Galactic Cosmic Rays and Anomalous Components. In: the Heliosphere: Pioneer 10/11, Ulysses, and IMP 8 Measurements. Paper presented at: *Proceedings 22nd International Cosmic Ray Conference, Dublin*, **3**, 378–381.

Lukasiak, L. Ferrando, P., McDonald, F. B. and Webber, W. R. (1994) Cosmic ray iotopic composition of C, N, O, Ne, Mg, Si nuclei in the energy range 50–200 MeV per nucleon measured by the voyager spacecraft during the solar minimum period. *Astrophys. J.* **426**, 366–372.

Mahoney, W. A., Ling, J. C., Wheaton, W. A. and Jacobson, A. S. (1984) HEAO 3 Discovery of ^{26}Al in the Interstellar Medium. *Astrophys. J.* **286**, 578–585.

McComas, D. (1999) private communication.

McDonald, F. B., Ferrando, P., Heber, B., Kunow, H., McGuire, R., Müller-Mellin, R., Paizis, C., Raviart, A. and Wibberenz, G. (1997) A comparative study of cosmic ray radial and latitudinal gradients in the inner and outer heliosphere. *J. Geophys. Res.* **102**, 4643–4651.

McKibben, R. B. (1985) 'Modulation of galactic cosmic rays in the heliosphere. Paper presented at: *The Sun and The Heliosphere in Three Dimensions (Proceedings XIXth ESLAB Symposium)*, edited by R. G. Marsden. Reidel, pp. 361–374.

McKibben, R. B. (1987) Cosmic ray modulation. Paper presented at: *Solar Wind 6*, edited by V. J. Pizzo, T. Holzer and D. G. Sime, NCAR Tech Note −306 + Proceedings pp. 615–633.

McKibben, R. B. (1989a) Cosmic rays in the local interstellar medium. In: Grzedzielski, S. and Page, D. E. (eds). *Physics of the Outer Heliosphere*, COSPAR Colloquia Series, Vol. 1, Pergamon, pp. 107–118.

McKibben, R. B. (1989b) Re–analysis and confirmation of positive latitude gradients for anomalous helium and galactic cosmic rays measured in 1975–1976 with Pioneer 11. *J. Geophys. Res.* **94**, 17,021–17,032.

McKibben, R. B. (1998) Three-dimensional solar modulation of cosmic rays and anomalous components in the inner heliosphere. *Sp. Sci. Rev.* **83**, 21–32.

McKibben, R. B. (1999) Jovian electrons and CIRs, in McKibben *et al.* Modulation of Cosmic Rays and Anomalous Components by CIRs, Report of Working Group 5. *Sp. Sci. Rev.* in press, 1999.

McKibben, R. B., Simpson, J. A., Zhang, M., Bame, S. and Balogh, A. (1995) Ulysses out-of-ecliptic observations of 27-Day variations in high energy cosmic ray intensity. *Sp. Sci. Rev.* **72**, 403–408.

McKibben, R. B., Connell, J. J., Lopate, C., Simpson, J. A. and Zhang, M. (1996) 'Observations of Galactic cosmic rays and the anomalous helium during the Ulysses passage from the south to the north solar poles', *Astron. Astrophys.* **316**, 547–554.

McKibben, R. B., Burger, R. A., Heber, B., Jokipii, J. R., McDonald, F. B. and Potgieter, M. S. (1998) Latitudinal structure of modulation in the heliosphere, pp. 188–193 in Fisk *et al.* Global Processes That Determine Cosmic Ray Modulation, Report of Working Group 1. *Sp. Sci. Rev.* **83**, 179–214.

Mewaldt, R. A. (1989) The abundances of isotopes in the cosmic radiation. In: Waddington, C. J. (ed.) *Cosmic Abundances of Matter*, AIP Press, New York, pp. 124–146.

Meyer, J.-P., Drury, L. O'C. and Ellison, D. C. (1997) Galactic cosmic rays from Supernova remnants. I. A cosmic ray composition controlled by volatility and mass-to-charge ratio. *Astrophys. J.* **487**, 182–196.

Page, D. E. (1985) Opening address. Paper preseented at: *The Sun and The Heliosphere in Three Dimensions (Proceedings XIXth ESLAB Symposium)*, edited by R. G. Marsden. Reidel, p. 1.

Paizis, C., Heber, B., Raviart, A., Ducros, R., Ferrando, P., Rastoin, C., Kuunow, H., Müller-Mellin, R., Sierks, H. and Wibberenz, W. (1995) Latitudinal effects of galactic cosmic rays observed onboard the Ulysses spacecraft. Paper presented at: Proceedings 24th International Cosmic Ray Conference, Rome, **4**, 756–759.

Paizis, C., Heber, B., Raviart, A., Potgieter, M. S., Ferrando, P. and Müller-Mellin, R. (1997) Compton-getting factor and latitude variation of cosmic rays. Paper presented at: *Proceedings 25th International Cosmic Ray Conference*, Durban, S.A., **2**, 93–96.

Paizis, C., Heber, B., Ferrando, P., Raviart, A., Falconi, B., Marzolla, S., Potgieter, M. S., Bothmer, V., Kunow, H., Müller-Mellin, R. and Posner, A. (1999) Amplitude evolution and rigidity dependence of the 26-day recurrent cosmic ray decreases. COSPIN/KET results, *J. Geophys. Res.* **104**, 28,241–28,247.

Parker, E. N. (1965) The passage of energetic charged particles through interplanetary space, *Plan. Sp. Sci.* **13**, 9–49.

Pizzo, V. J. (1994) Global, quasi-steady dynamics of the distant solar wind. 1. Origin of North-South Flows in the Outer Heliosphere, *J. Geophys. Res.* **99**, 4173–4183.

Potgieter, M. S. (1998) The modulation of galactic cosmic rays in the heliosphere: Theory and Models. *Sp. Sci. Rev.* **83**, 147–158.

Potgieter, M. S. and Moraal, H. (1985) A drift model for the modulation of gaoactic cosmic rays. *Astrophys. J.* **294**, 425–440.

Potgieter, M. S. and le Roux, J. A. (1989) The predictions of a time-dependent drift model compared with cosmic-ray intensity observations from 1976 1989. In: Grzedzielski, S. and Page, D. E. (eds) *Physics of the Outer Heliosphere*, COSPAR Colloquia Series, Vol. 1, Pergamon, pp. 139–142.

Potgieter, M. S. and Haasbroek, L. J. (1993) The simulation of base-line cosmic-ray modulation for the Ulysses trajectory. Paper presented at: Proceedings 23rd International Cosmic Ray Conference, Calgary, **3**, 457–460.

Potgieter, M. S. Haasbroek, L. J. Ferrando, P. and Heber, B. (1997) The modelling of the latitude dependence of cosmic ray protons and electrons in the inner heliosphere. *Adv. Sp. Res.* **19**(6), 917–920.

Rickett, B. J. and Coles, W. A. (1983) 'Solar cycle evolution of the solar wind in three dimensions', in *Solar Wind 5* (Neugebauer, M. ed), NASA Conf. Publication 2280, pp. 315–321.

Silberberg, R., Tsao, C. H. and Shapiro, M. M. (1998) A weak reacceleration model for cosmic rays that fits both heavy ions and electrons. In: Shapiro, M. M., Silberberg, R. and Wefel, J. P. (eds) *Towards the Millenium in Astrophysics*, Singapore: *World Scientific*, pp. 227–240.

Simnett, G. M. and Roelof, E. C. (1998) Low energy particles, pp. 240–247 in Simnett *et al.* Corotating Particle Events, Report of Working Group 2. *Sp. Sci. Rev.* **83**, 215–258.

Simpson, J. A. (1983) Elemental and isotopic composition of the Galactic cosmic rays. *Ann. Rev. Nucl. Part. Sci.* **33**, 323–381.

Simpson, J. A. (1999) Private communication.

Simpson, J. A. and Connell, J. J. (1998) Cosmic ray ^{26}Al and its decay in the galaxy. *Astrophys. J.* **497**, L85–88.

Simpson, J. A. and Garcia-Munoz, M. (1988) Cosmic ray lifetime in the galaxy: Experimental Results and Models. *Sp. Sci. Rev.* **46**, 205–XXX.

Simpson, J. A., Anglin, J. D., Balogh, A., Bercovitch, M., Bouman, J. M., Budzinski, E. E. Burrows, J. R., Carvell, J. J., Garcia-Munoz, M., Henrion, J., Hynds, R. J., Iwers, B., Jacquet, R., Kunow, H., Lentz, G., Marsden, R. G., McKibben, R. B., Müller-Mellin, R., Page, D. E., Perkins, M., Treguer, L., Tuzzolino, A. J., Wenzel, K.-P. and Wibberenz, G. (1992) The Ulysses cosmic ray and solar particle investigation. *Astron. Astrophys.* **92**, 365–399.

Simpson, J. A., Smith, D. A., Zhang, M. and Balogh, A. (1993) Jovian electron propagation in three dimensions of the heliosphere. *Sp. Sci. Rev.* **83**, 7–19.

Simpson, J. A., Anglin, J. D., Bothmer, V., Connell, J. J., Ferrando, P., Heber, B., Kunow, H., Lopate, C., Marsden, R. G., McKibben, R. B., Müller-Mellin, R., Paizis, C., Rastoin, C., Raviart, A., Sanderson, T. R., Sierks, H., Trattner, K. J., Wenzel, K. P., Wibberenz, G. and Zhang, M. (1995) Cosmic ray and solar particle investigations over the south polar regions of the sun, *Science*, **268**, 1019–1023.

Simpson, J. A., Zhang, M. and Bame, S. (1996) A solar polar north-south asymmetry for cosmic-ray propagation in the heliosphere: The Ulysses Pole-to-Pole Rapid Transit. *Astrophys. J.* **465**, L69–L72.

Smith, E. J. and Balogh, A. (1995) Ulysses observations of the radial magnetic field, *Geophys. Res. Lett.* **23**, 3317–3320.

Smith, E. J., Jokipii, J. R., Kóta, J., Lepping, R. P. and Szabo, A. (2000) Evidence of a north-south asymmetry in the heliosphere associated with a southward displacement of the heliospheric current sheet. *Astrophys. J.* **533**, 1084–1089.

Stone, E. C. (1987) Cosmic ray studies out of the ecliptic. Paper presented at: Proceedings 20th International Cosmic Ray Conferences, Moscow, **7**, 105–114.

Suess, S. T., Smith, E. J., Phillips, J., Goldstein, B. E. and Nerney, S. (1996) Latitudinal dependence of the radial IMF component – interplanetary impring. *Astron. Astrophys.* **316**, 304–312.

Thayer, M. R. (1997) An investigation into sulfur isotopes in the Galactic cosmic rays. *Astrophys. J.* **482**, 792–795.

Theilemann, F.-K., Nomoto, K., Iwamoto, K. and Brachwitz, F. (1997) Nucleosynthesis in SNe Ia and their impact on galactic evolution. In: Ruiz–Lapuente, P., Canal, R. and Isern, J (eds) *Thermonuclear Supernovae*, Kluwer, Boston, pp. 485–514.

Timmes, F. X., Woosley, S. E. and Weaver, T. A. (1995) Galactic chemical evolution: hydrogen through Zinc. *Astrophys. J. Supp.* **98**, 617–658.

Trattner, K. J., Marsden, R. G., Bothmer, V., Sanderson, T. R., Wenzel, K.-P., Klecker, B., and Hovestadt, D. (1995) The Ulysses south polar pass: Anomalous Component of Cosmic Rays, *Geophys. Res. Letters*, **22**, 3349–3352.

Trattner, K. J., Marsden, R. G., Bothmer, V., Sanderson, T. R., Wenzel, K.-P., Klecker, B., and Hovestadt, D. (1996) Ulysses COSPIN/LET latitudinal gradients of cosmic ray O, N, and Ne, *Astron. Astrophys.* **316**, 519–528.

Urch, I. H. and Gleeson, L. J. (1971) Radial gradients and anisotropies due to galactic cosmic rays, *Astrophys. Sp. Sci.* **16**, 55–74.

Wiedenbeck, M. E. (1983) The abundance of the radioactive isotope ^{26}Al in galactic cosmic rays. Paper presented at: *Proceedings 18th International Cosmic Ray Conference, Bangalore*, **9**, 147–150.

Wiedenbeck, M. E. and Greiner, D. E. (1980) A cosmic ray age based on the abundance of ^{10}Be. *Astrophys. J.* **239**, L139–L142.

Wiedenbeck, M. E., Binns, W. R., Christian, E. R., Cummings, A. C., Dougherty, B. L., Hink, P. L., Klarmann, J., Leske, R. A., Lijowski, M., Mewaldt, R. A., Stone, E. C., Thayer, M. R., von Rosenvinge, T. T., Yanasak, N. E. (1999) Constraints on the time delay between nucleosynthesis and cosmic ray acceleration from observations of ^{59}Ni and ^{59}Co. *Astrophys. J.* **523**, L61–L64.

Woosley, S. E. and Weaver, T. A. (1981) Anomalous composition of cosmic rays. *Astrophys. J.* **243**, 651–659.

Wuosmaa, A. H., Ahmad, I., Fischer, S. M, Greene, J. P., Hackman, G., Nanal, V., Savard, G., Schiffer, J. P., Wilt, P., Austin, S. M., Brown, B. A., Freedman, S. J. and Connell, J. J., (1998) Decay Partial Half–life of ^{54}Mn and cosmic ray chronometry, *Phys. Rev. Lett.* **80**, 2085–2088.

Zaerpoor, K., Chan, Y. D., DiGregorio, D. E., Dragowsky, M. F., Hindi, M. M., Isaac, M. C. P., Krane, H. S., Larimer, R. M., Macchiavelli, A. O., Macleod, R. W., Miocinovic, P. and Norman, E. B. (1997) Galactic confinement time of iron-group cosmic rays derived from the ^{54}Mn chronometer. *Phys Rev. Lett.* **79**, 4306–4309.

Zhang, M. (1997) A linear relationship between the latitude gradient and 26-day recurrent variations in the fluxes of galactic cosmic rays and anomalous nuclear components. I. observations. *Astrophys. J.* **488**, 841–853.

Zhang, M. (1999a) A path integral approach to the theory of heliospheric cosmic-ray modulation. *Astrophys. J.* **510**, 715–725.

Zhang, M. (1999b) A markov stochastic process theory of cosmic-ray modulation. *Astrophys. J.* **513**, 409–420.

Zhang, M. McKibben, R. B., Simpson, J. A., Staines, K., Anglin, J. D., Marsden, R. G., Sanderson, T. R. and Wenzel, K.-P. (1995) Impulsive bursts of energetic particles in the high-latitude dusk-side magnetosphere of Jupiter. *J. Geophys. Res.* **100**, 19497–19512.

9

Cosmic dust

Eberhard Grün, Harald Krüger and Markus Landgraf

9.1 INTRODUCTION

There are several methods to study cosmic dust. The traditional method to determine the global structure of the interplanetary dust cloud is by zodiacal light observations (e.g. see Leinert and Grun, 1990 for a review). Before *Ulysses*, zodiacal light observations provided the primary means of studying the out-of-ecliptic distribution of the interplanetary dust cloud. By inversion of these scattered light observations it has been derived that grain properties depend upon their elevation above the ecliptic plane (i.e. upon the inclination of their orbits, (Levasseur-Regourd, 1991)). Several three-dimensional models of the intensity, polarization, and colour of the zodiacal dust cloud have been developed (Giese *et al.*, 1985a; 1986).

Thermal infrared observations by *IRAS* (Hauser *et al.*, 1984) and especially *COBE* (Reach *et al.*, 1995) give further constraints on the zodiacal dust cloud outside and above the Earth's orbit. Interpretations of these observations, however, rely heavily on assumptions about the size distribution and material properties of the dust particles. More recently, the infrared spectral energy distribution and brightness fluctuations in the zodiacal cloud have been studied with ISO (Abraham *et al.*, 1999a; 1999b). These observations will also allow for better determination of the size distribution of interplanetary dust grains.

The dust experiments on board the *Pioneer 10* and *Pioneer 11* spacecraft provided information on the radial dependence of the spatial density of large dust grains (~20 µm diameter) outside Earth's orbit. Between 1 and 3.3 AU these experiments detected a decrease in dust abundance proportional to $r^{-1.5}$, r being the heliocentric distance (Humes *et al.*, 1974; Hanner *et al.*, 1976). The photometer on board *Pioneer 10* did not record any scattered sunlight above the background beyond the asteroid belt while the penetration experiment recorded dust impacts out to about 18 AU heliocentric distance at an almost constant rate (Humes, 1980).

Measurements of dust in the inner solar system with the *Pioneer 8*, *Pioneer 9*, and *Helios* space probes and the *HEOS-2* satellite showed that there are several populations of dust particles which possess different dynamical properties. A population of slow-moving, small (10^{-16} kg $<$ m $< 10^{-14}$ kg) particles has been observed by the *Pioneer 8/9* (Berg and Grün, 1973) and *HEOS-2* (Hoffman et al., 1975) dust experiments. The *Helios* dust experiment (Grün et al., 1980) confirmed these particles which orbit the Sun on low eccentric orbits ($e < 0.4$). These low angular momentum 'apex' particles were thought to originate from collisional break-up of larger meteoroids in the inner solar system (Grün and Zook, 1980). Another population which consists of very small particles ($m \sim 10^{-16}$ kg) has been detected by the *Pioneer 8/9* and *Helios* space probes arriving at the sensors from approximately the solar direction. These particles have been identified (Zook and Berg, 1975) as small grains generated in the inner solar system which leave the solar system on hyperbolic orbits due to the dominating effect of the radiation pressure force (β-meteoroids).

The orbital distribution of larger meteoroids is best known from meteor observations. Sporadic meteoroids move on orbits with an average eccentricity of 0.4 and an average semi-major axis of 1.25 AU at Earth's orbit (Sekanina and Southworth, 1975). *Helios* measurements allowed the identification of an interplanetary dust population (Grün et al., 1980) which consists of particles on highly eccentric orbits ($e > 0.4$) and with semi-major axes larger than 0.5 AU. *Pioneer 11 in-situ* data obtained between 4 and 5 AU are best explained by meteoroids moving on highly eccentric orbits (Humes, 1980). These particle populations resemble most closely the sporadic meteor population.

Before *Ulysses*, dust was observed in the magnetosphere of Jupiter, both by *in-situ* experiments on board *Pioneer 10* and *Pioneer 11*, and by remote-sensing instrumentation on board *Voyager 1* and *Voyager 2*. *Pioneer 10/11* measured a 1,000 times higher flux of micrometeoroids in the vicinity of Jupiter compared with the interplanetary flux. The *Voyager* imaging experiment detected a ring of particulates at distances out to 2.5 jovian radii, as well as volcanic activity on the jovian satellite Io.

One of the most important results of the *Ulysses* mission is the identification and characterization of a wide range of interstellar phenomena inside the solar system. A surprise was the identification and the dominance of the interstellar dust flux in the outer solar system. Before this discovery by *Ulysses* it was believed that interstellar grains are prevented from reaching the planetary region by electromagnetic interaction with the solar wind magnetic field. The interplanetary zodiacal dust flux was thought to dominate the near-ecliptic planetary region while at high ecliptic latitudes only a very low flux of dust released from long-period comets should be present. Therefore, the characterization of the interplanetary dust cloud was the prime goal of the *Ulysses* dust investigation.

Around Jupiter fly-by in 1992 *Ulysses* had discovered intense collimated streams of dust particles. The grains were detected at a distance out to 2 AU from Jupiter along the ecliptic plane and at high ecliptic latitudes (Grün et al., 1993). Modelling of the dust trajectories showed that the grains must have been ejected from the jovian system (Zook et al., 1996). Later measurements by the *Galileo* spacecraft within the jovian magnetosphere revealed Io as the source of the grains (Graps et al., 2000).

Sec. 9.1] Introduction 375

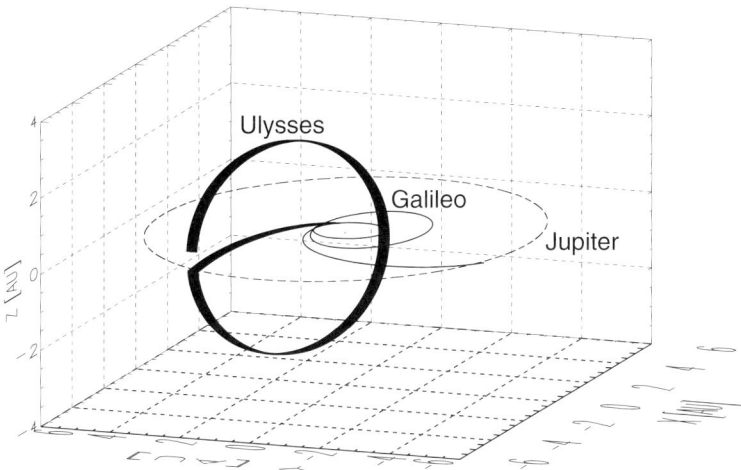

Figure 9.1. Trajectories of the *Ulysses* (heavy solid line) and the *Galileo* (thin solid line) spacecraft. The Sun is in the centre, Jupiter's (dashed) and *Galileo's* trajectories are in the ecliptic plane. The initial trajectory of *Ulysses* from Earth to Jupiter was also in the ecliptic plane. Subsequently, *Ulysses* was thrown into an orbit plane inclined by 79° to the ecliptic plane. Vernal equinox is towards the positive *x* direction and north is at the top. The *Ulysses* trajectory is shown until the end of 1997.

Here, we focus on measurements by the *in-situ* dust detectors on board the *Ulysses* and *Galileo* spacecraft to obtain information on the large-scale structure and the dynamics of the interplanetary dust cloud (Section 9.3), dust in the environment of Jupiter (Section 9.4), and interstellar dust (Section 9.5). The dust detectors measure dust along the the spacecraft's interplanetary trajectories. From these measurements a three-dimensional model of the spatial dust distribution can be constructed. The orbits of the *Galileo* and *Ulysses* spacecraft are displayed in Figure 9.1.

9.1.1 Dust objectives

Ulysses carried a high-sensitivity dust detector to the outer solar system for the first time. The detector is five orders of magnitude more sensitive than those on the *Pioneer 10* and *Pioneer 11* spacecraft (Humes, 1980). The objectives of the *Ulysses* dust experiment (as stated in the original proposal (Grün *et al.*, 1992)) were to:

1 determine the three-dimensional structure of the zodiacal cloud;
2 characterize its dynamical state; and
3 search for interstellar dust penetrating the solar system.

The prime objective was the determination of the three-dimensional dust distribution in the solar system. In order to reach this goal with the help of out-of-ecliptic *Ulysses* measurements, reference dust measurements had to be provided in the ecliptic plane.

This was almost ideally achieved by the *Galileo* mission that was launched a year earlier but took dust measurements for 6 years in interplanetary space near the ecliptic plane with a twin of the *Ulysses* dust detector. While *Ulysses* data are very good at determining the absolute latitude dependence (and inclination dependence) of meteoroids, measurements by *Galileo* near the ecliptic plane determine the radial and eccentricity dependence for low inclination orbits. The achievement of the third *Ulysses* objective (i.e. the identification of interstellar dust particles) required that interstellar dust had to be identified and separated from interplanetary dust. For measurements by both the *Galileo* and *Ulysses* detectors, this distinction was easy outside about 3 AU because it was found that the flux of interstellar dust grains dominated and differed significantly from prograde interplanetary dust, both in direction and speed (Grün *et al.*, 1993; 1994b; Baguhl *et al.*, 1995a; 1995b).

9.2 INSTRUMENTATION

Ulysses is a spinning spacecraft with its spin-axis coincident with the antenna-pointing direction. Its antenna usually points towards Earth. The *Ulysses* dust detector is mounted at an angle of 85° from the antenna direction. The field-of-view of the dust detector is a cone of 140° full angle and the spin-averaged effective sensor area for impacts varies with the angle between the impact direction and the antenna direction (Grün *et al.*, 1992). Because the detector is mounted almost perpendicular to the spacecraft spin-axis, the maximum spin-averaged sensitive area of $200\,\text{cm}^2$ is reached for impacts along a plane almost perpendicular to the antenna direction (the geometry of dust detection during Jupiter fly-by is sketched in Figure 9.8). The impact direction (rotation angle) in this plane is determined by the spin position of the spacecraft around its spin-axis at the time of a dust impact. The rotation angle is zero when the dust-sensor axis is closest to the ecliptic north direction. The rotation angle is measured in a right-handed system around the antenna direction.

The *Ulysses* dust detector (Grün *et al.*, 1992) is an impact ionization sensor which measures the plasma cloud generated upon impact of submicrometre and micrometre-sized dust particles onto the detector target (cf. Figure 9.2). Up to three independent measurements of the ionization cloud created during impact are used to derive both the mass and the impact speed of the dust grains (Grün *et al.*, 1995a). The detector mass threshold m_t is proportional to the positive charge component Q_1 of the plasma produced during the impact, which itself strongly depends on the impact speed v. Using the calibration parameters $(Q_1/m)_0$, m_0, v_0, and α, which have been approximated from detector calibrations (Grün *et al.*, 1995a), the corresponding mass threshold can be calculated:

$$m_t = \frac{Q_1}{(Q_1/m)_0} = m_0 \left(\frac{v_0}{v}\right)^\alpha \tag{9.1}$$

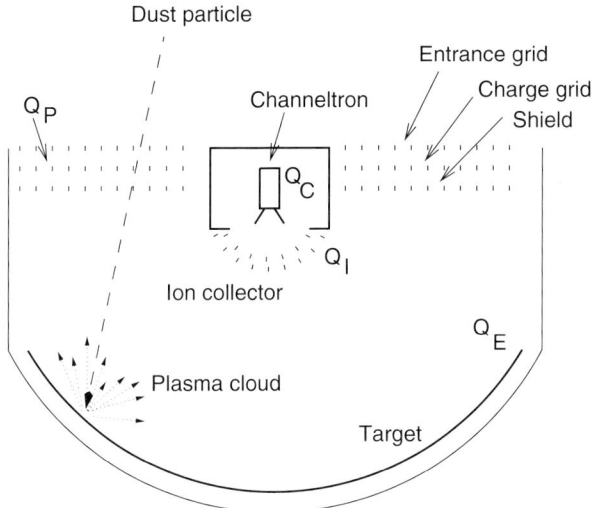

Figure 9.2. Schematic configuration of the *Ulysses* dust detector (GRU). Particles hitting the target create a plasma cloud and up to a three charge signals (Q_I, Q_E, Q_C) are used for dust impact identification.

with $\alpha \sim 3.5$. For example, an impact charge of $Q_I = 8 \times 10^{-14}$ C refers to a mass threshold $m_t = 3 \times 10^{-17}$ kg at $20\,\mathrm{km\,s^{-1}}$ impact speed.

The dynamic range of the impact charge measurement is 10^6 which is also the dynamic range of the mass determination for particles with constant impact speeds. The calibrated speed range of the instrument is $2\,\mathrm{km\,s^{-1}} \leq v \leq 70\,\mathrm{km\,s^{-1}}$ which corresponds to a calibrated mass range of $10^{-19\cdots-9}$ kg. Impact speeds can be determined with an accuracy of about a factor of 2 and the accuracy of the mass determination of a single particle is about a factor of 10.

Impact-related data such as up to three charge measurements, impact time – and rotation angle – are normally all transmitted to Earth for each impact. Thus, a complete record of impact charge, particle mass, impact velocity, impact time, and impact direction is available for each detected particle. Impact rates are derived from the number of detections within a given time interval.

In addition to *Ulysses*, the *Galileo* spacecraft carries a nearly identical dust detector on board. After being launched in 1989 *Galileo* traversed interplanetary space and was injected into a bound orbit about Jupiter in 1995. *Galileo* dust data supplement *Ulysses* measurements because *Galileo* measured dust along the ecliptic plane at times when *Ulysses* was at high ecliptic latitudes. This allowed for the determination of radial and latitudinal variation in interplanetary dust cloud (Grün et al., 1997). Dust data obtained with the detectors on board both – *Ulysses* and *Galileo* can be found in the literature (Grün et al., 1995b,c; Krüger, 1999b,c).

9.3 INTERPLANETARY DUST BACKGROUND IN THE ECLIPTIC PLANE AND ABOVE THE SOLAR POLES

The *Ulysses* dust measurements provide – as a function of time and spaceprobe position – the flux of particles as a function of impact direction, speed, and particle mass. Here we include data obtained by *Ulysses* from launch to the completion of the first inclined orbit. Shortly after launch the impact rate of 10^{-17} kg-and-smaller particles declined from a few impacts per day to about one impact per 3 days (Figure 9.3). Most of the time of interplanetary cruise until 1996 the dust impact rate stayed at this level. In late 1996 the impact rate decreased again by about a factor of 3. For about 1 year around Jupiter fly-by the impact rate of the smallest impacts increased by up to a factor of 1,000 for periods of 1 to several days. These dust streams are the second discovery of the *Ulysses* dust instrument (Grün *et al.*, 1993). Their characteristics will be discussed in Section 9.4.

At high ecliptic latitudes close to the solar poles a surprisingly large dust flux was recorded. Around the time of crossing the ecliptic plane the impact rate of big particles ($m > 10^{-15}$ kg) increased by about a factor of 10, whereas less massive

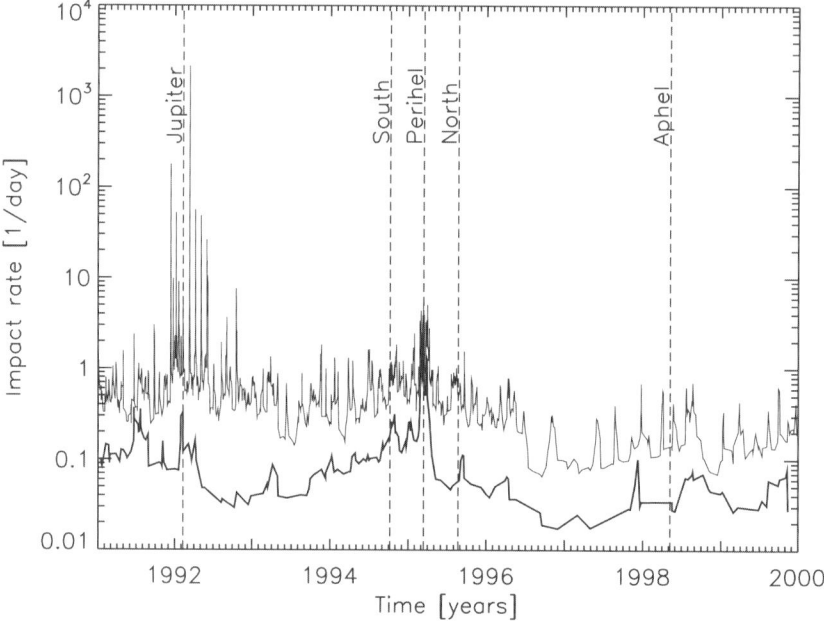

Figure 9.3. Dust-impact rates observed by *Ulysses*. The impact rates shown cover the periods from launch to Jupiter fly-by and the full out-of-ecliptic orbit from Jupiter over the poles of the Sun, through the ecliptic plane (perihelion) and out to aphelion at Jupiter distance. The impact rates are all impacts recorded (upper trace) and impacts of dust particles with $m > 10^{-15}$ kg (lower trace). The impact rates are sliding means always including six impacts.

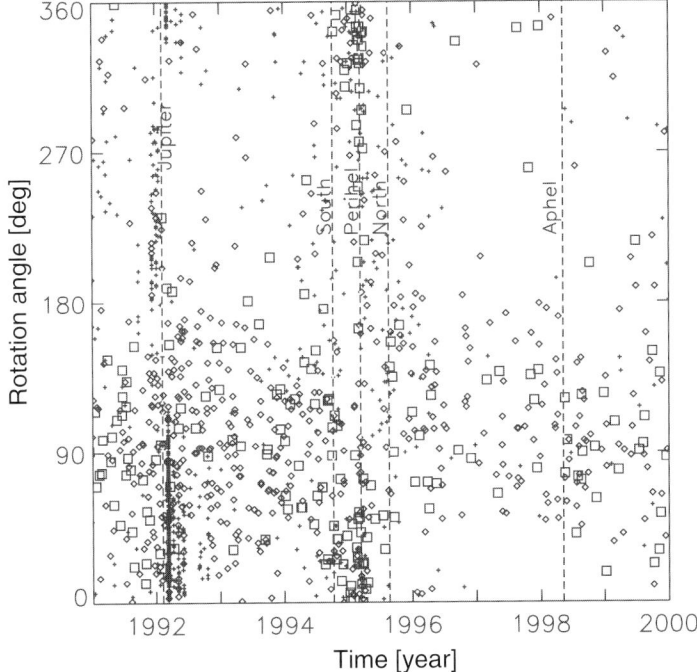

Figure 9.4. Dust-impact directions (rotation angle) observed by *Ulysses*. The rotation angle is defined in Figure 9.8. The rotation angle is 0° when the dust-sensor axis points closest to the ecliptic north pole. At rotation angle 90° the detector axis points parallel to the ecliptic plane in the direction of planetary motion, and at 180° it points closest to the south ecliptic pole. The data cover the periods from launch to Jupiter and a full out-of-ecliptic orbit. The smallest impact of particles with masses $m \leq 10^{-17}$ kg are denoted by crosses, medium massive particles (10^{-17} kg $< m \leq 10^{-15}$ kg) are denoted by diamonds, and most massive particles ($m > 10^{-15}$ kg) are denoted by squares.

particles showed a smaller enhancement. Particles smaller than 10^{-17} kg were recorded in abundance for some time shortly after launch, around Jupiter fly-by as jovian dust streams, and around the times of the polar passages. These events will be discussed in the sections on β-meteoroids (Section 9.3.2) and on Jupiter dust streams (Section 9.4).

The directional distribution of dust impacts (Figure 9.4) shows a similar separation in different categories as the impact rate. Here, we concentrate on particles bigger than 10^{-17} kg, smaller particles are discussed in their respective sections. The impact directions were concentrated at rotation angles of about 90° for most of the time. This direction is compatible with the flow direction of interstellar grains through the solar system (cf. Section 9.5). Only around the time of ecliptic plane crossing did the mean impact direction shift to about 0° rotation angle. This direction is compatible with the flux of interplanetary meteoroids around the Sun. Mostly during the passage from the north pole to *Ulysses*' aphelion were a few big

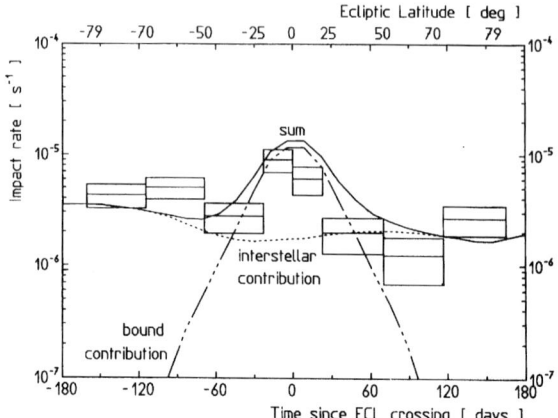

Figure 9.5. Dust-impact rate (impact charge $Q_I > 8 \times 10^{-14}$ C) observed by *Ulysses* during its south–north traverse around the time of ecliptic plane crossing (ECL). The top scale gives the spacecraft latitude. The boxes indicate mean impact rates and standard deviations. The observed impact rates are compared with models of the two major dust components in the solar system: interstellar dust and interplanetary dust on bound orbits about the Sun.

impacts were recorded from about 270° rotation angle (i.e. the direction one would expect particles on prograde bound orbits to arrive from).

9.3.1 Ulysses' south–north traverse

The *Ulysses* mission is especially well suited to obtain a latitudinal profile of the interplanetary dust cloud. In the distance range from 2.3 to 1.3 AU, *Ulysses* passed from close to the ecliptic south pole (−79° ecliptic latitude) through the ecliptic plane to the north pole (+79°). The outer portions of the out-of-ecliptic orbit (beyond 2.3 AU) are not suited for characterizing interplanetary dust because of the dominant interstellar dust population (Baguhl *et al.*, 1993).

The relation between the dust density at a given latitude and the inclination distribution is simply given by the fact that dust particles recorded at ecliptic latitude λ must have inclinations $i \geq \lambda$ in order to reach this latitude. Figure 9.5 shows the impact rate during the pole-to-pole traverse. The passages over the solar south and north poles occurred 170 days before and after ecliptic plane crossing, respectively. A total of 109 impacts (with impact charges $Q_I > 8 \times 10^{-14}$ C) were recorded during this time. The impact rate stayed relatively constant except for the maximum (9×10^{-6} s^{-1}) during ecliptic plane crossing. The impact rate at the northern leg is about a factor of 2 below that of the southern one. This is due to the varying spacecraft attitude which followed the direction to the Earth. This variation of spacecraft attitude is also reflected in the variation of rotation angles of the impacts which were detected during the south–north traverse. Over the south solar pole most large impacts (with impact charges $Q_I > 8 \times 10^{-14}$ C) occurred at rotation angles between 0° and 150° which includes the interstellar direction

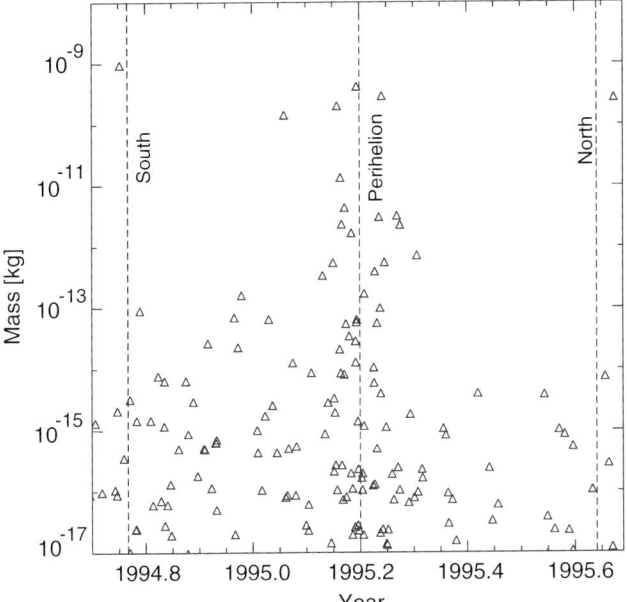

Figure 9.6. Masses of dust particles observed by *Ulysses* during its south–north traverse. Ecliptic plane crossing coincided with perihelion passage.

(cf. Figure 9.7). Closer to the ecliptic plane the rotation angle range of large impacts widened and moved further to the north direction (rotation angle $\sim 0°$). At ecliptic plane passage these impacts were recorded in a wide range around the north direction (0–100° and 200–360°). On the northern pass, rotation angles covered the whole range and above the north pole it ranged from 50°–200°, again including the interstellar direction.

Figure 9.6 shows the masses of dust particles detected during the south–north traverse. Twenty particles with masses greater than 10^{-13} kg were detected during the south–north traverse, fifteen of these particles were recorded during the 80 days when the spacecraft was close to the ecliptic plane ($-30° < \lambda < 30°$). This increased flux of big particles near the ecliptic plane is obviously due to the zodiacal dust population. The mass distribution during the south–north traverse seems to be composed of two distinct components: the interstellar dust component, which has peak masses between 10^{-14} and 10^{-15} kg and a bigger interplanetary dust component dominating the dust flux in the near-ecliptic region.

9.3.2 β-meteoroids

β-meteoroids are particles that leave the solar system on hyperbolic orbits because of radiation pressure and electromagnetic solar wind interactions (Hamilton *et al.*, 1996). They could only be observed by *Ulysses* for a few short periods during the

Ulysses mission: the early mission phase, both polar passes (Baguhl *et al.*, 1995a), and the perihelion ecliptic plane crossing. Wehry and Mann (1999) have identified β-meteoroids in the *Ulysses* dataset and discuss some properties.

9.3.3 Mass distribution of interplanetary particles

Although *Galileo* and *Ulysses* dust data cover a mass range $10^{-19}\,\text{kg} \leq m \leq 10^{-9}\,\text{kg}$, the statistically best and most complete measurements range only from $10^{-17}\,\text{kg} \leq m \leq 10^{-12}\,\text{kg}$. The mass distribution of the interplanetary dust is best known near the Earth. It has been determined from lunar crater counts and Earth-orbiting spacecraft (*Pegasus, Explorer 16* and *Explorer 23, Pioneer 8* and *Pioneer 9, LDEF*, and many others). Here we will use the mass distribution which Divine (1993) derived from the measurements presented by Grün *et al.*, 1985b). In addition we will use the radial dependence of the dust density that is based on zodiacal light observations by *Helios* (Leinert *et al.*, 1976).

9.3.4 Description of model distributions

Several dynamical meteoroid models have been developed in the last decades to describe the interplanetary meteoroid environment. The most comprehensive one so far is the model of Divine (1993) and its extension by Staubach (Staubach and Grün, 1995; Staubach, 1996) that synthesizes meteor data, zodiacal light observations, and *in-situ* measurements of interplanetary dust. The model describes dust concentrations and fluxes on the basis of meteoroid populations with distinct orbital characteristics. Gravitational (keplerian) dynamics and radiation pressure effects are employed in order to derive impact rates. Besides impact rates, impact directions and speeds can be modelled. The model includes dust populations on bound orbits around the Sun and an interstellar dust population that penetrates the solar system on unbound trajectories.

For the model description of interplanetary micrometeoroids we follow in principle the method of Divine (1993) with the extension introduced by Staubach (Staubach and Grün 1995; Staubach, 1996). The spatial dust density and the directional dust flux at each position in interplanetary space are synthesized from 'model' distributions of orbital elements. Following Divine (1993) the interplanetary meteoroid complex is represented by several populations of dust particles that are defined by their orbital elements and mass distributions. In order to simplify the meteoroid model, several assumptions have been made: (1) the zodiacal dust cloud is symmetric about the ecliptic plane, and (2) it has rotational symmetry about the ecliptic polar axis. Therefore, in the model the spatial dust density depends only on distance r from the Sun and height z above the ecliptic plane. Small asymmetries relative to the ecliptic plane (Leinert and Grün, 1990) or a small offset of the cloud centre from the Sun (Dermott *et al.*, 1994) are ignored. No time dependences are considered, although for the submicrometre dust flux a 22-year cycle has been suggested from theoretical considerations (Landgraf, 1998, 2000; Landgraf *et al.*,

2000) and found in the data (Morfill and Grün, 1979a,b; Gustafson and Misconi, 1979; Hamilton et al., 1996).

In addition to solar gravity, orbits of micrometre-sized dust grains are affected by solar radiation pressure. The ratio of radiation pressure force F_{rad}, to gravitational attraction by the Sun F_{grav} is defined by the factor $\beta = F_{rad}/F_{grav}$ (Burns et al., 1979). This β-value is strongly dependent on material composition and structure of the dust particles. In the dynamical dust model, β-values obtained by Gustafson (1994) from Mie calculations for homogeneous spheres are assumed. For particles with masses $m > 10^{-13}$ kg the influence of radiation pressure is negligible. At 10^{-14} kg, 10^{-16} kg, and 5×10^{-18} kg β-values of 0.3, 0.8, and 0.3, respectively, have been assumed. These values represent a continuous dependence and reflect the combination of the solar spectrum and particles size. For bound orbits $\beta < 1$ is required.

In order to evaluate the flux at position **r** (bold characters indicate vector quantities) the dust-particle velocity **v** has to be determined from the orbital elements, perihelion distance r_1, eccentricity e, and inclination i, and from the assumed β-value. The relative velocity $\mathbf{u}_D$ between a dust particle and the spacecraft with velocity $\mathbf{v}_{DB}$ is then $\mathbf{u}_D = \mathbf{v} - \mathbf{v}_{DB}$. The sensitivity of a detector can be expressed by its mass threshold m_t and its angular sensitivity. For spinning spacecraft like Galileo and Ulysses the effective sensitive area Γ is a function of the angle γ between the impact direction and the spacecraft axis (Grün et al., 1992).

Populations of interplanetary meteoroids are described by independent distributions of orbital inclination $p_i(i)$, eccentricity $p_e(e)$, perihelion distance $N_1(r_1)$, and particle mass $H_M(m)$, and the corresponding β-values. The detector threshold m_t, and the angular sensitivity Γ are used as weighting factors for the calculated flux (Divine, 1993; Grün et al., 1997). For any assumed dust population (defined by its distribution functions) and a given observation condition (spacecraft position $\mathbf{r}_{DB}$, velocity $\mathbf{v}_{DB}$, detector orientation $\mathbf{r}_D$, and sensitivity weighting factors), model fluxes and spatial densities can be calculated.

The distribution functions have been iterated by comparison with the corresponding measurement: the difference between the measurement and the sum of model fluxes of all dust populations is expressed by a residual. Each distribution function consists of a limited number of parameters (e.g. the inclination and eccentricity distributions are represented by seven parameters each) which were varied by an iteration algorithm in such a way that the residual was minimized.

Populations of particles on heliocentric bound orbits have been defined by the procedure described above. The populations are *ad-hoc* populations used for fitting purposes and do not correspond to specific sources of dust. Divine's core population (and the corresponding orbital distribution) has been found to fit micrometeoroids of masses $m > 10^{-13}$ kg that are not affected by radiation pressure. Three populations of smaller meteoroids, with $\beta > 0$, on bound heliocentric orbits fit the smaller particles detected by Galileo and Ulysses, mostly inside 3 AU distance from the Sun. Although all populations are defined over the whole mass range (10^{-21} kg $\leq m \leq 10^{-3}$ kg) each population dominates in a narrow mass interval for which a constant β-value has been assumed. The interstellar dust population is well

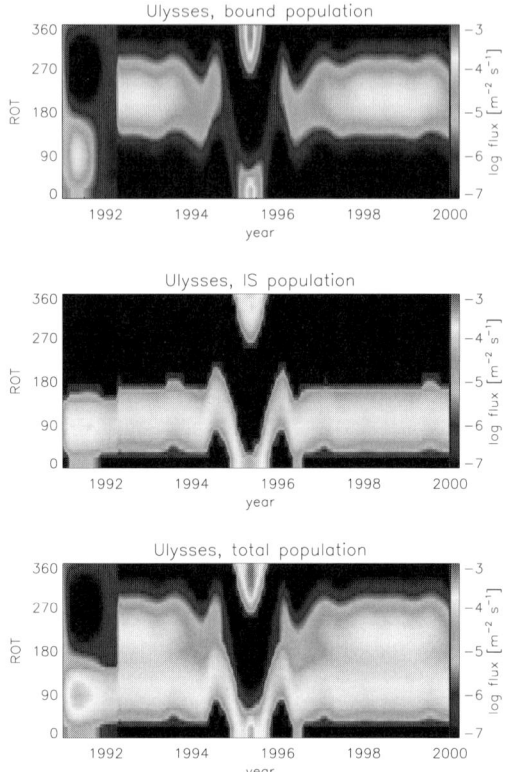

Figure 9.7. Models of directional impact rates (of particles with masses $m > 10^{-17}$ kg) onto *Ulysses* from launch to completion of its first out-of-ecliptic orbit. The grayscale coding of flux values is shown on the right side of each panel. Impact rates of interplanetary dust on bound orbits (upper panel), interstellar dust (middle panel), and the total impact rate (lower panel) are shown. The total impact rate can be compared with the observed directional impact rates in Figure 9.4 (the number of impacts – diamonds and squares – in a time and a rotation angle interval corresponds to the modelled impact rate).

represented by *Ulysses* and *Galileo* dust measurements at large heliocentric distances (Grün et al., 1993, 1994b; Baguhl, 1995a). The interstellar dust flow is, within the measurement uncertainties, aligned with the flow of interstellar gas through the heliosphere (Witte et al., 1993). Observed dust speeds are compatible with the 26 km s^{-1} unperturbed gas speed. About 60 per cent of the interstellar grains have masses between 10^{-17} kg and 10^{-15} kg. With an assumed value of $\beta = 1$ interstellar grains pass on straight trajectories through the planetary system. This β value is typical for the bulk of interstellar grains observed by *Ulysses* (for a detailed discussion see Landgraf et al., 1999). Model fluxes are calculated with these assumptions.

Figure 9.7 shows the calculated directional fluxes for the considered time period. The flux values are grayscale-coded and represent the flux in a phase-space interval, ΔROT and Δt. These model fluxes can be compared with the measurements shown in Figure 9.4 where the number of impacts in the same phase-space interval should be considered.

9.3.5 Comparison with zodiacal light

The model can be used to derive the global interplanetary dust distribution as represented by visible zodiacal light and infrared thermal-emission observations. The latitudinal density distribution was derived from zodiacal light observations (Leinert and Grün, 1990) or from infrared observations with *COBE* (Reach *et al*, 1995). Infrared brightnesses like those observed by *COBE* observations were taken in a band approximately perpendicular to the solar direction and, therefore, refer to dust outside Earth's orbit and do not represent dust close to the Sun (this is in contrast to zodiacal light observations which can be performed in almost any direction). The difference between the density distributions that fit the observed zodiacal and infrared brightnesses suggests that zodiacal dust in the inner solar system has a wider distribution (to which the zodiacal light measurements refer) than dust outside the Earth's orbit to which *COBE* data refer. Divine's core population approximates the latitudinal density function of zodiacal light observations for low latitudes ($\lambda > 10°$) and that of infrared observations for high latitudes ($\lambda > 20°$). The density distributions for the small particle populations have a wider latitudinal distribution and are closer to the zodiacal light distribution. The assumed constant density of interstellar dust provides a small but constant contribution at all latitudes.

From the dust populations, Grün *et al.*, (1997) have calculated model brightnesses of the zodiacal light as observed from Earth. By far the largest contribution comes from the core population with masses $m > 10^{-13}$ kg. Particle populations with masses $m < 10^{-13}$ kg contribute less than 1 per cent to the zodiacal light brightness at 1 AU. In order to obtain this result it was assumed that the albedo (ratio of visual light reflected by a grain to the amount of incident light) is $p = 0.05$ for the core population and $p = 0.02$ for the asteroidal population. We had to assume that the small particle populations consist of very dark particles (albedo of $p = 0.01$) in order for the sum of all populations not to exceed the observed brightness.

9.4 ELECTROMAGNETICALLY INTERACTING DUST: JUPITER DUST STREAMS

9.4.1 Observations

On 10 March 1992, about 1 month after Jupiter fly-by, perhaps one of the most unexpected findings of the *Ulysses* mission was detected: an intense stream of dust grains which was soon recognized to originate in the Jupiter system (Baguhl *et al.*, 1993). During this burst more than 350 impacts were detected within 26 hours. This exceeded by more than two orders of magnitude the typical impact rates of interplanetary and interstellar dust seen before. Ten more such dust streams could be

386 Cosmic dust [Ch. 9

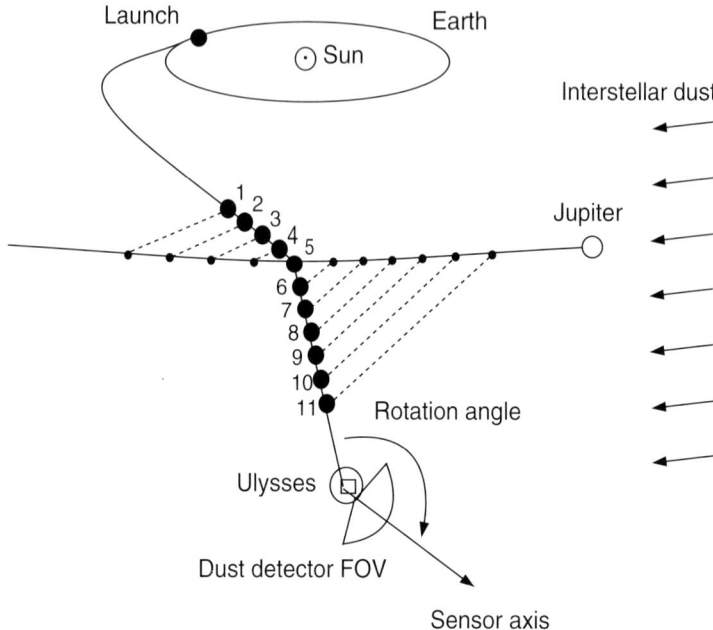

Figure 9.8. *Ulysses* trajectory and geometry of dust detection – oblique view from above the ecliptic plane. The position of the Sun and orbits of Earth and Jupiter (in the foreground) are also shown. The trajectory of *Ulysses* after Jupiter closest approach is deflected into an orbit inclined by 79° to the ecliptic plane going south. Numners along the trajectory refer to positions of *Ulysses* at which dust streams were detected, dotted lines point to Jupiter. The spacecraft spins about an axis which points towards Earth. The dust detector on board has a 140° conical FOV, and is mounted almost at right angle (85°) to the *Ulysses* spin axis. Radiant directions from which it can sense impacts therefore include the plane perpendicular to the spacecraft–Earth line. The rotation angle of the sensor axis at the time of a dust impact is measured from the ecliptic north direction. Arrows indicate the approach direction of interstellar dust.

revealed in the data when *Ulysses* was within 2 AU from Jupiter (Grün et al., 1993 Buguhl et al., 1993). The geometry of dust detection during *Ulysses'* Jupiter fly-by is shown in Figure 9.8 and the observed impact rate which shows the dust streams as individual spikes is displayed in Figure 9.9. The streams occurred at approximately monthly intervals (28 ± 3 days). No periodic dust phenomenon in interplanetary space was known before for small dust grains. Details of the streams are summarized in Table 9.1.

The impact direction (rotation angle) of the dust-stream particles is shown in Figure 9.10. The Jupiter system was the most likely source of the streams because of the following reasons:

1 the streams were narrow and collimated, which required a relatively close-by source (otherwise they should have been dispersed in space and time);

Sec. 9.4] **Electromagnetically interacting dust: Jupiter dust streams** 387

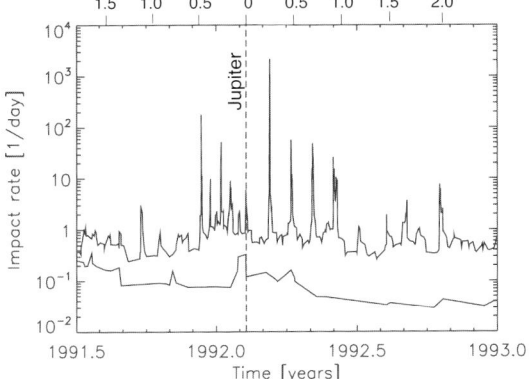

Figure 9.9. Impact rate of dust particles observed by *Ulysses* around Jupiter fly-by. The curves show all impacts recorded (upper curve) and impacts of dust particles with masses greater than 10^{-15} kg. The impact rates are means always including six impacts. The distance from Jupiter is indicated at the top.

Table 9.1. Parameters of the Jupiter dust streams adopted from Baguhl *et al.*, 1993. The time corresponds to the centre of the burst. Closest approach to Jupiter occurred on 92-039.5. Jupiter radius is $R_J = 71,492$ km.

Stream number	Days from Jupiter fly-by	Date (yr-day) Duration (h)	Number of particles	Mean rotation angle (deg)	Distance to Sun (AU)	Distance to Jupiter (R_J)
Stream 1	−136.7	91–267 76.0	6	182	4.30	2,356
Stream 2	−57.8	91–346 10.4	21	193	4.93	996
Stream 3	−45.8	91-358 18.0	7	295	5.08	796
Stream 4	−32.1	92–007 10.5	15	224	5.14	561
Stream 5	−20.4	92–019 57.1	10	259	5.29	367
Stream 6	31.2	92–071 34.3	327	49	5.4	549
Stream 7	59.4	92–099 60.9	28	47	5.39	1,107
Stream 8	87.4	92–126 55.6	33	29	5.38	1,484
Stream 9	116.7	92–156 107	28	25	5.36	1,965
Stream 10	205.4	92–244 215	12	29	5.32	3,418
Stream 11	254.3	92–293 103.4	13	57	5.2	4,200

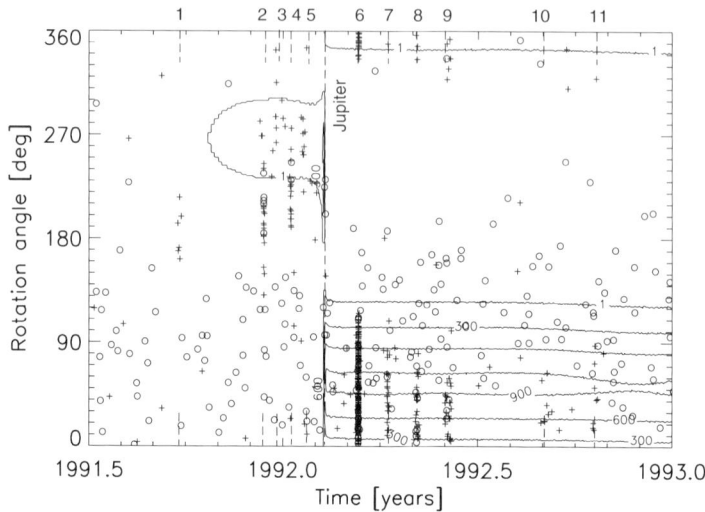

Figure 9.10. Dust impact directions observed around Jupier fly-by. Crosses denote particles with masses $m < 10^{-17}$ kg, circles denote particles with $m > 10^{-17}$ kg. The dust streams are indicated by dashed lines and are labelled 1–11. Closest approach to Jupiter is indicated by a long dashed line. The contours show the sensitive area of the dust sensor for particles approaching on straight lines from the line-of-sight to Jupiter.

2 the streams were concentrated near Jupiter and the strongest one was detected closest to the planet;

3 the streams before Jupiter fly-by approached *Ulysses* from directions almost opposite to the streams from after fly-by (All streams, however, radiated from close to the line-of-sight direction to Jupiter);

4 the observed periodicity suggested that all streams originated from a single source and seemed to rule out cometary or asteroidal origins of individual streams.

Comet *Shoemaker-Levy 9*, before its tidal disruption in 1992, was especially considered as a possible source (Grün et al., 1994a). Only two of the eleven streams, however, were compatible with an origin from the comet.

Confirmation of the Jupiter dust streams came from *Galileo*: dust 'storms' with up to 10,000 impacts per day were recorded about half a year before *Galileo's* arrival at Jupiter when the spacecraft was within 0.5 AU from the planet (Grün et al., 1996). Later, with *Galileo*, the dust stream particles were detected within Jupiter's magnetosphere (Grün et al., 1997, 1998a, b). The data showed a strong fluctuation with Jupiter's 10 h rotation period which demonstrated the electromagnetic interaction of the dust grains with the ambient magnetic field of Jupiter. The dust streams seen in interplanetary space are the continuation of the streams detected by *Galileo* within the magnetosphere.

9.4.2 Particle masses and speeds

The dust instruments on board *Ulysses* and *Galileo* have been calibrated in the laboratory. The calibrated range covers 3–70 km s^{-1} in speed and 10^{-9}–10^{-19} kg in mass (Grün *et al.*, 1995a). By applying these calibrations, masses and speeds of the stream particles were 1×10^{-19} kg to 9×10^{-17} kg and 20 to 56 km s^{-1} respectively. Uncertainties for the velocity were a factor of 2, and, for the mass, a factor of 10. Assuming an average density of 1 g cm^{-3} the derived sizes of the particles were 0.03–0.3 μm. These values, however, were challenged by later investigations.

Although the particles' approach directions were close to the line-of-sight to Jupiter, the approach direction of most streams deviated too much from the direction to Jupiter to be explained by gravitational forces alone. The deviation of the stream direction from the Jupiter direction was correlated with the magnitude and direction of the interplanetary magnetic field (especially its tangential component). Strong non-gravitational forces must have been acting on the grains to explain the observed impact directions. In numerical simulations Zook *et al.*, (1996) integrated the trajectories of many particles backward in time and away from *Ulysses* in various directions (Figure 9.11). Only particles which came close to Jupiter (<100 Jupiter radii) were considered compatible with a jovian origin. The dust particles were assumed to be electrically charged to +5 V by a balance between solar photo-electron emission and neutralization by solar wind electrons. The solar-wind speed was assumed to be 400 km s^{-1} Forces acting on the particles included

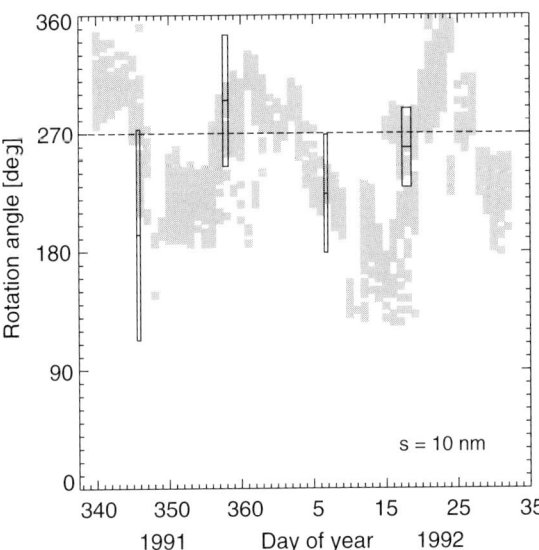

Figure 9.11. Simulated arrival directions ROT of 10 nm dust particles from Jupiter to *Ulysses* (shaded area) compared with directions from which dust streams were observed by the *Ulysses* dust sensor (boxes; from Zook, priv. comm.). The box sizes indicate the duration and range of impact directions.

390 Cosmic dust

Sun's and Jupiter's gravity and the interaction with the heliospheric magnetic field as observed by *Ulysses* (Balogh *et al.*, 1993). One important result from this analysis was that the particles were about 10^3 times less massive and 5–10 times faster than the values implied by the calibration of the dust instrument. Only particles with velocities in excess of $200\,\mathrm{km\,s^{-1}}$ and masses of the order of 10^{-21} kg were compatible with an origin in the Jupiter system. The corresponding particle radii were only 5–10 nm. Particles smaller than 5 nm were unable to travel from Jupiter to *Ulysses*. They were rather caught up by the solar wind and swept away. Particles significantly larger than 25 nm were not compatible with the measured impact directions because they did not interact strongly enough with the heliospheric magnetic field. For 10 nm sized particles the Lorentz force exceeds gravity by more than a factor of 1,000 and the trajectories of such particles are totally dominated by the interaction with the heliospheric magnetic field. Although the simulated impact directions can explain the observations, *Ulysses* should have detected particles in between the dust streams when no impacts were detected. This indicates that apart from interaction with the heliospheric magnetic field there must be other processes that modulate the particle trajectories.

The analysis by Zook *et al.* has demonstrated that the solar wind magnetic field acts as a giant mass-velocity spectrometer for charged dust grains. In particular, masses and impact speeds of the grains derived from the instrument calibration Grün *et al.*, 1995a) were shown to be invalid for the tiny Jupiter dust stream particles. Masses and speeds of stream particles derived from the *Galileo* measurements agreed very well with the values obtained for the streams detected by *Ulysses* (J. C. Liou and H. Zook, priv. comm.).

9.4.3 Dust source and particle acceleration in Jupiter's magnetosphere

What is the source of the dust streams in the Jupiter system, and which mechanism can accelerate the grains to sufficiently high speeds so that they are ejected into interplanetary and, possibly, interstellar space? One possible source was suggested even before the arrival of the two *Voyager* spacecraft at Jupiter in 1979: the volcanoes on Io (Johnson *et al.*, 1980; Morfill *et al.*, 1980). Small grains entrained in volcanic plumes follow ballistic trajectories and reach altitudes where they become exposed to the Io plasma torus. When grains encounter this high-density plasma region, they rapidly collect electrostatic charge and interact with the local magnetic field. For grains smaller than 0.1 µm, the resulting Lorentz force overcomes Io's gravity and these grains become injected into Jupiter's magnetosphere. The injection velocity is the relative velocity between the corotating magnetic field and Io (i.e. 57 km s). This production mechanism sets an upper limit on the size of the dust particles escaping from Io.

Dust grains which are positively charged and released at Io's distance ($r = 5.9\,\mathrm{R_J} = $ Jupiter radius $\mathrm{R_J} = 71,492$ km) are accelerated outward and leave the Jupiter system if their radii are between about 9 nm to 180 nm (Grün et al., 1998a). Smaller particles remain tied to the magnetic field lines and gyrate around them like ions do. Bigger grains move on gravitationally bound orbits which are

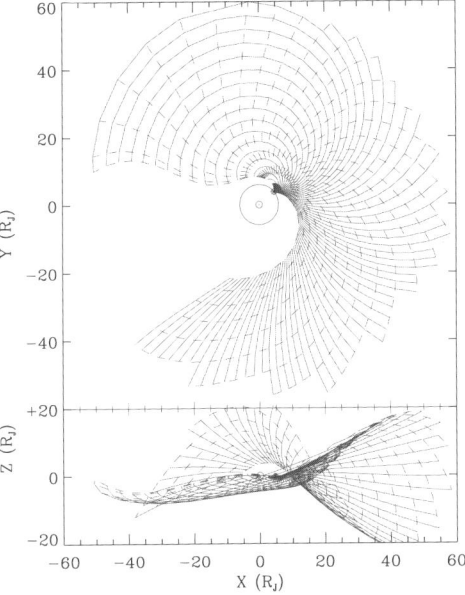

Figure 9.12. Jupiter's 'dusty ballerina skirt' formed by Jupiter stream particles moving away from the planet in a warped dust sheet (Grün *et al.*, 1998a). Jupiter is in the centre and the particles are released from a point source at Io's distance. The dust sheet is defined by synchrones (solid lines, positions of dust particles released at the same time) and syndynes (dashed lines, positions of dust particles which have the same charge-to-mass ratio, q/m). Adjacent synchrones are 1332 s apart. Syndynes of differently sized particles range from $s = 9.0$ nm to $s = 100$ nm, or, alternatively, at a surface potential of $U_g = +3$V to q/m-values between 984 and 8 C kg^{-1}. Particle trajectories have been followed only out to 50 R$_J$ which causes the ragged outer edge of the dust sheet. In steady state the dust sheet extends to much larger distances. The configuration of the dust sheet depends strongly on the phase of the magnetic field.

affected by the Lorentz force. The size range of expelled particles narrows down if the source is located closer to Jupiter, with the upper size limit staying approximately constant at 180 nm. For example, at the outer distance of the gossamer ring (3R$_J$) the lower size limit is 28 nm (Hamilton and Burns, 1993). This radial variation of the lower size limit of expelled particles is an indication of the source distance from Jupiter (Grün *et al.*, 1998a). Sizes of ejected grains of about 10 nm as determined by Zook *et al.*, (1996) are compatible with an Io source but contradict a source significantly closer to Jupiter (e.g. like the gossamer ring).

To a first order approximation, particles which are released from a source close to Jupiter's equatorial plane (e.g. Io or the gossamer ring) are accelerated along this plane. Beside this outward acceleration, however, there is a significant out-of-plane component of the electromagnetically induced force because Jupiter's magnetic field is tilted by 9.6° w.r.t. to the planet's rotation axis. Thus, the dust particles are deflected out of Jupiter's equatorial plane where they originated: Particles which

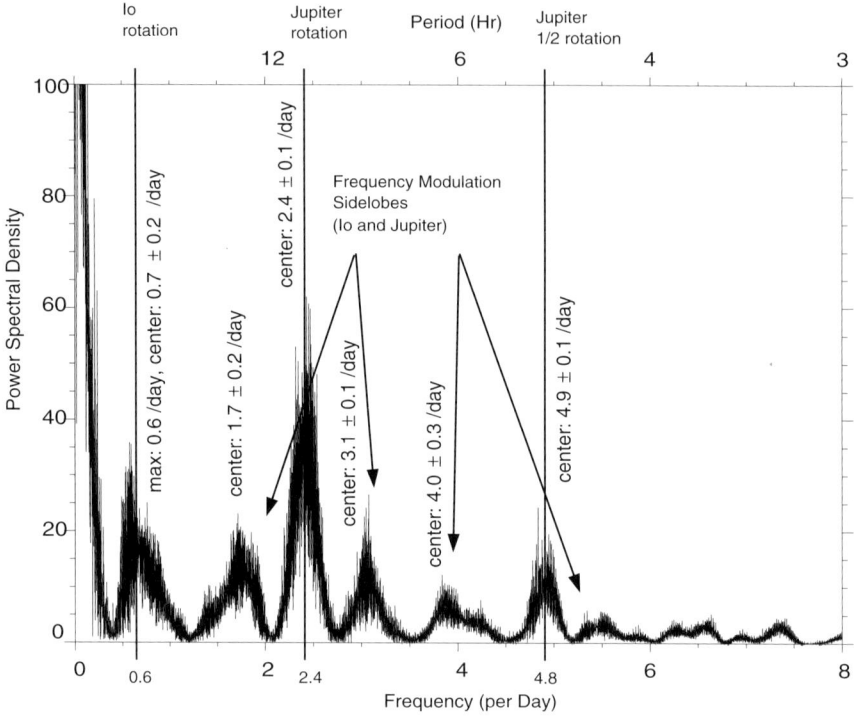

Figure 9.13. A Scargle-Lomb periodogram (Scargle, 1982) for two years of *Galileo* dust data (Grün *et al.*, 1996 and 1997; Graps *et al.*, 2000). A source with a 10 h period, Jupiter's magnetic field, is modulated by Io's frequency which leads to peaks at 13.9 ad 7.7 h. There are additional frequencies, a half period of Jupiter's magnetic field (4.9 h), Io as a single source (41.9h), and a low-frequency oscillation trend due to the spacecraft obital geometry.

are continuously released from a point source move away from Jupiter in a warped dust sheet, where particles are separated according to their size (or more appropriately to their charge to mass ratio) and time of generation. This scenario is depicted in Figure 9.12. A detector in Jupiter's equatorial plane detects an increased number of particles when this dust sheet passes over its position. Fluctuations in the dust impact rate with S and 10 h periods and impact directions observed by *Galileo* (Grün *et al.*, 1997, 1998b) can be explained by grains which are electromagnetically coupled to the magnetic field. However, only particles with a narrow size range of about 10 nm can explain the observed features. Modelling implies that larger or smaller particles have not been detected by the *Galileo* dust instrument. The 28 day periodicity detected by *Ulysses* in interplanetary space (cf. Figure 9.9) can be explained by a resonance between the rotation period of Jupiter (and, hence, its magnetic field) and Io's orbital period (Horányi *et al.*, 1993a). Although modelling results are compatible with Io being the source of the dust stream particles, modelling alone could not prove the source. In addition to the 5 h and 10 h periods which are compatible with Jupiter's rotation period, a modulation of the impact rate with

Io's orbital period (42 h) was found in the data (Krüger et al., 1999a). A periodogram which transformed 2 years of impact rate data into frequency space shows an amplitude modulation of Jupiter's magnetic field frequency and Io's orbital frequency (Figure 9.13; Graps et al., 2000). The periods of the magnetic field and Io modulate each other, and the only explanation is Io being the source of the 'carrier frequency'. Photometric observations of the Io plumes obtained with *Voyager* imply a size range of 5–15 nm (Collins, 1981) and recent Hubble Space Telescope observations constrained the particles to be smaller than 80 nm (Spencer et al., 1997). This is compatible with the results of Zook et al., (1996).

The characteristic signature of an amplitude modulation of oscillation frequencies is a carrier frequency (ω_0) with side frequencies, or 'modulation products,' ($\omega_1 + \omega_0$) and ($\omega_1 + \omega_0$). If Io's orbital frequency is a carrier frequency, and charged dust trapped in Jupiter's magnetic field is a frequency of Jupiter's rotation, that is amplitude-modulating Io's signal, then we would see a frequency at Jupiter's rotation period with side frequencies plus and minus Io's orbital frequency. The frequency transformed data obtained with *Galileo* directly display this amplitude-modulation scenario.

The dynamics of the charged nanometre-sized dust particles which are released from Io, pass the Io torus and are finally ejected out of Jupiter's magnetosphere is strongly affected by the geometry of the magnetic field. In a completely radially symmetric magnetic-field configuration no 10 h period should be evident in the impact rate, only the 5 h period should be seen. The prominent modulation of the rate with the 10 h period indicates that there must be a variation in the acceleration of the particles which is correlated with jovian local time. The Io dust stream particles will serve as test particles for future investigations of the jovian magnetic field environment.

9.5 INTERSTELLAR DUST

Prior to the *Ulysses* dust measurements, the abundance and even the mere existence of interstellar dust grains inside the solar system was controversial. Interstellar grains which are not bound to the Sun or any other star, but belong to the interstellar medium of our Galaxy, should pass by the Sun on its way through that medium. By using the density of the interstellar medium surrounding the solar system and the general assumption that 1 per cent of the mass of the interstellar medium is contained in dust grains (Holzer, 1989), it was postulated that the flux of these grains in the solar system has to be of the order of 10^{-3} m^{-2} s^{-1} This is higher than the total flux of cosmic dust observed at Earth orbit (Grün et al., 1985). This non-detection led to the conclusion that there has to be a physical process which removes interstellar dust from at least the inner solar system. Levy and Jokipii (1976) showed that the dominant force on small dust grains in the solar system is the Lorentz force caused by the solar wind magnetic field sweeping by the electrically charged dust grains. Electron emission due to solar UV photoeffect, in equilibrium with the sticking of electrons from the surrounding plasma environment, charges the grains to a positive surface potential. Since the Lorentz force acts perpendicular to the solar wind velocity vector, this seemed to be a mechanism for diverting the stream of interstellar dust grains out of the solar system. Nonetheless, it was argued (Gustafson and Misconi 1979; Morfill and Grün, 1979) that the Lorentz force not only diverts, but also concentrates interstellar dust, depending on the phase of the solar cycle. Furthermore, the Lorentz force is not as effective on larger grains as it is on small ones, because it accelerates grains proportional to their charge-to-mass ratio q/m. The charge-to-mass ratio, in turn, is inversely proportional to the grain radius s, if a constant surface potential and a spherical shape of the grain are assumed.

The size range $0.005\,\mu m \leq s \leq 0.25\,\mu m$ of classical interstellar dust grains, which have been postulated to explain the extinction of starlight (Mathis *et al.*, 1997), indicated that these grains should be heavily affected by interaction with the solar wind magnetic field. Before *Ulysses* the proposed properties of interstellar dust were

(a) there are no big ($s > 0.25\,\mu m$) dust grains in the diffuse interstellar medium surrounding the Sun, and
(b) the small ones are prevented from entering the planetary system by the solar wind magnetic field.

9.5.1 Discovery and identification of interstellar dust grains

After *Ulysses* had flown by Jupiter in February 1992, dust impacts were detected predominantly from a rotation angle between 0° and 180°. As can be seen in Figure 9.8, this corresponds to grains approaching from the retrograde direction. Significantly fewer impacts were detected from the prograde direction from which grains released from solar system bodies are expected, (cf Figure 9.4). Attempts to explain the measurements by a population of retrograde interplanetary grains failed because the impact rate of grains from the retrograde direction was constant while *Ulysses* was moving to higher and higher heliocentric latitudes (Baguhl *et al.*, 1996). Since the impact rotation angle of about 100° coincided with the upstream direction of gas from the local interstellar medium (Witte *et al.*, 1993) it was concluded that *Ulysses* detected particles belonging to a stream of dust grains originating from the interstellar medium (Grün *et al.*, 1993). In this respect the delay of the *Ulysses* mission and the resulting change in the final orbit turned out to be fortunate. On the initially planned orbit the Ulysses spin-axis would have pointed nearly parallel to the upstream direction of the interstellar dust most of the time. Since the dust detector points nearly perpendicular to the spin-axis, the impact rate of interstellar dust grains would have been strongly reduced and the identification of interstellar impacts would have been much more difficult.

To support the evidence for interstellar grains in the solar system, the impact velocities of grains detected from the retrograde direction after Jupiter fly-by have been analyzed. Most measured impact velocities, although uncertain by up to a factor of 2, indicated that the grains had heliocentric velocities in excess of the local solar system escape velocity, even if radiation pressure effects were neglected

(Grün et al., 1994b). Analysis of the data collected with the *Galileo* dust instrument further supported the interpretation of the *Ulysses* data as impacts of interstellar grains (Baguhl et al., 1995a). The discovery of interstellar grains in the solar system contradicted the considerations of the removal of interstellar dust summarized in points (a) and (b) above. This contradiction is resolved by the following considerations:

(A) interstellar grains with radii larger than 0.25 μm exist in the diffused medium. They cannot be detected by measuring the extinction of starlight because they scatter and absorb light independently of the wavelength of the incident radiation. Therefore they do not influence the shape of the interstellar extinction curve, and

(B) the solar wind magnetic field can indeed concentrate interstellar grains in the solar system. This depends on the phase of the solar cycle (Gustafson and Misconi, 1979).

After it was settled that the dataset gathered by the *Ulysses* dust detector contains interstellar grains and that they dominate the impacts detected in the outer solar system, the question was raised how we can distinguish them from impacts of interplanetary grains. In other words, how can we find a subset of the *Ulysses* dust data which contains only impacts from interstellar grains with high confidence?

As described above, the impact direction is the major characteristic that distinguishes interstellar from interplanetary grains in the outer solar system. Therefore, the identification criterion for interstellar impacts in the *Ulysses* dataset is:

Impacts that have been detected (a) after Ulysses left the ecliptic plane, and (b) with a rotation angle for which the interstellar gas upstream direction lies within the field of view of the dust detector (plus a 10° margin) are preliminarily identified as interstellar impacts.

The dataset defined in this way still contained a contribution from interplanetary impacts because during the short time of the first perihelion passage of *Ulysses* around March 1995, interplanetary grains hit the detector from the same direction as interstellar grains. Furthermore, Hamilton et al., (1996) suggested that very small ($s < 0.1$ μm) interplanetary grains can reach high latitudes when they are electromagnetically ejected from the solar system during favourable phases of the solar cycle. These grains can also contribute to the dataset defined above. Therefore, impacts have been removed from the preliminary dataset that a) have been measured around ecliptic plane crossing ($\pm 60°$ ecliptic latitude), or b) that have been measured above the solar poles with very small impact signals ($Q_1 < 10^{-13}$ C), indicating small masses.

In Figure 9.4 we show all dust impacts detected by *Ulysses* in a rotation-angle-vs-time diagram. The majority of particles in the intermediate mass range is compatible with an interstellar origin. They dominate in the outer solar system.

9.5.2 Mass distribution and cosmic abundances

As explained above, the existence of interstellar grains in the solar system can be explained by the fact that they are too big to be swept away by the solar wind magnetic field. Why can the dust models based on extinction measurements not simply be extended to account for big grains? The reason is that the mass of solid matter in the interstellar medium is limited by the *cosmic abundance of heavy elements*. The most abundant elements in the Galaxy are hydrogen and helium. In dust grains the amount of both of these elements is negligible as compared to heavier elements. The mass contained in solid dust grains is therefore limited by the available mass contained in elements heavier than helium. On average, 99 per cent of the mass of the local interstellar medium is carried by hydrogen and helium atoms (Holzer, 1989). Therefore, the total mass of dust grains cannot exceed 1 per cent of the overall mass of the medium.

If we calculate a mass distribution of the dust grains, the contribution of each grain mass interval to the total dust mass can be determined. The mass distribution for previous interstellar grain models postulates that most mass is contained in the large grains, although they are less abundant. If this grain mass distribution were extrapolated to infinitely high masses, the total mass contained in dust grains would be infinite. This would contradict cosmic abundance measurements. Therefore, the big interstellar grains found by *Ulysses* are important for understanding the composition of the interstellar medium.

We can quantify the contribution of the grains measured by *Ulysses* to the overall mass in dust in the interstellar medium by calculating the mass distribution for the *Ulysses* measurements and comparing it with the mass distribution of existing interstellar dust models. The mass distribution of interstellar grains measured by *Ulysses* is shown in Figure 9.14. The detected masses range from 10^{-18}–10^{-13} kg and most grains have masses between 3×10^{-16} and 1×10^{-15} kg (Landgraf *et al.*, 2000).

To compare the *in-situ* measurements with the mass distribution of dust models based on extinction measurements (Mathis *et al.*, 1997, MRN distribution hereafter), we calculate the contribution of each logarithmic mass interval to the overall mass density in the interstellar medium. The result is shown in Figure 9.15.

Comparing the *in-situ* measurements with the MRN distribution one finds two discrepancies:

- Small grains ($m < 10^{-16}$ kg) are deficient.
- The *in-situ* distribution extends to much larger masses than the MRN cut-off.

The first discrepancy was explained to be caused by the partial removal of small grains due to their interaction with the solar wind magnetic field (Grün *et al.*, 1994b; Landgraf 2000). In addition, interstellar grains with diameters around 0.4 μm are deflected by solar radiation pressure and are thus deficient in the inner solar system (Landgraf *et al.*, 1999). However, as explained above, the existence of *large* interstellar grains cannot be accounted for by dust models based on extinction measurements. Figure 9.15 shows that the large grains detected by *Ulysses* and *Galileo*

Sec. 9.5]	Interstellar dust	397

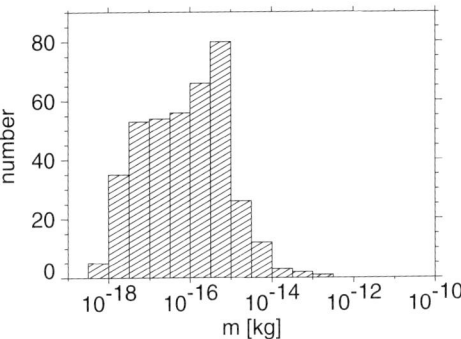

Figure 9.14. Histogram of the mass distribution of interstellar grains detected by the *Ulysses* dust instrument. The detection threshold for grains impacting with $20\,\mathrm{km\,s^{-1}}$ is $\sim 10^{-18}$ kg.

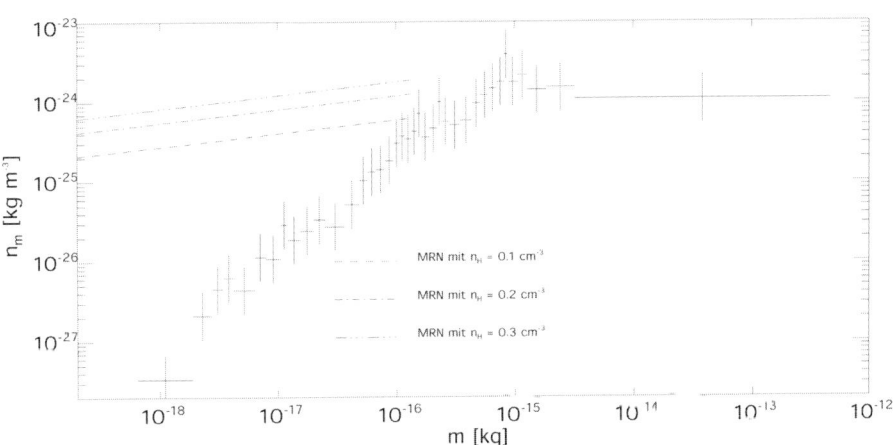

Figure 9.15. Mass density per logarithmic mass interval (Frisch et al., 1999). Crosses show the distribution of grains detected by *Ulysses* and *Galileo*. The broken lines represent the mass per logarithmic mass interval for the MRN distribution assuming three different overall densities of the local interstellar medium (measured by number n_H of hydrogen atoms per unit volume). The upper cut-off of the MRN distribution is 2×10^{-16} kg.

contribute significantly to the overall mass density of heavy elements (heavier than helium) in the local interstellar environment. Calculating the overall mass of the local interstellar medium locked up in grains and assuming that also smaller grains contribute, it was found (Frisch et al., 1999) that dust grains make up 2 per cent of the mass of interstellar matter surrounding the solar system. This is twice the amount of dust expected in the average medium of the Galaxy, which is not expected, since the local medium is less dense than average and already contains large amounts of heavy elements in the gas phase (Frisch et al., 1999).

9.6 SUMMARY AND CONCLUSIONS

The prime goal of the *Ulysses* mission with respect to dust was to determine the three-dimensional structure of the interplanetary dust cloud. This goal has been fully accomplished. *Ulysses* dust data is a crucial ingredient of the comprehensive dynamical model that describes the interplanetary dust cloud (Grün *et al.*, 1997).

Three more dust populations have been discovered that have not been seen before by any other *in-situ* dust experiment: jovian dust stream particles near Jupiter, electromagnetically controlled β-meteoroids above the poles of the Sun, and interstellar grains everywhere else.

The jovian dust streams discovered by *Ulysses* demonstrated the fact that dust may become an intimate player in magnetospheric processes. We are still at the beginning of our understanding of these interrelations but already now the importance of dust becomes evident. The dust streams are monitors of the volcanic plume activity on Io in a way that is not accomplished yet by any other observational method. So far, only occasional images from *Voyagers* and *Galileo* have shown the plumes which are not necessarily related to lava outpours on Io also observed from Earth. Electromagnetically coupled dust grains are probes of the plasma environment in the Io torus where they get their initial charge before they are emitted by Jupiter's magnetic field from the jovian system. Dust stream particles observed at different times originate from different portions of the Io torus. Therefore, observations tracing back to different local times in the Io torus may reveal local time variations in the torus that are predicted by theory. Tiny dust grains can transport material to regions of the magnetosphere where other known magnetospheric processes are unable to transport ions to. Glows of sodium and other elements may be examples for such a transport. Finally, dust stream particles will be used as carriers of information about the chemical composition in Io's volcanic plumes when *Cassini* flys by Jupiter in December 2000. The dust instrument on board may analyze their chemical composition. In February 2004 *Ulysses* will fly by Jupiter within a distance of 0.8 AU. A repetition of dust stream measurements in interplanetary space will be beneficial to test our understanding of this new phenomenon.

Yet another population of very small dust particles was observed at high ecliptic latitudes over the Sun's pole. This population has been suggested to be electromagnetically deflected β-meteoroids generated at low ecliptic latitudes and subsequently deflected by the contemporary interplanetary magnetic field to high latitudes. Long-term observations (over two solar cycles) of these particles may prove the prediction that during one solar cycle these particles are found to leave the solar system at high latitudes again while during the other cycle they escape near the ecliptic. This way, these particles are the Sun's analogue to Jupiter's dust streams. The study of these particles will provide one further piece in the puzzle of how much dust and from which region in the solar system solid particles are ejected to the interstellar medium.

The question if the interstellar dust grains measured in the solar system by *Ulysses* and *Galileo* are 'native' to the local interstellar medium (i.e. if they originate from the medium) or if they have been injected, is connected to their

Sec. 9.6] Summary and conclusions 399

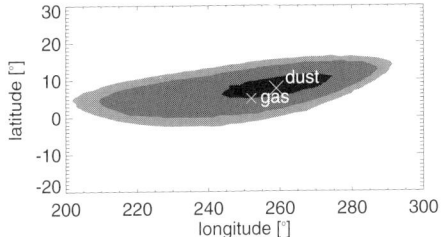

Figure 9.16. Ecliptic longitude and latitude of the upstream direction of interstellar dust grains measured by *Ulysses* after Jupiter fly-by. The contour plot of a χ^2 analysis shows $1\sigma, 2\sigma$ and 3σ confidence levels in black, dark grey, and light grey, respectively. The minimum χ^2 is achieved at 259° longitude and 8° latitude (indicated by the cross labelled 'dust'). The measurements of the helium upstream direction are indicated by the smaller cross labelled 'gas'

kinematic state with respect to the medium. Can we measure a systematic motion of the grains relative to the gas?

The velocity vector of neutral helium atoms with respect to the Sun has been determined with the *Ulysses* GAS experiment. The helium upstream direction was found to be 252° ecliptic longitude and 5° latitude, and the absolute relative velocity is 25.22 km s^{-1} (Witte et al., 1993). By determining the dust upstream direction and comparing it to the values for the helium atoms, we can establish the kinematic relationship between dust and gas. Since the *Ulysses* dust detector has a field of view of ±140°, the impact direction of an individual grain is not accurately determined and we have to apply a statistical analysis of the distribution of measured rotation angles. Figure 9.16 shows the result.

We find that the dust upstream direction coincides with the gas direction on the 1σ confidence level. This indicates, that interstellar gas and dust is in rest with respect to each other. However, since the determination of the dust upstream direction has considerable uncertainties, a velocity dispersion as large as 10 km s^{-1} between dust and gas cannot be ruled out (Landgraf et al., 1999). The *Ulysses* data indicate a dust component at rest with the local interstellar gas.

For interstellar dust it seems that *Ulysses* has raised a number of new questions by solving the question about the existence of interstellar dust grains in the solar system. These can by summarized as follows:

• Is the concept of cosmic elementary abundances correct on all length scales or are there small-scale inhomogeneities of the galactic chemistry?
• If the answer to the first question is negative, the interstellar grains we measure with *Ulysses* cannot originate from the interstellar gas of the surrounding medium. So how did they get there and what governs the dynamics of solid grains in the interstellar medium of our Galaxy?

To further confirm our findings and to improve the understanding of the interstellar medium surrounding our solar system, long term measurements are needed because the total flux of interstellar dust grains is low. *Ulysses* provides an ideal platform for these measurements, because of its fortunate orbit geometry.

9.7 ACKNOWLEDGEMENTS.

We thank the ESA/NASA *Ulysses* projects for effective and successful mission operations. This work has been supported by Deutsches Zentrum für Luft- und Raumfahrt e.V. (DLR) and by the Sonderforschungsbereich Sternentste- hung of the Deutsche Forschungsgemeinschaft (DFG).

9.8 REFERENCES

Abraham, P., Leinert, C., Acosta-Pulido, J., Schmidtobreick, L. and Lemke, D. (1999a) Zodiacal light observations with ISOPHOT, Proceedings of the conference. *The Universe as seen by ISO*, eds. P. Cox and M. F. Kessler, ESA SP-427, 145–148.

Abraham, P., Leinert, C. and Lemke, D. (1999b) Interplanetary dust as seen in the zodiacal light with ISO. In: d'Hendecourt, L. Joblin C. and Jones, A. (eds) *Solid Interstellar Matter: The ISO Revolution*, Les Houches No 11, *EDP Sciences*, Les Ullis pp. 3–18.

Baguhl, M., Grün, E., Linkert, D., Linkert, G. and Siddique, N. (1993) Identification of 'small' dust impacts in the Ulysses dust detector data. *Planet. Space Sci.* **41**, 1085–1098.

Baguhl M., Hamilton, D. P., Grün, E., Dermott, S. F., Fechtig, H., Hanner, M. S., Kissel, J., Lindblad, B. A.Linkert, D. Linkert, G. Mann, I. McDonnell, J. A. M. Morfill, G. E. Polanskey, C., Riemann, R. Schwehm, G. Staubach P. and Zook, H. A. (1995a) Dust measurements at high ecliptic latitudes. *Science*, **268**, 1016–1019.

Baguhl, M., Grün, E. Hamilton, D. P. Linkert, G. Riemann, R. and Staubach, P. (1995b) The flux of interstellar dust observed by Ulysses and Galileo. *Space Science Reviews*, **72**, 471–476.

Baguhl, M., Grün, E. and Landgraf, M. (1996) In situ measurements of interstellar dust with the Ulysses and Galileo spaceprobes (von Steiger, R., Lallement, R. and Lee, M. A. (eds)) *The Heliosphere in the Local Interstellar Medium. Space Science Reviews*, Kluwer Academic Publishers, vol. **78**, 165–172.

Balogh, A. Erdös, G., Forsyth, R. J. and Smith, E. J. (1993) The evolution of the interplanetary sector structure in 1992. *Geophys. Res. Lett.* **20**, 2331–2334.

Berg, O. E. and Grün, E. (1973) Evidence of hyperbolic cosmic dust particles. In: *Space Research* XIII, Akademie Verlag, Berlin, 1047–1055.

Burns, J. A. Lamy, P. L. and Soter, S. (1979) Radiation forces on small particles in the solar system. *Icarus*, **40**, 1–48.

Collins, S. A. (1981) Spatial color variations in the volcanic plume at Loki on Io. *J. Geophys. Res.* **86**, 8621.

Dermott, S. F. Jayaraman, S., Xu, Y. L., Gustafson, B. A. S. and Liou, J. C., (1993) A circumpolar ring of asteroidal dust in resonant lock with the Earth. *Nature*. **369**, 719–723.

Divine, N. (1993) Five populations of interplanetary meteoroids. *J. Geophys. Res.* **98**, 17029–17048.

Frisch, P. C., Dorschner, J., Geiß, J., Greenberg, J. M., Grün, E., Landgraf, M., Hoppe, P., Jones, A. P., Krätschmer, W., Linde, T. J., Morfill, G. E., Reach, W., Slavin, J., Svestka, J., Witt, A. and Zank, G. P. (1999) Dust in the local interstellar wind. *Astrophysical Journal.* **525**, 492–516.

Giese, R. H., Kinateder, G., Kneissel, B. and Rittich, U. (1985) Optical models of the three-dimensional distribution of interplanetary dust. In: Giese R. H. and Lamy, P. (eds) Properties and interactions of interplanetary dust, Reidel Publishing Co., Dordrecht, 255–259.

Giese, R. H., Kneissel, B. and Rittich, U. (1986) Three-dimensional zodiacal dust cloud: a comparative study. *Icarus.* **68**, 395–411.

Graps, A., Grün, E., Svedhem, H., Krüger, H., Horányi, M., Heck, A. and Lammers, S. (2000) Io as a source of the jovian dust streams. *Nature.* **405**, 48–50.

Grün, E. and Zook, H. A. (1980) Dynamics of micrometeoroids, in: Solid Particles in the Solar System, eds. I. Halliday and B. A. McIntosh, Reidel Publ. Co., Dordrecht, 293–298.

Grün, E., Pailer, N., Fechtig, H. and Kissel, J. (1980) Orbital and physical characteristics of micrometeoroids in the inner solar system as observed by Helios 1. *Planet. and Space Sci.* **29**, 333–349.

Grün, E., Zook, H. A., Fechtig, H. and Giese, R. H. (1985) Collisional balance of the meteoritic complex. *Icarus.* **62**, 244–272.

Grün, E., Fechtig, H., Giese, R. H., Kissel, J., Linkert, D., Maas, D., McDonnell, J. A. M., Morfill, G. E., Schwehm, G. and Zook, H.A. (1992) The Ulysses dust experiment. *Astron. Astrophys. Suppl. Ser.* **92**, 411–423.

Grün, E., Zook, H. A., Baguhl, M., Balogh, A., Bame, S. J., Fechtig, H., Forsyth, R., Hanner, M. S., Horányi, M., Kissel, J., Lindblad, B.-A., Linkert, D., Linkert, G., Mann, I., McDonnell, J. A. M., Morfill, G. E., Phillips, J. L., Polanskey, C., Schwehm, G., Siddique, N., Staubach, P., Svestka, J. and Taylor, A. (1993) Discovery of jovian dust streams and interstellar grains by the Ulysses spacecraft. *Nature.* **362**, 428–430.

Grün, E., Hamilton, D. P., Baguhl, M., Riemann, R., Horányi, M. and Polanskey, C. (1994a) Dust streams from comet Shoemaker-Levy 9? *Geophys. Res. Lett.* **21**, 1035–1038.

Grün, E., Gustafson, B. Å. S., Mann, I., Baguhl, M., Morfill, G. E., Staubach, P., Taylor, A. and Zook, H. A. (1994b) Interstellar dust in the heliosphere. *Astronomy and Astrophysics.* **286**, 915–924.

Grün, E., Baguhl, M., Fechtig, H., Hamilton, D.P., Kissel, J., Linkert, D., Linkert, G. and Riemann, R. (1995a) Reduction of Galileo and Ulysses dust data. *Planet. Space Sci.* **43**, 941–951.

Grün, E., Baguhl, M., Divine, N., Fechtig, H., Hamilton, D. P., Hanner, M. S., Kissel, J., Lindblad, B.-A., Linkert, D., Linkert, G., Mann, I., McDonnell, J. A. M., Morfill, G. E., Polanskey, C., Riemann, R., Schwehm, G., Siddique, N., Staubach P. and Zook, H. A. (1995b) Three years of Galileo dust data. *Planet. Space Sci.* **43**, 953–969.

Grün, E., Baguhl, M., Divine, N., Fechtig, H., Hamilton, D. P., Hanner, M. S., Kissel, J., Lindblad, B.-A., Linkert, D., Linkert, G., Mann, I., McDonnell, J. A. M., Morfill, G. E., Polanskey, C., Riemann, R., Schwehm, G., Siddique, N., Staubach, P. and Zook, H.A. (1995c) Two years of Ulysses dust data. Planet. *Space Sci.* **43**, 971–999.

Grün, E., Baguhl, M., Hamilton, D. P., Riemann, R., Zook, H. A., Dermott, S., Fechtig, H., Gustafson, B. A., Hanner, M. S., Horányi, M., Khurana, K. K., Kissel, J., Kivelson, M., Lindblad, B.-A., Linkert, D., Linkert, G., Mann, I., McDonnell, J. A. M., Morfill, G. E., Polanskey, C., Schwehm, G. and Srama, R. (1996) Constraints from Galileo observations on the origin of jovian dust streams. *Nature.* **381**, 395–398.

Grün, E., Staubach, P., Baguhl, M., Hamilton, D. P., Zook, H. A., Dermott, S., Gustafson, B. A., Fechtig, H., Kissel, J., Linkert, D., Linkert, G., Srama, R., Hanner, M. S., Polanskey, C., Horányi, M., Lindblad, B.-A., Mann, I., McDonnell, J. A. M., Morfill, G.E. and Schwehm, G. (1997) South-north and radial traverses through the zodiacal cloud. *Icarus.* **129**, 270–288.

Grün, E., Krüger, H., Dermott, S., Fechtig, H., Graps, A., Gustafson, B. A., Hamilton, D. P., Hanner, M. S., Heck, A., Horányi, M., Kissel, J., Lindblad, B.-A., Linkert, D., Linkert, G., Mann, I., McDonnell, J. A. M., Morfill, G. E., Polanskey, C., Schwehm, G., Srama, R. and Zook, H. A. (1998a) Dust measurements in the jovian magnetosphere. *Geophys. Res. Lett.* **24**, 2171–2174.

Grün, E., Krüger, H., Graps, A., Hamilton, D. P., Heck, A., Linkert, G., Zook, H. A., Dermott, S., Fechtig, H., Gustafson, B. Å., Hanner, M. S., Horányi, M., Kissel, J., Lindblad, B.-A., Linkert, D., Mann, I., McDonnell, J. A. M., Morfill, G. E., Polanskey, C., Schwehm, G. and Srama, R. (1998b) Galileo observes electromagnetically coupled dust in the jovian magnetosphere. *J. Geophys. Res.* **103**, 20011–20022.

Gustafson, B. Å. S. and Misconi, N. Y. (1979) Streaming of interstellar grains in the solar system. *Nature.* **282**, 276–278.

Hamilton, D. P. and Burns, J. A. (1993) Ejection of dust from Jupiter's gossamer ring. *Nature.* **364**, 695–699.

Hamilton, D. P., Grün, E. and Baguhl, M. (1996) Electromagnetic escape of dust from the solar system. In: Gustafson, B. Å. S. and Hanner, M. S. (eds) *Physics, Chemistry, and Dynamics of Interplanetary Dust. Astronomical Society of the Pacific Conference Series.* vol. **104**, pp. 31–34.

Hanner, M. S., Sparrow, J. G., Weinberg, J. L. and Beeson, D. E. (1976) Pioneer 10 observations of zodiacal light brightness near the ecliptic: Changes with heliocentric distance. In: Lecture Notes in Physics, **48**, *Interplanetary Dust and Zodiacal Light.*

Hauser, M. G., Gillett, F. C., Low, F. J., Gautier, T. N., Beichman, C. A., Neugebauer, G., Aumann, H. H., Band, B., Boggess, N., Emerson, J. P., Houck, J. R., Soifer, B. T. and Walker, R. G. (1984) IRAS observations of the diffuse infrared background. *Astrophys. J.* **278**, L15–L18.

Hoffmann, H.-J., Fechtig, H., Grün, E., Kissel, J. (1975) Temporal fluctuation and anisotropy of the micrometeroid flux in the Earth-Moon system. *Planet. and Space Sci.* **23**, 985–991.

Holzer, T. E. (1989) Interaction between the solar wind and the interstellar medium. *Annual Review of Astronomy and Astrophysics.* **27**, 199–234.

Horányi, M., Morfill, G. E. and Grün, E. (1993a) Mechnism for the acceleration and ejection of dust grains from Jupiter's magnetosphere. *Nature.* **363**, 144–146.

Horányi, M., Morfill, G. E. and Grün, E. (1993b) The dusty ballerina skirt of jupiter. *J. Geophys. Res.* **98**, 21,245–21,251.

Horányi, M., Grün, E., Heck, A. (1997) Modelling the Galileo dust measurements at Jupiter. *Geophys. Res. Let.* **24**, 2175–2178.

Humes, D. H., Alvarez, J. M., O'Neal, R.L. and Kinard, W. H. (1974) The interplanetary and near-Jupiter meteoroid environment. *J. Geophys. Res.* **79**, 3677–3684.

Humes, D. H. (1980) Results of Pioneer 10 and 11 meteoroid experiments: Interplanetary and near-Saturn. *J. Geophys. Res.* **85**, 5841–5852.

Johnson, T. V., Morfill, G. and Grün, E. (1980) Dust in Jupiter's magnetosphere: An Io source? *Geophys. Res. Lett.* **7**, 305–308.

Krüger, H., Grün, E., Graps, A. and Lammers, S. (1999a) Observations of electromagnetically coupled dust in the jovian magnetosphere. *Astrophys. and Space Sci.* **264**, 247–256.

Krüger, H., Grün, E., Landgraf, M., Baguhl, M., Dermott, S., Fechtig, H., Gustafson, B. Å., Hamilton, D. P., Hanner, M. S., Horanyi, M., Kissel, J., Lindblad, B., Linkert, D., Linkert, G., Mann, I., McDonnell, J. A. M., Morfill, G. E., Polanskey, C., Schwehm, G., Srama, R. and Zook, H. (1999b) Three years of Ulysses dust data: 1993 to 1995. *Planetary and Space Science.* **47**, 363–383.

Krüger, H., Grün, E., Hamilton, D. P., Baguhl, M., Dermott, S., Fechtig, H., Gustafson, B. A., Hanner, M. S., Horanyi, M., Kissel, J., Lindblad, B. A., Linkert, D., Linkert, G., Mann, I., McDonnell, J. A. M., Morfill, G. E., Polansey, C., Riemann, R., Schwehm, G., Srama, R., and Zook, H. (1999c) Three years of Galileo dust data: II. 1993 to 1995. *Planetary and Space Science.* **47**, 85–106.

Landgraf, M. (1998) Modellierung der Dynamik und Interpretation der in-situ Messung interstellaren Staubs in der Lokalen Umgebung des Sonnensystems, Ph.D. thesis, Ruprecht–Karls–Universität Heidelberg (in German).

Landgraf, M. (2000) Modeling the motion and distribution of interstellar dust inside the heliosphere. *Journal of Geophysical Research.* **105**, AS, 10303–10316.

Landgraf, M., Augustsson, K., Grün, E. and Gustafson, B. Å. S. (1999) Deflection of the local interstellar dust flow by solar radiation pressure. *Science.* **286**, 2319–2322.

Landgraf, M., Baggaley, W. J. and Grijn, E. (2000) Aspects of the mass distribution of interstellar dust grains in the solar system from in-situ measurements. *Journal of Geophysical Research.* **105**, AS, 10343–10352.

Leinert, C. and Grün, E. (1990) Interplanetary Dust. In: Schwenn, I. R. and Marsch, E. (eds) *Physics of the Inner Heliosphere*, Springer Verlag, Berlin, pp. 207–275.

Leinert, C., Richter, I., Pitz, E. and Planck, B. (1976) The zodiacal light from 1.0 to 0.3 A.U. as observed by the Helios space probes. *Astron. Astrophys.* **103**, 177–188.

Levasseur-Regourd, A.-C. (1991) The zodiacal cloud complex. In: Levasseur-Regourd, A. C. (ed.) *Origin and Evolution of Interplanetary dust.* Kluwer Academic Publishers, Dordrecht, 13–138.

Levy, E. H. and Jokipii, J. R. (1976) Penetration of interstellar dust into the Solar System. *Nature.* **264**, 423–424.

Mathis, J. S., Rumpi, W. and Nordsieck, K. H. (1997) The size distribution of interstellar grains. *Astrophysical Journal.* **280**, 425

Morfill, G. E. and Grün, E. (1979a) The motion of charged dust particles in interplanetary space – I. The zodiacal dust cloud. *Planet. Space Sci.* **27**, 1269–1282.

Morfill, G. E. and Grün, E. (1979b) The motion of charged dust particles in interplanetary space – II. Interstellar grains. *Planet. Space Sci.* **27**, 1283–1292.

Morfill, G., Grün, E. and Johnson, T. V. (1980) Dust in Jupiter's magnetosphere: Physical processes. *Planet. Space Sci.* **28**, 1087–1100.

Reach, W. T., Franz, B. A., Weiland, J. L., Hauser, M. G., Kelsall, T. N., Wright, E. L., Rawley, G., Stemwedel, S. W. and Spiesman, W. J. (1995) Observational confirmation of a circumsolar dust ring by the COBE satellite. *Nature*, **374**, 521–523.

Scargle, J. D. (1982) Studies in astronomical time series II.: Statistical aspects of spectral analysis of unevenly spaced data. *Astrophys. J.* **263**, 835–853.

Sekanina, Z. and Southworth, R. B. (1975) Physical and dynamical studies of meteors, meteor-fragmentation and stream-distribution studies, NASA CR-2615.

Spencer, J. R., Sartoretti, P., Bellester, G. E., McEwen, A. S., Clarke, J. T. and McGrath, M. A. (1997) The pele plume (Io): observations with the Hubble Space Telescope. *Geophys. Res. Lett.* **24**(20), 2471–2474.

Staubach P., and Grün, E. (1995) Development of an upgraded meteoroid model. *Advances in Space Res.* **16**, (11)103–(11)106.

Staubach, P. (1996) Numerische Modellierung der Dynamik von Mikrometeoroiden und ihre Bedeutung für interplanetare Raumsonden und geozentrische Satelliten, PhD. thesis, Ruprecht-Karls-Universität Heidelberg.

Wehry, A. and Mann, I. (1999) Identification of β-meteoroids from measurements of the dust detector on board the Ulysses spacecraft. *Astron. & Astophys.* **341**, 296.

Witte, M., Rosenbauer, H., Banaszkiewicz, H. and Fahr, H. (1993) The Ulysses neutral gas experiment – Determination of the velocity and temperature of the interstellar neutral helium. *Advances in Space Res.* **13**, (6)121–(6)130.

Zook, H. A. and Berg, O. E. (1975) A source for hyperbolic cosmic dust particles. *Planet and Space Sci.* **23**, 183–203.

Zook, H. A., Grün, E., Baguhl, M., Hamilton, D. P., Linkert, G., Liou, J.-C., Forsyth, R. and Phillips, J. L. (1996) Solar wind magnetic field bending of jovian dust trajectories. *Science.* **274**, 1501–1503.

Index

ACE (Advanced Composition Explorer) 16, 62, 322, 353, 363, 330
ACRs (Anomalous Cosmic Rays) 38, 272–275, 287
  see also Cosmic rays; Energetic particles
Age, turbulence 181
Alfvén decorrelation effect 189
Alfvén speed 75
Alfvén waves 23–24, 69–70, 96–97, 113, 167, 196
Alfvénic turbulence 24
Apollo 62, 67

Backward-propagating wave 110
BDE (Bi-Directional suprathermal Electrons) 137
  see also Electrons
Beta-meteoroids 374, 381–382, 398
  see also Cosmic dust
Bow shock 307
Bump-on-tail distribution 236

Carbon ions 39
Carrier frequency 393
Carrington rotation 47
Cassini 322
Challenger 16
Channeling 277–278
CIRs (Corotating Interaction Regions) 5, 107–120, 122–127, 129–133, 171, 262, 268, 299, 304
  northern hemisphere 120–121
  southern hemisphere 117, 119–120

Classical Orbital Elements 17–18
  argument of perihelion 17
  eccentricity 17
  inclination 17
  longitude of the ascending node 17
  mean anomaly 18
  mean distance 17
  true anomaly 18
Closed fields 4
CMEs (Coronal Mass Ejections) 5, 28, 54, 66, 107, 134–138, 140–158, 245–246, 265
COBE 373, 385
Composition signatures 121–122
Compressions, see Microstreams
Corona
  expansion 3
  holes 4, 43
  outer 3
  plasma 3
  structure 4
Corotating shocks 31, 127
Corotating structures 107
  CIR 3-D models 128–133
  CIR formation 108–116
  CIR observations 116–127
  see also CIRs
Cosmic abundance of heavy elements 396
Cosmic dust 373–375
  background in the ecliptic plane and above the solar poles 378–380
  beta-meteoroids 381–382

Cosmic Dust (cont)
  mass distribution of interplanetary particles 382
  model distributions 382–385
  south–north traverse by *Ulysses* 380–381
  zodiacal light comparison 385
  electromagnetically interacting 385–388
    particle masses and speed 389–390
    source and particle acceleration in Jupiter's magnetosphere 390–393
  instrumentation on board *Ulysses* 376–377
  interstellar 393–394
    discovery and identification 394–395
    mass distribution and abundances 396–397
  objectives of *Ulysses* mission 375–376
  see also Instrumentation
Cosmic rays 327–333
  adiabatic deceleration 334
  chronometry 359–360
  composition studies 351
    instrumentation for isotopic studies 351–357
    measurements of radioactive isotopes 357–361
  diffusion 334
  electron capture secondary isotopes 363
  flux 327–328
  in polar regions 24
  mean density 358
  modulation of electrons and atypical components 334
    26-day recurrent modulations 346–349
    implications of measurements for modulation theory 349–351
    modulation theory 334–336
    solar cycle modulation 336–346
  recurrent decreases 32
  regeneration 358
  Ulysses mission, future contributions 361
    composition studies 363–364
    modulation studies 361–363
  see also ACRs; Modulation
COSPIN (Cosmic and Solar Particle Investigations) 260, 331, 332
  see slso HET

Coulomb collisions 70
Counterstreaming electron events 127–128
CRIS (Cosmic Ray Isotope Spectrometer) 330
CRR (Corotating Rarefaction Region) 29
CRRES (Chemical Release and Radiation Effects Satellite) 331, 353

dE/dx versus residual-energy analysis 351
Degree of ionization 21
Discontinuities 195–198, 200–202, 206–222
  directional 26
  Directional Discontinuities (DDs) 195
    (CDs) Contact 195
    (RDs) Rotational 195
    (TDs) Tangential 195
    shocks 195
  hydromagnetic 26
  LB method (Lepping and Behannon) 196
  rotational 26
  tangential 26
  TS method (Tsurutani and Smith) 196
  waves at IP 246, 248–249
  see also Measurements by *Ulysses*;
Discovery 9
Dissipation 195
Drifts 335
DSN (Deep Space Network) 9, 15
Dust 36
  interstellar 393–397
  storms 388
  see also Cosmic dust

Electromagnetic lower hybrid waves 241
Electron heat flux 52, 88–90, 127
Electron to proton temperature ratio 33
Electrons 80–90
  core 80, 86
  halo 80, 87–88, 252
  photoelectrons 80
  strahl 81, 90–91
  suprathermal 127, 137
  temperature 83
Elsässer variables 171–174, 192
Energetic particles 259–263
  periodic structures in flux time series 278–280
  propagation and transport 275–276

charged particles in interplanetary structure 278, 276–277
latitudinal 276
solar–terrestrial connections 280–282
sources 263
anomalous cosmic rays 272–275
interplanetary 266–272
solar 263–266
Energetic particles associated with CIR 31
ESA (European Space Agency) 8
Explorer 16 382
Explorer 23 382
Extended inner source 291

Fast latitude scan 33–34
Fast wind 34, 63
structure and variation 25–26
Fermi shock acceleration mechanism 267
FES (Fast Envelope Sampler) 231
Field line random walk 178
Filtration 38
FIP (First Ionization Potential) 21, 311
bias 356
Fisk model 98
FIT (First Ionization Time) 62
Flow deflections 124–126
FLS (Fast Latitude Scan) 32, 50–51, 117
Fluctuations 167–175, 192–194, 199, 203–205
$1/f$ at large scales 171
$f^{-(5/3)}$ at small scales 171
Flux ropes 139–140, 145, 147, 153, 246
Forward Shock (FS) 112, 123, 131, 158
Forward-propagating wave 110
Freezing-in temperature 21, 59–60, 121–122
FS (Forward Shocks) 300, 301

g-mode oscillations 280
see also Gravity-wave excitations
Galileo 374–377, 382–384, 388, 393, 395–396, 398
Gas dynamic simulations 152
GAS instrument 27, 322
GCRs (Galactic Cosmic Rays) 36–37
see also Cosmic rays; Energetic particles
Generalized p model 189–191
Global merged interaction region 142
Gradient and curvature drifts 334
see also Drifts
Gradient, the 341
Gravity waves excitations 58

HCS (Heliospheric Current Sheet) 29, 46, 116, 120
warps 270
Heliopause 52, 307
Helios 8, 56, 68, 74–75, 85, 89–90, 110, 171, 173, 182, 184–185, 193, 373–374, 382
Helios 1 43, 88, 183, 229
Helios 2 43, 88, 229
Heliosphere 1–8, 10
origin and properties 3
Heliospheric medium 3
Helium 65–68, 149, 291–300, 302, 304
Helium-3 297, 299
Helium-4 295–297
HEOS-2 373–374
HET (High Energy Telescope) (COSPIN) 330, 331, 352
HI-SCALE instrument 262, 279
High-energy tails 73
HMF (Heliospheric Magnetic Field) 5
Hot model 292
Hydrogen 291–300, 302, 304

IAL (Ion Acoustic-Like) waves 241, 243, 250–251
see also Radio and plasma waves at high latitude; Waves
ICE 83–84, 88
ICMEs (Interplanetary CME) events 134, 136, 140, 264
and the heliospheric current sheet 146–147
composition of high latitude 149
multispacecraft observations 154–158
overexpansion of high latitude 149–154
solar origin 143–145
speeds at high latitudes 148–149
see also CMEs; Magnetic field
IMP 184
IMP-1 7
IMP-6 45, 83, 87, 239
IMP-7 45, 83, 87, 358
IMP-8 8, 16, 24, 45, 83, 87, 149, 154–155, 279, 331, 338, 358

Inner heliosphere 8, 349
Instability criterion 213
Instrumentation 231–233, 376–377
Interplanetary heating 68–69, 78
Interplanetary medium, see Heliospheric medium
Interplanetary meteoroids 383
Interplume wind 58
Interstellar constituents 35–36
   cosmic ray gas 35
Io 374, 390, 393, 398
Ion beams 73–78
Ion-heat flux 52
Ionization temperatures, see Freezing-in temperature
IP (Interplanetary) 229
IP shocks 240–245
IPM (Interplanetary Medium) 259
IPS (Interplanetary Scintillation) 44–45
IRAS 373
ISEE-1 233
ISEE-3 71, 76, 142, 146, 229, 244, 279, 331, 358
Isotope ratios 37
Isotopic measurements 330–331

Jokipii and Kóta model 97
JPL (Jet Propulsion Laboratory) 15
Jupiter 7–10, 17–18, 30, 35–36, 45, 52, 59, 95–96, 137, 142, 259, 267, 279, 333, 336, 338, 343, 350, 374–375, 378–379, 394, 398
Jupiter dust streams 385–392
   see also Cosmic dust

K41 theory 189
K65 theory 189
Kepler's equation 18
KET (Kiel Electron Telescope) 331, 338
KNLS (Kinetic Non-Linear Schrödinger) model 206
Kraichnan's model, see K65 model

Langmuir waves 236, 239, 242, 244, 246, 249, 278
   see also Radio and plasma waves at high latitude; Waves

LASCO (Large Angle Spectroscopic Coronagraph) 135
Latitudinal gradients 339–346
   rigidity depencence 344–346
LDEF 382
LET (Low Energy Telescope) 262, 331
LIC (Local Interstellar Cloud) 306–312
   see also Pickup ions
Linear holes 213
LISM (Local Interstellar Medium) 3, 5
Local type III emission 237
Lorentz force 390–391, 394
Lunik 1 and *Lunik 2* 7

Magnesium to oxygen ratio 21, 62, 64–65
Magnetic clouds 246
Magnetic dipole displacement 24
Magnetic field 5, 91–92
   and Alfvén waves 23–25
   azimuth angle 119
   compressions 120
   direction 94–98
   magnitude 119
   photosphere 22
   solar source and field polarity 92–93
   strength 93–94
   structure 119
   topology of ICMEs 145–146
   see also Alfvén waves; Measurements by *Ulysses*
Magnetic holes 26, 78, 212–214, 246, 248–249
Magnetic-field sector boundary, see HCS
Magnetosphere disturbances 281
Magnetosphere, Jovian 333
Mariner 85
Mariner 2 3, 7
Mariner 10 87
Mars 7, 267
MDLS (Modified Derivative Non-linear Schrödinger) equation 206
MDs (Magnetic Decreases) 209, 212–217
Measurements by *Ulysses* 167, 219–222
   discontinuities and Alfvén waves 195
      Alfvénic shocks 205
      DD radial and latitudinal gradients 196–199
      evolution of non-linear Alfvén waves

and rotational discontinuities 172, 206–207
  electron-capture isotopes 360
  high-mass resolution 352
    interrelationship 201–204
    MDs and magnetic holes 212–217
    north–south asymmetries 205
    slow shocks 217–218
    tangential discontinuities at high latitudes 209
    TDs versus RDs 207–211
    thicknesses 199–201
  polar heliosphere fluctuations 168–171, 173–175
    and solar-wind structure 192–193
    evolution, large scale 175–185
    turbulent processes, small scale 185–192
  radioactive isotopes 357–360
  stable isotopes 353–357
  uniqueness of data 167–168
Meteoroid models 382
Meteoroids, see Interplanetary meteoroids; beta-meteoroids
Microinstabilities 77
Microstreams 25, 56, 58, 69, 185, 193
Mini-sectors 194
Minimum variance directions 175, 178
Modulation 329–357, 361–364
  insensitivity to solar wind velocity 344
Modulation equation (8.1) 334–335
Modulation products 393

NASA (National Aeronautics and Space Administration) 8
Neutral helium 37
Neutral line 262
Normalized cross-helicity 181, 192–193
North–south asymmetry 205, 341–343
  see also Measurements by *Ulysses*

OESP (Out of Ecliptic/Solar Polar) 240
Open fields 4
Out-Of-Ecliptic mission (OOE) 2, 16
Overexpansion, see ICMEs

Page, D. E. 328
Parker Archimedean spiral 23, 92, 94–95, 174, 221, 328, 334, 350

Parker solar-wind model 23, 95, 97
Parker, Eugene 3, 5, 7, 334
Particle propagation 275–278
PBS (Pressure Balance Structure) 26, 55, 56, 58
Pegasus 382
PFR (Plasma Frequency Receiver) 231
Pickup ions 37–39, 153, 182, 272, 287–288
  charge states 290
  composition of the LIC 309–312
  hydrogen and helium 291–292
    velocities in the high-speed solar wind 292–299
    velocities in the in-ecliptic low-speed solar wind 299–304
  interaction with solar wind 319–321
  interstellar, mass heavier than helium 304–306
  ionization state of the LIC 306–309
  new population from an inner source 312–321
  production and measurements 288–291
  spatial distributions 290–291
Pioneer 111, 126, 344
Pioneer 8 373–374, 382
Pioneer 9 373–374, 382
Pioneer 10 7–8, 30, 36, 43, 68, 69, 292, 329, 373–375
Pioneer 11 7–8, 36, 43, 68, 69, 292, 329, 340, 350, 373–375
Pizzo's model 128–132
Plasma beta 213
Plasma bubbles 135
Plasma waves 33
Plumes 58
Polar pass 10
Polarity reversal 335
Polarization 202
Polytrope index 69, 83
Potential-field source-surface model 47
Principal Investigators 11, 14
Protons 292–295

QTN (Quasithermal Noise) 82
Quasi-periodic variations 279, 346
'Quiet' solar wind 249–253
  see also Waves

Radial gradients 338–339
Radio and plasma waves at high latitude 26–28
  Ion Acoustic Waves (IAWs) 27
  Langmuir waves (LWs) 27
  line-of-sight radio waves 27
  Whistler Mode Waves (WMWs) 27
Radio bursts, solar 233–240
Radio waves 244–245
RAR (Radio Astronomy Receiver) 231
Reverse Waves (RW) 113, 123, 131, 158, 154
  see Backward-propagating wave
ROID (Rate of Occurrence of Interplanetary Discontinuities) 196, 198–199
RS (Reverse Shocks) 300–301

Saturn 7–8
Seed particles 270
SEPs (Solar Energetic Particles) 28
Shock geometry 240
Shock pairs 28, 130, 134, 218–219
Shocks 110
Shoemaker-Levy 9 388
Sidereal solar rotation period 119
Simulation, gas dynamic 108–109
Skylab 4, 134
Slow shocks 195, 217–218
Slow wind 34, 63
SLS (Slow Latitude Scan) 33
Small damping 252
SMM (Solar Maximum Mission) 134
SN1987a 357
SOHO 16, 58, 61–62, 71, 76, 134, 143, 145, 322
Solar Cycle 23 263
Solar flares 136
Solar maximum polar pass 362
Solar wind 2–3, 43
  convection 334
  density and mass, momentum and energy fluxes 49–52
  electrons 80
  elements 62–63
  heavy ions 58–59
    abundance 62–65
    helium 65–68
  ionization states 59–62
  high- and low speed 43–44
  ion-distribution functions 68
    anisotropies 78–80
    high-energy tails 73
    ion beams 73–78
    temperatures 68–72
  low latitude 30
  measurements, *Ulysses* instruments 81–83
    core anisotropy 86
    electron-heat flux 88–90
    halo electrons and interplanetary potential 87–88
    radial temperature gradient and polytrope index 83–86
    strahl 90–91
  non-radial expansion 22–23
  polar 21–22
  speed 44–49
  structures in the high-latitude wind 52–58
  variability 4
  see also Cosmic dust; Instrumentation
Spartan 201 61
Spectral index 170, 177, 181, 190
Stanford source surface model 116–117
Strahl 90–91
Stream Interface (SI) 113
Stream–stream interactions 183
Streamer belt 29, 112, 125–126, 143, 192
Sun–Earth connections 280–282
SWICS (Solar Wind Ion Composition Spectrometer) 38, 58–59, 288, 321
SWOOPS 66, 68, 78, 81, 347
SXT (Soft X-ray Telescope) 143

TDs 248
  see also Discontinuities
Termination shock 5, 52, 288, 309
Thermal radio noise 27, 251–253
Three-dimensional hydrodynamic simulations 156
Transient flows
  CME origin 134–138, 140–142
  ICME observations 142–158
  see also CMEs; ICMEs
Transient shocks 136
Transition region 30–32
Transition scale 190
Turbulent processes, small scale

energy transfer 188–190
intermittency 185–188
transition scale 190, 192

Ulysses dataset 395
Ulysses mission 1–8
  characteristics 8
    data archiving 17
    history 15–17
    orbit 9–10, 17–19
    payload 10–14
    spacecraft operations 13, 15
  disruptions and delays 16
  dust experiment 375–376
  investigations, hardware 14
  objectives 8–9
  orbit 11–12
  pre-launch configuration 15
  scientific discoveries 20
    coupling between high- and low-latitude heliosphere 28–35
    high-latitude heliosphere 21–28
    interstellar constituents and medium outside the heliosphere 35–39
  timeline of events 13
  trajectory 338
Unified Radio and Plasma Wave Experiment 13
Unshocked regions 115
Uranus 8
URAP (Unified Radio and Plasma Wave Investigation) 82, 231, 249
UVCS (Ultra-Violet Coronagraph Spectrograph) 62

Vela 7
Velocity decline 192
Venus 7
Vlasov model 217
VLF waves 250
Voyager 30, 85, 126, 184, 279, 344, 398
Voyager 1 5, 8, 43, 229, 292, 309, 321, 329, 331, 340, 358, 374
Voyager 2 8, 43, 68–69, 229, 292, 309, 331, 340, 358, 374

Wavenumbers 179
Waves 229
  associated with IP shocks 240–241
    ion acoustic waves 241, 243–244
    Langmuir waves 244
    low-frequency electromagnetic waves 241–242
    radio waves 244–245
  at IP discontinuities and magnetic holes 246, 248–249
  in coronal mass ejections and magnetic clouds 245–246
  instrumentation for observations 231–233
  ion acoustic waves 250–251
  modes in the solar wind 229–231
  radio bursts caused by solar flares 233
    applications in remote study of the IPM 239–240
    type III theory 233–239
  thermal noise 251–253
  VLF and expanding regions of the solar wind 250
  whistler and heat-flux regulation 250
  'quiet' solar wind 249–250
Weighted slab model 356
WFA (Waveform Analyzer) 231, 232
Whistlers 250
  see also Electromagnetic lower hybrid waves; Radio and plasma waves at high latitude
WIND 16, 25, 27, 45, 62, 71, 73, 76, 142, 171, 229, 237, 239–240, 322, 343
WKB (Wentzel, Kramers, Brillouin) model 178
Wolf–Rayet ejecta 356
Wrong polarity 92

Yohkoh 136, 143, 145

Zodiacal dust cloud 382
Zodiacal light observations 373, 385